METHODS IN MOLECULAR BIOLOGY™

Series Editor
John M. Walker
School of Life Sciences
University of Hertfordshire
Hatfield, Hertfordshire, AL10 9AB, UK

For further volumes:
http://www.springer.com/series/7651

Allostery

Methods and Protocols

Edited by

Aron W. Fenton

Department of Biochemistry and Molecular Biology, The University of Kansas Medical Center, Kansas City, KS, USA

Humana Press

Editor
Aron W. Fenton
Department of Biochemistry and Molecular Biology
The University of Kansas Medical Center
Kansas City, KS, USA
afenton@kumc.edu

ISSN 1064-3745 e-ISSN 1940-6029
ISBN 978-1-4939-5864-1 ISBN 978-1-61779-334-9 (eBook)
DOI 10.1007/978-1-61779-334-9
Springer New York Dordrecht Heidelberg London

Printed on acid-free paper

Humana press is a part of Springer Science+Business Media (www.springer.com)

Dedication

I am convinced that all that is in the Universe revolves around my amazing wife; without her efforts, I could not do science. I have been in the fortunate position to have been trained by three mentors who are not only good people, but also believe in science at the highest caliber. Therefore, I would like to dedicate this book to these four individuals: Shellee Fenton and Drs. James B. Blair, Gregory D. Reinhart, and Gerald M. Carlson.

Preface

In the past 7 years, allostery has resurfaced as a major focus in understanding protein structure/function. Much of this rejuvenated interest has been driven by the ability of NMR to monitor protein dynamics and the potential of determining how these dynamics contribute to protein functions, including allostery (1–6). A second driving force for the recent interest is a growing appreciation that allosteric drugs offer safety advantages over conventional drugs (7–9). This renewed interest has resulted in several reviews on allostery (10–13).

At the onset of any discussion on allostery, it is beneficial to review the exact phenomenon included in the discussion. Shortly after the original use of "allosteric" (14), confusion over the definition of this term showed up in the literature. One source of confusion is whether "allostery" and "cooperative" should be treated as two synonyms to describe the same principle or if these words describe two different phenomena. To indicate similarities at the phenomenological level, it is now common to use "allostery" and "cooperative" interchangeably, with further definition as either "homotropic," to indicate energetic coupling when the two ligands are identical, or "heterotropic," to indicate energetic coupling when the two ligands are nonidentical. Even with these distinctions, the classification of homotropic and heterotropic as independent forms of regulation has been much debated. Alberto Sols articulated why these properties should be considered as related but independent properties by emphasizing that homotropic mechanisms require that the protein is an oligomer (15):

> ...because of confusion between two frequently linked but essentially independent concepts: (i) specifically regulatory sites and (ii) multiplicity of interacting equal sites in oligomeric proteins.... To compound the tendency to confusion, oligomerism is not only not necessarily linked to allosteric (heterotropic) effects but is not even the only basis for positive cooperativity (homotropic)....

By contrast, a purely thermodynamic view led Harvey Fisher and coworkers to express the similarities in these two properties (16):

> The term "cooperativity," or, more precisely, "heterotropic cooperativity" has been used occasionally to describe systems such as that shown ... (in an allosteric energy cycle)... in cases where the binding of one ligand either increases or decreases the affinity of a second, chemically distinguishable ligand. A majority of workers in the field, however, prefer to restrict the use of the term "cooperativity" to homotropic systems, and to refer to such effects in heterotropic systems as "positive and negative interactions." ....however, such a formal distinction between homotropic and heterotropic systems (implying as it does that the two classes of systems require totally different mechanisms to achieve what is essentially the same result) is an unwarranted assumption and one which may prove to be misleading.

Given this long standing historical debate, we have found the most productive approach is to define the type of regulation that is being described. However, one distinction that should be noted is the additional challenges associated with the study of homotropic systems since the concentrations of the two ligands cannot be varied independently. In this book, the majority of the chapters focus on studies of heterotropic systems. However, given the historical association between heterotropic and homotropic effects, techniques specific to the study of homotropic systems are also represented.

A second level of confusion is whether "allostery" includes any reference to a change in protein conformation. The original definition given by Monod et al. in 1963 (14) had no reference to conformational changes. Shortly thereafter, Monod and coworkers offered a plausible model to explain allostery derived from assumed conformational changes (17). The 1965 reference has been used to suggest that the recent introduction of dynamics into the discussion of allostery offers a "new view" of allostery (3, 18, 19). Others have relied on the 1963 definition to emphasize that the original definition of allostery placed no constraints on the molecular source of allosteric regulation and that dynamics were always accounted for in the description of this phenomena (11, 12, 20, 21).

In the Fenton laboratory, we use the word "allostery" to refer to heterotropic coupling events, with no implication that the mechanism for this through-protein communication is restricted to a change in protein conformation. Therefore, allosteric regulation is defined functionally as how a macromolecule binds one ligand differently when a second ligand is or is not prebound to the macromolecule. Since the definition of allostery influences what is expected as the "molecular source of allostery" (22), the use of the same definition has been strongly encouraged throughout all chapters in this volume (12). However, unifying the use of terms across all structure/function studies is an unrealistic goal, and even in several chapters of this volume, the influence of historical deviations of our favored definition is apparent.

Despite the semantic debates regarding classification, the common feature of allosteric systems is ligand-induced, through-protein changes. Therefore, any technique that can be used to study protein structure/function questions can be applied to the study of allostery. As such, the primary value of this book is the logic that is necessary to study this phenomenon, a phenomenon that is well recognized through the history of the life sciences and very poorly understood at the molecular level.

Kansas City, KS, USA ***Aron W. Fenton***

References

1. Tzeng, S. R., and Kalodimos, C. G. Protein dynamics and allostery: an NMR view, *Curr Opin Struct Biol.*
2. Smock, R. G., and Gierasch, L. M. (2009) Sending signals dynamically, *Science 324*, 198–203.
3. Kern, D., and Zuiderweg, E. R. (2003) The role of dynamics in allosteric regulation, *Curr Opin Struct Biol 13*, 748–757.
4. Lipchock, J. M., and Loria, J. P. Nanometer propagation of millisecond motions in V-type allostery, *Structure 18*, 1596–1607.
5. Bruschweiler, S., Schanda, P., Kloiber, K., Brutscher, B., Kontaxis, G., Konrat, R., and Tollinger, M. (2009) Direct observation of the dynamic process underlying allosteric signal transmission, *J Am Chem Soc 131*, 3063–3068.
6. Yan, J., Liu, Y., Lukasik, S. M., Speck, N. A., and Bushweller, J. H. (2004) CBFbeta allosterically regulates the Runx1 Runt domain via a dynamic conformational equilibrium, *Nat Struct Mol Biol 11*, 901–906.
7. Peracchi, A., and Mozzarelli, A. (2010) Exploring and exploiting allostery: Models, evolution, and drug targeting, *Biochim Biophys Acta.*
8. Groebe, D. R. (2006) Screening for positive allosteric modulators of biological targets, *Drug discovery today 11*, 632–639.
9. Groebe, D. R. (2009) In search of negative allosteric modulators of biological targets, *Drug discovery today 14*, 41–49.
10. Hilser, V. J. Biochemistry. An ensemble view of allostery, *Science 327*, 653–654.

11. Reinhart, G. D. (2004) Quantitative analysis and interpretation of allosteric behavior, *Methods Enzymol 380*, 187–203.
12. Fenton, A. W. (2008) Allostery: an illustrated definition for the 'second secret of life', *Trends Biochem Sci 33*, 420–425.
13. Lindsley, J. E., and Rutter, J. (2006) Whence cometh the allosterome?, *Proc Natl Acad Sci U S A 103*, 10533–10535.
14. Monod, J., Changeux, J. P., and Jacob, F. (1963) Allosteric proteins and cellular control systems, *J Mol Biol 6*, 306–329.
15. Sols, A. (1981) Multimodulation of Enzyme Activity, *Current Topics in Cellular Regulation 19*, 77–101.
16. Subramanian, S., Stickel, D. C., Colen, A. H., and Fisher, H. F. (1978) Thermodynamics of heterotropic interactions. The glutamate dehydrogenase . NADPH . glutamate complex, *J Biol Chem 253*, 8369–8374.
17. Monod, J., Wyman, J., and Changeux, J. P. (1965) On the Nature of Allosteric Transitions: a Plausible Model, *J Mol Biol 12*, 88–118.
18. Gunasekaran, K., Ma, B., and Nussinov, R. (2004) Is allostery an intrinsic property of all dynamic proteins?, *Proteins 57*, 433–443.
19. Swain, J. F., and Gierasch, L. M. (2006) The changing landscape of protein allostery, *Curr Opin Struct Biol 16*, 102–108.
20. Weber, G. (1972) Ligand binding and internal equilibria in proteins, *Biochemistry 11*, 864–878.
21. Cui, Q., and Karplus, M. (2008) Allostery and cooperativity revisited, *Protein Sci 17*, 1295–1307.
22. Fenton, A. W., Johnson, T. A., and Holyoak, T. (2010) The pyruvate kinase model system, a cautionary tale for the use of osmolyte perturbations to support conformational equilibria in allostery, *Protein Sci 19*, 1796–1800.

Acknowledgment

This work was supported in part by NIH grant DK78076.

Contents

Part III Monitoring Allosteric Changes in Protein Dynamics/Subpopulation Distribution

Part IV Macromolecular and Ligand Engineering Allosteric Functions

Part V Computational Methods/Aids in the Study of Allostery

Contributors

Olga Abian · *Institute of Biocomputation and Physics of Complex Systems (BIFI), Universidad de Zaragoza, Zaragoza, Spain; Departamento de Bioquímica y Biología Molecular y Celular, Universidad de Zaragoza, Zaragoza, Spain; Fundación ARAID, Diputación General de Aragón, Aragón, Spain*

Gary K. Ackers · *(Deceased) Department of Biochemistry & Molecular Biology, Washington University School of Medicine, St. Louis, MO, USA*

David L. Bain · *Department of Pharmaceutical Sciences, University of Colorado, Aurora, CO, USA*

Dorothy Beckett · *Department of Chemistry and Biochemistry, University of Maryland, College Park, MD, USA*

Sabine Bouguet-Bonnet · *Methodologie RMN (CRM2; UMR 7036, UHP-CNRS), Faculté des Sciences et Techniques, Nancy-Université, Vandoeuvre-les-Nancy, France*

Matthias Buck · *Department of Physiology and Biophysics, Case Comprehensive Cancer Center, Center for Proteomics and Bioinformatics, Case Western Reserve University School of Medicine, Cleveland, OH, USA; Department of Neurosciences, Case Comprehensive Cancer Center, Center for Proteomics and Bioinformatics, Case Western Reserve University School of Medicine, Cleveland, OH, USA; Department of Pharmacology, Case Comprehensive Cancer Center, Center for Proteomics and Bioinformatics, Case Western Reserve University School of Medicine, Cleveland, OH, USA*

Gerald M. Carlson · *Department of Biochemistry and Molecular Biology, University of Kansas Medical Center, Kansas City, KS, USA*

José C. Clemente · *Oncology Research & Development, GlaxoSmithKline, Upper Providence, PA, USA*

Dennis A. Dougherty · *Division of Chemistry and Chemical Engineering, California Institute of Technology, Pasadena, CA, USA*

Aron W. Fenton · *Department of Biochemistry and Molecular Biology, The University of Kansas Medical Center, Kansas City, KS, USA*

Harvey F. Fisher · *Department of Biochemistry and Molecular Biology, University of Kansas Medical Center, Kansas City, KS, USA*

Anthony A. Fodor · *Department of Bioinformatics and Genomics, University of North Carolina at Charlotte, Charlotte, NC, USA*

Stuart A. Forman · *Department of Anesthesia Critical Care & Pain Medicine, Massachusetts General Hospital, Boston, MA, USA*

Gabriel A. Frank · *Departments of Structural Biology and Chemical Physics, Weizmann Institute of Science, Rehovot, Israel*

Shawnalea J. Frazier · *Division of Biology, California Institute of Technology, Pasadena, CA, USA*

David P. Giedroc · *Department of Chemistry, Indiana University, Bloomington, IN, USA*

NICHOLAS E. GROSSOEHME · *Department of Chemistry, Indiana University, Bloomington, IN, USA; Department of Chemistry, Physics and Geology, Winthrop University, Rock Hill, SC, USA*
HEIDI E. HAMM · *Department of Pharmacology, Vanderbilt University Medical Center, Nashville, TN, USA*
GILAD HARAN · *Department of Chemical Physics, Weizmann Institute of Science, Rehovot, Israel*
PETR HERMAN · *Faculty of Mathematics and Physics, Institute of Physics, Charles University, Prague, Czech Republic*
JO M. HOLT · *Oro Valley, AZ, USA*
AMNON HOROVITZ · *Department of Structural Biology, Weizmann Institute of Science, Rehovot, Israel*
ANDREI Y. ISTOMIN · *Department of Bioinformatics and Genomics, University of North Carolina at Charlotte, Charlotte, NC, USA*
T.M. IVERSON · *Departments of Biochemistry and Pharmacology, Vanderbilt University Medical Center, Nashville, TN, USA*
DONALD J. JACOBS · *Department of Physics and Optical Science, University of North Carolina at Charlotte, Charlotte, NC, USA*
EILEEN K. JAFFE · *Fox Chase Cancer Center, Philadelphia, PA, USA*
ALI I. KAYA · *Department of Pharmacology, Vanderbilt University Medical Center, Nashville, TN, USA*
JAMES K. KRANZ · *Biopharmaceuticals Research & Development, GlaxoSmithKline, Upper Merion, PA, USA*
KYLE E. KRETH · *Department of Bioinformatics and Genomics, University of North Carolina at Charlotte, Charlotte, NC, USA*
SARAH H. LAWRENCE · *Fox Chase Cancer Center, Philadelphia, PA, USA*
J. CHING LEE · *Department of Biochemistry and Molecular Biology, University of Texas Medical Branch, Galveston, TX, USA*
HENRY A. LESTER · *Division of Biology, California Institute of Technology, Pasadena, CA, USA*
DENNIS R. LIVESAY · *Department of Bioinformatics and Genomics, University of North Carolina at Charlotte, Charlotte, NC, USA*
AARON L. LUCIUS · *Department of Chemistry, The University of Alabama at Birmingham, Birmingham, AL, USA*
MARTA MARTINEZ-JULVEZ · *Institute of Biocomputation and Physics of Complex Systems (BIFI), Universidad de Zaragoza, Zaragoza, Spain; Departamento de Bioquímica y Biología Molecular y Celular, Universidad de Zaragoza, Zaragoza, Spain*
MILAGROS MEDINA · *Institute of Biocomputation and Physics of Complex Systems (BIFI), Universidad de Zaragoza, Zaragoza, Spain; Departamento de Bioquímica y Biología Molecular y Celular, Universidad de Zaragoza, Zaragoza, Spain; Fundación ARAID, Diputación General de Aragón, Aragón, Spain*
DENGMING MING · *Department of Physiology and Biophysics, School of Life Science, Fudan University, Shanghai, China*

AMIE D. MOODY • *Department of Pharmaceutical Sciences, University of Colorado, Aurora, CO, USA*
JAMES M. MOTTONEN • *Department of Physics and Optical Science, University of North Carolina at Charlotte, Charlotte, NC, USA*
OWEN W. NADEAU • *Department of Biochemistry and Molecular Biology, University of Kansas Medical Center, Kansas City, KS, USA*
CHARULATA B. PRASANNAN • *Department of Biochemistry and Molecular Biology, The University of Kansas Medical Center, Kansas City, KS, USA*
ANITA M. PREININGER • *Department of Pharmacology, Vanderbilt University Medical Center, Nashville, TN, USA*
GREGORY D. REINHART • *Department of Biochemistry and Biophysics, Texas A&M University and Texas AgriLife Research, College Station, TX, USA*
JAMES P. ROBBLEE • *Department of Pharmaceutical Sciences, University of Colorado, Aurora, CO, USA*
MARTIN K. SAFO • *Department of Medicinal Chemistry, School of Pharmacy & Institute for Structural Biology and Drug Discovery, Virginia Commonwealth University, Richmond, VA, USA*
JAI A. P. SHANATA • *Division of Chemistry and Chemical Engineering, California Institute of Technology, Pasadena, CA, USA*
RYAN P. STAFFORD • *Department of Chemistry, The University of Alabama at Birmingham, Birmingham, AL, USA*
DEIRDRE STEWART • *Department of Anesthesia Critical Care & Pain Medicine, Massachusetts General Hospital, Boston, MA, USA*
QINGLING TANG • *Department of Biochemistry and Molecular Biology, The University of Kansas Medical Center, Kansas City, KS, USA*
TARJANI M. THAKER • *Department of Biochemistry, Vanderbilt University Medical Center, Nashville, TN, USA*
CUIJUAN TIE • *Department of Biochemistry and Biophysics, Texas A&M University and Texas AgriLife Research, College Station, TX, USA*
ASHUTOSH TRIPATHI • *Department of Medicinal Chemistry, School of Pharmacy & Institute for Structural Biology and Drug Discovery, Virginia Commonwealth University, Richmond, VA, USA*
SONIA VEGA • *Institute of Biocomputation and Physics of Complex Systems (BIFI), Universidad de Zaragoza, Zaragoza, Spain; Departamento de Bioquímica y Biología Molecular y Celular, Universidad de Zaragoza, Zaragoza, Spain; Fundación ARAID, Diputación General de Aragón, Aragón, Spain*
ADRIAN VELAZQUEZ-CAMPOY • *Institute of Biocomputation and Physics of Complex Systems (BIFI), Universidad de Zaragoza, Zaragoza, Spain; Departamento de Bioquímica y Biología Molecular y Celular, Universidad de Zaragoza, Zaragoza, Spain; Fundación ARAID, Diputación General de Aragón, Aragón, Spain*
DEEPTAK VERMA • *Department of Bioinformatics and Genomics, University of North Carolina at Charlotte, Charlotte, NC, USA*
P. KEITH VERONESE • *Department of Chemistry, The University of Alabama at Birmingham, Birmingham, AL, USA*

OLEG K. VOROV · *Department of Physics and Optical Science, University of North Carolina at Charlotte, Charlotte, NC, USA*
MICHAEL E. WALL · *Computer, Computational, and Statistical Sciences Division, Center for Nonlinear Studies, Los Alamos National Laboratory, Los Alamos, NM, USA*
LIQUN ZHANG · *Department of Physiology and Biophysics, Case Western Reserve University School of Medicine, Cleveland, OH, USA*

Part I

Monitoring Allosteric Function

Chapter 1

Binding Techniques to Study the Allosteric Energy Cycle

James K. Kranz and José C. Clemente

Abstract

Thermodynamic principles of cooperativity and allostery have long been used as a starting point to begin understanding the interplay between ligand binding events. Understanding the nature of allosteric effects requires an experimental technique that can be used to quantify ligand binding energies and simultaneously give experimental insights into the conformational dynamics at play upon ligand binding. CD spectroscopy provides macroscopic information about the relative secondary and tertiary structures present in a protein. Here, we use this spectroscopic technique with thermal shift assays wherein ligand binding constants can be quantified based on their stabilizing effect against thermally induced protein denaturation. Binding constants for two ligands are used to determine a pairwise coupling free energy which defines the shared energy that favors or opposes binding of the second ligand binding event in an allosteric system. In CD-based thermal shift assays, temperature is the driving force for protein unfolding and can also influence protein conformational dynamics present in the unbound protein or ligand-bound proteins. Dihydrofolate reductase (DHFR) and glutamate dehydrogenase (GDH) are proposed as example test systems. NADP and methotrexate bind DHFR with positive cooperativity. Mammalian GDH exhibits negative cooperativity with respect to binding of NAD and NADPH coenzyme molecules, activation by ADP, and inhibition by GTP.

Key words: Ligand binding, Allosteric energy cycle, Allostery, Thermal shift, Circular dichroism

1. Introduction

It is challenging to succinctly summarize the term "allostery" given its more than 60-year history applied to innumerable systems, not only because of the large volume of publications on the subject but also due to the system-specific nature of allosteric interactions. As Gregorio Weber pointed out in 1972 (1):

> In considering the possible classes of chemical equilibria in proteins in solution, we are naturally led to distinguish the following possibilities: (a) the binding equilibria between the protein and added ligands, usually small molecules; (b) the internal, first-order equilibria between parts of the same peptide chain; (c) the second- and higher order equilibria among the several covalent chains that constitute the multi-chain proteins, or among the macromolecules in single-chain proteins. Evidently no one of these equilibria can be treated in isolation from the others.

Aron W. Fenton (ed.), *Allostery: Methods and Protocols*, Methods in Molecular Biology, vol. 796,
DOI 10.1007/978-1-61779-334-9_1, © Springer Science+Business Media, LLC 2012

We now have a wealth of experimental data indicating proteins naturally exist in a constant state of motion, described by populations of conformations. This is in fact the source of ligand-induced protein folding, or local folding coupled to binding, and can be the source of great specificity (2–5) or enormous plasticity (6–8) in ligand recognition. Motions coupled to enzyme catalysis have been long hypothesized as the source of enormous rate accelerations relative to their solution counterparts; the change in dynamics of a functioning enzyme has recently comes to light (9, 10). Ligands can shift the population of protein conformational states in unpredictable ways; in general, binding induces rigidity in a complex manner (8, 11, 12), but at least one example of an enzyme shows an increase in conformational heterogeneity upon substrate binding wherein key conformers become populated only under steady-state conditions (13).

An allosterically modulated system is one in which the native, unliganded state of the protein has multiple conformations that are accessible and that ligand binding causes a shift in the population of states owing to an energetic preference for the ligand to bind to a particular structural form of the protein. In other words, the allostery arises from the interdependence of ligand binding events that modulates protein conformational heterogeneity. Likewise, allostery in enzymatic function arises from the interdependence of substrate binding, catalysis, and product release steps each with their own unique relationship to the population of conformational states available to the protein.

The term "allostery" typically is associated with the complex functions of a multidomain protein, where the fractional saturation by endogenous ligands is characterized by site-resolved binding or kinetic constants that are nonequivalent. By definition, heterotropic allosteric regulation (i.e., allostery) must exhibit two defining characteristics: (1) binding of the effector ligand to the macromolecule elicits a change in a functional property of the macromolecule, either enzyme catalysis or binding of a second ligand; and (2) the effector binds to the macromolecule at a site topographically distinct from the functional site. This definition applies to monomeric proteins as well, where the native state is characterized by conformational dynamic regions that change upon ligand binding.

Returning again to Weber (1), the concept of allostery lends itself to an intuitive approach employing the notion of "apparent" or "conditional" energies. For the simplest case of a monomeric protein and a single ligand, the protein exists in a number of conformations, $A_1 \ldots A_N$, that are not equally populated depending on the energetic stability of each state (14), represented in the schematic in Fig. 1. Interconversion between any two states, i and j, is described by a first-order equilibrium constant, $k^0{}_{ij}$ and $k^1{}_{ij}$, where the binary subscript 0 denotes a transition in the unliganded state of the protein and 1 denotes the same transition taking place between liganded forms of the protein. The ligand, X, has an affinity for every conformation of the protein, given by $K_i(\mathrm{X})$. The observation of a binding event by a macroscopic analytical technique yields only an apparent dissociation constant, K_{app}, that encapsulates the interconversion between the unbound manifold of states and the bond manifold of states:

$$K_{\mathrm{app}}(\mathrm{X}) = \frac{\left(\sum_{i=1}^{N}[\mathrm{A}_i]\right)[\mathrm{X}]}{\left(\sum_{i=1}^{N}[\mathrm{A}_i\mathrm{X}]\right)}. \quad (1)$$

The ensemble of conformations available to the protein in either the unbound or X-liganded manifolds defined by $K_{\mathrm{app}}(\mathrm{X})$ represents the measurable quantity that is the free energy of binding, $\Delta G_{\mathrm{app}}(\mathrm{X}) = RT\ln K_{\mathrm{app}}(\mathrm{X})$. Of course, not all states (or microstates) are equally probable (15), and the probability of each state can change upon ligation.

The schematic described in Fig. 1 also extends to a system containing dual ligands X and Y, which is described by a thermodyanamic cycle (Fig. 2). Equilibrium constants for binding either X or Y to an unliganded protein, $K^0_{\mathrm{app}}(\mathrm{X})$ or $K^0_{\mathrm{app}}(\mathrm{Y})$, respectively, are equivalent to apparent binding constants in Eq. 1. Equilibrium dissociation constants for binding the second ligand once a first binding event has occurred are $K^1_{\mathrm{app}}(\mathrm{X})$ for binding X to a complex of AY, or $K^1_{\mathrm{app}}(\mathrm{Y})$ for binding Y to a complex of AX. We maintain the formalism that the protein exists in a distribution of states to emphasize that any binding event can alter the relative population of states for any part of the thermodynamic cycle.

In any thermodynamic cycle, additivity is required for transitions described as a state function, generally as a free energy of binding, ΔG, though enthalpy and entropy, ΔH and $(-T\Delta S)$ are

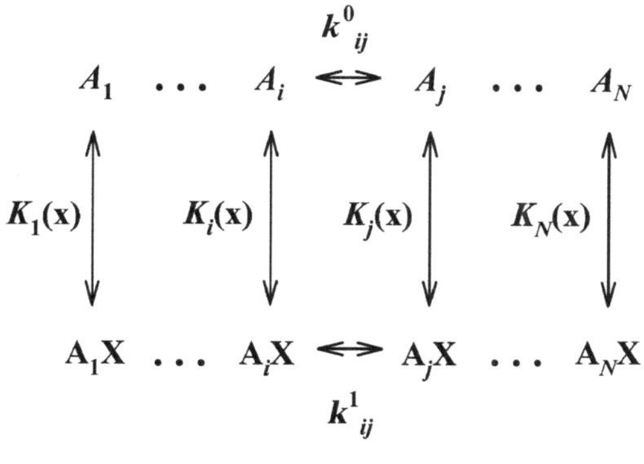

Fig. 1. A_1, A_i, A_j, and A_N chain conformations for hypothetical protein. First-order equilibrium constant of unliganded conformations *i* and *j*, k^0_{ij}; first-order equilibrium constant of X-liganded conformations *i* and *j*, k^1_{ij}. Second-order dissociation constants of X binding to any individual protein conformation, $K_i(\mathrm{X})$.

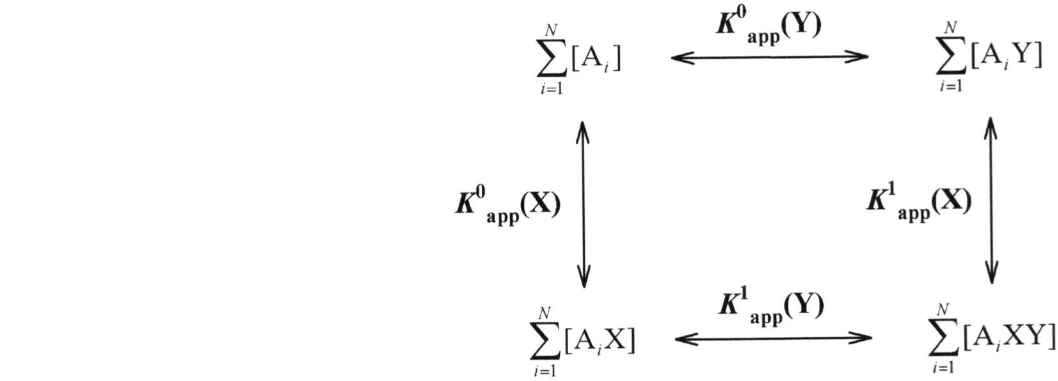

Fig. 2. Thermodynamic cycle describing a population of states, A_i, for a hypothetical protein, and binding of either X or Y ligands, or both X and Y. Apparent equilibrium constants describe binding of the first ligand, K^0_{app}(X) or K^0_{app}(Y), or binding of the second ligand to a complex of the protein and the other ligand, K^1_{app}(X) or K^1_{app}(Y).

also state functions and thus can be used to further dissect ΔG in a similar manner. The overall free energies of the system are described by $\Delta G_{app}(X)$, $\Delta G_{app}(Y)$, and $\Delta G_{app}(XY)$, where both ligands bind simultaneously, and at topographically nonoverlapping sites. From Fig. 2, free energy conservation requires that:

$$\Delta G_{app}(X) = RT \ln K^0_{app}(X) = \Delta G^0_{app}(X), \tag{2}$$

$$\Delta G_{app}(Y) = RT \ln K^0_{app}(Y) = \Delta G^0_{app}(Y), \tag{3}$$

$$\begin{aligned}\Delta G_{app}(XY) &= RT \ln K^0_{app}(X) + RT \ln K^1_{app}(Y) \\ &= \Delta G^0_{app}(X) + \Delta G^1_{app}(Y),\end{aligned} \tag{4a}$$

$$\begin{aligned}&= RT \ln K^0_{app}(Y) + RT \ln K^1_{app}(X) \\ &= \Delta G^0_{app}(Y) + \Delta G^1_{app}(X),\end{aligned} \tag{4b}$$

$$= \Delta G^0_{app}(X) + \Delta G^0_{app}(Y) + \Delta G_c, \tag{4c}$$

$$\Delta G_c = \Delta G^1_{app}(X) - \Delta G^0_{app}(X), \tag{5a}$$

$$= \Delta G^1_{app}(Y) - \Delta G^0_{app}(Y). \tag{5b}$$

Free energy terms in Eqs. 2–4c are only “apparent” in the sense that they encompass ligand binding to a distribution of protein conformational states that are not defined. They are precisely the ΔG values one measures experimentally as the binding energy of X, Y or both X and Y. Moreover, the thermodynamic cycle is additive, but not necessarily symmetric, as defined by a coupling free energy term ΔG_c (1, 16–18). As shown in Eq. 5a, 5b, ΔG_c is measured experimentally from either $\Delta G^0_{app}(X) - \Delta G^0_{app}(X)$, or from ΔG^1_{app}

(Y)− ΔG^0_{app}(Y). Positive cooperativity is observed when ΔG^1_{app}(X) < ΔG^0_{app}(X) or when ΔG^1_{app}(Y) < ΔG^0_{app}(Y).

Practically speaking, cooperativity is a thermodynamic description of interdependent binding events. Heterotrophic allostery further describes the relationship between multiple ligand binding events that modulate the affinity of each other via conformational changes; i.e., binding results in a change of the relative populations among possible protein conformational states. Understanding the underlying mechanisms that are described by cooperativity and allostery requires a deep investigation using multiple approaches, thus beyond the scope of any one technique. What is perhaps most important from a practical sense is defining: (1) that the system of interest has two different ligands that are functionally relevant; (2) they are either competitive or bind non-competitively, in topographically distinct locations; and (3) an experimental method that provides some insights as to whether or not allostery exists in the system of interest.

Circular dichroism (CD) spectroscopy has been used extensively to characterize protein folding and unfolding, both as a function of denaturants and temperature. It is a truly general technique that is not limited to constraints of molecular size or chemical composition, is a "label-free" approach that does not require secondary conjugation or chemical modification to obtain data on protein structural composition. CD has unique resolving power in estimating the relative fraction of secondary structural elements of a protein, as described extensively by Greenfield (19–23) as well as many others, too numerous to elaborate. Moreover, modern improvements in CD instrumentation and data fitting makes it straightforward to collect full spectra as a function of temperature, providing a means to quantitate the presence of subtle fluctuations in protein structure that occurs at temperatures below the primary protein unfolding transition.

What is lacking in the literature, however, is the use of CD to measure ligand binding; in particular, the combination of CD as a tool to monitor ligand-induced thermal shifts. Experimentally, replicate protein unfolding experiments are performed at different concentrations of ligands; the T_m value representing the mid-point unfolding temperature is related to the total ligand concentration, from which a K_D is obtained. When coupled with additional information on the composition of the ligands and protein, binding affinities from thermal shift analyses can be coupled to insights on the changes of protein structure in response to temperature or ligand binding, supporting a hypothesis of structural changes linked to binding in reference to model systems. Thus, for a system that is thought to be allosteric, CD provides a useful first tool to fully characterize binding of both ligands independently, to establish the presence of cooperative binding when both ligands are present via thermodynamic cycles, and to provide a first approximation of allosteric interactions that can be further elaborated by other techniques.

2. Materials

1. Test protein of interest, such as bovine liver dihydrofolate reductases, bDHFR (Sigma-Aldrich D6385), or bovine glutamate dehydrogenase, bGDH (Sigma-Aldrich G7882) in a suitable assay buffer.
2. Ligands of interest, such as NADPH (Sigma: N9660, or equivalent), methotrexate (Sigma: M9929, or equivalent), ADP (Fluka 01897, or equivalent), GTP (Fluka 01897, or equivalent).
3. CD spectrophotometer, such as the Olis DSM CD or equivalent, outfitted with a Peltier-based thermal cell holder and appropriate quartz cuvettes.

3. Methods

The essentials of CD spectroscopy have been exhaustively elaborated in the literature (see Greenfield (20–27) and references therein) and will not be detailed here. Historically, measurements of CD signal at single wavelengths have been employed to follow the kinetics and thermodynamics of protein folding, the major advantage being rapid data collection and simplified data analysis. The limitation of this approach is the loss of information content that is present in monitoring unfolding over a wide range of wavelengths. Modern improvements in both instrumentation and data analysis allow researchers to employ methods such as singular value decomposition (SVD) (28) or similar methods (20), toward the interpretation of CD spectra. The advantage of full spectral analyses applied to allosteric binding is the ability to uniquely quantify the fractional change in α-helix, β-sheet, and random coil upon ligand binding. Here, we encourage the use of full CD spectra in monitoring thermal shift assays for the purpose of gaining some insights into protein conformational changes that may be coupled to binding, though a single wavelength is sufficient for K_D determination in a CD-based thermal shift assay (Table 1).

DHFR represents a suitable test system for exploring positive cooperativity via CD-based thermal shift assays. DHFR is a well-characterized system with ligand-induced conformational changes

Table 1
Features of protein CD spectra

Far-UV CD	
Range	(190–240 nm)
Source of signal	Peptide bond in asymmetric environment; protein secondary structure
α-Helix	Negative at 208 nm Negative at 222 nm Positive at 192 nm
β-Sheet	Negative at 218 nm Positive at 196 nm
Random coil	Positive at 212 nm Negative at 195 nm
Near-UV CD	
Range	(260–320 nm)
Source of signal	Aromatics in asymmetric environment; protein tertiary structure

and positive cooperativity in binding of NADPH and the antifolate compound methotrexate (29). Likewise, GDH represents a suitable test system for exploring negative cooperativity via CD-based thermal shift assays. GDH is allosterically regulated by GTP, ADP, and coenzymes NADH or NADPH (30, 31), and is known to change protein conformation in a ligand-dependent manner (32). Though DHFR and GDH are presented as specific examples for sample preparation and data collection, experiments will be discussed generally without specific analysis of either system.

3.1. Sample Preparation and Data Collection

1. Compound solutions: Test ligands are prepared in DMSO (or aqueous buffer if solubility is sufficiently high) at a 50- to 100-fold concentrated solution, generally in the 10–100 mM range. Compounds may be diluted in DMSO serially, then added to protein in a uniform dilution. A typical thermal shift assay employs 8–12 different concentrations of a test compound with a single negative control (DMSO alone), spanning a final concentration range from 0.05 to 200 μM, for example.
2. Protein solution: The protein is diluted from a concentrated stock to a working concentration of ~0.5–20 μM protein into a suitable assay buffer; sufficient volume should be prepared

for a unique CD sample at each test concentration, then aliquoted into microcentrifuge tubes. The exact concentrations of protein and dye are defined by experimental assay development studies.

3. Compound addition/equilibration: compounds are diluted uniformly by a factor of 50- to 100-fold into the aliquoted protein, mixing via pipettor or gentle vortexing.
4. Individual samples are placed into CD cuvettes, and thermally denatured. Either a single wavelength or spectra are collected as a function of temperature (see Notes 1–3).

3.2. Thermal Shift Assay Data Analysis

The details of thermal shift analysis for ligand binding have been described for calorimetric studies (33, 34) and for fluorescent-based probe studies (35–37), and the thermodynamic analysis also applies to CD experiments. To measure the strength of ligand binding from a thermal shift assay, the data are fit in two stages. These are described by Eqs. 6 and 7a, 7b.

$$y(T) = y_{F,T_m} + m_F(T - T_m) + \frac{y_{U,T_m} - y_{F,T_m} + m_U(T - T_m)}{1 + e^{\Delta H_{U,T_r} + \Delta C_{p,U}(T - T_r) - T(\Delta S_{U,T_r} + \Delta C_{p,U}\ln(T/T_r))/RT}}. \quad (6)$$

The temperature-dependent CD signal, $y(T)$, is used to determine the protein thermal stability, T_m, at any given concentration of ligand (see Note 4). The protein folding free energy, $\Delta G_U(T)$ must be replaced by the Gibbs–Helmholtz relationships (38), and is composed of a temperature-dependent enthalpy, $\Delta H_U(T)$, and a temperature-dependent entropy, $\Delta S_U(T)$; a heat capacity for protein unfolding, $\Delta C_{p,U}$, defines the temperature dependence of $\Delta H_{U,Tr}$ and $\Delta S_{U,Tr}$. Temperature-dependent baselines for fully folded and fully unfolded protein are described by linear functions: $y_F(T) = y_{F,T_m} + m_F(T - T_m)$ and $y_U(T) = y_{U,T_m} + m_U(T - T_m)$, for folded and unfolded, respectively. The assumption of linear baselines may not be valid when protein fluctuations occur as a function of temperature.

Figure 3a shows example protein unfolding data for human GDH alone and in the presence of ADP. Protein melting was monitored by CD at 220 nm, fit to Eq. 6 by nonlinear least squares minimization algorithm to estimate the parameters of y_{F,T_m} and m_F, y_{U,T_m} and m_U, $\Delta H_{U,T_r}$, $\Delta S_U(T)$, and T_m for each individual sample. $\Delta C_{p,U}$ is always highly correlated with $\Delta H_{U,T_r}$ in parameter estimation, and is thus difficult to determine independently from $\Delta H_{U,T_r}$ with nonlinear least squares regression analysis. The value of $\Delta C_{p,U}$ was held fixed near its value measured by other techniques, generally estimated from differential scanning calorimetry (DSC), or is estimated based on protein composition (39).

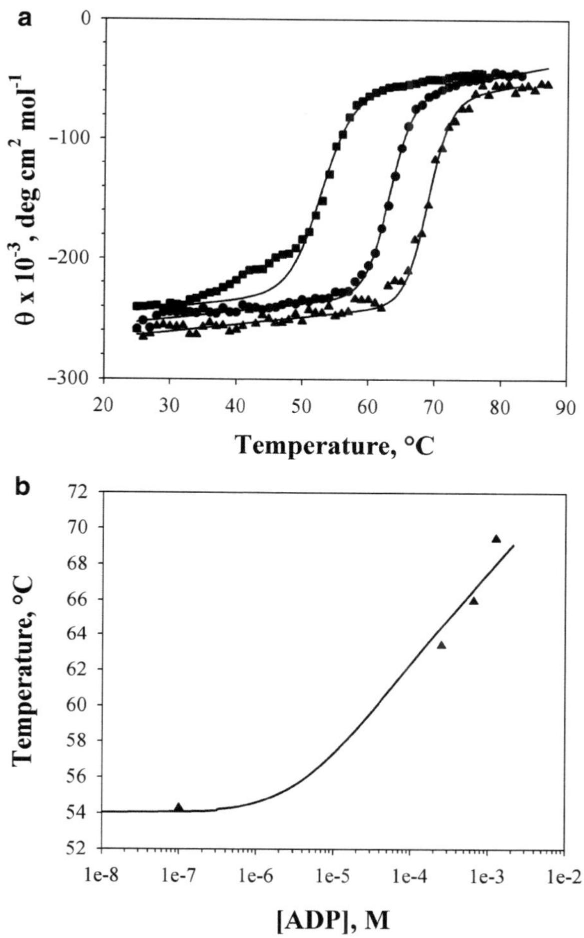

Fig. 3. (**a**) Thermally induced unfolding of (10 μM) *bovine* glutamate dehydrogenase (Sigma-Aldrich), monitored at 220 nm. Unfolding was performed on the apo protein (*squares*) or in the presence of either 250 μM ADP (*circles*) or 1 mM ADP (*triangles*). Data are fit to Eq. 7a, 7b, which gave $T_m = 54.2\ (\pm 0.1)$°C, $\Delta H_{app} = 90\ (\pm 15)$ kcal/mol for apo DHFR; baseline parameters were $y_f = -227{,}000$ (deg cm^2 mol^{-1}) and $m_f = 620$ (deg cm^2 mol^{-1}°C^{-1}) for the native state, and $y_u = -65{,}400$ (deg cm^2 mol^{-1}) and $m_u = -980$ (deg cm^2 mol^{-1}°C^{-1}) for the denatured state. Similar baseline values were obtained for ADP-bound transitions. (**b**) Plots of T_m versus ligand concentration, used to fit a binding constant of ADP to GDH.

Figure 3a also shows thermal unfolding of GDH in the presence of either 250 μM ADP or 1 mM ADP. As expected, GDH is stabilized against unfolding with an elevation of T_m in proportion to increasing [ADP]. What is more interesting is the behavior of GDH protein conformation in the pretransition region. At temperatures below ~30°C, both the apo and ADP-bound forms (250 μM and 1 mM) of GDH show common baseline behavior, with similar slopes and similar CD spectra (data not shown). However, in the 35–50°C range, the apo form deviates from linear behavior of ADP-bound GDH, with a change in elipticity at 220 nm consistent with an increase in random coil character. This observation is suggestive of a temperature-dependent change in the distribution of GDH conformational states in the 35–50°C range, specific to the unliganded form. Binding of ADP induces a change in protein structure that is consistent with the CD characteristics of the apo state near room temperature. Here, we can see the limitations of using single wavelength CD data in studying protein conformations in CD-based thermal melts. Further evaluation using CD spectral analyses of temperature-dependent conformational changes (see Greenfield (20–27)) and references therein) would be needed to ascertain a connection to allosteric ligand binding.

3.3. Binding Constants from Thermal Shift Assays

In a second stage of thermal shift data analysis, the T_m value of the protein at each concentration of ligand is related to the expected effect for a ligand with a given binding affinity, K_b or K_D. The relationship between total ligand concentration, total protein concentration, and the two equilibrium constants for protein unfolding and ligand binding is defined by Eq. 7a, in terms of equilibrium constants; data must be fit using Eq. 7b which expresses equilibrium constants in terms of their temperature-dependent forms:

$$L_t = (1 - K_U)\left(\frac{P_t}{2} + \frac{1}{K_U K_b}\right), \tag{7a}$$

$$\begin{aligned} L_t = \Bigg(& 1 - e^{-\left(\Delta H_{U,T_r} + \Delta C_{p,U}(T_m - T_r) - T_m\left(\Delta S_{U,T_r} + \Delta C_{p,U}\ln(T_m/T_r)\right)\right)/RT_m} \Bigg) \\ & \times \left(\frac{P_t}{2} + \frac{1}{e^{-\left(\Delta H_{U,T_r} + \Delta C_{p,U}(T_m - T_r) - T_m\left(\Delta S_{U,T_r} + \Delta C_{p,U}\ln(T_m/T_r)\right)\right)/RT_m}} \right. \\ & \left. \times\, e^{-\left(\Delta H_b(T_0) + \Delta C_{p,b}(T - T_0) - T\left(\Delta S_b(T_0) + \Delta C_{p,b}\ln(T/T_0)\right)\right)/RT} \right) \end{aligned} \tag{7b}$$

P_t is the total protein concentration (sum of U, N, and NL_b for unfolded, native, or ligand-bond forms). L_t is the total ligand concentration (sum of L_f and NL_b for free or protein-bound forms). K_U is the protein unfolding equilibrium constant in both Eqs. 6 and 7a, 7b; K_b or K_D is described in terms of an enthalpy, $\Delta H_b(T_0)$, an entropy, $\Delta S_b(T_0)$, and heat capacity, $\Delta C_{p,b}$, of ligand binding in Eq. 7b.

Figure 3b shows an example of binding constant determination for ADP binding to GDH. Protein thermal melting temperatures, T_m, are plotted as a function of ligand concentration, and fit to Eq. 7a, 7b. ADP stabilized the protein against unfolding: T_m = 63.4°C with 250 μM ADP, T_m = 66.0°C with 650 μM ADP, and T_m = 69.4°C with 1 mM ADP for the three data points shown as an example. Binding constants are estimated at an arbitrary reference temperature, generally at 25 or 37°C; here, K_D(37°C) = 3 μM and ΔG(37°C) = −7.8 kcal/mol for a binding free energy.

Each ligand should be assayed for binding by thermal shift assays independently, to determine ΔG^0_{app} (X) and ΔG^0_{app} (Y). Generally, protein thermal denaturation should be performed at several different concentrations of ligand ($n > 6$, preferably $n > 10$) for each K_D measurement. Theoretical curves are generated to demonstrate how the dependence of T_m on ligand concentration is affected by several important parameters in the thermal shift assay. The theoretical behavior of weak to tight binding ligands based on variations in theoretical affinity is shown (Fig. 4a). The K_D greatly

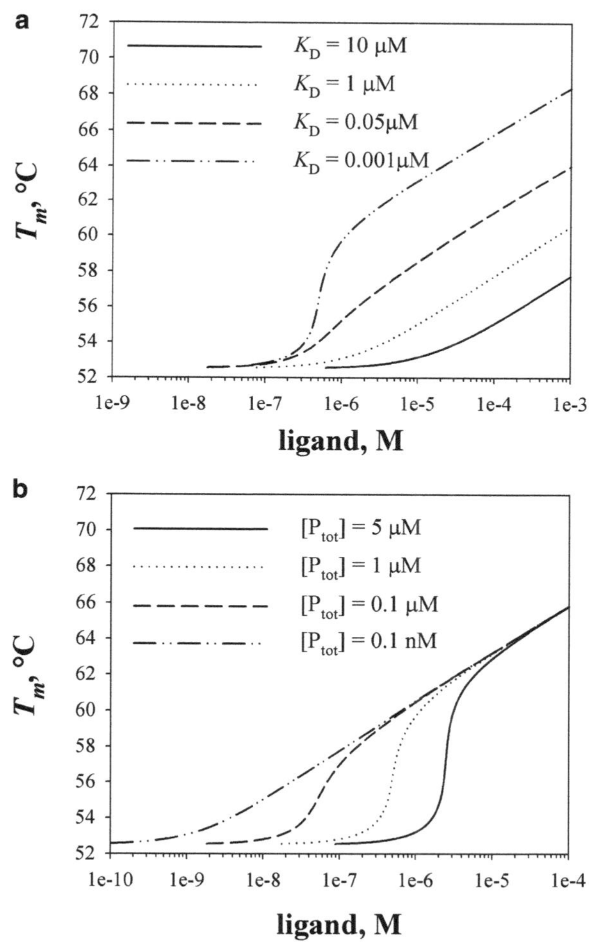

Fig. 4. Simulations of thermal shift concentration response curves using Eq. 7a, 7b. The thermodynamic parameters describing protein stability were held constant for all simulations. (**a**) Effect of differing ligand binding affinities ranging from $K_D = 0.001–10$ μM, with a constant protein concentration, $P_t = 1$ μM. (**b**) Effect of varying protein concentrations from $P_t = 0.0001–10$ μM, with a constant ligand binding affinity, $K_D = 0.001$ μM.

affects the magnitude of ΔT_m at a given concentration of added ligand. For the two weakest simulated binding curves, there is no change in T_m at ligand concentrations below their respective ligand affinities. Simulated curves for tight binding inhibitors highlight the influence of the protein concentration on the shape of the dose–response curve. In these simulations, a sigmoidal dependence of T_m on ligand concentration is observed just below the protein concentration, consistent with a saturation-binding experiment. This aspect of thermal shift assays is a unique consequence of the Gibbs–Helmholtz equation (33, 34), the result of which is a continuing shift in the equilibrium between native and denatured protein as more and more ligand is added.

Simulations in Fig. 4b further explore the affect of protein concentration on the shape of the dose–response curve, assuming a constant high affinity of $K_D = 1$ nM. Simulations varied the protein concentration from $[P_t] = 0.1$, 1, or 10 μM as well as the theoretical limit where the protein concentration is infinitely dilute ($[P_t] << K_D$). In each of the examples where the protein concentration exceeds the K_D, a sigmoidal dependence of T_m on ligand concentration is observed until protein saturation is achieved. This phenomenon also gives insight into the binding stoichiometry between ligand and protein.

Following determination of K_D values for X and Y ligands individually, a concentration of either X or Y is selected for subsequent cooperativity studies; at tenfold above the K_D, the protein should be sufficiently saturated with either X or Y to test cooperative binding of the alternative ligand, though iterative testing at different concentrations may be required. Samples are prepared again, including a common constant concentration of the secondary ligand. For example, in testing binding of X at different concentrations, all samples contain Y at a fixed concentration. Note that the coupling free energy can be determined from either combination of $K^0_{app}(X)$ and $K^0_{app}(Y)$ and $K^1_{app}(Y)$; it is not necessary to determine all four (though it is recommended to confirm additivity around the thermodynamic cycle). All binding constants are converted to a free energy of binding, and the coupling free energy is determined.

Spectral analysis of the CD data should be performed to ascertain fluctuations in the native state upon binding either X or Y, and importantly with both ligands bound. In addition, the temperature dependence of native state structure and of ligand-bound states can provide additional evidence of changes in protein conformational states. In order to interpret a CD change as a change in the dynamics of the protein, full spectral analysis is strongly encouraged. Finally, we wish to emphasize the use of this method as a potential litmus test for the presence of cooperativity and possible allostery; the data should inspire follow-up experimentation that is better suited for characterizing the precise mechanism of allosteric binding.

4. Notes

1. No two CD spectrophotometers have the same dynamic range and operational restrictions; consult the instrument manual or manufacturer for specifics.
2. For reference, typical wavelengths used in CD spectroscopy are shown in Table 1. For a protein with significant α-helical structure, unfolding may be monitored at 222 and 208 nm. Recommended wavelength ranges for analyzing secondary structure vary (20), but typically encompass 200–240 nm.
3. For thermal shift assays, the experimental temperature is steadily increased at such a rate as to allow thermal equilibrium to be maintained throughout the experiment. Typical temperature ramp rates range from 0.1 to 10°C/min (commonly 1°C/min). The spectra or individual wavelength data are collected at regular intervals, 0.2–1°C/image, over a temperature range spanning the typical protein unfolding temperatures of 25–95°C.
4. Equations in Subheading 3.2 used for fitting protein thermal stability values, T_m, assume a linear dependence on temperature. If there is a temperature-dependent change in protein conformation or distribution of conformations, the native state baseline may not adhere to a linear temperature dependence. A rigorous analysis of CD spectra as a function of temperature, as suggested above (but is beyond the scope of the current paper), is necessary to lend interpretation to temperature-dependent structural changes.

References

1. Weber, G. (1972) Ligand binding and internal equilibria in proteins. *Biochemistry* **11**, 864–878
2. Kranz, J. K., and Hall, K. B. (1999) RNA recognition by the human U1A protein is mediated by a network of local cooperative interactions that create the optimal binding surface. *J Mol Biol* **285**, 215–231
3. Showalter, S. A., and Hall, K. B. (2004) Altering the RNA-binding mode of the U1A RBD1 protein. *J Mol Biol* **335**, 465–480
4. Showalter, S. A., and Hall, K. B. (2002) 5. A functional role for correlated motion in the N-terminal RNA-binding domain of human U1A protein. *J Mol Biol* **322**, 533–54
5. Shoemaker, B. A., Portman, J. J., and Wolynes, P. G. (2000) Speeding molecular recognition by using the folding funnel: the fly-casting mechanism. *Proc Natl Acad Sci U S A* **97**, 8868–887
6. Lee, A. L., and Wand, A. J. (2001) Microscopic origins of entropy, heat capacity and the glass transition in proteins. *Nature* **411**, 501–504
7. Brokx, R. D., Lopez, M. M., Vogel, H. J., and Makhatadze, G. I. (2001) Energetics of target

peptide binding by calmodulin reveals different modes of binding. *J Biol Chem* **276**, 14083–14091

8. Frederick, K. K., Marlow, M. S., Valentine, K. G., and Wand, A. J. (2007) Conformational entropy in molecular recognition by proteins. *Nature* **448**, 325–329
9. Henzler-Wildman, K., and Kern, D. (2007) Dynamic personalities of proteins. *Nature* **450**, 964–972
10. Nashine, V. C., Hammes-Schiffer, S., and Benkovic, S. J. (2010) Coupled motions in enzyme catalysis. *Curr Opin Chem Biol* **14**, 644–651
11. Tsai, C. J., del Sol, A., and Nussinov, R. (2008) Allostery: absence of a change in shape does not imply that allostery is not at play. *J Mol Biol* **378**, 1–11
12. Wand, A. J. (2001) Dynamic activation of protein function: a view emerging from NMR spectroscopy. *Nat Struct Biol* **8**, 926–931
13. Larion, M., Salinas, R. K., Bruschweiler-Li, L., Bruschweiler, R., and Miller, B. G. (2010) Direct evidence of conformational heterogeneity in human pancreatic glucokinase from high-resolution nuclear magnetic resonance. *Biochemistry* **49**, 7969–7971
14. Hilser, V. J., Garcia-Moreno, E. B., Oas, T. G., Kapp, G., and Whitten, S. T. (2006) A statistical thermodynamic model of the protein ensemble. *Chem Rev* **106**, 1545–1558
15. Vertrees, J., Wrabl, J. O., and Hilser, V. J. (2009) Energetic profiling of protein folds. *Methods Enzymol* **455**, 299–327
16. Di Cera, E. (1998) Site-specific analysis of mutational effects in proteins. *Adv Protein Chem* **51**, 59–119
17. Jencks, W. P. (1981) On the attribution and additivity of binding energies. *Proc Natl Acad Sci U S A* **78**, 4046–4050
18. Weber, G. (1975) Energetics of ligand binding to proteins. *Adv Protein Chem* **29**, 1–83
19. Greenfield, N., and Fasman, G. D. (1969) Computed circular dichroism spectra for the evaluation of protein conformation. *Biochemistry* **8**, 4108–4116
20. Greenfield, N. J. (1996) Methods to estimate the conformation of proteins and polypeptides from circular dichroism data. *Anal Biochem* **235**, 1–10
21. Greenfield, N. J. (2006) Determination of the folding of proteins as a function of denaturants, osmolytes or ligands using circular dichroism. *Nat. Protocols* **1**, 2733–2741
22. Greenfield, N. J. (2006) Analysis of the kinetics of folding of proteins and peptides using circular dichroism. *Nat. Protocols* **1**, 2891–2899
23. Greenfield, N. J. (2006) Using circular dichroism collected as a function of temperature to determine the thermodynamics of protein unfolding and binding interactions. *Nat. Protocols* **1**, 2527–2535
24. Adler, A. J., Greenfield, N. J., and Fasman, G. D. (1973) Circular dichroism and optical rotatory dispersion of proteins and polypeptides. *Methods Enzymol.* **27**, 675–735
25. Greenfield, N. J. (2004) Circular dichroism analysis for protein-protein interactions. *Methods Mol. Biol.* **261**, 55–78
26. Greenfield, N. J. (2004) Analysis of circular dichroism data. *Methods Enzymol.* **383**, 282–317
27. Greenfield, N. J. (2006) Determination of the folding of proteins as a function of denaturants, osmolytes or ligands using circular dichroism. *Nat. Protocols* **1**, 2733–2741
28. Hennessey, J. P., Jr., and Johnson, W. C., Jr. (1981) Information content in the circular dichroism of proteins. *Biochemistry* **20**, 1085–1094
29. Bystroff, C., and Kraut, J. (1991) Crystal structure of unliganded Escherichia coli dihydrofolate reductase. Ligand-induced conformational changes and cooperativity in binding. *Biochemistry* **30**, 2227–2239
30. Frieden, C. (1965) Glutamate dehydrogenase. VI. survey of Purine Nucleotide and other effects on the enzyme from various sources. *J Biol Chem* **240**, 2028–2035
31. George, A., and Bell, J. E. (1980) Effects of adenosine 5′-diphosphate on bovine glutamate dehydrogenase: diethyl pyrocarbonate modification. *Biochemistry* **19**, 6057–6061
32. Banerjee, S., Schmidt, T., Fang, J., Stanley, C. A., and Smith, T. J. (2003) Structural studies on ADP activation of mammalian glutamate dehydrogenase and the evolution of regulation. *Biochemistry* **42**, 3446–3456
33. Brandts, J. F., and Lin, L. N. (1990) Study of strong to ultratight protein interactions using differential scanning calorimetry. *Biochemistry* **29**, 6927–6940
34. Shrake, A., and Ross, P. D. (1990) Ligand-induced biphasic protein denaturation. *J Biol Chem* **265**, 5055–5059
35. Kranz, J. K., and Schalk-Hihi, C. (2011) Protein thermal shifts to identify low molecular weight fragments. *Methods Enzymol* **493**, 277–98

36. Matulis, D., Kranz, J. K., Salemme, F. R., and Todd, M. J. (2005) Thermodynamic stability of carbonic anhydrase: measurements of binding affinity and stoichiometry using ThermoFluor. *Biochemistry* 44, 5258–5266
37. Zhang, R., and Monsma, F. (2010) Fluorescence-based thermal shift assays. *Curr Opin Drug Discov Devel* **13**, 389–402
38. Privalov, P. L. (1979) Stability of proteins: small globular proteins. *Adv Protein Chem* **33**, 167–241
39. Murphy, K. P., and Gill, S. J. (1991) Solid model compounds and the thermodynamics of protein unfolding. *J Mol Biol* **222**, 699–709

Chapter 2

Kinetic Trapping of a Key Hemoglobin Intermediate

Jo M. Holt and Gary K. Ackers

Abstract

The complete binding cascade of human hemoglobin consists of a series of partially ligated intermediates. The individual intermediate binding constants cannot be distinguished in O_2 binding curves, however, each constant can be determined from the O_2-induced change in assembly constant for the $\alpha_2\beta_2$ tetramer from its constituent $\alpha\beta$ dimers. The characterization of these O_2 binding constants has shown the Hb cascade to be asymmetric in nature, with binding dependent upon the specific distribution of O_2 among the four hemesites. A stopped-flow approach to measuring the dissociation constant of a key doubly ligated intermediate, that in which one dimer is oxygenated and the other is not, is described. The intermediate is transiently formed in the absence of O_2 and then allowed to dissociate in the presence of O_2. The free dimers thus released are trapped by the plasma protein haptoglobin, the rate limiting step being that of tetramer dissociation. The kinetic constant observed for the dissociation of this intermediate confirms the value for its equilibrium O_2 binding constant, previously determined under equilibrium conditions by subzero isoelectric focusing.

Key words: Hemoglobin, Cooperativity, Allostery, Thermodynamic linkage, Oxygen binding, Haptoglobin, Stopped flow

1. Introduction

Human hemoglobin (Hb) is composed of four heme-containing subunits, two α and two β, each of which binds a single O_2 ligand. The tetramer readily dissociates to its constituent $\alpha\beta$ dimers, indicating that the dimer–dimer interface is weaker than the intra-dimer interface. Interaction among the four subunits is manifested in a strong positive cooperativity of O_2 binding and release. Fourteen partially ligated intermediates are generated in the course of O_2 binding, each with a different configuration of bound ligand among the four hemesites (Fig. 1). The binding constant for the first oxygen ligand, K_1, has the same value for all subunits, α^1, β^1, α^2, and β^2. However, binding the second O_2 to yield two ligands on a single $\alpha\beta$ dimer (tetramer species *21*) is approximately tenfold

Aron W. Fenton (ed.), *Allostery: Methods and Protocols*, Methods in Molecular Biology, vol. 796,
DOI 10.1007/978-1-61779-334-9_2, © Springer Science+Business Media, LLC 2012

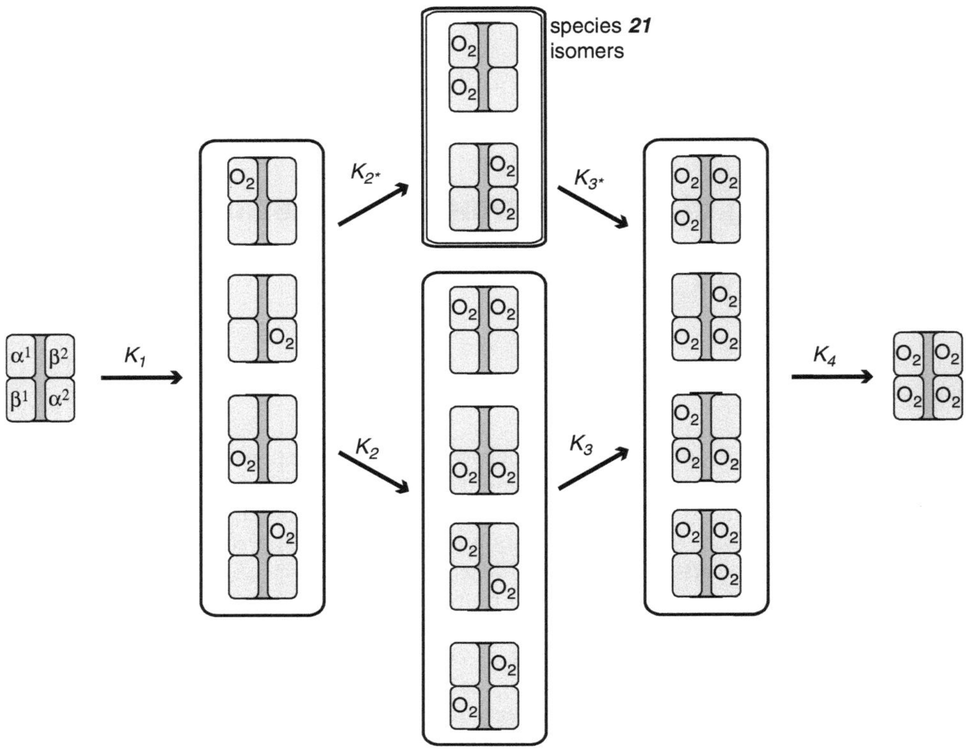

Fig. 1. The Hb binding cascade, showing all possible distributions of bound O_2 among the four subunits, including isomeric forms. The species *21* intermediate is that in which one $\alpha\beta$ dimer has no ligands and the partner dimer is fully oxygenated.

more favorable than binding the second O_2 to yield one ligand on each dimer ($K_{2*} = 10\ K_2$, Fig. 1) (1). This difference demonstrates that the Hb subunits do not respond uniformly to oxygenation, contrary to what has historically been accepted (2).

An experimental approach to measuring the intermediate binding constants for Hb (which are not accessible from analysis of traditional O_2 binding curves) was developed using linkage thermodynamics, capitalizing on the strict correlation between the O_2 binding constants K_1, K_2, K_{2*}, K_3, K_{3*}, and K_4, and the $\alpha\beta$ dimer $\rightarrow$ $\alpha_2\beta_2$ tetramer assembly constants for each tetramer (3). Because O_2 is an inherently labile Hb ligand, it is necessary to fix the configuration of bound ligand in the tetramer, and this is accomplished through the use of nonlabile hemesite analogs, such as the substitution of Zn(II) heme for the native Fe(II) heme (1). The intermediate binding constants have been characterized with a variety of hemesite analogs, in differing solution conditions, and with both symmetric and asymmetric modifications, providing a quantitative description of the entire binding cascade (4, 5).

The Zn(II) heme analog does not bind ligands such as O_2 and has proven to be an excellent deoxy hemesite analog (6–8). Although direct O_2 binding to the asymmetric doubly ligated species *21* is not experimentally feasible, even with the Zn(II) heme analog, subzero isoelectric focusing has permitted its dimer → tetramer equilibrium constant, and thus its O_2 binding constant, to be determined via thermodynamic linkage analysis (8, 9). The difference between the assembly free energy, ΔG_2, of species *21* and a singly ligated species is equal to the difference in free energy of O_2 binding, a value referred to as the cooperative free energy. When added to the intrinsic (or noncooperative) free energy of O_2 binding (observed with the free αβ dimer), the cooperative free energy yields the O_2 binding free energy for any step in the cascade.

In a kinetic experiment described here, the Zn/FeO_2 species *21* tetramer is transiently formed and then allowed to dissociate. The free dimers thus formed are trapped by the plasma protein haptoglobin, with the rate limiting step being that of tetramer dissociation (10). The kinetic constant for dissociation of the Zn/FeO_2 species *21* tetramer is measured in this manner, confirming the value for the equilibrium O_2 binding constant.

2. Materials

2.1. Preparation of Native FeO_2 Tetramers

1. Normal saline solution: 0.9% w/v NaCl, i.e., 9 g NaCl in 1 L.
2. Buffer exchange and concentrators: Sephadex G25 column, stirred-cells (Amicon), disposable ultrafiltration tubes.
3. Cation exchange chromatography: HPSP Sephadex (Amersham), FPLC (Pharmacia).
4. Chromatography Buffer A: 0.01 M NaH_2PO_4, 1 mM EDTA, pH adjusted to 6.8. Store at 4°C.
5. Chromatography Buffer B: 0.02 M Na_2HPO_4, 1 mM EDTA, pH adjusted to 8.3. Store at 4°C.
6. Mixed-bed ion exchange resin (BioRad AG501 ×8).

2.2. Preparation of Zn Tetramers

1. Acetone (high purity).
2. Concentrated HCl.
3. Ethylene glycol.
4. Methanol.
5. A low-temperature water bath with external circulation, using tubing that is resistant to methanol and ethylene glycol.
6. Jacketed reaction vessel (2-L), glass only, acid washed.
7. Dialyzate: 1 mM sodium bicarbonate, 0.1 mM dithiothreitol.
8. Zn(II) protoporphyrin IX (Frontier Scientific, Logan, UT).
9. Syringe filters: 0.45 μm cellulose acetate.

2.3. Preparation of Zn/Fe Asymmetric Hybrid Species 21u

1. Anaerobic chamber (Coy Laboratory Products, Inc.).
2. Deoxygenation procedures were carried out with humidified ultra high purity N which was further purified by O -removing cartridges (QC2+ panel/Agilent Technologies, Supelco), connected to the chamber by stainless steel or copper tubing (Swagelok).
3. Glass gas-tight syringes (Hamilton).
4. Sodium dithionite. Store in a dessicator in the freezer.
5. Standard buffer: 0.1 M Tris (Trizma base, SigmaUltra), 0.1 M NaCl (SigmaUltra), 1 mM Na_2EDTA (SigmaUltra) titrated at 21.5°C with concentrated HCl at pH 7.4 to give a total [Cl^-] of 0.18 M.

2.4. Dimer Trapping with Haptoglobin

1. Stopped-flow spectrophotometer (Applied Photophysics SX.18MV), customized with an anaerobic chamber (Coy Laboratory Products, Inc.) enclosing the sample handling unit of the spectrophotometer.
2. Human haptoglobin type 1:1 (Sigma). Store frozen.

3. Methods

The absorbance decrease in the Soret region that accompanies the formation of the haptoglobin-deoxy dimer complex has been used to measure the rate constant of tetramer dissociation, k_{off}, for both modified and native Hbs over a range of solution conditions (11–15). In combination with the dimer → tetramer assembly rate constant, $k_{on} = 1.1 \times 10^6\ M^{-1}\ s^{-1}$ (2, 15), the equilibrium assembly constant can be obtained. The value for k_{on} has been shown to be invariant among a wide range of modifications and mutations to the deoxy tetramer (15).

The species *21* tetramer can be formed by hybridizing parent tetramers ZnHb and native FeO_2Hb, however, the maximum fraction of *21* hybrid formed at equilibrium is only 3% of the mixture, as measured by subzero isoelectric focusing (8). However, under anaerobic conditions, the unligated species *21u* tetramer comprises nearly 30% of the hybrid mixture after 24 h of equilibration (Fig. 2). Addition of O_2 to this anaerobic mixture results in rapid formation of species *21* from *21u*, followed by tetramer dissociation and disproportionation to the parent species, such that the species *21* fraction of the hybrid mixture falls from 30 to 3% within a few seconds. The reaction of haptoglobin with fully ligated species FeO_2Hb is rapid ($t_{1/2} = 1$ s), but does not result in an appreciable change in absorbance. Reaction of haptoglobin with the parent species

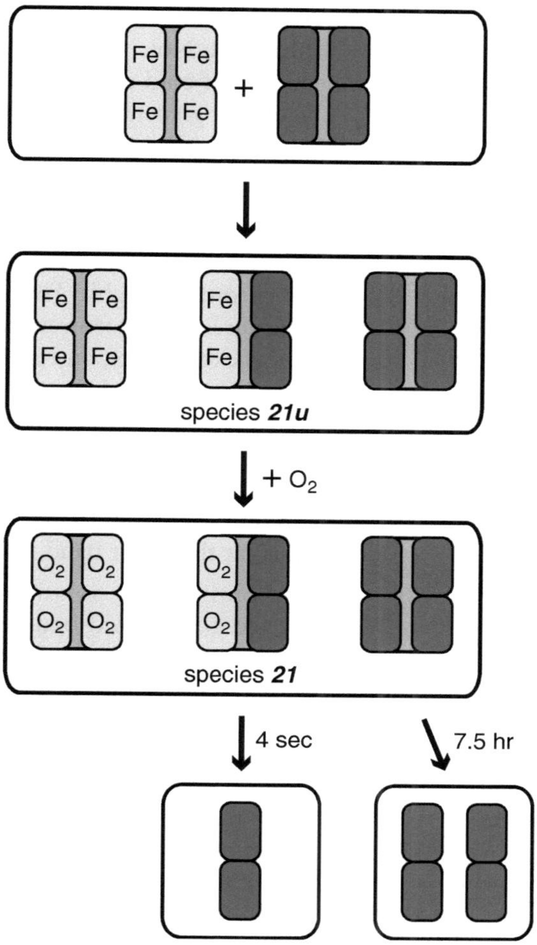

Fig. 2. Hybridization and dissociation of species *21*. Native Hb, which contains the Fe(II) heme group, is mixed with Hb containing Zn(II) heme (dark gray subunits). Through the naturally occurring process of dimer exchange, a hybrid tetramer is formed (species *21u*) in which the subunits of one dimer contain native Fe(II) heme and the subunits of the partner dimer contain Zn(II) heme. O_2 binds solely to the Fe(II) heme, creating the species *21* hybrid as well as an oxygenated parent tetramer (all four subunits ligated). Only the Zn(II) heme-containing dimers generate an absorbance change when bound to haptoglobin, and only those dimers originating from species *21* contribute to the signal over the first 500 s.

ZnFe and the Zn/FeO_2 hybrid *21* does produce a change in absorbance. The reaction with ZnFe does not contribute significantly to the kinetic profile, as it occurs with a very slow dissociation rate ($t_{1/2} = 7.5$ h) (16). The absorbance change over 500 s is thus due solely to the dissociation of the Zn/FeO_2 hybrid tetramer.

3.1. Preparation of Native FeO_2 Tetramers

1. Normal adult human blood is collected in EDTA-coated tubes (see Note 1). Plasma is removed by adding a threefold volume of cold normal saline solution followed by low-speed (1,600 × *g*) centrifugation for 30 min at 4°C (see Note 2).
2. Red blood cells are lysed by the addition of distilled water at room temperature for 1 h, followed by high-speed centrifugation at 4°C to pellet the red cell membranes.
3. The supernatant is exchanged into the starting buffer, Buffer A, for chromatography. The eluant is then re-concentrated, but to no greater concentration than 50 mg/mL (the limits of the cation exchanger), and applied to a cation exchange column. An SP Sepharose HP column volume of 250 mL can process 2 g of Hb per run. HbA_0 is the major band eluted with a linear gradient of 15–100% Buffer B over ten column volumes (9). Concentration is measured by visible absorbance using an extinction coefficient of 131 m M^{-1} cm^{-1} (on a per heme basis) at the Soret band maximum of 415 nm.
4. The purified Hb solution is concentrated to 3–4 mM and applied to a mixed-bed resin (1 g resin per 20 mL Hb) to strip all organic and inorganic phosphates, resulting in a deionized Hb solution. The concentration is measured prior to storage in liquid N_2 (see Note 3).

3.2. Preparation of Zn Tetramers

1. $ZnHbA_0$ is prepared from deionized native $FeHbA_0$ (0.65 mM) by first removing the Fe hemes using acid-acetone precipitation (17) (see Note 4). Acetone (1.5 L) is acidified with 750 μL concentrated HCl and equilibrated to −20°C in a 2-L jacketed reaction vessel cooled by an ethylene glycol/methanol (1:1 v/v mixture) circulating bath in a cold room (4–7°C). The reaction vessel is placed on a magnetic stirrer. Approximately 300 mL of the acid-acetone solution is transferred to a beaker and kept at −20°C in the circulating bath, along with several 5 mL glass pipettes and a tube rack with four 50 mL glass centrifuge tubes. With vigorous stirring, 300–400 mg of 0.65 mM Hb is added dropwise to the remaining 1.1 L of cold acid-acetone in the reaction vessel. Stirring is continued for 10–15 min, then stopped, and the globin is allowed to settle for 15–30 min. The colored supernatant is removed by aspiration. The remaining globin solution is pelleted at 1,200 × *g* for 20 s at −10°C, the supernatant is discarded, and the centrifuge tubes returned to the −20°C bath (see Note 5). Multiple additions of the globin-acetone suspension to the same four centrifuge tubes are required.
2. The globin precipitate in each tube is then washed by gentle resuspension with 15 mL of the unused acid-acetone solution

followed by centrifugation. This wash is performed 4–5 times. At 4°C, the globin pellet is dissolved in 3–5 mL of cold distilled water, transferred to a dialysis bag (see Note 6), and dialyzed against distilled water (2 L at 4°C) for 6–10 h with one change of dialyzate. The dialyzate is changed to sodium bicarbonate, and dialysis is continued until a precipitate of denatured globin appears, typically 24 h. At this point, the dialyzate is changed to the starting buffer for cation exchange, Buffer A, and dialysis continues for at least 6 h (see Note 7). Any precipitate remaining in the dialysis bag is removed by centrifugation (1,400 × *g* for 1 min at 4°C).

3. The globin is now reconstituted with Zn^{2+} heme under reduced or red light, due to the light sensitivity of Zn heme in the presence of O_2. 20 mL of globin solution in Buffer A and 100 mM HCl are prepared and placed on ice. A 1.5-fold equivalent amount of Zn heme is dissolved in 0.2 mL of 0.1 M NaOH, diluted tenfold with distilled water, and placed on ice. The Zn heme solution is slowly added dropwise to the gently stirring globin on ice. The pH is monitored with pH strips, and the cold, dilute HCl is added as needed to maintain the pH between 8 and 9. After the heme addition is complete, stirring is continued for 1 h. Then, the solution is allowed to warm to 20°C while stirring, and then stirring is continued for 10 min at 20°C. Typically, a precipitate appears and is removed by centrifugation (1,400 × *g*, 1 min). The ZnHb supernatant is cooled to 4°C and titrated with the cold, dilute HCl to pH 6.7. The solution is then filtered with a syringe filter prior to chromatography.
4. Chromatography is carried out on SP Sepharose with a gradient of 15–100% Buffer B over 11 column volumes. The reconstituted Hb should have a concentration not greater than 50 mg/mL when applied to SP Sepharose. The major peak is collected and concentration is determined ($\varepsilon_{373} = 22.8\ mM^{-1}\ cm^{-1}$). The ratio A_{468}/A_{373} should be 0.05. The ZnHb is then concentrated to 1–4 mM in heme and frozen in liquid N_2 for storage.
5. ZnHb thus prepared is analyzed by electrospray mass spectometry to verify its molecular weight. Purity is initially verified by the presence of a single band on isoelectric focusing (Pharmacia Phast System with ampholytes 6.7–7.7).

3.3. Preparation of the Zn/Fe Asymmetric Hybrid Species 21u

1. Hybridization of the Zn Hb tetramer with the native deoxyFe Hb tetramer is carried out in an anaerobic chamber purged continuously by a low flow of N_2 at room temperature. It is typically necessary to flush the chamber for 24 h prior to use (see Note 8). All glassware and solutions in this procedure are anaerobic, having been N_2-flushed in the

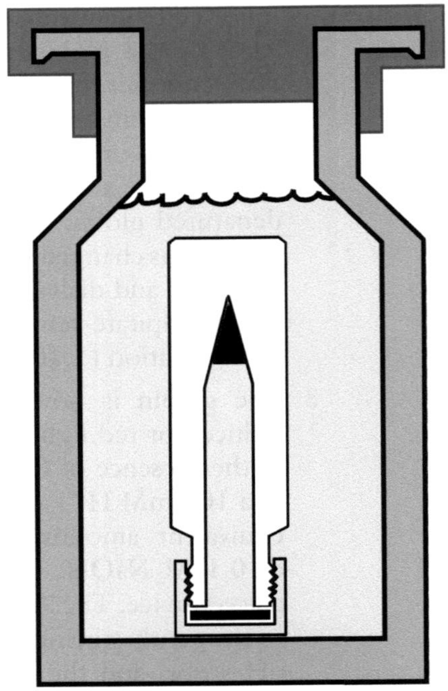

Fig. 3. The hybridization vial within a vial. The inner (upside down) vial contains the hybridization sample, and is submerged in a solution of sodium dithionite. The dithionite solution is under a positive pressure of purified N_2 which is maintained by constant flow via needles inserted through the rubber septum of the outer vial. The vial assembly is constructed in an anaerobic chamber, but can be removed from the chamber for incubation over hours or days.

chamber for a minimum of 1 h prior to use. All transfers of solutions are carried out with gas-tight syringes.

2. Small volumes (e.g., 10–30 μL) of unligated parent species ZnHb and FeHb in standard buffer are placed in septum-sealed vials. Equimolar amounts (0.5 mM each) are mixed in a small septum-sealed vial which is then submerged in a larger vial containing 0.1% dithionite (Fig. 3) (see Note 9) (10).
3. The hybridization vial is then incubated at 21.5°C in the anaerobic chamber or placed in a bath outside the chamber. The incubation continues for 24–36 h.

3.4. Dimer Trapping with Haptoglobin

1. The anaerobic chamber surrounding the sample handling unit of the stopped-flow spectrophotometer is continuously flushed with purified N_2 for at least 48 h. The sample chamber is cooled by an external water bath to 21.5°C.
2. A sample of haptoglobin (16 μM) in a septum-sealed vial is equilibrated with 100% O_2 by flushing for 0.5–1 h.

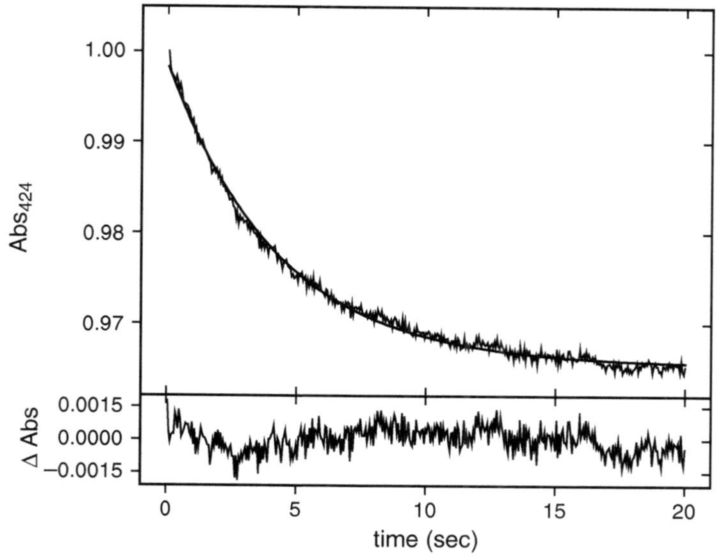

Fig. 4. Stopped-flow results. Absorbance decrease at 424 nm upon haptoglobin association with Zn(II) $\alpha\beta$ dimers after 24 h of anaerobic hybridization. The reaction time course is fit with a single exponential. The residual of this fit is shown below the panel.

3. Upon completion of hybridization, deoxygenated standard buffer is added to the hybridization mixture to give a final Hb concentration of 16 μM, followed by an anaerobic transfer to one of the drive syringes of the stopped flow. The pathlength is set to 2 mm and the dead time to 1 ms. The haptoglobin solution is transferred to the other drive syringe of the stopped-flow instrument. The change in absorbance upon mixing is monitored at 424 nm.
4. The absorbance decay follows a single exponential (Fig. 4), which can be fit either with TableCurve or SigmaPlot (Systat Software), both of which provide an accurate propagation of errors. The data analyzed should be averages of at least triplicates with calculation of standard deviations. The rate constant for dissociation of tetramer to free dimer thus observed is $k_{off} = 0.20 \pm 0.02\ s^{-1}$. Combined with the consensus dimer to tetramer association rate constant, k_{on}, to calculate the tetramer assembly free energy ΔG_2 ($\Delta G_2 = -RT \ln(k_{on}/k_{off})$) yields a value of $-9.1 \pm .1$ kcal/mol for Zn/FeO_2 (15).
5. Assignment of the observed absorbance change to dissociation of the hybrid tetramer, rather than to a parent tetramer, can be addressed by substituting the hybridization mixture with (a) the ZnHb parent, (b) the FeHb parent, or (c) a hybridization mixture at various times after mixing. The dependence of amplitude on incubation time can be

compared to the timecourse for species *21u* formation measured independently by subzero isoelectric focusing (8), to confirm that the increase in amplitude with incubation time is consistent with the formation of species *21u*. In another control experiment, the anaerobic hybrid mixture is reacted with an oxygenated buffer solution which contains no haptoglobin.

4. Notes

1. Procedures involving whole human blood require institutional preauthorization and added precautions, similar to those employed for handling infectious materials.
2. Blood is ideally collected fresh from known volunteers and used immediately to minimize the degree of auto-oxidation of the heme Fe. Outdated blood from blood banks typically has significant levels of oxidized heme (Fe(III)), which complicates purification procedures. The collection vacutainer should be EDTA-coated (lavender top), not heparin-coated. Unopened vacutainers can be stored upright at 4°C overnight, during which time the plasma and red blood cells will separate. The plasma can be carefully removed with a Pasteur pipette, greatly simplifying the washing process.
3. For both short-term and long-term storage, Hb is best stored in the absence of salts, which promote oxidation of the heme Fe. Hb acts as its own buffer. Small volumes can be frozen in cryo-vials, but for large volumes it is preferable to form beads by dropwise addition to liquid N_2 using a Pasteur pipette. The beads are then transferred to vials or bottles for storage in a dewar.
4. The concentration of Hb is very important at this stage. The efficiency of heme removal decreases dramatically as Hb concentration is increased. All concentrations listed are on a per heme basis.
5. Do not permit the globin pellet to warm during the transfer to and from the centrifuge. Denaturation of the globin is irreversible and will occur at 8°C or warmer.
6. Dialysis tubing should have a molecular weight cutoff of 10,000 or less, below the molecular weight of the individual subunits.
7. A pH below neutrality is important for maintenance of globin solubility.
8. Anaerobic work requires attention to detail and, above all, patience. Plastic vials and tubing absorb O_2 and time

is needed for flushing with purified N_2. Although the N_2 in the tank may be very pure, O_2 can leak at the regulator and at any of the fittings connected to the regulator, therefore, the N_2 should be scrubbed prior to entering the anaerobic chamber.

9. Sodium dithionite crystals are notoriously impure, and it is not unusual to detect a sulfurous odor when opening a new bottle. It is necessary to try different batches from different suppliers, and sometimes to purify by re-crystallization.

References

1. Ackers, G. K., Doyle, M. L., Myers, D., and Daugherty, M. A. (1992) Molecular code for cooperativity in hemoglobin, *Science* **255**, 54–63.
2. Ackers, G. K. (1998) Deciphering the molecular code of hemoglobin cooperativity, *Advances in Protein Chemistry* **51**, 185–253.
3. Smith, F. R., and Ackers, G. K. (1985) Experimental resolution of cooperative free energies for the ten ligation states of human hemoglobin, *Proc. Natl. Acad. Sci. USA* **82**, 5347–5351.
4. Ackers, G. K., and Holt, J. M. (2006) Asymmetric cooperativity in a symmetric tetramer: Human hemoglobin, *J. Biol. Chem.* **281**, 11441–11443.
5. Ackers, G. K., Dalessio, P. M., Lew, G. H., Daugherty, M. A., and Holt, J. M. (2002) Single residue modification of only one dimer within the hemoglobin tetramer reveals autonomous dimer function, *Proc. Natl. Acad. Sci. USA* **99**, 9777–9782.
6. Huang, Y., Doyle, M. L., and Ackers, G. K. (1996) The oxygen-binding intermediates of human hemoglobin: evaluation of their contributions to cooperativity using zinc-containing hybrids, *Biophys. J.* **71**, 2094–2105.
7. Huang, Y., Yonetani, T., Tsuneshige, A., Hoffman, B. M., and Ackers, G. K. (1996) Heterometallic hybrids of homometallic human hemoglobins, *Proc. Natl. Acad. Sci. USA* **93**, 4425–4430.
8. Ackers, G. K., Holt, J. M., Huang, Y., Grinkova, Y., Klinger, A. L., and Denisov, I. (2000) Confirmation of a unique intra-dimer cooperativity in the human hemoglobin $\alpha^1\beta^1$ half-oxygenated intermediate supports the symmetry rule model of allosteric regulation, *Proteins: Struct. Func. Gen.* ***Suppl.*** **4**, 23–43.
9. Ackers, G. K., Holt, J. M., Burgie, E. S., and Yarian, C. S. (2004) Analyzing intermediate state cooperativity in hemoglobin, In *Methods in Enzymology* (Holt, J. M., Johnson, M.J., and Ackers, G.K., Ed.), pp 3–28, Elsevier, San Diego, CA.
10. Holt, J. M., Klinger, A. L., Yarian, C. S., Keelara, V., and Ackers, G. K. (2005) Asymmetric distribution of cooperativity in the binding cascade of normal human hemoglobin. 1. Cooperative and noncooperative oxygen binding in Zn-substituted hemoglobin, *Biochemistry* **44**, 11925–11938.
11. Chu, A. H., Turner, B. W., and Ackers, G. K. (1984) Effects of protons on the oxygenation-linked subunit assembly in human hemoglobin, *Biochemistry* **23**, 604–617.
12. Huang, Y., and Ackers, G. K. (1995) Enthalpic and entropic components of cooperativity for the partially ligated intermediates of hemoglobin support a "Symmetry Rule" mechanism, *Biochemistry* **34**, 6316–6327.
13. Huang, Y., and Ackers, G. K. (1996) Transformation of cooperative free energies between ligation systems of hemoglobin: resolution of the carbon monoxide binding intermediates, *Biochemistry* **35**, 704–718.
14. Huang, Y., Koestner, M. L., and Ackers, G. K. (1996) Heterotropic effects of chloride on the ligation microstates of hemoglobin at constant water activity, *Biophys. J.* **71**, 2106–2116.
15. Turner, G. J., Galacteros, F., Doyle, M. L., Hedlund, B., Pettigrew, D. W., Turner, B. W., Smith, F. R., Moo-Penn, W., Rucknagel, D. L., and Ackers, G. K. (1992) Mutagenic dissection of hemoglobin cooperativity: effects of amino acid alteration on subunit assembly of oxy and deoxy tetramers, *Proteins: Struct. Funct. Genet.* **14**, 333–350.
16. Ip, S. H. C., Johnson, M. L., and Ackers, G. K. (1976) Kinetics of deoxyhemoglobin subunit dissociation determined by haptoglobin binding: estimation of the equilibrium constant from forward and reverse rates, *Biochemistry* **15**, 654–660.
17. Scholler, D. M., Wang, M. Y. R., and Hoffman, B. M. (1978) Metal-substituted hemoglobin and other hemoproteins, *Meth. Enzymol.* **52**, 487–493.

Chapter 3

Allosteric Coupling Between Transition Metal-Binding Sites in Homooligomeric Metal Sensor Proteins

Nicholas E. Grossoehme and David P. Giedroc

Abstract

Intracellular concentrations of transition metal ions are controlled at the transcriptional level by a panel of metalloregulatory proteins that collectively allow the cell to respond to changes in bioavailable metal concentration to elicit the appropriate cellular response, e.g., upregulation of genes coding for metal export or detoxification proteins in the event of metal excess. These proteins represent a specialized class of allosteric regulators that are ideal for studying ligand-mediated allostery in a comprehensive way due to the size, stability, reactivity, and the spectroscopic properties of transition metal ions as allosteric ligands. In addition to the commonly studied heterotropic regulation of metal binding and DNA binding, many of these proteins exhibit homotropic allostery, i.e., communication between two or more identical metal (ligand) binding sites on an oligomer. This chapter aims to guide the reader through the design and execution of experiments that allow quantification of the thermodynamic driving forces (ΔG_C, ΔH_C, and ΔS_C) that govern both homotropic and heterotropic allosteric interactions in metal sensor proteins as well as the steps required to remove the influence of complex speciation from the measured parameter values.

Key words: Metalloregulation, Metal sensor protein, Metals in biology, Isothermal titration calorimetry, ITC, Allosteric coupling free energy

1. Introduction

Allostery is the simple idea that binding of an effector molecule can influence the chemistry or reactivity at another, often spatially distinct, site (1, 2). This communication has evolved as a necessary feature of a wide variety of biological macromolecules and is essential for proper cellular function. Cellular sensory machinery, e.g., transcriptional regulatory proteins, takes advantage of this phenomenon by enabling communication between an effector binding site and a DNA binding interface such that occupancy of the effector site influences the affinity of the protein for its DNA operator. While other chapters in this book focus on other facets

Aron W. Fenton (ed.), *Allostery: Methods and Protocols*, Methods in Molecular Biology, vol. 796,
DOI 10.1007/978-1-61779-334-9_3, © Springer Science+Business Media, LLC 2012

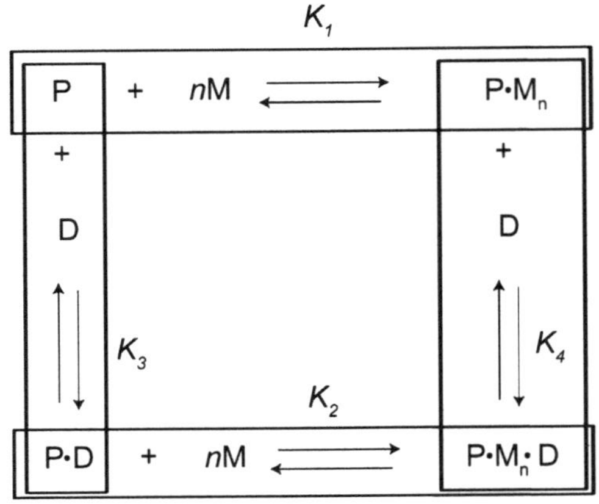

Fig. 1. Heterotropic allosteric coupling scheme. This general four-state thermodynamic cycle accounts for the possible states that a theoretical metalloregulator P can adopt. The vertical and horizontal boxes are the two equilibrium pairs that can be measured to determine the overall allosteric coupling energy, ΔG_C, as described in the text and the four equilibrium constants, along with concentration, dictates the population of each state. The horizontal reactions are defined by n metal (M) ions interacting with the apoprotein (P) or protein–DNA complex (P·D) to form a metallated protein (P·M$_n$) or the ternary metal–protein–DNA complex (P·M$_n$·D), respectively, while the vertical equilibria show the possible interactions the DNA (D) can have with P and P·M$_n$.

of measuring allostery ranging from specific techniques to mapping the communication pathway, this chapter develops a general approach to quantify the thermodynamic forces that govern communication between one binding site and another. We specifically focus on metalloregulatory or "metal sensor" proteins, a specialized class of allosteric transcriptional regulators, which are characterized by many features that make these proteins ideal models for understanding allosteric communication (3). These proteins, like other families of transcriptional regulators, provide an opportunity to observe two types of allosteric communication: heterotropic, which is the communication between the metal regulatory site and the DNA binding interface, and homotropic, which derives from the observation that identical metal-binding sites on homooligomeric proteins are characterized by distinct thermodynamic properties.

1.1. Heterotropic Allosteric Coupling

The widely accepted and applied model of allostery (4) that is drawn upon for this review is depicted in Fig. 1. This model is characterized by closed thermodynamic cycle ($\sum_{i=1}^{4} \Delta X_i = 0$, where X is any thermodynamic state function) with the four corners representative of the four states that a homooligomeric protein (P) can theoretically adopt. When the allosteric protein is a metalloregulator, these are the apoprotein, P, the fully metal coordinated protein with n metal-binding sites filled, P·M$_n$, an operator DNA associated state for the apoprotein, P·D, and the ternary complex, P·M$_n$·D, where the protein is fully coordinated by both metal and DNA ligands. The magnitude of the equilibrium constants, K_i, that describes the transition between these four states dictates the biological role of a metalloregulator. For example, if $K_4 > > K_3$, the P·M$_n$·D and P states are stable in solution and the protein represses metal uptake upon binding to the cognate metal ion, M. Alternatively, in the cases that $K_3 >> K_4$, the P·M$_n$ and P·D states are biologically relevant and the regulator is likely involved in derepression of the transcription of genes that encode proteins involved in efflux or detoxification of the cognate metal ion, M (3, 5, 6).

This scheme provides a general means to determine the degree to which an effector molecule influences the affinity of a metal sensor protein for the DNA operator, i.e., the degree to which the regulatory protein allosterically couples the two distinct ligand binding sites. When considering the discussion above, it can be concluded that this coupling free energy, ΔG_C, is related to the relative magnitudes of these two equilibrium constants. Indeed, the coupling energy is extracted from the ratio

$$K_C = \frac{K_4}{K_3} = \frac{K_2}{K_1} = \frac{[P][P \cdot M_n \cdot D]}{[P \cdot D][P \cdot M_n]} \tag{1}$$

and describes the ligand exchange equilibrium

$$P{\cdot}M_n + P{\cdot}D \rightleftharpoons P + P{\cdot}M_n{\cdot}D \tag{2}$$

This is perfectly consistent with the example given above. When $K_4 >> K_3$, the products of Eq. 2 are favored. As Eq. 1 shows, this comparison can also be made between the overall metal-binding equilibria, K_1 and K_2, which suggests that quantifying the allosteric coupling between metal binding and DNA binding can be accomplished if either pair of equilibria (vertical or horizontal boxes in Fig. 1) can be measured. It then follows that the coupling free energy, ΔG_C, is simply calculated from the thermodynamic relationship in Eq. 3 and can be further described according to other fundamental thermodynamic relationships (Eqs. 4–5).

$$\Delta G_C = -RT \ln K_C, \tag{3}$$

$$\Delta H_C = \Delta H_4 - \Delta H_3 = \Delta H_2 - \Delta H_1, \tag{4}$$

$$\Delta S_C = \Delta S_4 - \Delta S_3 = \Delta S_2 - \Delta S_1, \tag{5}$$

In the simplified scheme developed above, K_1 and K_2 describe the overall reaction and are, therefore, products of the stepwise equilibrium constants, $K_a K_b$,...K_n defined by units of M^{-n}. Thus, to explicitly consider stepwise binding constants, K_a, K_b,....K_n, the simplified cycle in Fig. 1 needs to be expanded to include these additional equilibria, as shown in Fig. 2. Note that this scheme with $n = 2$ contains two overlapping $n = 1$ allosteric cycles, corresponding to the first and second metal-binding events, which are highlighted by the gray and white boxes, respectively. It then follows that *if* the two sequential metal-binding events are considered a single overall event (β_{M2} and β_{MD2}, where $\beta_{M2} = K_{M1} \cdot K_{M2}$ and β_{M2} and β_{MD2}, where $\beta_{MD2} = K_{MD1} \cdot K_{MD2}$), the coupling cycle collapses to that described in Fig. 1 (see Note 1). Calculating the allosteric energies for the individual cycles (gray and white boxes in Fig. 2) can be accomplished exactly as described for the simplified $n = 1$ case, with the inclusion of the additional equilibria in the overall cycle in Fig. 2

$$K_C = \frac{K_{DNA3}}{K_{DNA1}} = \frac{\beta_{MD2}}{\beta_{M2}} = \frac{[P][P \cdot M_2 \cdot D]}{[P \cdot D][P \cdot M_2]}, \quad (6)$$

where K_C describes the ligand exchange reaction

$$P \cdot M_2 + P \cdot D \rightleftharpoons P + P \cdot M_2 \cdot D \quad (7)$$

and the coupling enthalpy and entropy are calculated with

$$\begin{aligned} \Delta H_C &= \Delta H_{DNA3} - \Delta H_{DNA1} \\ &= (\Delta H_{MD1} + \Delta H_{MD2}) - (\Delta H_{M1} + \Delta H_{M2}), \end{aligned} \quad (8)$$

$$\begin{aligned} \Delta S_C &= \Delta S_{DNA3} - \Delta S_{DNA1} \\ &= (\Delta S_{MD1} + \Delta S_{MD2}) - (\Delta S_{M1} + \Delta S_{M2}). \end{aligned} \quad (9)$$

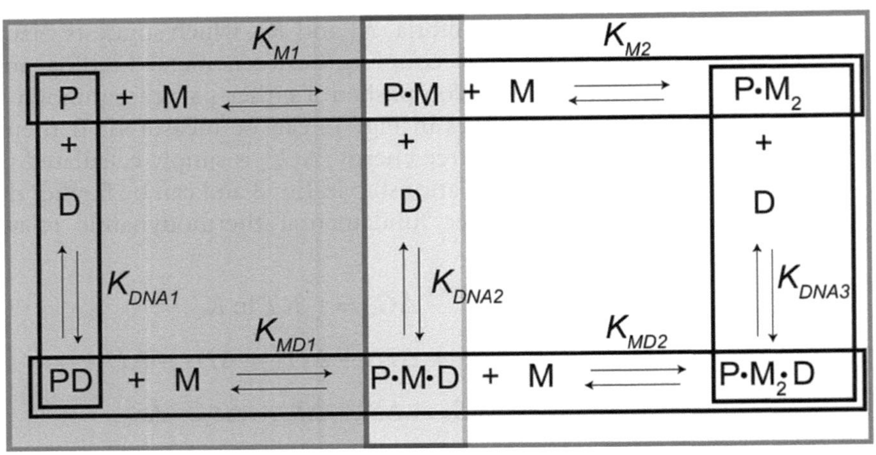

Fig. 2. Expanded heterotropic allosteric coupling scheme. This general six-state thermodynamic cycle accounts for the four allosteric "end" states a homodimeric metalloregulatory protein (P) can hypothetically adopt: apo (P), metal-bound ($P \cdot M_2$), DNA-bound apoprotein (P·D) and a "ternary" protein–metal–DNA complex ($P \cdot M_2 \cdot D$). In addition, two intermediate states are shown corresponding to the singly metallated protein (P·M) and "ternary" complex (P·M·D). The *outer vertical and horizontal boxes* are the two equilibrium pairs that can be measured to determine the overall allosteric coupling energy, ΔG_C, as described in the text. The *gray and white boxes* highlight the thermodynamic cycles that can be used to determine the stepwise coupling energies for the individual metal ions.

1.2. Homotropic Allosteric Coupling

Homotropic allostery in transcriptional regulators is specific to proteins, commonly homodimeric, that contain at least two identical effector binding sites, as represented in the two horizontal equilibria in Fig. 2. If the binding of the first ligand influences the thermodynamic properties of the second binding event, then this system exemplifies homotropic allostery. Measuring this form of allostery is particularly convenient since it can be observed in a single metal-binding experiment, as can be seen in the representative titration in Fig. 3. This titration clearly shows two metal ions,

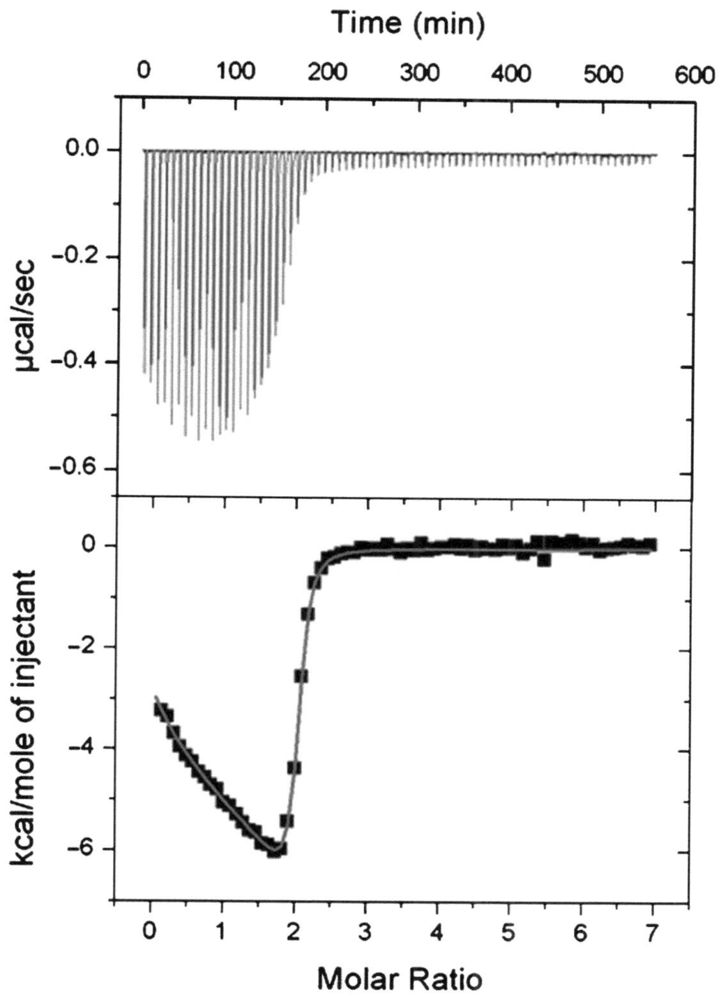

Fig. 3. Representative ITC titration indicative of negative allosteric communication between two identical binding sites on a homodimer. 1.4 mM Zn^{2+} titrated into 34 μM CzrA dimer in 50 mM ACES and 400 mM NaCl at 25°C and pH 7.0 (15). The *top panel* show the raw ITC data with each peak corresponding to an individual 3 μL injection (first injection is 1 μL) plotted as power vs. time. The *bottom panel* show the integrated, concentration normalized data plotted as ΔH vs. the Zn^{2+}-CzrA dimer molar ratio. Here, we see that each CzrA dimer binds two nonequivalent Zn^{2+} ions with negative cooperativity; however, this system is interestingly characterized by a homotropic coupling enthalpy, ΔH_C, that opposes ΔG_C. This feature has been observed in other allosteric regulators (31).

M_1 (Zn_1) and M_2 (Zn_2), binding to a protein (CzrA) with distinct nonequivalent thermodynamics, exemplifying homotropic allosteric communication. Since standard fitting procedures do not account for this necessary statistical feature, it is essential to correct the macroscopic binding constants, K_{M1} and K_{M2} (or K_{MD1} and K_{MD2}). This is accomplished by casting the K_{M1} and K_{M2} in terms of a single microscopic binding event, k, and a cooperativity term, ω, with a factor of 2 included to account for the two statistically equivalent metal-binding sites, as demonstrated in Eqs. 10–12 (7).

$$K_{M1} = 2k, \tag{10}$$

$$K_{M2} = \frac{\omega k}{2}, \tag{11}$$

$$\Delta G_C = -RT \ln \omega = -RT \ln \frac{4K_{M2}}{K_{M1}}. \tag{12}$$

According to this analysis, ω is the microscopic coupling constant and its magnitude dictates the degree to which two homotropic binding events are coupled. When $\omega > 1$, the binding of the first ligand increases the affinity of the 2nd ligand (positively cooperative) and, conversely, when $\omega < 1$, the system is characterized by negative cooperativity. The enthalpic and entropic coupling energies for homotropic interactions can be calculated from Eqs. 13 and 14, respectively.

$$\Delta H_C = \Delta H_{M2} - \Delta H_{M1}, \tag{13}$$

$$\Delta S_C = \frac{-\Delta G_C + \Delta H_C}{T}. \tag{14}$$

2. Materials

1. The ultimate goal of the approach outlined here is to quantify the global thermodynamics that drive allosteric processes. Therefore, a sensitive microcalorimeter is necessary. Two common commercial sources for this instrument are MicroCal, LLC. (Northampton, MA) and TA Instruments (Lindon, UT), respectively. In addition to the standard 1.4 or 1 mL reaction cell volumes, each of these suppliers also offer a more contemporary model that minimizes sample volume (~200 μL) and increases absolute sensitivity (8) (see Note 2).
2. Reagents of the highest purity are strongly recommended since even modest contamination can significantly influence the solution properties of metal ions. High-purity standard biological reagents (buffers, etc.) are typically available

from a variety of suppliers (e.g., Sigma, Fisher, VWR). Most metal salts are available in ultra high purity grade from Alfa Aesar.

3. If anaerobic conditions are necessary, appropriate oxygen free chambers are necessary for sample preparation and calorimetric measurement.
4. DNA binding and metal into protein/DNA complex titrations requires duplex DNA corresponding to the native operator sequence. Single stranded DNA oligonucleotides are commercially available in high purity from a number of commercial sources including Operon (Huntsville, AL) and IDT (Coralville, IA). Following purification (see Note 3), accurately determine ssDNA concentration using the appropriate extinction coefficient (http://www.idtdna.com/analyzer/Applications/OligoAnalyzer/). Anneal the two strands by mixing equimolar concentrations in an eppendorf tube and heating to 95°C followed by slowly cooling to room temperature, with DNA duplex formation confirmed by native polyacrylamide gel electrophoresis. Care must be taken to avoid fold-back intramolecular DNA hairpin structures that might arise from the palindromic or nearly palindromic and, therefore, self-complementary nature of the individual ssDNA strands (see Note 4). DNA duplexes prepared in this way at high strand concentrations are stored at −20°C and are stable indefinitely.
5. Protein preparation should be carried out by standard protocols (see Note 5). It is recommended that samples estimated to be $\geq$ 95% pure by SDS-PAGE are used.

3. Methods

3.1. Solution Condition Considerations

Conditional variability can appreciably influence the heat measured by bulk thermodynamic techniques such as ITC. It is, therefore, of paramount importance to select appropriate solution conditions. Listed below are a number of conditional variables that should be specifically addressed. Note that this is not an exhaustive list and additional considerations should be identified on an individual experimental basis. In general, it is recommended to select solution conditions that generate quantifiable speciation to enable robust mathematical analysis.

1. *pH.* In addition to the obvious effects of pH on biological macromolecules (e.g., acid/base catalyzed hydrolysis), pH can significantly influence the solution properties of cationic metals. For example, at basic or neutral pH, $Fe(OH)_3$ is

sparsely soluble ($K_{sp} = 2.64 \times 10^{-39}$ M^2 (9)). The ideal pH would mimic the conditions that dictate the chemistry in vivo. Note that the pH can significantly influence the apparent affinity, as discussed later.

2. *Buffer.* The obvious criterion for buffer selection is to maintain a constant pH. However, when transition metal ions are of interest, metal-buffer equilibria must be considered explicitly since nearly all buffers associate with metal ions to some degree (10, 11), although a number of these are very weakly coordinating (12). It is also notable that some buffers promote redox activity through stabilization of one oxidation state relative to another, e.g., copper (13). As noted above, it is recommended to use buffers that are characterized by quantifiable metal-buffer affinity and speciation so as to limit unquantifiable metal solution chemistry. A number of resources are available to guide the reader to buffers with metal chemistries rigorously quantified (10, 11). If the experiment necessitates a specific buffer and thermodynamic information is not available for this system, these values may be determined calorimetrically using the guidelines provided below (14–17). Some experiments require the presence of a strong metal chelator to enable measurement of a quantifiable binding curve (vida infra); in these cases the metal-buffer interaction may become insignificant.

3. *Ionic Strength.* The ionic strength (I) of a solution directly influences the activity coefficient, and hence the measurable equilibrium constant. Eq. 15 shows Debye–Hückel relationship which indicates that I is a function of the total ionic content and scales with the square of the valency (z_i) for each species (i).

$$I = 1/2 \sum_i z_i^2 [i] \quad (15)$$

In theory, all ions in solution should be included in this calculation; however, under typical experimental conditions (50–500 mM monovalent salt concentration (see Note 6)), the contribution from metal salts and ligand are commonly negligible, although it may be necessary to consider the influence of the buffer at elevated concentrations. High concentrations of DNA, as well as its counter ion, can have a very significant impact on ionic strength; however, the simple relationship presented in Eq. 15 cannot describe polyelectrolyte contributions appropriately; a nonlinear Poisson–Boltzmann analysis is necessary for accurately accountability (18). Owing to the difficulty of these calculations, and typically minor influence it has on the data, they are commonly ignored. 100 mM monovalent salt has commonly been used for metal-binding experiments (11) and provides a standard for direct comparisons between different systems.

4. *Temperature.* Equilibrium constants are inherently temperature-dependent and most biological processes are characterized by $\Delta C_p \neq 0$ resulting in ΔH variance with temperature. Since K_i and ΔH are the two directly measurable variables in an ITC experiment, temperature is a very important experimental variable. Fortunately, modern calorimeters maintain a constant temperature (~4–80°C) over the course of very long experiments. Therefore, one only needs to decide the most appropriate temperature for the system of interest, with 25 and 37°C commonly employed.
5. *Oxygen Sensitivity.* Owing to the largely reducing potential of most intracellular environments, metal ions and surface cysteines tend to be in a reduced state, i.e., Cu^+ and Cys-SH or Cys-S^- as compared to Cu^{2+} and Cys-S–S-Cys. For systems that are susceptible to metal and/or ligand oxidation, anaerobic preparation and experimentation is strongly advised. The use of "pseudo-anaerobic" or chemically reducing conditions is fraught with serious shortcomings. Most common reducing agents interact with metal ions with significant affinities, with dithiothreitol (DTT) being the dramatic particularly notable case as many metal ions make very high-affinity metal–DTT complexes (11, 19). These complexes outcompete the desired metal–protein interactions, particularly in light of the large molar excess relative to protein that is necessary for these experiments. Further, comprehensive control experiments are required to verify that the unavoidable redox chemistry in an aerobic environment, e.g., reducing agent oxidation, does not influence the net heat flow (i.e., the measurable variable). For these reasons, it is strongly advised that oxygen-sensitive experiments be conducted under strictly anaerobic conditions in the absence of chemical reductants (see Subheading 3).

3.2. Metal-Free Buffers

As this guide is geared toward quantifying metal-ion-driven chemistries, it is necessary to ensure that all materials are prepared in a way to minimize contamination. For systems not focusing on metals, the guidelines below are still worthwhile, as they ensure minimal interference from extraneous sources.

1. *Preparation of Glassware.* Standard silicate laboratory glassware is very susceptible to metal contamination. This is particularly true for "hard" metals (i.e., Fe^{3+} and Mg^{2+}), which form strong electrostatic interactions with anions

(20). Metalloregulatory proteins, which can possess very high affinities for their cognate metal ion (3, 21), can, therefore, leach metals from contaminated glass surfaces. This is easily avoided by ensuring that all glassware is soaked in 1% nitric acid (HNO_3) enabling protons to outcompete metal ions at the surface. Following acid treatment, the glassware should be rinsed exhaustively (≥3×) with metal-free water (see below) to avoid an unwanted change in the pH of buffer solutions.

2. *Metal-Free Water.* Standard RODI water purification, common in most research laboratories, is not sufficient to remove metal ions to the degree required for quantitative metal-binding experiments. Further purification can be provided by numerous standard purification systems that are capable of deionization to a resistance ≥18 MΩ cm. Alternatively, strong metal chelators conjugated to solid styrene beads are commercially available, i.e., Chelex, and can be used to treat laboratory-grade RODI water to produce operationally defined "metal-free" water. This can be accomplished by passing water through a vertical column containing chelating resin and collecting it in an acid-washed container. Alternatively, the resin can be added directly into the water and shaken for several hours. Incubation at elevated temperatures can expedite this procedure. Separate the phases by centrifugation and careful decanting.
3. *Buffer Preparation.* Buffer salts, as provided by the manufacturer, are commonly contaminated with small amounts of divalent metal ions. Removal of metal ions can easily be accomplished by treating the prepared buffer with Chelex, as described just above. Note that Na^+ or H^+ ions (depending on the regeneration protocol used for the Chelex resin) replace the metals to maintain electrostatic neutrality and, depending on the amount of metal removed from the buffer, this may be significant. pH and conductivity measurements of the post-Chelex solution are strongly suggested.

3.3. Anaerobic Preparations

For cases that require extra care to minimize oxidative conditions as discussed in Subheading 1, additional steps must be taken to ensure a rigorously anaerobic environment, since thoroughly deoxygenated buffers and solutions are required by these experiments.

1. Prepare the buffer solution using metal-free water and remove residual metal as necessary (see Subheading 2).
2. Two standard degassing protocols are available to ensure that all solutions are free of oxygen. The first method is more

rapid, but has the potential for mild increase in salt and buffer concentration due to water evaporation (see Note 7), while the second is much more thorough and minimizes solvent evaporation (see Note 8).

3. Stock metal solutions should be prepared and stored under an inert atmosphere. The simplest method is to dissolve a known mass of metal salt in an anaerobic chamber (Vacuum Atmospheres or Coy). If an anaerobic chamber is not available, deoxygenation can be accomplished suboptimally by extensive bubbling of argon or nitrogen from a cylinder of compressed gas through a metal stock solution. The concentration of this stock solution should be verified by standard metal quantification techniques prior to use.
4. Completely buffer exchange the purified protein into an oxygen-free buffer. This is accomplished by concentrating the protein stock to ~12 mL, transferring to an anaerobic chamber and dialyzing at least 4 h in 500 mL of the buffer to be used to metal-binding experiments. The dialysis buffer should be exchanged four times to ensure complete removal of oxygen and metal chelators or reducing agents that may have been used during protein purification.

3.4. Metal-Binding Assays

In general, many experimental design considerations are discussed in detail in instrument manuals (22) and any user should be familiar with these publications. It is recommended that titrations are designed that involve well-defined metal–chelator interactions, as direct titrations involving "free" metal ions pose a number of potential problems. Some common examples of useful chelators include nitrilotriacetic acid (NTA), imidazole, triethylenetetramine (trien), ethylenediamine (en), and Tris buffer; however, note that different degrees of competition are elicited by each of these(10, 23) (vida infra). As mentioned above, unknown solution chemistry may be occurring that cannot be appropriately accounted for. Further, it is highly unlikely that metal ions are present in a cellular environment not associated with a cellular chelator. Second, it is quite likely that a direct titration of metal → protein lies outside of the experimental window that allows for a robust fit to a unique binding constant. An example titration of Zn^{2+} into CzrA, a homodimeric Zn^{2+}/Co^{2+} sensor from *Staphylococcus aureus*, is shown in Fig. 3.

1. Ensure that the calorimeter is well maintained, calibrated, and operating properly. Refer to the instrument manual for specific directions.
2. Prepare at least 2 mL of 10–100 μM protein in a predetermined experimental buffer. Ensure that the buffer is metal free as described in Subheading 2. At least four rounds

of dialysis are recommended. The dialysis buffer from the last round should be saved to prepare the metal solution and used in the reference cell of the calorimeter.

3. Prepare at least 1 mL of 1 mM (see Note 9) metal salt from a stock solution (see Note 10) in the appropriate experimental buffer. It has been our experience that preparing extra (~10 mL) metal titrant solution is ideal because it enables concentration determination (via AAS or ICP-MS) of this solution eliminating error arising from dilution. To ensure that the buffer exactly matches the protein buffer, the buffer from the final protein dialysis step may be used.
4. Ensure that the calorimeter has been thoroughly cleaned (see Note 11).
5. Rinse the sample cell and titration syringe with the experimental buffer.
6. Load the titration syringe, sample cell, and reference per manufacturer's recommendations.
7. Insert the appropriate experimental parameters. An iterative process is likely required to determine the ideal parameter values. While the experiment is running, pay close attention to the injection volumes, as large injections may mask data inflections, and make fitting impossible or inaccurate (see Note 12). If the instrument cycle is complete before the reaction has reached completion, it may be possible to refill the syringe and continue the experiment (this can be done multiple times if necessary, refer to instrument specifications). Once all data are acquired, the files can be concatenated. MicroCal has developed software to automate this procedure.
8. Repeat the experiment at least two more times. Make appropriate adjustments to the experimental parameters.

If the goal is to quantify the heterotropic coupling energy, this set of experiments must be repeated with the protein–DNA complex. Duplex DNA preparation is described in Subheading 2. Note that this measurement may be very challenging unless very high concentrations can be reached in cases where the affinity of the ternary P·M_n·D complex is low. It is strongly urged that the investigator show that all of the protein be quantitatively contained in the protein–DNA complex after the completion of the metal-binding experiment (15).

1. Select a gel filtration column capable of separating dsDNA and the metal-bound or apo protein complex from the protein–DNA complex. Pass each of these solutions across the column to generate a retention volume profile for the system of interest.

2. Collect the contents of the reaction cell once the reaction has reached completion. Take all necessary precautions regarding specific techniques necessary for the system (i.e., anaerobic).
3. Load the sample on the calibrated gel filtration column, making sure to monitor the absorbance at 260 nm and 280 nm. If the protein of interest has a low molar extinction coefficient, monitor at 240 nm as well.
4. Collect the appropriate fractions.
5. Quantify the total metal concentration contained in these two samples.
6. Calculate the total metal–protein stoichiometry for the two samples, ensuring an accounting for dilution.

3.5. Accounting for Speciation

3.5.1. Adjusting the Binding Constant

Metal speciation, which comprises metal interactions with other solutes to form multiple metal–solute complexes, is unavoidable under the solution conditions necessary to conduct these experiments. Standard ITC data fitting packages cannot account for these competing equilibria. It is, therefore, necessary to correct the optimized thermodynamic parameters obtained by fitting the experimental data to a standard physical model(s). Provided the experiment is carefully designed, all competing interactions with these solutes, which can be any experimental buffer salt (e.g., Tris) or low-affinity metal chelator (e.g., NTA), collectively referred to as "B" here (see Note 2), are known and readily accounted for as described below. For the purpose of this exercise, it is assumed that the metal interacts with the experimental buffer (B) to form two metal-buffer species, MB and MB_2, where B is the form of the buffer that complexes with the metal ion, likely the fully deprotonated form, and the higher affinity ligand, L, is present as a single molecular species. This approach can easily be extended to more complex systems that explicitly consider ligand (L) protonation chemistry that is likely to occur over the biological pH range (see Note 13) as described below (Eq. 24 is a representative equilibrium).

Consider the following expression:

$$(1 - \alpha_{MB} - \alpha_{MB2})M + \alpha_{MB}MB + \alpha_{MB2}MB_2 + L \rightleftarrows ML + (\alpha_{MB} + 2\alpha_{MB2})B \qquad (16)$$

In this expression, α_{MB} and α_{MB2} are the stoichiometric coefficients associated with MB and MB_2, respectively, and determined from the individual equilibrium constants, K_i or β_i (see Notes 1 and 14).

$$\alpha_{MB} = \frac{[MB]}{C_M} = \frac{K_{MB}[B]}{\alpha_{Buffer}}, \qquad (17)$$

$$\alpha_{\mathrm{MB2}} = \frac{[\mathrm{MB}]}{C_{\mathrm{M}}} = \frac{\beta_{\mathrm{MB2}}[\mathrm{B}]^2}{\alpha_{\mathrm{Buffer}}}, \tag{18}$$

$$\alpha_{\mathrm{Buffer}} = 1 + K_{\mathrm{MB}}[\mathrm{B}] + \beta_{\mathrm{MB2}}[\mathrm{B}]^2. \tag{19}$$

The standard *one-site* fitting model assumes a direct interaction between metal (M) and ligand (L) according to

$$\mathrm{M} + \mathrm{L} \rightleftharpoons \mathrm{ML}, \tag{20}$$

$$K_{\mathrm{cal}} = \frac{[\mathrm{ML}]}{[\mathrm{M}][\mathrm{L}]}. \tag{21}$$

This assumption ensures that all metal ions are present as the free hydrated metal (M) or associated with the macromolecular ligand L such that

$$K_{\mathrm{cal}} = \frac{[\mathrm{ML}]}{(C_{\mathrm{M}} - [\mathrm{ML}])(C_{\mathrm{L}} - [\mathrm{ML}])} = \frac{[\mathrm{ML}]}{[\mathrm{M}][\mathrm{L}]}, \tag{22}$$

where C_{M} and C_{L} are the total metal and ligand concentrations, respectively. This, of course, is not the case for the chemical system described by Eq. 16. Therefore, Eq. 22 must be expanded to account for the additional chemical species that are present in the reaction.

$$\begin{aligned} K_{\mathrm{cal}} &= \frac{[\mathrm{ML}]}{(C_{\mathrm{M}} - [\mathrm{ML}])(C_{\mathrm{L}} - [\mathrm{ML}])} \\ &= \frac{[\mathrm{ML}]}{\left(1 + K_{\mathrm{MB}}[\mathrm{B}] + \beta_{\mathrm{MB2}}[\mathrm{B}]^2\right)[\mathrm{M}][\mathrm{L}]} = \frac{K_{\mathrm{ML}}}{\alpha_{\mathrm{Buffer}}}. \end{aligned} \tag{23}$$

It then follows that the apparent binding constant, K_{cal}, can be converted to the competition-independent value, K_{ML}, by simply multiplying it by the metal-buffer reaction quotient, α_{Buffer}. If it is necessary to account for acid–base equilibria of the ligand, L, a corresponding competition value may be added to this expression (see Note 13).

3.5.2. Adjusting the Enthalpy

To extract the enthalpy change associated with metal–ligand interactions of interest, and the corresponding protein structural or dynamic response (Eq. 20), collectively ΔH_{ML}, the heats associated with the coupled chemical events must be subtracted from the measured calorimetric enthalpy, ΔH_{cal}. To accomplish this, Hess' Law is applied to the overall equilibrium which enables the isolation of individual equilibria that contribute to the net evolved heat. For this example, we extend the equilibrium in Eq. 16 to include a single protonation/deprotonation event on the ligand since subtraction of proton-buffer heats is required.

$$(1-\alpha_{MB}-\alpha_{MB2})M+\alpha_{MB}MB+\alpha_{MB2}MB_2+(1-\alpha_{HL})L+\alpha_{HL}HL \rightleftarrows ML+(\alpha_{MB}+2\alpha_{MB2}-\alpha_{HL})B+\alpha_{HL}HB, \quad (24)$$

$$\alpha_{HL}=\frac{[HL]}{C_L}=\frac{K_{HL}[H]}{\alpha_{Proton}}, \quad (25)$$

$$\alpha_{Proton}=1+K_{HL}[H]. \quad (26)$$

To begin the disassembly of Eq. 24, we consider three categories of equilibria: those involving metal–buffer (MB_i) interactions (Eqs. 27–28), those involving proton flow (Eqs. 29–30), and finally the desired metal–ligand interaction (Eq. 31). Note that these equations are written such that they sum up the overall equilibrium and Eqs. 27–29 are the reverse of the association reaction; therefore, a negative sign must be placed in front of standard state functions to describe the equations as written, as indicated.

$$\{MB \rightleftharpoons M+B\}\alpha_{MB} \quad -\alpha_{MB}\Delta H_{MB}, \quad (27)$$

$$\{MB_2 \rightleftharpoons M+2B\}\alpha_{MB2} \quad -\alpha_{MB2}\Delta H_{MB2}, \quad (28)$$

$$\{HL \rightleftharpoons L+H\}\alpha_{HL} \quad -\alpha_{HL}\Delta H_{HL}, \quad (29)$$

$$\{B+H \rightleftharpoons HB\}\alpha_{HL} \quad -\alpha_{HL}\Delta H_{HB}, \quad (30)$$

$$M+L \rightleftharpoons ML \quad -\Delta H_{ML}. \quad (31)$$

According to this scheme, the desired heat associated with Eq. 31 (ΔH_{ML}) can be calculated by subtracting the heat from Eq. 27 through Eq. 30 from the overall measured enthalpy, ΔH_{cal}.

$$\Delta H_{ML}=\Delta H_{cal}+\alpha_{MB}\Delta H_{MB}+\alpha_{MB2}\Delta H_{MB2}+\alpha_{HL}\Delta H_{HL}-\alpha_{HL}H_{HB}. \quad (32)$$

However, this necessitates a priori knowledge of all these values. While this is readily done for ΔH_{HB} from literature sources, the specific set of conditions influences all of the other competing equilibria. Therefore, it is necessary to experimentally determine each of these. As a representative example, the protocol below employs the well-characterized metal–EDTA interactions (11) to determine the total metal-buffer heat, which is the sum of the individual metal–buffer interactions, as shown in Eq. 33.

$$\Delta H_{MBtot}=-(\alpha_{MB}\Delta H_{MB}+\alpha_{MB2}\Delta H_{MB2}). \quad (33)$$

1. Ensure cleanliness and proper operation of the calorimeter (see Note 11).
2. Prepare at least 2 mL of 100 μM (see Note 10) EDTA in an identical buffer used for the metal–protein titration.
3. Prepare at least 1 mL of 1 mM (see Note 9) metal salt in identical experimental buffer.
4. Conduct the titration using the appropriate instrument settings. It is recommended to use the same settings as that for the metal–protein titration. Ideally, a step function is observed; consider a tighter binding ligand if this is not the case.
5. Repeat at least twice.
6. Average the measured heats. Propagate error appropriately (24).
7. Calculate α_{HL} and α_{H2L}, and α_{Proton} (see Note 15) from $K_{HL} = 10^{9.52}$ and $\beta_{H2L} = 10^{15.65}$ for EDTA (the ligand (L) in this case).
8. Calculate ΔH_{MBtot} according to

$$\Delta H_{MBtot} = \Delta H_{cal} - \alpha_{HL}(\Delta H_{HB} - \Delta H_{HL}) - -\Delta H_{ML}. \quad (34)$$

This can then be used to subtract ΔH_{MBtot} from any reaction under identical conditions (i.e., Eq. 32). Further, this set of equations and approach is completely generic and can be applied to any metal–ligand interaction assuming that adjustments are made in each category for additional (or fewer) species.

3.5.3. Determining Protons Displaced Calorimetrically

As described above, due to the dramatic influence buffer–proton interactions can have on the overall ΔH, it is necessary to subtract this value from ΔH_{cal}. To accomplish this correctly, the net number of protons displaced or taken up (n_{H+}) must be known.

$$n_{H+} = \sum_i i\alpha_{HiL}. \quad (35)$$

If all relevant pK_as are known, this value can easily be calculated according to Eq. 29 and the experimental pH; however, for biological macromolecules, it is likely that it needs to be experimentally determined. While this can be accomplished by other methods (25), determining this value calorimetrically provides additional thermodynamic data and makes the optimized parameter values more statistically robust. This method (26, 27) is based on the fact that under identical conditions of pH, salt concentration and type, and temperature, changing the buffer (B) leads only to a change in the heat associated with the buffer chemistry; for the case at hand, this includes metal–buffer and proton–buffer interactions.

1. Select three or more buffers that maintain a constant pH at the experimental pH of interest, with each characterized by a significantly different ΔH_{HB} (11).
2. Prepare the protein and buffer taking the necessary precautions to ensure otherwise identical solution conditions and using the same batch of protein stock as before if possible.
3. Determine $\Delta H_{\mathrm{MBtot}}$, as described in Subheading 5.2. Note that some metal–buffer pairs can influence the ligand exchange kinetics; therefore, keenly monitor the first few injections as well as the injections nearing stoichiometric equivalence to be sure that equilibrium is attained after each injection (see Note 12).
4. Conduct metal → protein titrations in triplicate in the each buffer, again taking care to ensure that all experiments are carried out at the same temperature while monitoring the progress of successive injections (see Note 12).
5. Repeat this procedure with more buffers as needed. At least three buffers are required for a robust determination of the $n_{\mathrm{H+}}$.

Once the data are collected and the metal–buffer interactions are subtracted, a plot of ΔH_{HB} vs. $\Delta H_{\mathrm{cal}} - \Delta H_{\mathrm{MBtot}}$ generates a straight line with the slope corresponding to the number of protons displaced (or consumed), $n_{\mathrm{H+}}$, under this specific set of solution conditions, according to

$$(\Delta H_{\mathrm{cal}} - \Delta H_{\mathrm{MBtot}}) = n_{\mathrm{H+}}\Delta H_{\mathrm{HB}} + (\Delta H_{\mathrm{ML}} + \Delta H_{\mathrm{HLtot}}), \quad (36)$$

where $\Delta H_{\mathrm{HLtot}}$ is the total enthalpy associated with proton–ligand interactions and $\Delta H_{\mathrm{ML}} + \Delta H_{\mathrm{HLtot}}$ is the heat associated with the metal–protein interaction and coupled proton release, both of which are constant throughout this series of experiments.

4. Notes

1. Standard equilibrium nomenclature defines β as an overall binding constant and is, therefore, the product of sequential equilibrium constants (e.g., $\beta_{\mathrm{MB2}} = K_{\mathrm{MB}}K_{\mathrm{MB2}}$). Ensure that the equilibrium constants used for these calculations describe the association equilibria. For example, in the generic reaction

$$\mathrm{X} + \mathrm{Y} \rightleftharpoons \mathrm{XY}$$

$$K = [XY]/[X][Y]$$

2. Standard nomenclature for ITC experiments, as used in MicroCal, LLC VP-ITC user manual (23), refers to the molecule in the injection syringe as the *ligand* (L) and the molecule in the reaction cell as the *macromolecule* (M). However, to remain consistent with common metal speciation nomenclature (11), in this manuscript we explicitly use M to refer to the *metal ion* and L refers to the *ligand* that binds to the metal ion in which the heat of association is measured. In our examples, we use L to indicate either protein or EDTA and include B as a representative molecule that interacts with the metal ion at a lower affinity than the *ligand* of interest in a competitive manner.
3. Large quantities of DNA are needed for these experiments, particularly given the standard 1.0 or 1.4 mL reaction volumes of the nanoITC (TA) or VP-ITC (MicroCal), respectively. Several 1 μmol scale DNA syntheses are likely needed to generate adequate material for these experiments. Single-stranded DNAs are routinely purified by denaturing PAGE followed by electroelution ($l > 20$ nts) or high resolution anion exchange chromatography ($l \leq 20$ nts), and ethanol precipitation. In the case of denaturing PAGE-purified DNAs, complete removal of acrylamide and urea is ensured by a final reverse phase clean up step using prepacked C18 columns (Alltech) and elution with 50% methanol. Dry to completeness with a SpeedVac.
4. Some duplex DNA operator sequences are palindromic or twofold symmetric or nearly so. As a result, a ssDNA hairpin may be thermodynamically favored over the intermolecular dsDNA. Strand annealing under conditions of high monovalent salt concentration (0.5–1 M NaCl) promotes duplex formation. Additionally, increased strand concentration may be needed to favor the intermolecular duplex formation. Note that rapid cooling should be avoided, as this process favors hairpin formation.
5. Since metal-binding events are of particular interest, it is strongly advised that purification by His-tags is avoided. Obviously, high-affinity interactions between these motifs and metal ions can obscure the metal-binding events of functional interest.
6. Elevated monovalent salt concentration in these systems is often required to enhance protein solubility. Further, if the protein–DNA interactions are of interest, high salt is often needed to ensure that the affinities are within the measurable range due to the sizeable electrostatic contribution to K_a for essentially all protein–DNA interactions (28–30). For a direct

comparison of binding affinities between different systems, the solution conditions must be identical.

7. Transfer/prepare the buffer in a 2–3 L round bottomed vacuum flask (a reaction flask from Kontes works well for this purpose). Attach the flask to a dual line manifold with one dedicated vacuum line and the other attached to a cylinder of argon. Situate the flask on a magnetic stirring mechanism and stir under high vacuum for at least 1 h/l of buffer (2 /l is recommended). Back-fill with argon. Stir for 1 h/l ensuring that the vessel is sealed. Repeat. Transfer the sealed vessel to an anaerobic chamber. Note that this method leads to a small increase in buffer concentration as a result of unavoidable solvent evaporation.
8. Prepare the buffer in a vacuum flask (typically available up to 500 mL) leaving at least 1/3 of the flask volume empty. Submerge the flask into liquid nitrogen or an isopropanol–dry ice slurry until completely frozen. While still frozen, expose to a high vacuum for 10–20 min. Close the flask and warm until completely melted. Submersion in tepid water can help this process; however, caution is urged to avoid fracturing the glassware as a result of a rapid temperature change. Repeat this process three times followed by backfilling the flask with Argon for transfer to an anaerobic chamber.
9. The concentration can be adjusted if necessary to account for very large or small measured heats or other experimental considerations.
10. Some metals are sparingly soluble at high concentrations unless the solution is acidic (e.g., Fe^{3+}). At a given pH, the soluble metal concentration can be easily calculated from solubility products (9). We tend to prepare ≥100 mM stock metal concentrations under neutral conditions if possible for use as titrants. Higher stock concentrations minimize the buffer dilution that occurs upon sample preparation; however, as long as >100 mM concentrations are used, this dilution is insignificant. Regularly verify the stock concentration of metal titrants by atomic absorption spectroscopy or ICP-MS.
11. Although several methods are available for cleaning and ITC, we recommend a protocol consisting of soaking the sample cell and titration syringe in 10 mM EDTA, 1 mM DTT prepared in 0.1% detergent (such as Micro-90) at 65°C for 4 h followed by a thorough rinse with metal-free water (1–2 L)
12. Although monitoring all injection aliquots is a good idea, it is particularly appropriate to monitor the first several injections to verify that enough time is allowed for the signal to return to baseline. Further, the injections leading up to the stoichiometric

equivalence or inflection point should also be monitored due to the reduced concentration of ligand and resulting slower reaction. If enough time is not allowed to verify reaction equilibrium, the data are unreliable.

13. If one desires to account for ligand protonation speciation, a competition term is easily derived: $\alpha_{\text{Proton}} = \sum_{i=0}^{n} \beta_i [H]^i$. However, it should be noted that unless the exact pK_a values are known or can be determined for all titratable groups on the ligand, accounting for acid–base chemistry is largely an approximation. Further, while the heat associated with amino-acid protonation chemistry has been quantified, these values are specific for the free amino acids and, therefore, neglect any local microenvironment, i.e., charge stabilization or dielectric constant. For this reason, these equilibria can be ignored and the results then become pH-specific (15). However, the heat associated with buffer protonation is readily determined and should be accounted for (see Subheadings 5.2 and 5.3)
14. It is assumed that the buffer is in large excess and the concentration is effectively constant.
15. $$\alpha_{\text{Proton}} = 1 + 10^{9.52}[\text{H}] + 10^{15.65}[\text{H}]^2,$$
$$\alpha_{\text{H2L}} = \frac{10^{15.65}[\text{H}]^2}{\alpha_{\text{Proton}}}.$$

References

1. Monod, J., Wyman, J., and Changeux, J.-P. (1965) On the nature of allosteric transitions: A plausible model, *J Mol Biol* ***12***, 88–118.
2. Monod, J., Changeux, J.-P., and Jacob, F. (1963) Allosteric proteins and cellular control systems, *J Mol Biol* ***6***, 306–329.
3. Ma, Z., Jacobsen, F. E., and Giedroc, D. P. (2009) Coordination chemistry of bacterial metal transport and sensing, *Chem Rev* ***109***, 4644–4681.
4. Reinhart, G. D. (2004) Quantitative analysis and interpretation of allosteric behavior, *Methods Enzymol* ***380***, 187–203.
5. Giedroc, D. P., and Arunkumar, A. I. (2007) Metal sensor proteins: nature's metalloregulated allosteric switches, *Dalton Trans*, 3107–3120.
6. Busenlehner, L.S., Pennella, M.A., and Giedroc, D.P. (2003) The SmtB/ArsR family of metalloregulatory transcriptional repressors: structural insights into prokaryotic metal resistance, *FEMS Microbiol Rev* ***27***, 131–143.
7. Lee, S., Arunkumar, A. I., Chen, X., and Giedroc, D. P. (2006) Structural insights into homo- and heterotropic allosteric coupling in the zinc sensor *S. aureus* CzrA from covalently fused dimers, *J Am Chem Soc* ***128***, 1937–1947.
8. Peters, W. B., Frasca, V., and Brown, R. K. (2009) Recent developments in isothermal titration calorimetry label free screening, *Combinatorial Chemistry; High Throughput Screening* ***12***, 772–790.
9. *CRC Handbook of Chemistry and Physics*, 75 ed., (1994) CRC Press, London.
10. Magyar, J. S., and Godwin, H. A. (2003) Spectropotentiometric analysis of metal binding to structural zinc-binding sites: accounting quantitatively for pH and metal ion buffering effects, *Anal Biochem* ***320***, 39–54.
11. NIST Standard Reference Database 46, Version 7.0.
12. Yu, Q., Kandegedara, A., Xu, Y., and Rorabacher, D. B. (1997) Avoiding interferences from good's buffers: a contiguous series of noncomplexing tertiary amine buffers

covering the entire range of pH 3–11, *Anal Biochem* ***253***, 50–56.
13. Hegetschweiler, K., and Saltman, P. (1986) Interaction of copper(II) with N-(2-hydroxyethyl)piperazine-N′-ethanesulfonic acid (HEPES), *Inorg Chem 25*, 107–109.
14. Grossoehme, N. E., Akilesh, S., Guerinot, M. L., and Wilcox, D. E. (2006) Metal-binding thermodynamics of the histidine-rich sequence from the metal-transport protein IRT1 of *Arabidopsis thaliana*, *Inorg Chem* ***45***, 8500–8508.
15. Grossoehme, N. E., and Giedroc, D. P. (2009) Energetics of allosteric negative coupling in the zinc sensor *S. aureus* CzrA, *J Am Chem Soc* ***131***, 17860–17870.
16. Grossoehme, N. E., Mulrooney, S. B., Hausinger, R. P., and Wilcox, D. E. (2007) Thermodynamics of Ni2+, Cu2+, and Zn2+ binding to the urease metallochaperone UreE, *Biochemistry* ***46***, 10506–10516.
17. Grossoehme, N. E., Spuches, A. M., and Wilcox, D. E. (2010) Application of isothermal titration calorimetry in bioinorganic chemistry, *J Biol Inorg Chem* ***15***, 1183–1191.
18. Sharp, K. A., and Honig, B. (1990) Calculating total electrostatic energies with the nonlinear Poisson-Boltzmann equation, *J Phys Chem* ***94***, 7684–7692.
19. Krezel, A., Lesniak, W., Jezowska-Bojczuk, M., Mlynarz, P., Brasuñ, J., Kozlowski, H., and Bal, W. (2001) Coordination of heavy metals by dithiothreitol, a commonly used thiol group protectant, *J Inorg Biochem* ***84***, 77–88.
20. Cotton, A.W., Geoffrey, M., Carlos, A., Bochman, M. (1999) Advanced Inorganic Chemistry, 6 ed., Wiley, New York.
21. Eicken, C., Pennella, M. A., Chen, X., Koshlap, K. M., VanZile, M. L., Sacchettini, J. C., and Giedroc, D. P. (2003) A metal-ligand-mediated intersubunit allosteric switch in related SmtB/ArsR zinc sensor proteins, *J Mol Biol* ***333***, 683–695.
22. MicroCal. (2002) MicroCalorimeter User's Manual, Northampton, MA.
23. O'Halloran, T. (1993) Transition metals in control of gene expression, *Science* ***261***, 715–725.
24. Taylor, J. R. (1997) in An Introduction to Error Analysis 2nd ed., University Science Books, California.
25. Yang, X., Chen-Barrett, Y., Arosio, P., and Chasteen, N. D. (1998) Reaction paths of iron oxidation and hydrolysis in horse spleen and recombinant human ferritins, *Biochemistry 37*, 9743–9750.
26. Doyle, M.L., Louie, G., Dal Monte, P.R., and Sokoloski, T.D. (1995) Tight binding affinities determined from thermodynamic linkage to protons by titration calorimetry, ***259,*** 183–194.
27. Baker, B. M., and Murphy, K. P. (1996) Evaluation of linked protonation effects in protein binding reactions using isothermal titration calorimetry, *Biophys J 71*, 2049–2055.
28. Record, M.T., Jr., Ha, J.H., and Fisher, M.A. (1991) Analysis of equilibrium and kinetic measurements to determine thermodynamic origins of stability and specificity and mechanism of formation of site-specific complexes between proteins and helical DNA, *Methods Enzymol* ***208***, 291–343.
29. Chen, X., Agarwal, A., and Giedroc, D.P. (1998) Structural and functional heterogeneity among the zinc fingers of human MRE-binding transcription factor-1, *Biochemistry 37*, 11152–11161.
30. Arunkumar, A.I., Campanello, G.C., and Giedroc, D.P. (2009) Solution structure of a paradigm ArsR family zinc sensor in the DNA-bound state, *Proc Natl Acad Sci USA*, ***106***, 18177–18182.
31. Popovych, N., Sun, S., Ebright, R. H., and Kalodimos, C. G. (2006) Dynamically driven protein allostery, *Nat Struct Mol Biol* ***13***, 831–838.

Chapter 4

Studying the Allosteric Energy Cycle by Isothermal Titration Calorimetry

Marta Martinez-Julvez, Olga Abian, Sonia Vega, Milagros Medina, and Adrian Velazquez-Campoy

Abstract

Isothermal titration calorimetry (ITC) is a powerful biophysical technique which allows a complete thermodynamic characterization of protein interactions with other molecules. The possibility of dissecting the Gibbs energy of interaction into its enthalpic and entropic contributions, as well as the detailed additional information experimentally accessible on the intermolecular interactions (stoichiometry, cooperativity, heat capacity changes, and coupled equilibria), make ITC a suitable technique for studying allosteric interactions in proteins. Two experimental methodologies for the characterization of allosteric heterotropic ligand interactions by ITC are described in this chapter, illustrated with two proteins with markedly different structural and functional features: a photosynthetic electron transfer protein and a drug target viral protease.

Key words: Titration calorimetry, Protein interactions, Thermodynamics, Allostery, Heterotropic interactions, Cooperativity

1. Introduction

Although somewhat imprecise, allostery may be defined as the control of protein conformational equilibrium by ligand binding (1–3). Thus, the statistical population of the different protein conformational states in a protein will depend on the presence of the ligand. Those conformational states able to bind the ligand will be stabilized and populated in the presence of the ligand, whereas those states unable to bind will be depopulated. The different conformational states will differ in their biological properties, in particular, in the ability to interact with another ligand. Therefore, the interaction of the protein with a given ligand will be modulated by the binding of another ligand (i.e., the binding parameters for ligand A will be a function of the binding parameters of ligand B and its concentration). The modulation of the conformational equilibrium of the

Aron W. Fenton (ed.), *Allostery: Methods and Protocols*, Methods in Molecular Biology, vol. 796,
DOI 10.1007/978-1-61779-334-9_4,

protein represents the energetic and structural basis for the cooperative interaction between two different ligands binding to a protein. As it will be shown below, the effect of both ligands will be reciprocal and equivalent, but there will be a concentration effect that will depend on their respective binding affinities.

Although, etymologically, allostery means "another site," the thermodynamic energy cycles used to describe allostery do not require the ligands involved to bind to different protein sites. In fact, competitive ligand binding (excluding ligands) may be considered as an extreme case of maximal negative cooperativity. Thus, allostery may be considered as the general phenomenon in which the binding of a ligand affects the thermodynamic binding parameters for the binding of a second ligand (3).

Allosteric interactions may correspond to positive cooperativity or negative cooperativity. In the first case, the binding of a ligand will increase the binding affinity of the second ligand, whereas in the second case, the binding of a ligand will decrease the binding affinity of the second one. It is important to point out here that, unless both ligands are competitive or the first ligand binding event is compulsory for the second one, the change in binding affinity elicited by the binding of a previous ligand is not usually dramatic; the Gibbs energy of binding is a thermodynamic potential rather insensitive to changes in the protein environment and molecular details, due to enthalpy–entropy compensation effects. However, the first derivatives of the Gibbs energy, enthalpy, and entropy are much more susceptible and informative, allowing better interpretation (4). Then, the enthalpy or the entropy of binding may exhibit considerable changes elicited by the binding of a previous ligand. Because isothermal titration calorimetry (ITC) is the only technique able to directly measure the enthalpy (and the entropy) of binding, with no assumptions or models implied, ITC is the appropriate technique to detect the often subtle changes observed in cooperative interactions, to assess binding cooperativity in protein allosteric regulation extracting information on the interatomic driving forces.

If the two ligands involved in the allosteric or cooperative phenomenon are identical, the ligand interactions are termed homotropic; if the two ligands are nonidentical, the ligand interactions are termed heterotropic. A well-known case of heterotropic interactions is the coupling between proton binding and ligand binding, which is the basis for the pH dependency of the ligand binding parameters. Homotropic interactions are typical from (homo)oligomeric proteins (e.g., hemoglobin), whereas heterotropic interactions are not restricted to oligomeric proteins. The analysis of homotropic interactions is easier than that of heterotropic interactions, as discussed elsewhere (5).

A description of ITC is not provided in this chapter. The basics of the technique (equipment, experimental design, data analysis, information available through ITC, advantages, and drawbacks) can be found elsewhere (6–11). This chapter focuses specifically on the use of ITC in order to dissect the interactions underlying the allosteric regulation in a protein binding two different ligands in a cooperative fashion. Two calorimetric approaches are addressed: an exact analysis and an approximate analysis. A full thorough description of the methodology can be found elsewhere (12–14).

1.1. Heterotropic Interactions

When a protein M presents two binding sites for two different ligands, A and B, an energy cycle involving four protein species must be considered (Fig. 1). Two binary complexes and one ternary complex coexist in equilibrium:

$$\begin{aligned}[\mathrm{MA}] &= K_{\mathrm{A}}[\mathrm{M}][\mathrm{A}],\\ [\mathrm{MB}] &= K_{\mathrm{B}}[\mathrm{M}][\mathrm{B}],\\ [\mathrm{MAB}] &= K_{\mathrm{A/B}}K_{\mathrm{B}}[\mathrm{M}][\mathrm{A}][\mathrm{B}] = K_{\mathrm{B/A}}K_{\mathrm{A}}[\mathrm{M}][\mathrm{A}][\mathrm{B}],\end{aligned} \quad (1)$$

where $K_{A/B}$ and $K_{B/A}$ are the association constants for ligand A and B binding to the macromolecule already bound to ligand B and A, respectively, and K_A and K_B are the association constants for ligand A and B binding to the free macromolecule. If the binding of one ligand has an influence on the binding of the other ligand, the $K_{A/B}$ and $K_{B/A}$ are different from K_A and K_B. It follows from Eq. 1 that:

$$K_{\mathrm{B}}K_{\mathrm{A/B}} = K_{\mathrm{A}}K_{\mathrm{B/A}}, \quad (2)$$

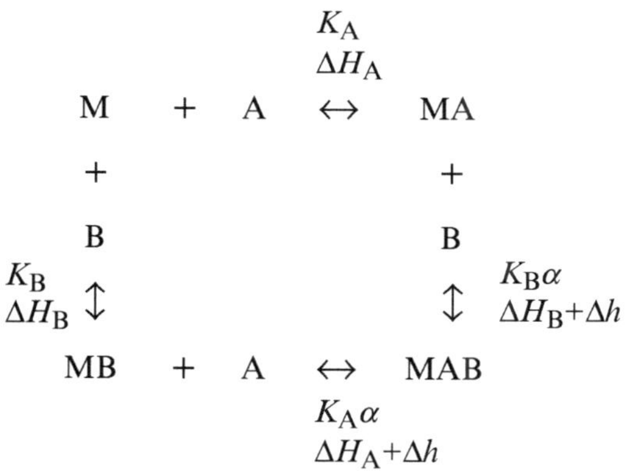

Fig. 1. Scheme representing the equilibrium processes and the complexes involved in the allosteric cycle of a protein binding two different ligands. The thermodynamic characterization of the full energy cycle entails determining the ligand binding parameters (K_A, ΔG_A, ΔH_A, $-T\Delta S_A$, K_B, ΔG_B, ΔH_B, $-T\Delta S_B$) and the cooperative interaction parameters (α, Δg, Δh, $-T\Delta s$). In addition, experiments under different experimental conditions (temperature, pH, etc.) will provide further information from the dependency of the thermodynamic parameters on such experimental variables.

which is in fact an expression of the energy conservation principle and similar to that of conditional probability. If an interaction or cooperativity constant, α, is introduced for the binding of ligand A when ligand B is bound to the macromolecule (2, 15, 16), then:

$$\begin{aligned} K_{\mathrm{A/B}} &= \alpha K_{\mathrm{A}}, \\ K_{\mathrm{B/A}} &= \alpha K_{\mathrm{B}}. \end{aligned} \tag{3}$$

Therefore, the influence between the two ligands is reciprocal: if the binding of ligand A modifies the binding affinity of ligand B, the binding of ligand B modifies the binding affinity of ligand A in the same extent.

The cooperativity interaction parameter α determines if the formation of the ternary complex MAB is more or less favorable than in the case of independent binding. If $\alpha = 0$, the formation of the ternary complex is not possible because the binding of one type of ligand blocks the binding of the other type (maximal negative cooperativity or competitive ligands). If $\alpha < 1$, the binding of one type of ligand lowers the affinity of binding of the other type of ligand (negative cooperativity). If $\alpha = 1$, the binding of one type of ligand does not have any influence on the affinity of binding of the other type of ligand (no cooperativity or independent ligands). If $\alpha > 1$, the binding of one type of ligand raises the affinity of binding of the other type of ligand (positive cooperativity).

The cooperative interaction parameter, α, is related to the cooperative Gibbs energy, Δg, as well as to the Gibbs energy for ligand A and B binding to the protein already bound to ligand B and A, respectively:

$$\Delta g = -RT \ln \alpha = \Delta G_{\mathrm{A/B}} - \Delta G_{\mathrm{A}} = \Delta G_{\mathrm{B/A}} - \Delta G_{\mathrm{B}}. \tag{4}$$

In addition, the cooperative interaction enthalpy, Δh, and entropy, Δs, may be defined. They are related to the cooperative interaction constant, and the corresponding parameters for ligand A and B binding to the protein already bound to ligand B and A, respectively:

$$\begin{aligned} \Delta h &= -T^2 \frac{\partial(\Delta g / T)}{\partial T} = RT^2 \frac{\partial \ln \alpha}{\partial T} = \Delta H_{\mathrm{A/B}} - \Delta H_{\mathrm{A}} = \Delta H_{\mathrm{B/A}} - \Delta H_{\mathrm{B}}, \\ \Delta s &= -\frac{\partial \Delta g}{\partial T} = R\left(\ln \alpha + T \frac{\partial \ln \alpha}{\partial T}\right) = \Delta S_{\mathrm{A/B}} - \Delta S_{\mathrm{A}} = \Delta S_{\mathrm{B/A}} - \Delta S_{\mathrm{B}}. \end{aligned} \tag{5}$$

The cooperativity interaction parameters (interaction constant, Gibbs energy, enthalpy, and entropy, besides the heat capacity and other additional experimentally accessible parameters) are a reflection of the long or short range interactions between both ligands (direct ligand–ligand interaction or coupled through a protein conformational change), and also solvent molecules, underlying the heterotropic allosteric phenomenon. The description of the methodology will only be based on affinity (or Gibbs energy) and enthalpy, because these parameters are directly determined by ITC, with entropy calculated in a straightforward manner from both.

1.2. Exact Analysis of Heterotropic Interactions

The power of this approach is the cost saving (time and material); at the same time, data analysis is more complex than in the approximate methodology (see next subheading).

Three experimental titrations must be performed: two binary titrations (titrating ligand A or ligand B into protein) in order to determine the binding parameters for ligands A and B (K_A, K_B, ΔH_A, and ΔH_B), and a ternary titration (injecting ligand A into protein + ligand B) in order to determine the cooperative binding parameters (α, Δh).

The binary titrations correspond to a standard calorimetric titration of a single ligand into a protein; data analysis is straightforward (see Note 1) and is described in the next subheading or elsewhere (6, 10, 11).

Data analysis for the ternary titration is more complex. Protein and ligand B (at saturating or subsaturating conditions) will be placed into the calorimetric cell, while ligand A will be placed into the injecting syringe. The concentration of each of the reactants in the calorimetric cell after any injection i is given by:

$$[\mathrm{M}]_{\mathrm{T},i} = [\mathrm{M}]_0 \left(1 - \frac{v}{V}\right)^i,$$
$$[\mathrm{A}]_{\mathrm{T},i} = [\mathrm{A}]_0 \left(1 - \left(1 - \frac{v}{V}\right)^i\right),$$
$$[\mathrm{B}]_{\mathrm{T},i} = [\mathrm{B}]_0 \left(1 - \frac{v}{V}\right)^i, \qquad (6)$$

where $[\mathrm{M}]_0$ is the initial concentration of the macromolecule in the calorimetric cell (see Note 2), $[\mathrm{A}]_0$ is the concentration of ligand A in the syringe, $[\mathrm{B}]_0$ is the initial concentration of ligand B in the calorimetric cell, v is the injection volume, and V is the cell volume. From the mass balance for the ternary system, the following set of equations (in three unknowns, the concentrations of free species M, A, and B), is obtained:

$$[\mathrm{M}]_{\mathrm{T}} = [\mathrm{M}] + K_A[\mathrm{M}][\mathrm{A}] + K_B[\mathrm{M}][\mathrm{B}] + \alpha K_A K_B[\mathrm{M}][\mathrm{A}][\mathrm{B}],$$
$$[\mathrm{A}]_{\mathrm{T}} = [\mathrm{A}] + K_A[\mathrm{M}][\mathrm{A}] + \alpha K_A K_B[\mathrm{M}][\mathrm{A}][\mathrm{B}],$$
$$[\mathrm{B}]_{\mathrm{T}} = [\mathrm{B}] + K_B[\mathrm{M}][\mathrm{B}] + \alpha K_A K_B[\mathrm{M}][\mathrm{A}][\mathrm{B}]. \qquad (7)$$

This set of equations can be solved numerically (e.g., Newton–Raphson method). Once the values of the free concentration of reactants are known, the concentration of the three different complexes, [MA], [MB], and [MAB], can be evaluated. Finally, the heat effect, q_i, associated with the injection i can be evaluated as follows:

$$q_i = V\left(\Delta H_A\left([MA]_i - [MA]_{i-1}\left(1-\frac{v}{V}\right)\right)\right.$$
$$+ \Delta H_B\left([MB]_i - [MB]_{i-1}\left(1-\frac{v}{V}\right)\right)$$
$$\left. +(\Delta H_A + \Delta H_B + \Delta h)\left([MAB]_i - [MAB]_{i-1}\left(1-\frac{v}{V}\right)\right)\right), \quad (8)$$

which indicates that the heat associated with injection i is related to the change in the concentration of each complex after that injection. The ternary titration allows estimating the cooperative interaction parameters (α and Δh) by nonlinear fitting analysis of the experimental data using Eq. 8 in just one experiment, knowing the binary binding parameters (K_A, K_B, ΔH_A, and ΔH_B) (see Note 3). It is advisable to include an adjustable parameter q_d in Eq. 8 taking into account incomplete saturation or a nonzero heat of dilution. In the data analysis, the heat effect q_i is normalized per mole of ligand injected in each injection, i.e., $Q_i = q_i/(v[A]_0)$, in order to get the differential isotherm.

1.3. Approximate Analysis of Heterotropic Interactions

The power (and the weakness, at the same time) of this approach consists of reducing the ternary system (M/A/B) to an equivalent binary system (M/A) with apparent binding parameters, which are dependent on the concentration of ligand B (see Notes 3 and 4).

From the general scheme shown in Fig. 1, the apparent association constant, the apparent Gibbs energy, and the apparent enthalpy for ligand A binding to the protein in the presence of ligand B (at a certain concentration) are given in refs. 1, 2, 12, 15 (see Note 5):

$$K_A^{app} = K_A \frac{1+\alpha K_B[B]}{1+K_B[B]},$$
$$\Delta G_A^{app} = \Delta G_A - RT\ln\frac{1+\alpha K_B[B]}{1+K_B[B]},$$
$$\Delta H_A^{app} = \Delta H_A - \Delta H_B\frac{K_B[B]}{1+K_B[B]} + (\Delta H_B + \Delta h)\frac{\alpha K_B[B]}{1+\alpha K_B[B]}. \quad (9)$$

Therefore, the ternary equilibrium problem can be described through a *quasi*-simple equilibrium, in which the effect of the presence of ligand B on the thermodynamic parameters of the binding of ligand A is accounted for by considering a set of modified apparent thermodynamic parameters dependent on ligand B (Eq. 9) (see Note 6).

Several titrations of ligand A into protein at different total concentrations of ligand B must be made. These titrations are considered to conform to the standard single ligand binding site model (6–11) (see Note 7). Briefly, protein and ligand B (at saturating or subsaturating conditions) will be placed into the calorimetric cell, while ligand A will be placed into the injecting syringe. The concentration of each of the reactants in the calorimetric cell after any injection i is given by:

$$[\mathrm{M}]_{\mathrm{T},i} = [\mathrm{M}]_0\left(1 - \frac{v}{V}\right)^i,$$
$$[\mathrm{A}]_{\mathrm{T},i} = [\mathrm{A}]_0\left(1 - \left(1 - \frac{v}{V}\right)^i\right). \quad (10)$$

From the mass balance for the ternary system, the following set of equations is obtained:

$$[\mathrm{M}]_\mathrm{T} = [\mathrm{M}] + K_\mathrm{A}^{\mathrm{app}}[\mathrm{M}][\mathrm{A}],$$
$$[\mathrm{A}]_\mathrm{T} = [\mathrm{A}] + K_\mathrm{A}^{\mathrm{app}}[\mathrm{M}][\mathrm{A}]. \quad (11)$$

This set of equations can be solved numerically or analytically. Once the values of the free concentration of reactants are known, the concentration of the complex [MA] can be evaluated. Finally, the heat effect, q_i, associated with the injection i can be evaluated as follows:

$$q_i = V\left(\Delta H_\mathrm{A}^{\mathrm{app}}\left([\mathrm{MA}]_i - [\mathrm{MA}]_{i-1}\left(1 - \frac{v}{V}\right)\right)\right), \quad (12)$$

which indicates that the heat associated with injection i is related to the change in the concentration of the complex after that injection. The thermodynamic apparent binding parameters for ligand A ($K_\mathrm{A}{}^{\mathrm{app}}$, $\Delta H_\mathrm{A}{}^{\mathrm{app}}$) are estimated at each ligand B concentration by nonlinear least squares regression analysis of the experimental data using the standard model for a single ligand binding to a protein (Eqs. 10–12). The cooperative interaction parameters, α and Δh, are obtained by nonlinear fitting analysis of the apparent binding parameters for ligand A as a function of ligand B (Eq. 9) (12, 17, 18) (see Notes 4, 8, 9, and 10).

There is a special situation in which ligand B is employed at a saturating concentration ($K_\mathrm{B}[\mathrm{B}] >> 1$, $\alpha K_\mathrm{B}[\mathrm{B}] >> 1$). Then, the apparent binding parameters for ligand A are simplified to:

$$K_\mathrm{A}^{\mathrm{app}} = K_\mathrm{A}\alpha,$$
$$\Delta G_\mathrm{A}^{\mathrm{app}} = \Delta G_\mathrm{A} - RT\ln\alpha,$$
$$\Delta H_\mathrm{A}^{\mathrm{app}} = \Delta H_\mathrm{A} + \Delta h. \quad (13)$$

Thus, performing only two titrations (one binary and one ternary), it is possible to obtain the binding parameters for ligand A binding to the protein (K_A, ΔH_A) from the binary titration (Eqs. 10–12), as well as the apparent binding parameters for ligand A binding to the protein (K_A^{app}, ΔH_A^{app}) in the presence of saturating ligand B concentration from the ternary titration. Then, the cooperative interaction parameters (α and Δh) are calculated using Eq. 13, without the need for determining the binding parameters for ligand B. Traditionally, this has been the standard methodology employed (16, 19) (see Notes 11 and 12).

2. Materials

1. Buffer solutions (see Note 13).
   Buffer A: 50 mM Tris–HCl, pH 8.
   Buffer B: 20 mM Tris–HCl, pH 7, 2 mM DTT (dithiothreitol), 2% w/v CHAPS (3-[(3-cholamidopropyl)dimethylammonio]-1-propanesulfonate), 50% w/v glycerol.
   Buffer C: 20 mM Tris–HCl, pH 7, 2 mM DTT.
2. Dialysis membrane of 10 kDa MWCO.
3. Isothermal titration calorimeter.

   Several isothermal calorimeters, with the required sensitivity for monitoring biological binding reactions, are commercially available (e.g., those manufactured by MicroCal or Calorimetry Science Corporation). The protocols described here may be performed with any of them.
4. Biological systems.

   Anabaena ferredoxin-$NADP^+$ reductase (FNR).

   Anabaena ferredoxin (Fd).

   $NADP^+$ (lyophilized powder, from Sigma, ref. N5755).

   Hepatitis C virus NS3 protease S139A (NS3 PR, inactive mutant).

   NS3 protease activator peptide pNS4A (lyophilized powder, from Anaspec, ref. 24491).

   NS3 protease substrate (lyophilized powder, from Bachem, ref. M-2235).

   Two different systems have been selected for illustrating the two different approaches in the analysis of heterotropic interactions: FNR for the exact analysis and NS3 PR for the approximate analysis.

   The FAD-dependent FNR catalyzes the reduction of $NADP^+$ by reduced Fd in the last step of the photosynthetic electron transfer chain, leading to the production of reducing power in the form of NADPH. When all the components of the reaction, Fd, FNR, and $NADP^+$, are in the oxidized state catalysis cannot take place, but interactions among the components of these electron transfer chain are still produced and equilibrium binding parameters can be determined. FNR binding sites for $NADP^+$ and Fd partially overlap, but it is an appropriate experimental example given the wealth of experimental data available in the literature for this ternary system (20, 21). In this system, Fd has been considered as ligand A and $NADP^+$ as ligand B (see Note 14).

NS3 PR is a viral serine protease considered as a target for drug development. It is an allosteric enzyme, in which the binding of the accessory viral protein NS4A elicits a conformational change enhancing the catalytic efficiency of the enzyme by increasing both the catalytic rate constant and the affinity for the substrate (22, 23). The binding site of NS4A is located in the N-terminal domain, while the substrate cleft is formed by both N- and C-terminal domains. The substrate is a modified fluorogenic peptide derivative (Ac-DED(EDANS)EE-Abu-L-lactoyl-SK(DABCYL)-NH_2). A lysine-flanked central fragment of NS4A (H-*KK*GSVVIVGRIILS*K*-OH), instead of the full-length NS4A, is used experimentally as activator allosteric ligand. In order to do binding experiments in the absence of enzymatic catalysis, an inactive mutant (S139A) is employed. In this system, the substrate has been considered as ligand A and pNS4A as ligand B (see Note 14).

3. Methods

3.1. Sample Preparation

3.1.1. FNR System

1. FNR and Fd can be overexpressed in Escherichia coli (E. coli) as previously described (24, 25).
2. FNR and Fd must be dialyzed in buffer A. $NADP^+$ must be dissolved in buffer A.
3. Protein concentrations can be estimated by UV absorbance using extinction coefficients of $\varepsilon_{458nm} = 9.4\ mM^{-1}\ cm^{-1}$ for FNR and $\varepsilon_{423nm} = 9.7\ mM^{-1}\ cm^{-1}$ for Fd (see Note 15).
4. $NADP^+$ concentration can be estimated by UV absorbance using an extinction coefficient of $\varepsilon_{259nm} = 16.9\ mM^{-1}\ cm^{-1}$.
5. For the binary titrations, prepare 20 μM FNR solution (5 mL), 300 μM Fd solution (0.6 mL), and 300 μM $NADP^+$ solution (0.6 mL).

6. For the ternary titration, prepare 20 μM FNR + 50 μM $NADP^+$ solution (2.5 mL), and 300 μM Fd solution (0.6 mL).

3.1.2. NS3 PR System

1. NS3 protease can be overexpressed in *E. coli* as previously described (26).
2. NS3 protease must be dialyzed in buffer C.
3. Protein concentration can be estimated by UV absorbance using an extinction coefficient of $\varepsilon_{280nm} = 18.5\ mM^{-1}\ cm^{-1}$.
4. pNS4A may be dissolved in water or buffer C at high concentration (~15 mM).
5. Substrate may be dissolved in DMSO at high concentration (~3 mM).
6. For the binary titration, prepare 10 μM NS3 PR solution (2.5 mL) and 150 μM substrate solution (0.6 mL). Buffers B and C must be mixed appropriately in order to get final glycerol and CHAPS concentrations of 25% and 1% w/v, respectively.
7. For the ternary titration, prepare 10 μM NS3 PR + 200 μM pNS4A solution (2.5 mL) and 150 μM substrate solution (0.6 mL). Buffers B and C must be mixed appropriately in order to get final glycerol and CHAPS concentrations of 25% and 1% w/v, respectively. Given the dissociation constant for the NS3-pNS4A interaction ($K_D = 10$ μM), almost 100% of the NS3 protease will be bound to pNS4A in the calorimetric cell.
8. Because substrate was dissolved in DMSO, DMSO must be added conveniently in order to get the same final DMSO concentration in both cell and syringe solutions (not more than 5% v/v, depending on the substrate stock solution concentration) (see Note 16).

3.2. Experimental Methodology: Exact Analysis

1. Perform first binary titration: titrate $NADP^+$ solution into FNR solution. Data analysis considering the standard model with a single ligand binding site (Eqs. 10–12) will provide the binding parameters for ligand B (K_B, ΔH_B).
2. Perform second binary titration: titrate Fd solution into FNR solution. Data analysis considering the standard model with a single ligand binding site (Eqs. 10–12) will provide the binding parameters for ligand A (K_A, ΔH_A).
3. Perform ternary titration: titrate Fd solution into FNR + $NADP^+$ solution. Data analysis considering the ternary model (Eqs. 6–8) will provide the cooperativity binding parameters (α, Δh).
4. Results should be consistent with the information shown in Fig. 2 (see Notes 17–24).

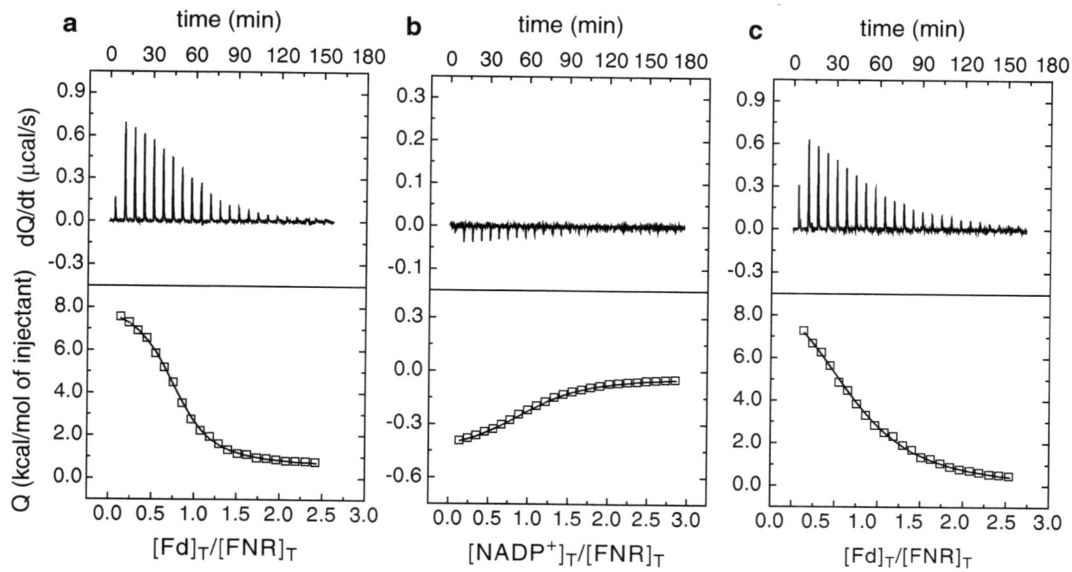

Fig. 2. Calorimetric titrations illustrating the determination of the cooperative heterotropic interactions in the FNR system, applying the exact analysis. (**a**) Fd titrated into FNR ($K_A = 6.8 \pm 0.4 \times 10^5\ M^{-1}$), $\Delta H_A = 7.8 \pm 0.2$ kcal/mol, (**b**) $NADP^+$ titrated into FNR ($K_B = 2.6 \pm 0.2 \times 10^5 M^{-1}$, $\Delta H_A = -0.4 \pm 0.2$ kcal/mol), and (**c**) Fd titrated into FNR with $NADP^+$ ($\alpha = 0.16 \pm 0.01$, $\Delta h = 4.5 \pm 0.2$ kcal/mol). The continuous lines are the best fits considering a single ligand binding site model (**a**, **b**), or the exact ternary model with two different ligand binding sites (**c**). $NADP^+$ and Fd binding shows negative cooperativity, with a cooperative Gibbs energy ($\Delta g = 1.1$ kcal/mol) partitioned into an unfavorable enthalpic contribution ($\Delta h = 4.5$ kcal/mol) and a favorable entropic contribution ($-T\Delta s = -3.4$ kcal/mol).

3.3. Experimental Methodology: Simple Approximate Analysis

1. Perform binary titration: titrate substrate solution into NS3 PR solution. Data analysis considering the standard model with a single ligand binding site (Eqs. 10–12) will provide the binding parameters for ligand A (K_A, ΔH_A).
2. Perform ternary titration: titrate substrate solution into NS3 PR + pNS4A solution. Data analysis considering the standard model will provide the apparent binding parameters for substrate in the presence of activator pNS4A peptide (K_A^{app}, ΔH_A^{app}).
3. The cooperative binding parameters (α, Δh) may be calculated using Eq. 13.
4. Results should be consistent with the information shown in Fig. 3 (see Notes 17–24).

4. Notes

1. Data analysis with the software provided by MicroCal (Origin) is straightforward and sufficient for simple systems (e.g., single ligand binding site). However, for the analysis of more complex systems, such as the exact ternary equilibrium presented here, it is necessary to create additional fitting custom routines.

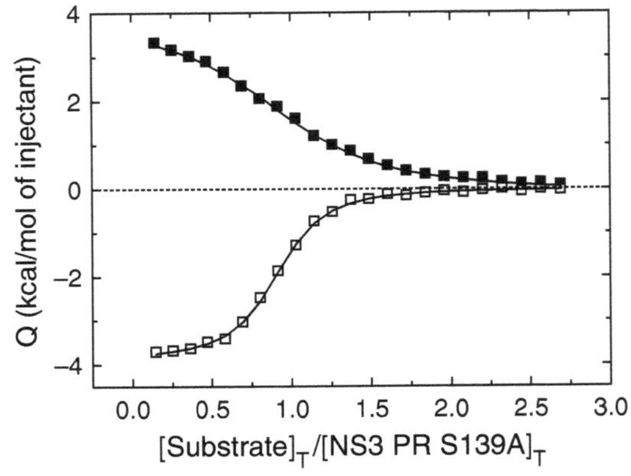

Fig. 3 Calorimetric titrations illustrating the determination of the cooperative heterotropic interactions in the NS3 PR system, applying the approximate analysis. Substrate titrated into inactive NS3 PR (*closed squares*; K_A = 8.1 ± 0.6 × 10^5 M^{-1}, ΔH_A = 3.9 ± 0.2 kcal/mol), and substrate titrated into inactive NS3 PR with pNS4A at saturating concentration (*open squares*; K_A^{app} = 4.2 ± 0.3 × 10^6 M^{-1}, ΔH_A^{app} = −3.9 ± 0.2 kcal/mol). The continuous lines are the best fits considering a single ligand binding site model (Eqs. 10–12). Values for the cooperative interaction constant α of 5.2 ± 0.8 and the cooperative interaction enthalpy Δh of −7.8 ± 0.4 kcal/mol were obtained using Eq. 13. Substrate and activator peptide pNS4A binding shows positive cooperativity, with a cooperative Gibbs energy (Δg = −1.0 kcal/mol) partitioned into a favorable enthalpic contribution (Δh = −7.8 kcal/mol) and an unfavorable entropic contribution ($-T\Delta s$ = 6.8 kcal/mol).

2. When defining custom fitting routines, it is advisable to introduce a parameter "*n*" representing the stoichiometry of the binding, even if a stoichiometry 1:1 is expected. This parameter would appear multiplying $[M]_0$ in Eq. 6. In that case, fractional values of *n* will account for a percentage of inactive or nonbinding-competent protein.
3. It is assumed that the interaction constant α is the same at any concentration of ligand B. However, α might depend on the concentration of ligand B, and, therefore, the interaction parameter might exhibit different values at low and high concentration of ligand B. The exact ternary model allows determining the cooperativity binding parameters at specific ligand B concentrations.
4. In general, the values of the interaction parameters estimated applying the approximate methodology are approximate and might be characterized by a significant error. It is important to point out that all the equations included in the approximate analysis are exact. The approximations are introduced when the free ligand B concentration to be applied in those equations is estimated and when those equations are applied.

5. The apparent binding parameters in the approximate analysis are not equal to the binding parameters defined in Eqs. 1–3. In particular, the apparent association constant, K_A^{app}, is not equal to the association constant $K_{A/B}$. Thus, contrary to $K_{A/B}$, K_A^{app} is ligand B concentration-dependent. Both K_A^{app} and $K_{A/B}$ will coincide in two limit cases: (1) when α is equal to one; (2) when the product $K_B[B]$ is sufficiently high (protein is saturated with ligand B) (*see* Eq. 13). Therefore, the ratio K_A^{app}/K_A is not, in general, equal to α. Likewise, the apparent binding enthalpy, ΔH_A^{app}, is not, in general, equal to the binding enthalpy $\Delta H_{A/B}$, and the difference $\Delta H_{A/B} - \Delta H_A$ is not equal to Δh.
6. An inconvenience in applying the approximate methodology is that, in general, the approximate binary system is not completely equivalent to the ternary system. The reason is that the free concentration of ligand B is not constant throughout the titration (as it is assumed in Eq. 9), and the apparent association constant and the apparent binding enthalpy for ligand A are not true constants (Eqs. 9). The discrepancy between the ternary system and its binary equivalency diminishes when the binding affinity of ligand B is low and/or its concentration is much higher than that of the protein.
7. The single ligand binding site model (Eqs. 10–12) represents a special case of the ternary model with two ligand binding sites (Eqs. 6–8) in which $[B]_0 = 0$, $K_B = 0$, $\Delta H_B = 0$, $\alpha = 0$, $\Delta h = 0$. Therefore, both binary and ternary titrations can be analyzed using the same fitting routine.
8. The concentration of free ligand B is not known accurately in a titration experiment, unless $[B]_T$ is much higher than $[M]_T$ and the free concentration of B can be approximated by the total concentration of B. A reasonable good *a priori* operational estimation of the free ligand B concentration is: (1) the difference between the total concentration of ligand B and the concentration of macromolecule in the calorimetric cell at the beginning of the experiment, if the concentration of ligand B is higher than the concentration of macromolecule; (2) the total concentration of ligand B, if the concentration of ligand B is lower than the concentration of macromolecule.
9. Because, usually, the calorimetric experiment is performed at constant cell volume, during the titration experiment the concentration of any molecule in the calorimetric cell decreases as the experiment progresses due to the injection of titrant solution from the syringe. Therefore, even if the binding of the two ligands is independent ($\alpha = 1$), the concentration of ligand B is not constant. One way to avoid this

particular problem is adding ligand B in the syringe at the same concentration than in the calorimetric cell (and modifying Eq. 6 accordingly). However, the binding cooperativity will still make the free ligand B concentration nonconstant.

10. The approximate method has the possibility of avoiding the binary titration with ligand B. Thus, either the parameters for ligand B binding to the protein can be obtained together with the cooperative binding parameters through the nonlinear fitting analysis (Eq. 9) (approximate method) (18) or they are not required for obtaining the cooperative binding parameters (Eq. 13) (simplified approximate method). Then, this methodology is appropriate when the ligand B exhibits very low affinity (18).
11. It might be impossible to achieve a saturating concentration of ligand B (e.g., it may exhibit a very low binding affinity or it may precipitate). In the case of maximal or very high negative cooperativity (α equal to zero or very small), high saturating concentrations of ligand B may cause a reduction in affinity for ligand A so large that might render the experiment useless. In these two cases, nonsaturating concentrations will not provide the right interaction parameters when using the simplified approximate analysis.
12. Experiments at fixed nonsaturating concentration of ligand B may provide more information than experiments at buffered or excess ligand B concentration (27).
13. The experimental conditions and buffer solutions should be carefully selected for a given protein system, depending on the characteristics (thermodynamic and kinetic stability, solubility, propensity to interact with organic molecules, etc.) of the interacting macromolecules.
14. Given that the cooperativity or heterotropic effect is reciprocal, there is no bias as to which ligand must be considered as ligand A or B. However, the ligand with larger binding enthalpy (larger signal) and with moderate affinity (not too high/low) will be more appropriate as ligand A, since it will be easier to observe changes in its binding parameters elicited by the presence of the other ligand.
15. Errors in reactant concentrations will propagate causing the estimated binding parameters to have significant errors.
16. A perfect match in composition between the cell and the injector solution is crucial for avoiding experimental artifacts stemming from dilution effects. Therefore, solutions must be carefully prepared when using organic co-solutes (e.g., glycerol, DMSO).

17. There are no general rules about the design of a given experiment or about limit values for the binding and cooperativity parameters in order to detect cooperativity. The effect of ligand B on the binding of ligand A depends on: the binding parameters for ligand B binding, the interaction cooperativity parameters, and the concentration of ligand B. In general, ligand B concentration must be large enough to observe an effect on the binding parameters for ligand A. Solubility limits or ligand exclusion may preclude the application of the simplified approximate method in which ligand B is employed at saturating concentration (Eq. 13) (see Note 12).
18. The methodologies described here are valid for $0 \leq \alpha < +\infty$. Practical considerations will impose limitations based on the apparent association constant for ligand A (K_A^{app}) being within the experimentally accessible range (10^3 M^{-1} < K_A^{app} < 10^9 M^{-1}). Fortunately, two binding parameters can be directly measured (K_A^{app}, ΔH_A^{app}), and both of them contain information on the cooperativity binding parameters. If K_A^{app} is too large, ΔH_A^{app} is still accurately determined; if K_A^{app} is too small, it will not be affected by large errors, contrary to ΔH_A^{app}. Therefore, more information is available using ITC, compared to other binding techniques, and there is a way to overcome the difficulties of having a very small/large binding affinity.
19. If $\alpha \neq 1$, cooperativity will be detected even if Δh is close to zero, because the binding affinity for ligand A will be modified by the presence of ligand B. Inversely, if $\Delta h \neq 0$, cooperativity will be detected even if α is close to 1, because the binding enthalpy for ligand A will be modified by the presence of ligand B. In this last scenario, the enthalpy–entropy compensation phenomenon arises (28, 29). These ideas underscore the power of ITC for characterizing allosteric interactions, with the possibility of determining simultaneously the affinity and the enthalpy of binding.
20. Even if the two ligands are competitive ($\alpha = 0$), in practice, a value slightly different from zero might be estimated from nonlinear fitting analysis, either with the exact or the approximate method. As a rule of thumb, as long as the value of the product $K_A\alpha$ is much lower than 1 (i.e., ligand A does not bind in the presence of ligand B), the value of α can be considered equal to zero.
21. The methodology presented here can be applied to any allosteric protein binding two different ligands. Besides, it can be readily generalized and extended to systems with different stoichiometries (30).

22. This methodology represents a very quick procedure to determine the cooperativity interaction parameters. The structural and energetics determinants of the cooperative interaction underlying the allosteric regulation can be established by carefully designing mutants of the protein and/or functional variants of the ligands. Thus, it is possible to identify which residues in the protein are responsible for the allosteric regulation; or which regions in a given ligand are important for the allosteric coupling, which are, in principle, not identical to those required for binding affinity (31–33). Therefore, it is possible to apply this methodology to the design of allosteric inhibitors.
23. Any event taking place in the calorimetric cell will add to overall energetics. In order to extract valid conclusions from the values of the thermodynamic binding parameters using molecular structure–energy correlations, contributions from extrinsic sources must be removed (e.g., contribution of buffer ionization to the observed binding enthalpy).
24. Additional experiments can be done under different experimental conditions (temperature, pH, ionic strength, co-solutes, mutations, or functional modifications) in order to obtain additional information on the cooperative heterotropic interactions and the allosteric regulation from the dependency of the cooperative binding parameters on those experimental variables (heat capacity, proton linkage, etc.).

Acknowledgments

Supported by grants CP07/00289 (O.A.) from the Instituto de Salud Carlos III (Spanish Ministry of Science and Innovation); grants SAF2004-07722 and BFU2010-19451 (A.V.-C.), and BIO2007-65890-C02-01 and BIO2010-14983 (M.M.) from the Spanish Ministry of Science and Innovation; grants PI044/09 and PI078/08 (A.V.-C.) from Diputación General de Aragón, Spain; and grant UZ2009-BIO-05 (A.V.-C.) from Universidad de Zaragoza. O.A. was supported by a Miguel Servet Research Contract from the Instituto de Salud Carlos III. A.V.-C. was supported by research contract from Fundación ARAID (Diputación General de Aragón, Spain).

References

1. Wyman J, Gill SJ (1990) Binding and linkage: Functional Chemistry of Biological Macromolecules. University Science Books, California, USA.
2. Reinhart GD, Hartleip SB, Symcox MM (1989) Role of coupling entropy in establishing the nature and magnitude of allosteric response. Proc Natl Acad Sci USA 86:4032–4036.
3. Fenton AW (2008) Allostery: An illustrated definition for the 'second secret of life'. Trends Biochem Sci 33:420–425.
4. Hinz HJ (1983) Thermodynamics of protein-ligand interactions: Calorimetric approaches. Annu Rev Biophys Bioengineer 12:285–317.
5. Freire E, Schön A, Velazquez-Campoy A (2009) Isothermal titration calorimetry: General formalism using binding polynomials. Methods Enzymol 455:127–155.
6. Wiseman T, Williston S, Brandts JF, Lin LN (1989) Rapid measurement of binding constants and heats of binding using a new titration calorimeter Anal Biochem 179:131–137.
7. Freire E., Mayorga OL, Straume M (1990) Isothermal titration calorimetry. Anal Chem 62:950A-959A.
8. Doyle ML (1997) Characterization of binding interactions by isothermal titration. Curr Opin Biotechnol 8:31–35.
9. Jelessarov I, Bosshard HR (1999) Isothermal titration calorimetry and differential scanning calorimetry as complementary tools to investigate the energetics of biomolecular recognition. J Mol Recognit 12:3–18.
10. Velazquez-Campoy A, Leavitt SA, Freire E (2004) Characterization of protein-protein interactions by isothermal titration calorimetry. Methods Mol Biol 261:35–54.
11. Velazquez-Campoy A, Freire E (2005) ITC in the post-genomic era...? Priceless. Biophys Chem 115:115–124.
12. Velazquez-Campoy A, Goñi G, Peregrina JR, Medina M (2006) Exact analysis of heterotropic interactions in proteins: Characterization of cooperative ligand binding by isothermal titration calorimetry. Biophys J 91:1887–1904.
13. Martinez-Julvez M, Medina M, Velazquez-Campoy A (2009) Binding thermodynamics of ferredoxin:NADP$^+$ reductase: two different protein substrates and one energetics. Biophys J 96:4966–4975.
14. Houtman JC, Brown PH, Bowden B, Yamaguchi H, Appella E, Samelson LE, Schuck P (2007) Studying multisite binary and ternary protein interactions by global analysis of isothermal titration calorimetry data in SEDPHAT: application to adaptor protein complexes in cell signaling. Protein Sci 16:30–42.
15. Weber G (1975) Energetics of ligand binding to proteins. Adv Prot Chem 29:1–83.
16. Subramanian S, Stickel DC, Colen AH, Fisher HF (1978) Thermodynamics of heterotropic interactions: The glutamate dehydrogenase-NADPH-glutamate complex. J Biol Chem 253:8369–8374.
17. Du W, Liu W-S, Payne DJ, Doyle ML (2000) Synergistic inhibitor binding to Streptococcus pneumoniae 5-enolpyruvylshikimate-3-phosphate synthase with both monovalent cations and substrate. Biochemistry 39:10140–10146.
18. Edgcomb SP, Baker BM, Murphy KP (2000) The energetics of phosphate binding to a protein complex. Protein Sci 9:927–933.
19. Fisher HF, Subramanian S, Stickel DC, Colen AH (1980) The thermodynamics of a negatively interacting allosteric effector system: The glutamate dehydrogenase-NADPH-ADP complexes. J Biol Chem 255:2509–2513.
20. Medina M, Gomez-Moreno C (2004) Interaction of ferredoxin-NADP$^+$ reductase with its substrates: Optimal interaction for efficient electron transfer. Photosynth Res 79:113–131.
21. Medina M (2009) Structural and mechanistic aspects of flavoproteins: Photosynthetic electron transfer from photosystem I to NADP$^+$. FEBS J 276:3942–3958.
22. Bianchi E, Urbani A, Biasiol G, Brunetti M, Pessi A, De Francesco R, Steinkühler C (1997) Complex formation between the hepatitis C virus serine protease and a synthetic NS4A cofactor peptide. Biochemistry 36:7890–7897.
23. Urbani A, Biasiol G, Brunetti M, Volpari C, Di Marco S, Sollazzo M, Orru S, Dal Piaz F, Casbarra A, Pucci P, Nardi C, Gallinari P, De Francesco R, Steinkühler C (1999) Multiple determinants influence complex formation of the hepatitis C virus NS3 protease domain with its NS4A cofactor peptide. Biochemistry 38:5206–5215.
24. Medina M, Martinez-Julvez M, Hurley JK, Tollin G, Gomez-Moreno C (1998) Involvement of glutamic acid 301 in the catalytic mechanism of ferredoxin-NADP$^+$ reductase from *Anabaena* PCC 7119. Biochemistry 37:2715–2728.

25. Faro M, Hurley JK, Medina M, Tollin G, Gomez-Moreno C (2002) Flavin photochemistry in the analysis of electron transfer reactions: role of charged and hydrophobic residues at the carboxyl terminus of ferredoxin-NADP$^+$ reductase in the interaction with its substrates. Bioelectrochemistry 56:19–21.

26. Abian O, Neira JL, Velazquez-Campoy A (2009) Thermodynamics of zinc binding to hepatitis C virus NS3 protease: a folding by binding event. Proteins 77:624–636.

27. Robert CH, Gill SJ, Wyman J (1988) Quantitative analysis of linkage in macromolecules when one ligand is present in limited total quantity. Biochemistry 27:6829–6835.

28. Fisher HF, Tally J (1997) Isoergonic cooperativity in glutamate dehydrogenase complexes: a new form of allostery. Biochemistry 36:10807–10810.

29. Tlapak-Simmons VL, Reinhart GD (1998) Obfuscation of allosteric structure-function relationships by enthalpy-entropy compensation. Biophys J 75:1010–1015.

30. Pulido NO, Salcedo G, Perez-Hernandez G, Jose-Nuñez C, Velazquez-Campoy A, Garcia-Hernandez E (2010) Energetic effects of magnesium in the recognition of adenosine nucleotides by the F1-ATPase β subunit. Biochemistry 49:5258–5268.

31. Fenton AW, Paricharttanakul NM, Reinhart GD (2003) Identification of substrate contact residues important for the allosteric regulation of phosphofructokinase from Eschericia coli. Biochemistry 42:6453–6459.

32. Williams R, Holyoak T, McDonald G, Gui C, Fenton AW (2006) Differentiating a ligand's chemical requirements for allosteric interactions from those for protein binding. Phenylalanine inhibition of pyruvate kinase. Biochemistry 45:5421–5429.

33. Fenton AW, Hutchinson M (2009) The pH dependence of the allosteric response of human liver pyruvate kinase to fructose-1,6-bisphosphate, ATP, and alanine. Arch Biochem Biophys 484:16–23.

Chapter 5

Detecting "Silent" Allosteric Coupling

Harvey F. Fisher

Abstract

Using isothermal calorimetry (ITC), we have found one case where a well-characterized allosteric activator showed no sign of allostery in its $\Delta G°$ of binding to successive sites on multiple subunits and another case where successive binding showed no $\Delta G°$ binding allostery but did show large entropy-compensated flip-flopping enthalpy changes. This behavior, which we have termed "isoergonic cooperativity" and others have referred to as "silent coupling" is quite simply explained by basic linkage theory when reactions are considered beyond the $\Delta G°$ level. Thus, direct calorimetric determination of all thermodynamic parameters including $\Delta H°$, $\Delta S°$, $\Delta G°$, $\Delta C_p°$, and $d(\Delta C_p°/dt)$ provides a more informative depiction of a ligand binding event and its consequences than does the mere measurement of $\Delta G°$ alone. We further discuss the benefits and limitations of methods that have previously been used to study silent coupling. In particular, ITC is free of the numerous pitfalls inherent in the application of van't Hoff and Årrhenius plots to allosteric phenomena. Aside from having a 30-fold advantage in precision, ITC is capable of measuring changes in enthalpy directly at five more levels of mathematical differentiation than are available to van't Hoff type approaches.

Key words: Isothermal calorimetry, Silent allostery, van't Hoff plots, Heat capacity, Coupled harmonic oscillators

...our concern... impelled us
To purify the dialect of the tribe

(T.S. Eliot – Little Gidding)

1. Introduction

1.1. Isoergonic (Silent) Cooperativity

We have discovered two very different cases of ligand binding, neither of which betrays evidence of allosteric behavior as judged by free-energy criteria, but do show pronounced anomalous phenomena in their enthalpic binding plots (1, 2).

The bovine liver glutamate dehydrogenase (blGDH) molecule is a hexamer consisting of two trimers of subunits stacked vertically in a somewhat eclipsed manner. Ligand-binding Isothermal Titration Calorimetry (ITC) titration curves for blGDH (represented as

Aron W. Fenton (ed.), *Allostery: Methods and Protocols*, Methods in Molecular Biology, vol. 796,
DOI 10.1007/978-1-61779-334-9_5,

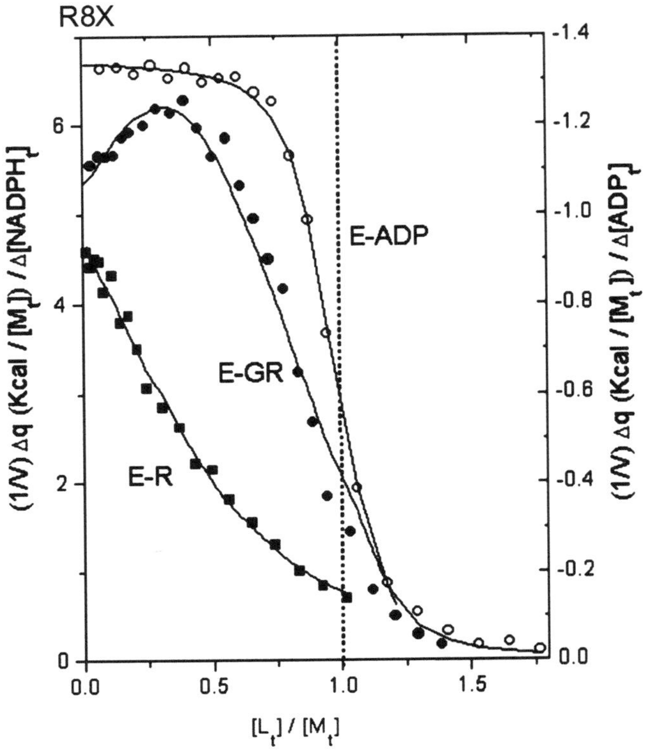

Fig. 1. Isothermal titrations of bovine liver glutamate dehydrogenase, with ADP (*right-hand ordinate*), with NADPH and with NADPH (*left-hand ordinate*) in the presence of 50 mM L-glutamate. Data collected at 25°C in 0.1 M phosphate buffer, pH 7.6. The ADP titration is fit to an independent binding site equation, and NADPH titrations are fit to the three-step interactive equation of Wiseman et al. (28).

"E") complexes are shown in Fig. 1. The binding of NADPH (R) requires a three step interactive isothermal equation, independent of whether L-glutamate (G) is or is not present. By contrast, the binding of ADP to a blGDH molecule is well fitted by and independent site binding isotherm, suggesting that ADP binds to each of the six subunits independently. However, the same dataset collected over a temperature range can be equally well fitted by a three-step interactive equation leading to a series of alternating entropy enthalpy flip-flops resembling the behavior of the ER complexes in Fig. 2. (It is important to note that the value of ΔH_1 shown in Fig. 2 represents the properties of the binding of the first ligand molecule to an enzyme binding site since it is calculated from the $\lim_{t\to 0} d(\Delta H)/d[L]$. However, ΔH_2 and ΔH_3 represent only the subsequent phenomenological behavior of the system as a whole and cannot properly assigned specifically to individual sites.) Both of the ER complexes appear to involve sequential binding to each subunit of a single trimer, but there is

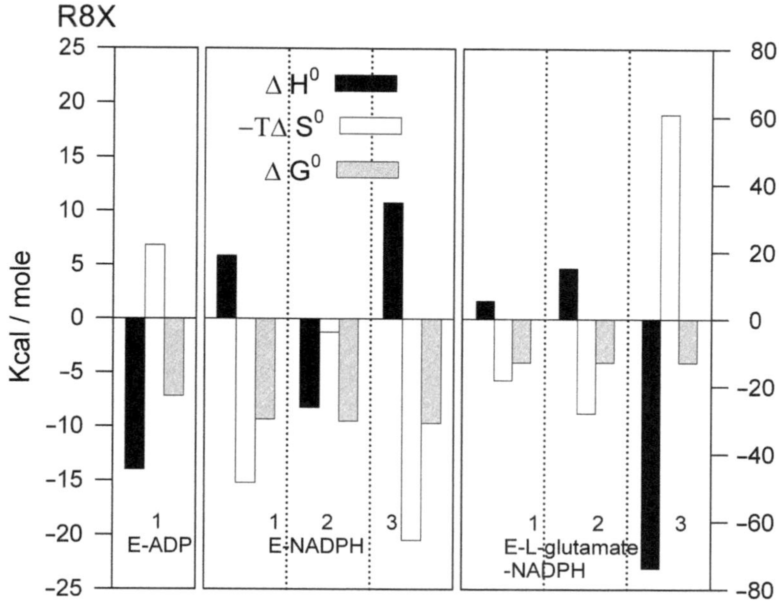

Fig. 2. Thermodynamic parameters for the independent single site binding of the ADP complex and of the three successive sites for the NADPH and the L-glutamate-NADPH complexes of bovine liver glutamate dehydrogenase calculated from the fitted curves shown in Fig. 1.

Table 1
Binding constants calculated from single and multiple interacting site analysis (1)[a]

Complex	$K_1 \times 10^4 (M^{-1})$	$K_2 \times 10^4 (M^{-1})$	$K_3 \times 10^4 (M^{-1})$
EGR	8.8 ± 1	8.1 ± 1	8.8 ± 1
ER	0.4 ± 0.05	0.4 ± 0.06	0.5 ± 0.1
E-ADP	45 ± 3	–	–

[a]Calculations were carried out using the Microcal Orgin ITC analysis software based on theory described in ref. 28

no indication of any such interaction between the two trimers. Table 1 shows that the intrinsic binding constants for each of the three binding steps are identical in both of these reactions and as such are reflected by ΔG° bars of identical height in Fig. 2. From the identity of these ΔG° values, one would conclude that these successive steps occur without any discernible interaction and thus could not be classified as allosteric. However, it can be seen in

Fig. 2 that the ΔH° values of successive binding steps and their opposing $T\Delta S^{\circ}$ values of successive binding steps alternate in sign with differences in magnitude of as much as 25 kcal M^{-1}. Hence, we have defined a system in which sequential binding steps show substantial differences in enthalpy unaccompanied by any significant change in their corresponding binding constants as "*Isoergonic Cooperativity.*" We find it useful to include in the definition of isoergonic/silent coupling cases where structural and kinetic observations prove the existence of classic allosteric behavior without the appearance of any corresponding free-energy phenomena.

This phenomenon is not easily explained by many of the versions of allosteric theory (e.g., two-state models) currently in use. It is, however, easily understood in terms of classic thermodynamic theory of linkage of multiple ligand-to-protein binding reactions. In order to provide such an explanation of this phenomenon and of allosteric phenomena in general, we next summarize the elements of linkage theory required for the task.

In the beginning, Gregorio Weber said (3):

- "The binding of a ligand to a protein cannot be considered as an event apart from the changes in that protein induced by that binding."
- "We *never* deal with the binding of a single ligand to a protein."
- "Therefore, it is impossible for the binding of two ligands to the same protein to be really independent."
- "The study of the interactions of proteins and small ligands provides a means of relating biological function to fundamental physical chemical principles."
- "Structural information must be accompanied by corresponding energetic studies if it is to lead to appropriate physical interpretation."

Stimulated by these axioms, an enormous amount of both theoretical and experimental work by a large number of contributors has led to substantial (but not unanimous) agreement on the following points:

1. Every biological reaction is initiated by a protein–ligand binding step.
2. Such reactions never involve the binding of only a single ligand or a single step.
3. The binding of two ligands to the same protein always involves a mutual interaction.
4. The product of a ligand-binding reaction is a new entity in itself; its structure and its properties may differ substantially from the simple sum of those of its initial components.

5. The concept of G. Weber's "thermodynamic square" permits the evaluation of a variety of interaction parameters.
6. Patterns of interaction parameters provide a basis for exploring ligand-binding energy transduction.
7. Measurement of ligand binding heat by techniques using higher levels of mathematical differentiation provide more reliable and more detailed dissection into parameters such as enthalpy, entropy, heat capacity, and their corresponding interaction terms.
8. The "fluctuating protein" concept suggests that protein reactions proceed as multiple traces on multidimensional free-energy landscapes.
9. Consequences of the thermodynamic complexity of ligand binding suggest new views of such processes as enzymatic catalysis, signaling events, and evolutionary aspects.
   Evidence supporting these conclusions has been summarized by Eftink and Biltonen (4) and by Fisher (5).

Wyman, in 1948 proposed a linkage theory expressed in terms of chemical potential theory (6). Weber replaced the elaborate chemical potential binding treatment of Wyman with the much simpler but still rigorous Gibbs free energy (ΔG) approach, based on the relationship, $\Delta G^{\circ} = -RT\ln K$, where K is the equilibrium constant for a ligand-binding event (3).

The Wyman and Weber mathematical approaches, appropriately applied, both lead to the same precisely thermodynamic conclusions. Because the Weber ΔG° approach leads more directly to experimentally determined and more easily comprehended parameters, we employ it exclusively here.

1.2. The Thermodynamic Square

The consequences of Weber's axioms are most clearly expressed in the form of the classic "thermodynamic square":

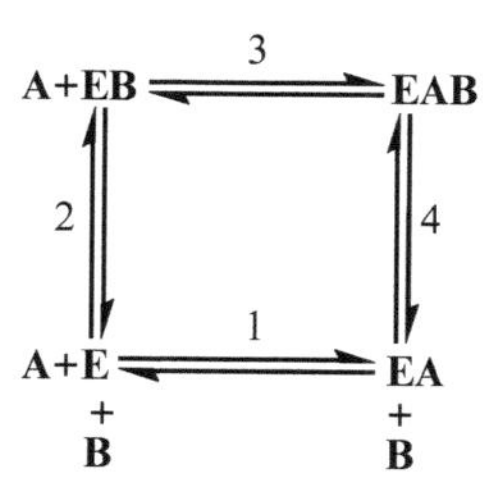

(1)

where equilibrium constants K_1, K_2, K_3, and K_4, correspond to the corresponding numbered steps, and

$\Delta G^{\circ}{}_n = -RT\ln\ K_n$ is again defined for each step.

Defining an interaction parameter ΔG_{I} as the algebraic difference between the free energy of formation of the ternary complex from E, A, and B and that of the sum of the free energies of formation of the two binary complexes:

$$\Delta G_{\mathrm{I}}^{\circ} = \Delta G_{1}^{\circ} + \Delta G_{4}^{\circ} - \left(\Delta G_{1}^{\circ} + \Delta G_{2}^{\circ}\right). \quad (2)$$

Since

$$\Delta G_{1}^{\mathrm{o}} + \Delta G_{4}^{\mathrm{o}} = \Delta G_{2}^{\mathrm{o}} + \Delta G_{3}^{\mathrm{o}} \quad (3)$$

and

$$\Delta G_{\mathrm{I}}^{\mathrm{o}} = \Delta G_{4}^{\circ} - \Delta G_{2}^{\circ} = \Delta G_{3}^{\circ} - \Delta G_{1}^{\circ}.$$

Substituting $-RT \ln K_n$ for each ΔG_n and rearranging, we define the free-energy interaction parameter, ΔG_{I} as:

$$\Delta G_{\mathrm{I}}^{\mathrm{o}} = -RT \ln\left(\frac{K_4}{K_2}\right) = -RT \ln\left(\frac{K_3}{K_1}\right). \quad (4)$$

Thus, $\Delta G_{\mathrm{I}}^{\mathrm{o}}$ is the difference in the free energy of A binding to the EB complex over that of A binding to free E.

Of course, the effect of the presence of bound B on the binding of A must be identical in sign and magnitude to that of the presence of bound A on the binding of B. It is apparent that $\Delta G_{\mathrm{I}}^{\mathrm{o}}$ may be either positive or negative depending on whether A and B mutually decrease or increase each other's affinity for the protein. In addition to the expression of ligand binding of cooperativity embodied by ΔG_{I}, Subramanian et al. extended the concept to include the additional interaction parameters; ΔH_{I}, ΔS_{I}, and ΔC_{PI} (7). As we see below, these new parameters are of considerable use in characterizing the nature of a given ligand–ligand or protein–ligand interaction, providing a sound fundamental basis for the theoretical and experimental exploration of coupled reactions, multiple-step linkage, and allosteric effects.

1.3. The Temperature Dependence of Protein–Ligand Binding

The overwhelming majority of studies on protein–ligand interactions (and particularly those expressed in the special dialect of linkage theory called "allostery") have been based on Gibb's Free-energy (ΔG°) measurements. Such entities as equilibrium constants (K), rate constants (k), and their corresponding quasi-thermodynamic counterparts $\Delta G^{\ddagger}$, $K^{\ddagger}$, and $k^{\dagger}$ are all free-energy parameters. Since, by definition, $\Delta G^{\circ} = \Delta H^{\circ} - T\Delta S^{\circ}$, it can be seen by inspection that ΔG° is a temperature-dependent parameter. Lumry and Rajender showed that ΔG° values are generally quite insensitive measurements of protein–ligand interactions due to the compensation between the ΔH° and $T\Delta S^{\circ}$ terms inherent in the definition of ΔG° itself (8). Edsall and Gutfreund (9) had pointed out that reactions involving substantial ΔS° values can change sign with a relatively modest change in temperature. Thus, it is quite possible that an inhibitory binding event observed at one temperature may become an activation event at another.

Therefore, a free-energy value observed at a single temperature is physically and mechanistically meaningless.

The temperature dependence of protein–ligand interactions is further complicated by the fact that $\Delta H°$ is itself generally temperature dependent. All such reactions when studied appropriately exhibit ΔC_p values in the range of 200–600 cal/deg M^{-1} (10) where $\Delta C_p = \Delta H°/\Delta T$. On this basis, Hinz and Jaenicke (11) stated that a $\Delta H°$ measured only at a single temperature is physically meaningless. Fine-grained ITC measurements by Fisher et al. (12) showed that ΔC_p is itself a temperature dependant function. Thus, the Hinz and Jaenicke caution must be extended to the realization that even a ΔC_p measured only from the $\Delta\Delta H$ between a single pair of temperatures has no real physical significance. It is evident that the proper thermodynamic characterization of the formation of a ternary complex requires the determination of the values of $\Delta G°$, $\Delta H°$, $\Delta S°$, and ΔC_p over a substantial temperature range for each of the four steps of the thermodynamic square portrayed in Eq. 1.

2. Methods of Measurement of ΔH

In general, any chemical reaction involves the production of a certain amount of heat designated as the quantity q. The lower case symbol is used to indicate that q is not a state property since it is strongly dependent on the implicit and quite variable contributions of ΔS and its related function, ΔC_p, and thereby on temperature. Thus, q is separated from $\Delta H°$ mathematically by five levels of integration. As a result, the determination of ΔH is strongly dependent on the experimental method used. Currently, there are two classes of experimental methods available for this purpose: (1) the various forms of the van't Hoff equation in which ΔH is determined by the temperature dependence of the binding constant K, and (2) direct calorimetric measurements.

2.1. Limitations of van't Hoff and Årrhenius Plots

The thermodynamic state properties which define the temperature dependence of a reaction include ΔH, ΔC_p, dC_p/dt, and $d^2\Delta C_p/dt$. This series of parameters involves a progression of successive degrees of mathematical derivatization. The range of phenomena that can be observed in any given experiment, however, depends on the method employed. In Table 2 we have listed varying cases of temperature dependence along with the phenomena that can be observed with both calorimetric and van't Hoff-based approaches. The table shows that van't Hoff plots can evaluate only the unrealistic class 1 and 2 cases and, at best, can indicate the presence of case 3 phenomena. By contrast,

Table 2
A comparison of levels of enthalpic measurement by two different experimental approaches redrawn from ref. 5

Case		Measurement		
	Method:	van't Hoff	Calorimetric	Differential scanning calorimeter
	Parameter:	ln K vs. $1/T$	ΔH vs. T	ΔC_P^o vs. T
1 $\Delta H^o = 0$ $\Delta C_P^o = 0$ $\frac{d\Delta C_P^o}{dT} = 0$				
2 ΔH^o – finite $\Delta C_P^o = 0$ $\frac{d\Delta C_P^o}{dT} = 0$				
3 ΔH^o – finite ΔC_P^o – finite $\frac{d\Delta C_P^o}{dT} = 0$				
4 ΔH^o – finite ΔC_P^o – finite $\frac{d\Delta C_P^o}{dT}$ – finite $\frac{d^2\Delta C_P^o}{dT^2} = 0$				
5 ΔH^o – finite ΔC_P^o – finite $\frac{d\Delta C_P^o}{dT}$ – finite $\frac{d^2\Delta C_P^o}{dT^2}$ – finite				

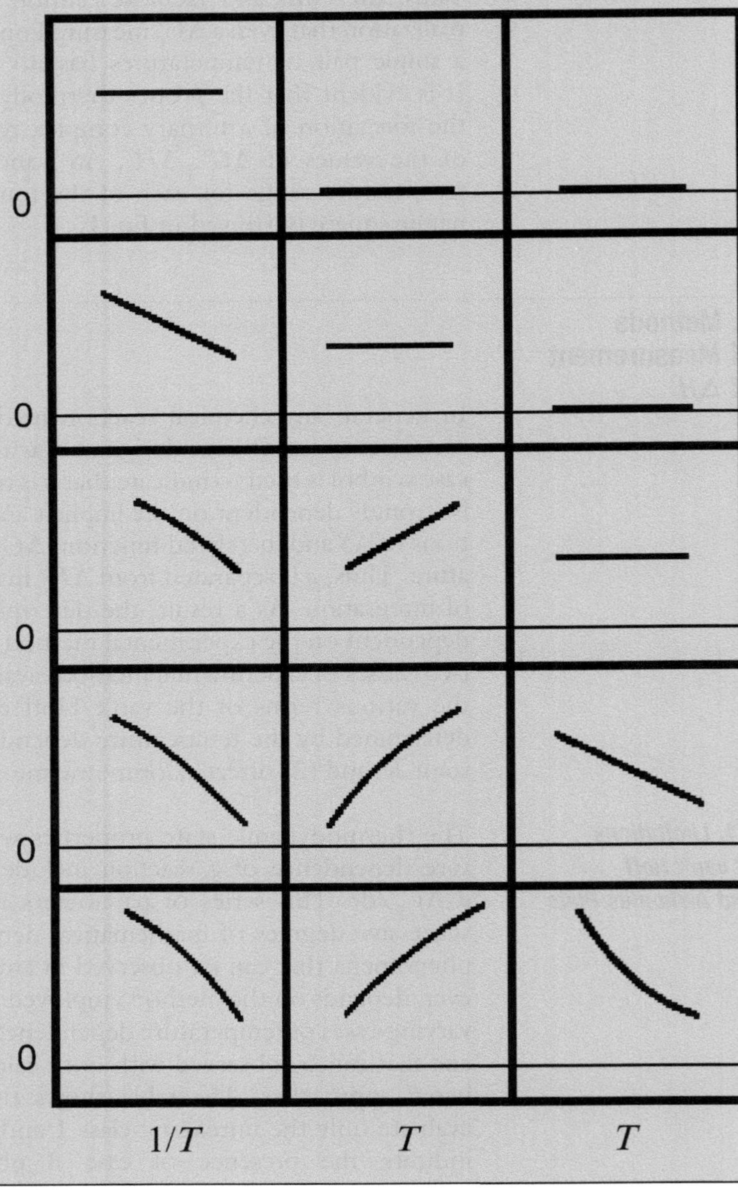

calorimetry, which measures the output or uptake of heat during a reaction directly, evaluates case 3, 4, and 5 parameters unequivocally, operating as it does at two-to-three levels of integration above that of the van't Hoff plot. In addition, the probable error in a calorimetric measurement is 15- to 20-fold less than the three kcal mol^{-1} error that characterizes even a carefully done van't Hoff experiment.

The term "enthalpy–entropy compensation" was first introduced by Lumry in 1970 (8), although such behavior in chemical reactions had been noted by Leffler and Grunwald as early as 1963 (13) and is, as we show, well supported by calorimetric studies. Much of the literature on the concept has relied on van't Hoff- and Årrhenius-based plots. After a decade or so of great interest and widespread (and not uncommonly inappropriate) application, the compensation concept fell into some disrepute. Exner pointed out that the attempt to extract two parameters (ΔH and ΔG) from temperature dependence data inevitably led to the false impression of enthalpy–entropy compensation even in a system in which such a relationship did not exist (14). Throughout the years, the same (quite correct) argument has been rehashed using various forms of statistical exposition (15, 16). More recently, Sharp has revisited the controversy in an article entitled "Entropy–enthalpy compensation: fact or artifact?"(17) and Cornish–Bowden has now published a denouncement of the concept under the title "Enthalpy–entropy compensation: a phantom phenomenon (18)." All of these attacks are absolutely correct within their scope but their criticisms are focused on inherently flawed studies. Admittedly, there is no shortage of flawed interpretations subject to this criticism, but the statistical artifact argument applies only to data obtained from an oversimplified form of the van't Hoff equation. The general form of the van't Hoff equation is:

$$\Delta H^{\circ} = -R\left(\frac{\partial \ln K}{\partial T}\right)_p. \quad (5)$$

This equation is exact and covers all five cases listed in Table 2. However, a simplified form of Eq. 5 is generally used to fit data presented on a van't Hoff plot:

$$\Delta \ln K = -\frac{\Delta H^{\mathrm{o}}}{RT}. \quad (6)$$

Equation 6, which has been used in every case referred to in the many denunciations of compensation, tacitly assumes that ΔH^{o} is itself independent of temperature. Klotz has commented that the unwarranted assumption implicit in Eq. 6 that ΔH^{o} must be independent of temperature "can lead to wildly incorrect conclusions" (19). Van Holde has concluded that values obtained by this oversimplified form of the van't Hoff equation "yields nothing more than the ΔH° of a reaction that does not exist" (20). A particular large and unavoidable source of error in this approach is the requirement of an extrapolation that is 25-fold longer than the range of the dataset itself. The argument has now hopefully been put to rest by a well-argued demonstration by Horn et al. that a "properly carried out" usage of a van't Hoff plot can indeed yield a reasonably correct value of the ΔH° of a reaction if, and only if, a calculated ΔC_p term is added to Eq. 6 (21).

While all of these critical essays were directed at van't Hoff plot based evidence of enthalpy–entropy compensation, they actually imply a broader generalization – that all ΔH values derived from van't Hoff analysis are at least suspect, if not simply incorrect. This heavy shadow of doubt, thus, must lie over the vast number of allosteric phenomena characterized using van't Hoff or Årrhenius plots. Even the basic data to which Jonsson et al. (22) and Limbach et al. (23) have applied their elegantly derived arguments claiming the occurrence of certain quantum-tunneling events should be subject to serious reevaluation from this point of view.

A substantial number of publications dealing with allosteric phenomena report on the results of investigations based not on the equilibrium approach of the van't Hoff plot, but rather on the temperature dependence of steady-state kinetic parameters as expressed by the Årrhenius plot which is of the same mathematical form as Eq. 5. Aside from carrying the same doubts raised by the statistical skeptics on the validity of the van't Hoff plot for multi-step reactions, the kinetic approaches to measuring a supposed enthalpy of activation ($\Delta H^{\ddagger}$), requires a further assumption which leads not to increased experimental error, but to totally invalid conclusions. Invariably, such studies make the totally incorrect assumption that the Michaelis constant, K_M is equal to (or roughly approximates) K_S, the dissociation constant of the substrate. The field of enzyme kinetics abanded this simplistic notion a half-century ago. Cook and Cleland (24) have stated that a steady-state K_M is in fact:

$$K_M = \frac{\sum [\text{Unliganded forms}]}{\sum [\text{Liganded forms}]}. \tag{7}$$

Gutfreund (25) has shown that even for the oversimplified and quite unrealistic reaction:

$$\mathrm{E} + \mathrm{S} \underset{K_{21}}{\overset{K_{12}}{\rightleftharpoons}} \mathrm{ES} \underset{K_{32}}{\overset{K_{23}}{\rightleftharpoons}} \mathrm{EP} \underset{K_{13}}{\overset{K_{31}}{\rightleftharpoons}} \mathrm{E} + \mathrm{P} \tag{8}$$

KM is defined by,

$$K_M = \frac{K_{12}K_{32} + K_{21}K_{31} + K_{23}K_{31}}{K_{12}(K_{23} + K_{32} + K_{31})}. \tag{9}$$

More realistic 6- to 12-step mechanisms would require expressions which would differ from $K_M = K_{21}/K_{12}$ in an even more unpredictably complex manner.

Given the sheer number, diversity and mathematical rigor of these attacks on the application of van't Hoff or Årrhenius plots to reactions involving more than a single step, investigators who continue to use these approaches would do well to include a thorough analysis of the cumulative error effects on their results.

2.2. Isothermal Titration Calorimetry

The criticisms of the practice of equating the results of van't Hoff or Årrhenius plots with thermodynamic parameters which we have just discussed do not apply in any sense to those obtained from methods in which ΔH is measured directly. The phenomena exhibited by either of these approaches are compared in Table 2. Upon examination of this table, it can be seen that the van't Hoff approach is limited to determining that a given reaction has a temperature dependence of a nondiscernable shape, source, or magnitude, while the more direct calorimetric approaches can both detect and measure features at two to three higher levels of integration. We show here that phenomena at these higher levels do indeed occur quite generally in protein–ligand binding reactions and that their recognition and interpretation provide the basis of resolving many of the complicated phenomena which comprise allostery into combinations of more easily interpretable and simply grasped physical–chemical concepts.

Calorimetry has now emerged from the period in which it was practiced by only a few specialists using traditional batch or flow calorimeters to a widely applied technique readily available to any investigator and employed extensively by the pharmaceutical industry. Indeed, it is now considered the optimal approach for the determination of the various thermodynamic parameters of any ligand-binding reaction due to its theoretical validity, its experimental accuracy, its wide breadth of applicability to reactions that do not provide optical signals, its simplicity and speed of operation, and its availability due to reasonably priced apparatus. This revolution came about as the direct result of the continuous isothermal titration calorimeter invented by John Brandts and marketed by Microcal Corp. as the "Omega" apparatus. Its origin is easily traced to the stimulus of Rufus Lumry whose vision revealed the need for such instrumentation to explore the phenomena predicted by a small and still largely unrecognized group of theoreticians who some 40 years ago began to view the protein as a mobile entity fluctuating rapidly between roughly equivalent isoergonic conformational states. Their conclusions, presented in a prophetic book entitled "The Fluctuating Enzyme" (27) were in sharp contrast to the view of a protein current at the time as that of a compact and relatively rigid structure. As we discuss in a later section, the field known as "allostery" is now some 40 years later

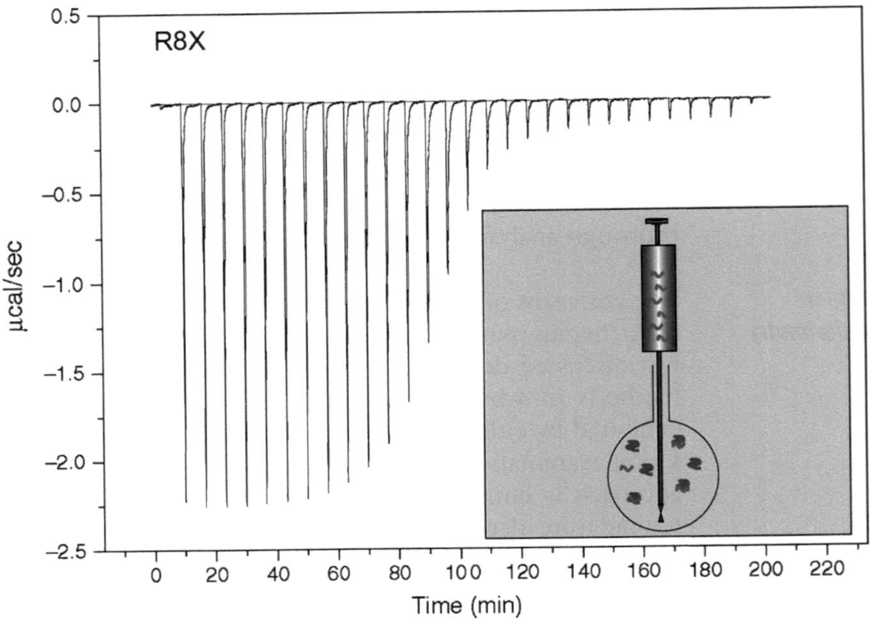

Fig. 3. A typical ITC experiment. The experiment shown corresponds to the titration of a phosphotyrosine peptide (TEGQpYQPQPA) with the SH2 domain of Lck (S. Leavitt, E. Freire, unpublished data). See Note 1.

beginning to rediscover the complex set of concepts originally predicted by R. Lumry, G.R. Welch, L. Blumenfield, S. Damjanovic, B. Gavish, R. Gregory, B. Somogyi, and M. Volkenstein.

The apparatus, experimental technique, and the basic analytic treatment of results have been described in detail by Wiseman et al. (26) Fisher and Singh (28) Indyk and Fisher (29), and Leavitt and Freire (30). We briefly summarize these topics.

In an isothermal titration calorimeter small incremental additions of a ligand-containing solution are delivered to a well-stirred and precisely thermostatted solution containing the protein solution. Δq the incremental heat developed from each addition, $\Delta[L_T]$ is measured. The phenomenological equation to which the resulting data are fitted (the "Wiseman equation") is:

$$\frac{\mathrm{d}q}{\mathrm{d}[L_T]} = \frac{\mathrm{d}}{\mathrm{d}([L_T]/[M_T])} \times \left[\frac{[M_T]\Delta H_1 K_1[\mathrm{L}] + \begin{pmatrix} \Delta H_1(\Delta H_1 + \Delta H_2)K_1K_2[L]^2 \\ +\Delta H_1 + \Delta H_2 + \cdots + \Delta H_n \end{pmatrix} K_1K_2 \ldots K_n[L]^n}{1 + K_1[L] + K_1K_2[L]^2 + K_1K_2 \ldots K_n[L]^n} \right]. \tag{10}$$

A sketch of the configuration of an ITC cell and the results of a typical titration experiment are shown in Fig. 3 (see Note 1) (30).

It can be seen that the sequence of incremental additions of ligand approaches $\mathrm{d}q/\mathrm{d}[L_\mathrm{T}]$ and that the integrated area (q_T/M) provides the ΔH° for the reaction.

The isotherms of many systems, including most multisubunit proteins and those containing allosteric modifier sites deviate significantly from the single-site equations. In such cases, the data must be fitted to the differential form of the general multisite interactive equation:

$$\frac{q}{[M_\mathrm{T}]} = \frac{(\Delta H_1)K_1[L] + (\Delta H_1 + \Delta H_2)K_1K_2[L]^2 + (\Delta H_1 + \Delta H_2 + \cdots + \Delta H_n)K_1K_2\ldots K_n[L]^n}{1 + K_1[L] + K_1K_2[L]^2 + K_1K_2\ldots K_n[L]^n}. \tag{11}$$

The differential forms for such equations become increasingly complex for cases involving more than two interacting sites and may not even be capable of being written explicitly. The Microcal software does, however, calculate and fit such data quite easily using digital differentiation.

3. Isothermal Calorimetric Experimental Results

Measurements of the individual steps involved in the formation of the ternary dead-end inhibitor complex of blGDH (E)–NADPH (R)–L-glutamate (G) (E–R–G) are conveniently considered in the form of the general thermodynamic square shown in Eq. 1.

$$\begin{array}{ccc} \mathrm{ER} & \rightleftharpoons & \mathrm{ERG} \\ \upharpoonleft\!\downharpoonright & & \upharpoonleft\!\downharpoonright \\ \mathrm{E} & \rightleftharpoons & \mathrm{EG} \end{array} \tag{12}$$

The results of ITC titrations of steps around the square obtained at 15° and at 25° are shown in Fig. 4. It can be seen by inspection that the ΔH of the formation of the ternary complex is by no means independent of temperature, showing a substantial ΔC_p. It is, therefore, represented by at least a case 3 level in Table 2 and its measurement by a simple van't Hoff equation would lead to invalid results. The values of the interaction parameters $\Delta G_\mathrm{I}^\circ$, $\Delta H_\mathrm{I}^\circ$ $T\Delta S_\mathrm{I}^\circ$, and $\Delta C_\mathrm{p,I}$ for this and for the E–NADH–ADP allosterically modified complex are shown in Fig. 5.

The obvious feature of the comparison of the patterns of the interaction parameters of these two complexes is the fact that in both cases a large negative $\Delta H_\mathrm{I}^\circ$ is largely offset by a large positive $\Delta S_\mathrm{I}^\circ$ leaving a relatively small $\Delta G_\mathrm{I}^\circ$ as an almost trivial result.

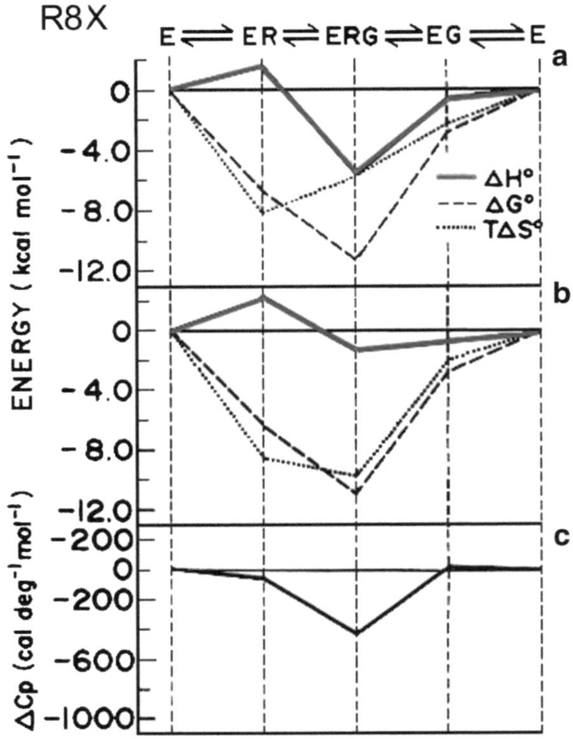

Fig. 4. Thermodynamic profiles for the formation of the E–NADPH–L-glutamate. Complex (7). It should be noted that $\Delta G°$ and $\Delta H°$ are plotted in the positive sense while $T\Delta S°$ is plotted in the negative sense so that downward changes in parameters reflect increased binding tendencies. (**a**) 25°C; (**b**) 15°C; (**c**) ΔC_p over the range 15–20°C I. R = NADPH; G = L-glutamate.

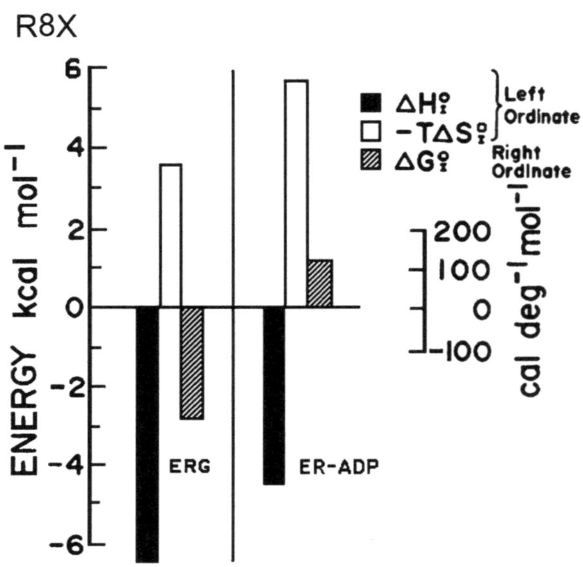

Fig. 5. Interaction parameters for the enzyme–NADPH–L-glutamate complex (a positive interaction) and for the enzyme–NADPH–ADP complex (a negative interaction). Data are from refs. 7 and 49.

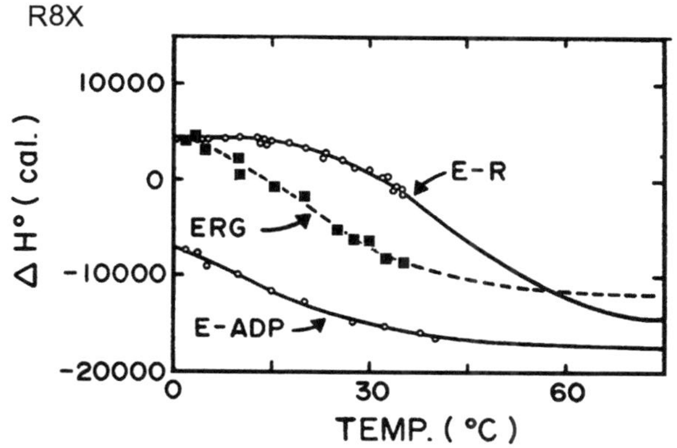

Fig. 6. The temperature dependence of the observed ΔH of formation of three complexes of glutamate dehydrogenase. All measurements were carried out in 0.1 M phosphate buffer pH 7.6 as described in ref. 12.

The observation is of course merely a clearly established case of enthalpy–entropy compensation. An interesting facet of this comparison is that while the overall patterns of thermodynamic parameters of these two complexes appear to be quite similar, yet one complex is kinetically inhibiting and the other is activating. Another particularly significant example of the use of ITC in dissecting complex allosteric phenomena is elucidation of the sequence of intersubunit interactions in hemoglobin by Ackers (31).

3.1. A Unifying Theory of the Thermodynamics of Coupled Ligand-Binding Steps

It is evident that the phenomena shown in Fig. 6 involve at the very least the complexity of behavior required by case 5 of Table 2. As such, they exceed the capability of detection by the van't Hoff approach by several orders of magnitude. Indeed, while the direct measurement of ΔH° of binding at two different temperatures is sufficient to demonstrate the requirement of case 5 behavior, the more extensive fine-grained temperature dependence of the ΔH° of binding of several enzyme–ligand complexes shown in the Fig. 6 clearly require even a higher level of theory to account for their quite disparate behavior. A quite simple unified theory developed by ourselves (5) and a nearly identical exposition by Eftink and Biltonen (4) has been proven able to account quantitatively for the peculiar curves shown in that Fig. 6. This model requires only the single (and by now generally accepted) notion that a protein may be involved in an equilibrium between more than one conformational form, E and E′, and that any given ligand, L, may bind unevenly to each form.

In such a case, the binding of a ligand to its preferred protein form necessarily shifts the $E \rightleftharpoons E'$ equilibrium in that direction. If the intrinsic enthalpies of the two forms differ, then shifting that equilibrium must lead to either an uptake or release of some certain amount of heat. Conversely, increasing the temperature in itself forces a shift in the E ⇌ E' equilibrium from the lower enthalpy form toward the higher enthalpy form. (Of course, the same two factors would be expected to affect an EL ⇌ E′L equilibrium if the ligand can bind to some degree with both forms. However, these two equilibria may be expected to exhibit quite different effects on both their free-energy (ΔG) and temperature (ΔH) parameters). Assuming for simplicity a case of mandatory binding in which a ligand "L" can bind only to the E form and that the resulting EL complex cannot isomerize to an E′L form, we express the reaction simply as:

$$\mathrm{E} + \mathrm{L} \rightleftharpoons \mathrm{EL} \tag{13}$$

being unaware of the actual two-step reaction that is occurring:

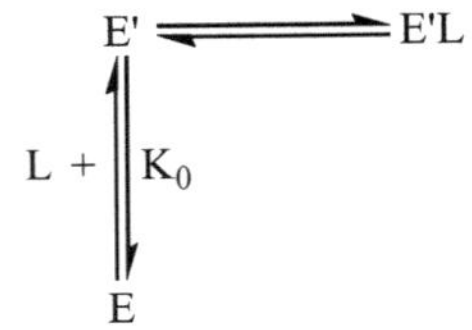

(14)

In such a case, a calorimetric measurement of the temperature dependence of the ΔH° of binding is expressed by the following:

$$\Delta H(T) = \frac{\Delta H_0^{\circ}}{1 + K_0}, \quad \text{where } K_0 = \exp\left[\Delta H_0^{\circ}\left(\frac{T - T_0}{RTT_0}\right)\right], \tag{15}$$

T_0 is the temperature at which $K_0 = 1$ and $\Delta G_0^{\circ} = 0$. The calorimetrically determined ΔH appears as shown in Fig. 7. As the figure shows, at temperatures well above T_0 the enzyme occupies the favored E′ high-enthalpy form and the only heat measured is ΔH_0°, the intrinsic heat of the E′ + L ⇌ EL reaction. At temperatures very much lower than T_0, however, the free enzyme is largely in the low-enthalpy E form, and must $E \rightleftharpoons E' \rightarrow E'L$ now undergo the two-step reaction. The observed ΔH_0° in this case is the sum of the ΔH values of the binding step and the isomerization step. In our derivation thus far, we have assumed that no step in the reaction has an intrinsic ΔC_p. However, Fig. 7 shows that where $T = T_0$ an apparent ΔC_p is observed. The magnitude of this apparent ΔC_p over a temperature range is:

$$\Delta C_\mathrm{p}(T) = \frac{\partial \Delta \mathrm{H}_0}{\partial \mathrm{T}} = \frac{\mathrm{K}_\mathrm{o}(\Delta \mathrm{H}_0^{\circ})^2}{(1 + \mathrm{K}_0)^2 R\mathrm{T}^2}. \tag{16}$$

It can be seen that by recognizing the two-step nature of most (if not all) protein–ligand binding reactions the resulting Eq. 15 which constitutes our alternative explanation, ΔC_p has completely disappeared as have all of the interaction parameters. This fact may require some explanation. It is clear that the ΔC_p terms that appear in the single-step binding scheme of Eq. 1 (and which

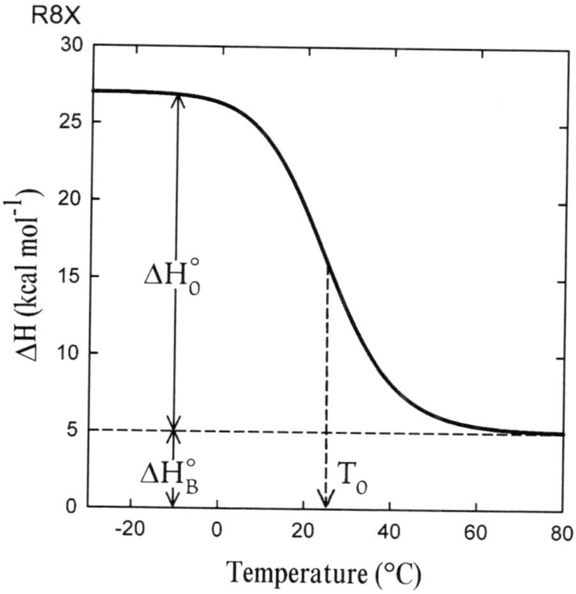

Fig. 7. The temperature dependence of the observed ΔH for a coupled hidden two-state system; $E \overset{\Delta H_0^o}{\rightleftharpoons} E' \overset{\Delta H_1^o}{\rightleftharpoons} EL$. It is assumed that [L]is present at saturating conditions.

customarily appear in protein–ligand binding reactions in general) represents nothing more than the inadequacy of the scheme to properly represent the more complex reaction which actually takes place. We suggest that this finding may apply to all chemical and physical reactions – a ΔC_p represents nothing more than our ignorance at any point in time of the full complexity of the phenomena we observe in a given experiment. As our understanding increases, the ΔC_p term decreases correspondingly and finally disappears completely.

The full expression of Eq. 15 and the phenomenon shown in Fig. 7 require a span of 80°, a range not usually attainable in protein solutions. Indeed, if T_0 happens to lie in the 35° to 45° range, then ΔH_0 is quite linear over a span of 40° and thus appears as in case 3 in Table 2 rather than the case 5 behavior shown in the full curve of Fig. 7.

3.2. Experimental Support for the Unifying Theory

In Fig. 4 we show the ΔH^o of the formation of three bovine liver glutamate dehydrogenase complexes determined by ITC (1). It can be seen that all three reactions exhibit class 5 behavior as defined in Table 2. Although the behavior of these three complexes appears to differ significantly, all three sets of data can be fitted precisely by Eq. 15 as shown by the solid lines in the figure. While the temperature-independent ΔH_0^o terms differ (as would be expected), the ΔH_0^o term, representing the heat of the protein conformational change, is nearly the same in all three cases having the surprisingly high value of 22 ± 1 kcal M^{-1}. (It is interesting to note that Kodama (32) found this same value for a conformational change in the calcium-myosin ATPase complex). Therefore, the pronounced difference in shape of the three curves is due to the wide variations in their T_0 values (it should be remembered that T_0, as we have defined it, is a ΔG function). We found T_0 values of 52°C for ADP, an allosteric modifier, 27°C for the positively cooperative binding of L-glutamate to the ER binary complex and −15°C for the formation of the ER complex itself.

3.3. Mechanistic Interpretation of the Isoergonic Phenomena

While the E-ADP isotherm in Fig. 1 requires only the simple independent three site binding equation, those of the ER and E–R–G complex clearly requires expansion of the equation to a three site interactive case using the model-free general equation of Wiseman:

$$\frac{q}{[M_\mathrm{T}]} = \frac{(\Delta H_1)K_1[L] + (\Delta H_1 + \Delta H_2)K_1K_2[L]^2 + (\Delta H_1 + \Delta H_2 + \cdots + \Delta H_n)K_1K_2\ldots K_n[L]^n}{1 + K_1[L] + K_1K_2[L]^2 + K_1K_2\ldots K_n[L]^n}, \quad (17)$$

where ΔH_n is the molar enthalpy of the binding to each site and K_n is the binding constant (M^{-1}) for each individual binding site.

This phenomenological equation fits the data quite well as shown by the solid lines in Fig. 1. The isothermal expression of Eq. 17, however, provides only the thermodynamic envelope into which any proposed mechanism must fit. A clue to such a viable mechanism is provided by the two-step binding event shown in Eq. 14 based on the isothermal ligand titrations whose results are shown in Fig. 4. Incorporating this feature into the interactive multisite equation of Wiseman, we proposed the following mechanism for the binding of R to the first two interactive binding sites:

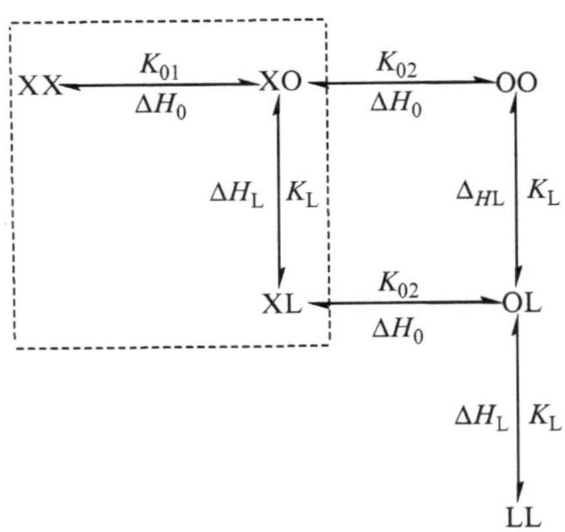

(18)

Here, we assume that the single ligand binding site on each enzyme subunit may adopt one of two possible conformations, a low enthalpy closed form (X) or a high enthalpy open form (O). We also assume that the ligand, L, cannot bind to any closed form,

X, but can bind to any open form, O, equally well with the same binding constant, K_L and the same enthalpy, ΔH_L. We assume that the enthalpy of the X $\rightleftharpoons$ O transition, ΔH_o, is identical for all transitions at any given temperature. The variable parameters are the equilibrium constants K_{01} or K_{02} which depend on the number of remaining open sites, and which thus express the intersubunit interactions.

The phenomenological equation for the two-site scheme shown in Eq. 18 is as follows:

$$\frac{1}{V}\left(\frac{q}{[M_T]}\right) = \frac{\begin{array}{l}[(\Delta H_o)K_{01} + (2\Delta H_o)K_{01}K_{02} + (\Delta H_o + \Delta H_L)K_{01}K_L[L] \\ + (2\Delta H_o + \Delta H_L)K_{01}K_{02}K_L[L] + (2\Delta H_o + 2\Delta H_L)K_L^2 K_{01}K_{02}[L]^2] \\ - [(\Delta H_o)K_{01} + (2\Delta H_o)K_{01}K_{02}]\end{array}}{1 + K_{01} + K_{01}K_{02} + K_L K_{01}[L] + K_{01}K_{02}K_L[L] + K_{01}K_{02}K_L^2[L]^2}. \quad (19)$$

This two-site scheme and its equation can be extended to include larger numbers of subunit binding sites, but the explicit equations become increasingly cumbersome. The Wiseman computer program can, however, solve such cases digitally, and we have shown that the data for the ER and ERG cases of Fig. 1 are very well fitted by a three-site interactive version of Eq. 11. The basic elements of this admittedly still crude mechanistic scheme are well supported by other thermodynamic, kinetic and structural studies described in ref. 5. It is enough for our purpose here to have proved theoretically and demonstrated experimentally that the high enthalpy two-state transitions that provide the linkage required by our isoergonic theory do in fact exist.

4. Other Observations of Silent Coupling

We have now demonstrated two examples of isoergonic coupling in blGDH. Reinhart has noted similar effects in the phosphofructokinase and carbamoyl-phosphate synthetase systems which he has termed "silent allostery" (33, 34). His experimental evidence is based solely on kinetic measurements expressed as Årrhenius plots which assume a temperature independent $\Delta H^{\ddagger}$ despite some nonlinear behavior in several plots. Thus, his interpretation is subject to all of the now voluminous criticism of the validity of van't Hoff plots. Nevertheless, the kinetic results are interpreted rigorously, and the sheer magnitude of the effects he has observed suggests that his conclusions are at least qualitatively correct. Therefore, "isoergonic" and "silent" cooperativity are equivalent terms, both expressing the conclusion that free-energy-based functions such as binding constants or rate constants are insufficient indicators of allosteric behavior.

5. Perspective

5.1. Results

The phenomenon we describe here entitled "isoergonic cooperativity" and the analogous findings of Reinhart, which he has labeled "silent allostery," are difficult to explain in the customary terms of allostery. However, the overall ΔG° is the quite small resultant of the two large opposing enthalpic and entropic forces in the majority of ligand–protein binding reactions. As a result, relatively small changes in these forces may change the algebraic sign of ΔG°. Therefore, it should not come as a complete surprise to find occasional cases where the resulting ΔG° is zero. Thus, from the viewpoint of basic thermodynamic linkage theory, such results are trivial. However, their real significance lies in that fact itself. One important aspect of these findings is the realization that ΔG functions may not be the most sensitive way of discovering and characterizing allosteric phenomena.

5.2. Allostery

A remarkable book written in 1930 by J.B.S Haldane laid the foundations of the new fields of enzymology in a fashion which has remained relevant to this very day (35). In describing the various classes of inhibition based on Lineweaver–Burk plots, Haldane defined noncompetitive inhibition as the case where the inhibitor binds in some way to another "form" (now understood as "place") than that of the substrate. In Greek, "another place" translates into "allosteric."

In 1963 Monod and coworkers introduced the term, "allostery." However, in 1965 Monod, Wyman, and Changeux proposed a highly specific mechanism for the sequential binding of oxygen to hemoglobin (36). Since their model was based on the rigorous chemical potential linkage theory of Wyman (6), their general notion of allostery was sound. The specific mechanistic details the authors proposed for this particular case were at least reasonable at the time. Unfortunately, it appears that the subsequent development of this new field took as their starting point these highly restrictive mechanistic details, which were later proved to be too simplistic in place of the rigorous linkage theory. As a result, the field has now developed into a language, further confused by several different dialects. Fenton has discussed the rather tangled sets of allosteric nomenclature and varied concepts of allostery (37). For example, the treatment of homotropic and heterotropic binding phenomena as if they were members of disparate branches of science, each with its own separate set of concepts and varied terminology, overemphasizes the strictly local idiosyncratic functional group interactions at the cost of obscuring the major mystery they hold in common – the transduction of the energy of binding between two allosterically related sites.

Possibly, as a result of this terminology, allostery seems to have separated itself from the sound thermodynamic basis from which it arose. As a result, the findings of many well-done and otherwise meaningful studies are now forced into this linguistic straight-jacket, and constructive scientific debate has been replaced by semantic argument. The substitution of special esoteric new terms for otherwise well-understood physical–chemical concepts has further increased the isolation of this important field from the general body of physics and chemistry.

Nowhere is this disjunction more apparent than in the allosteric application of kinetics to ligand-binding induced reactions (see Note 2). The allostery literature is replete with references to "K effectors" and "V effectors," terms hearkening back to Haldane's 1930 original definitions. The extensive progress of the last 80 years in the field of steady-state kinetics led by Cleland's formulation of a system of rigorous equations applicable to the most complex multistep reaction mechanisms (38) has been largely ignored (see Note 2). In the few cases, it has been mentioned that it has been generally misapplied. We have described the misuse of K_m in an earlier section. The apologetic use of such terms as "apparent dissociation constant" adds nothing to remedy the invalid assumptions employed. Cleland has shown that the parameter V/K_m is a definable and mechanistically valid parameter. Its use in the field of ligand–protein interactions is recommended.

The reader may notice that we have interpreted the results of our ITC ligand titrations strictly in terms of well founded classic thermodynamic concepts without the use of or any need of the phenomenological terms currently used in the allosteric literature. In urging our colleagues to abandon epistemological adventures and to interpret their results in rigorously proved thermodynamic (e.g., enthalpy–entropy compensation) and kinetic terms, I herewith offer my own contribution to that end by withdrawing the term "isoergonic cooperativity" from further use.

5.3. Coupling

Richard Feynman's epic work on statistical mechanics must certainly rank as one of the most significant events in the history of science (39). It provides a unifying explanation of all energetic interactions of matter from the subatomic to the known and unknown levels of the universe. Kim, in his review of Feynman's work concluded that every facet of it could be reduced to the interactions of coupled harmonic oscillators (40). Coupled systems exhibit phenomena not shown in any of their isolated components. The thermodynamic coupling factors of protein–ligand binding reactions are expressed in both Wyman chemical potential linkage theory and in the corresponding Weber free energy formalism. In the latter case, they appear explicitly as thermodynamic interaction parameters such as those shown in Fig. 5. At this point we must note a limitation to Weber's initial point that two ligands bound to the same protein molecule must interact with one another. While formally true, such an interaction occurs with a significant magnitude only to the extent that the two binding reactions are dynamically coupled. The required coupling in terms of ligand energy transduction through the protein structure is currently a matter of intensive investigation by a wide variety of experimental techniques. Lumry, following a suggestion by Pauling (41) suggested that ligand-binding energy could be transducted to an enzyme active site if the free-energy reaction coordinates of the protein conformation and that of the chemical reaction were complementary to one another (42). Calorimetric measurements such as those shown in Figs. 4 and 5 here appear to support that concept (43).

The kinetics of coupled harmonic oscillators has been fully developed in physics (as in coupled pendulums, for example) in quantum mechanics, and especially in electrical circuits, based on Kirkhoff's laws. No such general theory applicable to the kinetics of complex chemical reactions is available at present, however. Katchalski has made a start on the problem, applying Kirkhoff's laws treating complex mechanisms as electrical circuits (44). Oster has continued the development of this approach (45). Its completion would be of great help in understanding coupling in chemical and biological systems.

5.4. Energetic Landscapes

The free-energy coordinate of a chemical reaction is conventionally represented as a two dimensional curve as shown in panel A of Fig. 8. Based on the accumulating evidence of the ability of a protein molecule to occupy any of a large number of conformational states at a given point in time combined with the continual discovery of longer and more complex enzyme reaction mechanisms, Swint-Kruse and Fisher (46) have suggested that such reactions may be better represented on a three-dimensional landscape as shown in panel B of Fig. 8. Here, the free-energy profile of the various conformations available to a given enzyme form is plotted on one axis and the sequence of reaction steps on the other. The distribution of occupied conformations is seen to change on each succeeding step. The continuous bold traces denote coupled steps with rates indicated by the heights of their activation barriers. Gaps in the traces indicate the lack of such coupling. Thus, multiple pathways may exist and if so they may or may not be coupled. If reaction conditions could be identified that enabled only one contiguous pathway (bold line) to predominate on the landscape, then a representation like that in (a) would be a flattened version of a single line in (b). However, this relationship would not be true if multiple pathways are simultaneously accessible to an enzyme. When analyzing composite data with alternative energetic models, results for (a) do not represent a continuous pathway on (b) because of the nonadditivity of thermodynamic and

R8X

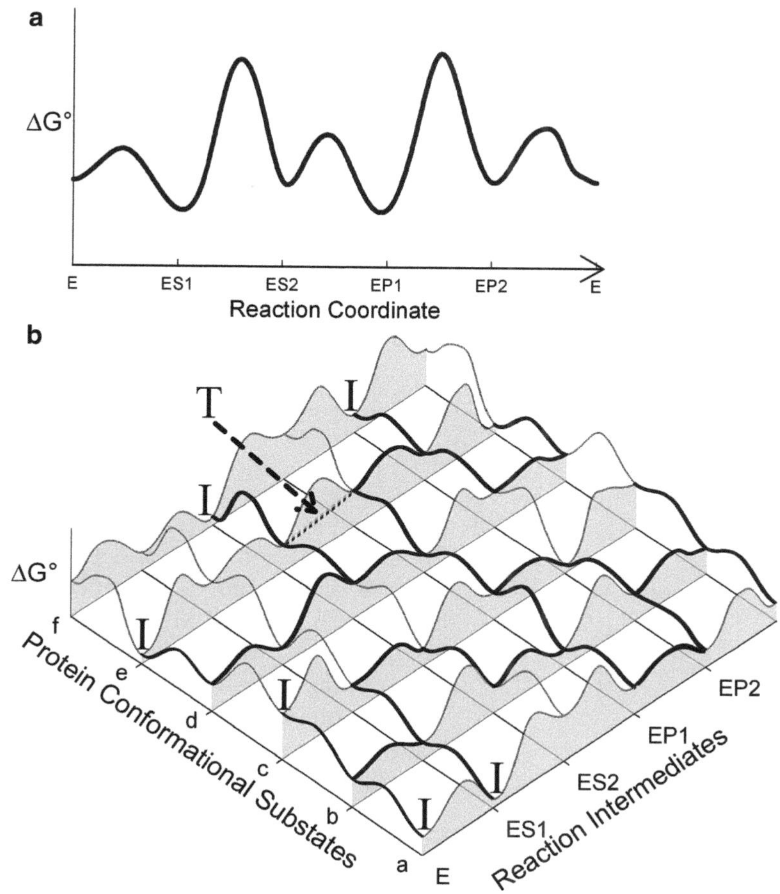

Fig. 8. Alternative depictions of enzymatic reaction pathways. (**a**) The conventional two-dimensional reaction coordinate for a simple enzymatic reaction, where the standard Gibbs free energy (ΔG°) is plotted against the reaction intermediates. (**b**) A three-dimensional patch of the potential energy landscape for a simple enzyme reaction, where ΔG° is plotted against the protein conformational substrates and the reaction intermediates. The *black lines* trace energetic barriers that exist between alternative protein conformations and between subsequent steps of the chemical reaction. Note that traces that parallel the protein conformation axis show changes in the population distribution as the reaction proceeds. *Bold lines* indicate the low-energy, more-probable reaction routes that are accessible to an enzyme population at a given pH, temperature and other solvent conditions. The *thin lines* connect steps whose transition barriers are too high to permit their occurrence at measurable rates, thus restricting the number of available reaction routes. Dead-end, inhibited forms are indicated by the symbol "I." A quantum "tunneling" event is indicated by the symbol "T" and the *dashed line*. In both graphs, E denotes enzyme, ES1 and ES2 are two different enzyme–substrate complexes, and EP1 and EP2 represent alternative enzyme–product complexes (46).

kinetic coupling of alternative pathways in the landscape. Evidence in support of this concept can be found in Rose's tracer studies on fumerase (47) and ITC combined with transient-state studies on α-amino acid dehydrogenases by Swint-Kruse and Fisher (46). Benkovic et al. have also developed the same concept based on studies of dihydrofolate reductase (48). Recalling our earlier comment that coupled systems may exhibit behavior not observed in that of their isolated components, this viewpoint may be of some use in interpreting otherwise puzzling allosteric observations. Thus, allostery may best be defined as coupled interactions between the binding of two ligands to separate sites on the same protein. Further taxonomic elaboration of this concept is neither necessary nor warranted.

6. Notes

1. This experiment was performed in 10 mM Pipes, pH 7.5 at 15°C. [SH2] = 81 μM and [TEGQpYQPQPA] = 0.4 mM. Analysis of the data, as described in the text, yields a binding affinity of $5.8 \times 10^{6}\ M^{-1}$ and a ΔH of −13.5 kcal/mol. The inset illustrates the configuration of an ITC reaction cell. The cell volume is 1.4 mL and is filled with the protein solution (gray splotches). The injection syringe, which also stirs the solution to assure proper mixing, is filled with the ligand solution (also gray). At specified time intervals, a small volume (typically 10 μL) of the ligand solution is injected into the cell, giving rise to the characteristic titration heat effects. Once the protein is saturated, the residual heat effects originate from dilution of the peptide and also from mechanical effects associated with the injection. These effects need to be subtracted before thermodynamic analysis. In an ITC experiment, the quantity measured and displayed on the *y*-axis is the time dependence of the electric power (μcal/s) necessary to maintain constant the temperature difference between the reaction and reference cells after each injection of reactant. The area under each peak is the heat (microcalories) associated with the process.
2. Editor Note: Although K_m should not be used as a binding constant, there have now been several descriptions of the proper use of steady-state derived K_m values to determine thermodynamic coupling constants (24, 50, 51). Dr. Cleland is an author of one of these descriptions (24).

References

1. Fisher, H. F. and Tally, J. (1997) Isoergonic cooperativity in glutamate dehydrogenase complexes: a new form of allostery. *Biochemistry* **36**, 10807–10810.
2. Fisher, H. F. and Tally, J. (1998) Isoergonic cooperativity: a novel form of allostery. *Methods Enzymol.* **295**, 331–349.
3. Weber, G. (1975) Energetics of ligand binding to proteins. *Adv. Protein Chem.* **29**, 1–83.
4. Eftink, M. R. and Biltonen, R. L. (1980) Thermodynamics of Interacting Biological Systems. In *Biological Calorimetry* (Breezer, A. E., Ed.) pp 343–408, Academic, New York.
5. Fisher, H. F. (1988) A unifying model of the thermodynamics of formation of dehydrogenase- ligand complexes. in *Advances in Enzymology and Related Areas of Molecular Biology* (Meister, A., Ed.) 61 ed., pp 1–46, Wiley, NJ.
6. Wyman, J., Jr. (1948) Heme proteins. *Adv. Protein Chem.* **4**, 407–531.
7. Subramanian, S., Stickel, D. C., Colen, A. H., and Fisher, H. F. (1978) Thermodynamics of heterotropic interactions. The glutamate dehydrogenase . NADPH . glutamate complex. *J. Biol. Chem.* **253**, 8369–8374.
8. Lumry, R. and Rajender, S. (1970) Enthalpy-entropy compensation phenomena in water solutions of proteins and small molecules: a ubiquitous property of water. *Biopolymers* **9**, 1125–1227.
9. Edsall, J. T. and Gutfreund, H. (1983) *Bio-Thermodynamics*, p 219, Wiley, New York.
10. Rialdi, G. and Biltonen, R. L. (1975) Thermodynamics and Thermochemistry. In *MTP International Review of Science. Physical Chemistry* (Skinner, H. A., Ed.) Series 2 ed., pp 147, Butterworths, London.
11. Hinz, H. J. and Jaenicke, R. (1975) Thermodynamics of complex formation between nicotinamide adenine dinucleotide and pig skeletal muscle lactate dehydrogenase. *Biochemistry* **14**, 24–27.
12. Fisher, H. F., Colen, A. H., and Medary, R. T. (1981) Temperature-dependent delta C_P generated by a shift in equilibrium between macrostates of an enzyme. *Nature* **292**, 271–272.
13. Leffler, J. E. and Greenwald, E. (1963) *Rates of Equilbria of Organic Reactions*, Wiley, London.
14. Exner, O. (1964) On the enthalpy-entropy relationship. *Collect. Czech. Chem. Commun.* **29**, 1094–1113.
15. Krug, R. R., Hunter, W. G., and Grieger, R. A. (1976) Enthalpy-entropy compensation. 1. Some fundamental statistical problems associated with the analysis of van't Hoff and Arrhenius data. *J. Phys. Chem.* **80**, 2335–2341.
16. Keleti, T. (1983) Errors in the evaluation of Arrhenius and van't Hoff plots. *Biochem. J.* **209**, 277–280.
17. Sharp, K. (2001) Entropy-enthalpy compensation: fact or artifact? *Protein Sci.* **10**, 661–667.
18. Cornish-Bowden, A. (2002) Enthalpy-entropy compensation: a phantom phenomenon. *J. Biosci.* **27**, 121–126.
19. Klotz, I. M. (1997) *Ligand-Recptor Energetics – A Guide for the Perplexed*, Wiley, New York.
20. Van Holde, K. E. (1985) *Physical Biochemistry*, Prentice-Hall, Englewood Cliffs, NJ.
21. Horn, J. R., Russell, D., Lewis, E. A., and Murphy, K. P. (2001) Van't Hoff and calorimetric enthalpies from isothermal titration calorimetry: are there significant discrepancies? *Biochemistry* **40**, 1774–1778.
22. Jonsson, T., Glickman, M. H., Sun, S., and Klinman, J. P. (1996) Experimental evidence for extensive tunneling of hydrogen in the lipoxygenase reaction: implications for enzyme catalysis. *J. Am. Chem. Soc.*, **118**, 10319–10320.
23. Limbach, H., Lopez, J. M., and Kohen, A. (2006) Arrhenius curves of hydrogen transfers: tunnel effects, isotope effects and effects of pre-equilibria. *Phil. Trans. R. Soc. B* **361**, 1399–1415.
24. Cook, P. and Cleland, W. W. (2007) *Enzyme Kinetics and Mechanism*, Garland Science Publishing, New York.
25. Gutfreund, H. (1998) *Kinetics for the Life Sciences*, U.K. Press Cambridge, Cambridge.
26. Fisher, H. F. and Singh, N. (1995) Calorimetric methods for interpreting protein-ligand interactions. *Methods Enzymol.* **259**, 194–221.
27. (1986) *The Flucuating Enzyme*, Wiley, London.
28. Wiseman, T., Williston, S., Brandts, J. F., and Lin, L. N. (1989) Rapid measurement of binding constants and heats of binding using a new titration calorimeter. *Anal. Biochem.* **179**, 131–137.
29. Indyk, L. and Fisher, H. F. (1998) Theoretical aspects of isothermal titration calorimetry. *Methods Enzymol.* **295**, 350–364.

30. Leavitt, S. and Freire, E. (2001) Direct measurement of protein binding energetics by isothermal titration calorimetry. *Curr. Opin. Struct. Biol.* **11**, 560–566.
31. Ackers, G. K., Doyle, M. L., Myers, D., and Daugherty, M. A. (1992) Molecular code for cooperativity in hemoglobin. *Science* **255**, 54–63.
32. Kodama, T. (1985) Thermodynamic analysis of muscle ATPase mechanisms. *Physiol. Rev.* **65**, 467–551.
33. Braxton, B. L., Tlapak-Simmons, V. L., and Reinhart, G. D. (1994) Temperature-induced inversion of allosteric phenomena. *J. Biol. Chem.* **269**, 47–50.
34. Tlapak-Simmons, V. L. and Reinhart, G. D. (1998) Obfuscation of allosteric structure-function relationships by enthalpy-entropy compensation. *Biophys. J.* **75**, 1010–1015.
35. Haldane, J. B. S. (1930) *Enzymes*, Green and Co., Longmans.
36. Monod, J., Wyman, J., Changeux, J. P. (1965) On the nature of allosteric transitions: a plausible model. *J. Mol. Biol.* **12**, 88–118.
37. Fenton, A. W. (2008) Allostery: An illustrated definition for the 'second secret of life'. *Trends Biochem. Sci.* **33**,420–425.
38. Cleland, W. W. (1963) The kinetics of enzyme-catalyzed reactions with two or more substrates or products. I. Nomenclature and rate equations. *Biochim. Biophys. Acta* **67**, 104–137.
39. Feynman, R. P., Kislinger, M., and Ravndal, F. (1971) Phenomenological model for diffractive excitation of Hadron resonances. *Phys. Rev. D* **3**, 2706.
40. Kim, Y. S. (2007) Coupled oscillators and Feynman's three papers. *J. Phys. Conf. Ser.* **70**, 1–19.
41. Pauling, L. (1946) Molecular architecture and biological reactions. *Chem. Eng. News* **24**, 1375.
42. Lumry, R. (1986) Free-Energy Management in Protein Reactions: Concepts, Complications and Compensation. In *The Fluctuating Enzyme* (Welch, G. R., Ed.) pp 32, Wiley, London.
43. Fisher, H. F. and Singh, N. (1991) Transduction of enzyme-ligand binding energy into catalytic driving force. *FEBS Lett.* **294**, 1–5.
44. Oster, G. F., Perelson, A. S., and Katchalsky, A. (1971). Network thermodynamics. *Nature* **234**, 393–397.
45. Oster, G. F. and Perelson, A. S. (1973). Systems, circuits and thermodynamics. *Israel J. Chem.* **11**, 445–478.
46. Swint-Kruse, L. and Fisher, H. F. (2008) Enzymatic reaction sequences as coupled multiple traces on a multidimensional landscape. *Trends Biochem. Sci.* **33**, 104–112.
47. Rose, I. A. (1998) How fumarase recycles after the malate -> fumarate reaction. Insights into the reaction mechanism. *Biochemistry* **37**, 17651–17658.
48. Benkovic, S. J., Hammes, G. G., and Hammes-Schiffer, S. (2008) Free-energy landscape of enzyme catalysis. *Biochemistry* **47**, 3317–3321.
49. Fisher, H. F., Subramanian, S., Stickel, D. C., and Colen, A. H. (1980) The thermodynamics of a negatively interacting allosteric effector system. The glutamate dehydrogenase . NADPH . ADP complexes. *J. Biol. Chem.* **255**, 2509–2513.
50. Reinhart, G. D. (1983) The determination of thermodynamic allosteric parameters of an enzyme undergoing stead-state turnover. *Arch. Biochem. Biophys.* 224, 389–401.
51. Reinhart, G. D. (2004) Quantitative analysis and interpretation of allosteric behavior. *Methods Enzymol.* **380**, 187–203.

Chapter 6

Using Mutant Cycle Analysis to Elucidate Long-Range Functional Coupling in Allosteric Receptors

Jai A.P. Shanata, Shawnalea J. Frazier, Henry A. Lester, and Dennis A. Dougherty

Abstract

Functional coupling of residues that are far apart in space is the quintessential property of allosteric receptors. Data from functional studies of allosteric receptors, such as whole-cell dose–response relations, can be used to determine if mutation to a receptor significantly impacts agonist potency. However, the classification of perturbations as primarily impacting binding or allosteric function is more challenging, often requiring detailed kinetic studies. This protocol describes a simple strategy, derived from mutant cycle analysis, for *e*lucidating *l*ong-range *f*unctional *c*oupling in *a*llosteric *r*eceptors (ELFCAR). Introduction of a gain-of-function reporter mutation, followed by a mutant cycle analysis of the readily measured macroscopic EC_{50} values can provide insight into the role of many physically distant targets. This new method should find broad application in determining the functional roles of residues in allosteric receptors.

Key words: Allostery, Signal transduction, Conformational change, Coupling, Structure–function study, Ion channel, Nicotinic receptor, Amino acids

1. Introduction

In allosteric receptors, high-affinity binding of a specific small molecule(s) or ion(s) to a particular binding site initiates a functional response, termed activation (1, 2). Typically, this functional response is most prominent in a region that is spatially distinct from the binding site, and the binding event conveys changes to structural and/or electronic properties of the protein over distances of tens of angstroms or more. This is a hallmark of allosteric receptors – the ability to propagate specific conformational changes over long distances across a macromolecule. A natural consequence of this effect is the categorization of functional domains or specific amino acids of the protein into two

Aron W. Fenton (ed.), *Allostery: Methods and Protocols*, Methods in Molecular Biology, vol. 796,
DOI 10.1007/978-1-61779-334-9_6,

classes: those that are primarily involved in binding the small molecule and those that contribute to the functional response. In a typical characterization of such a receptor, a dose–response curve based on a readout of receptor function produces, for example, an EC_{50}. When a specific perturbation is introduced – often a side chain mutation – any significant shift in EC_{50} could reflect changes in either agonist binding or functional coupling, as EC_{50} is a composite number that can be influenced by both these properties of an allosteric receptor (3). Which property the perturbation affects, however, cannot be resolved from the dose–response curve alone. Here, we describe a strategy based on mutant cycle analysis (4, 5) that, in favorable cases, provides a simple way to make such a classification.

In mutant cycle analysis, two perturbations are made to the system, first in isolation and then in tandem. The key metric is Ω, the coupling parameter, here given in terms of the functional metric EC_{50} (Fig. 1a). If the two perturbations are independent of one another, Ω should be ~1. That is, for independence, the individual mutations' effects on functional metrics should be multiplicative (additive when placed on a relative energy scale). If Ω deviates significantly from 1, an interaction between the perturbed sites is established. Typically, the method has been used to establish proximity between a pair of amino acids, often implicating a stabilizing interaction such as a hydrogen bond or an ion pair (6–8). However, in an allosteric protein, two residues can be *functionally* coupled even if they are quite far apart, and such behavior has been commonly seen (9).

In the present system, we define one perturbation as the "reporter." We confine this discussion to cases where the reporter is a well-characterized, gain-of-function mutation that is remote from the binding site of the agonist or allosteric modulator. The reporter mutation is then used to evaluate a target (Fig. 1b). Typically, the target is a particular amino acid whose role in receptor function is unknown, although in principle the method could also be used to probe the consequences of chemical changes to an agonist or allosteric modulator. In the present discussion, we focus on amino acid targets. A mutation to the target is then paired with the reporter mutation, with the goal of determining whether the target can be classified as primarily contributing to agonist binding or to the functional response. Our reporter-based application of mutant cycle analysis is termed *e*lucidating *l*ong-range *f*unctional *c*oupling in *a*llosteric *r*eceptors (ELFCAR) (10).

The key to ELFCAR analysis is the composite nature of the functional metric, EC_{50}. Consider a minimal kinetic scheme for an allosteric receptor such as this three-state model.

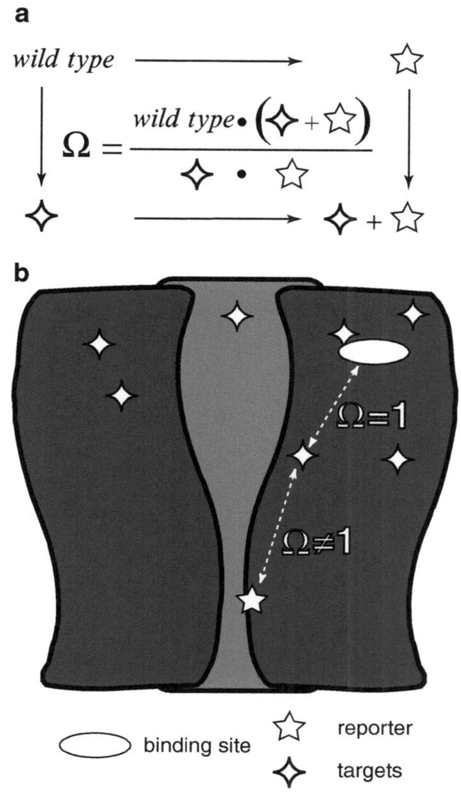

Fig. 1. Schematic representation of an allosteric receptor and mutant cycle analysis. The five-point star is the gain-of-function reporter mutation and the four-point stars are the distant targets being altered throughout the allosteric receptor. In these studies, the specific functional metric used is EC_{50}. (**a**) Definition of Ω, the coupling parameter, in terms of a generic functional metric for receptors with altered target, reporter, and both. (**b**) ELFCAR enables the assignment of functional coupling of physically distant regions of allosteric receptors ($\Omega \neq 1$) versus binding mutations ($\Omega = \sim 1$) when mutant cycle analysis is performed on wild-type receptors with high K_{act} and using reporter(s) that further significantly increase K_{act}.

$$R + A \underset{k_-}{\overset{[A]k_+}{\rightleftharpoons}} AR \underset{k_{deact}}{\overset{k_{act}}{\rightleftharpoons}} AR^*$$

For this scheme, R is an unactivated allosteric receptor, A is agonist, and R* is an activated allosteric receptor. This gives rise to the EC_{50} value defined by Eq. 1, where $K_D = k_-/k_+$ and $K_{act} = k_{act}/k_{deact}$.

$$EC_{50} = \frac{K_D}{1 + K_{act}}. \tag{1}$$

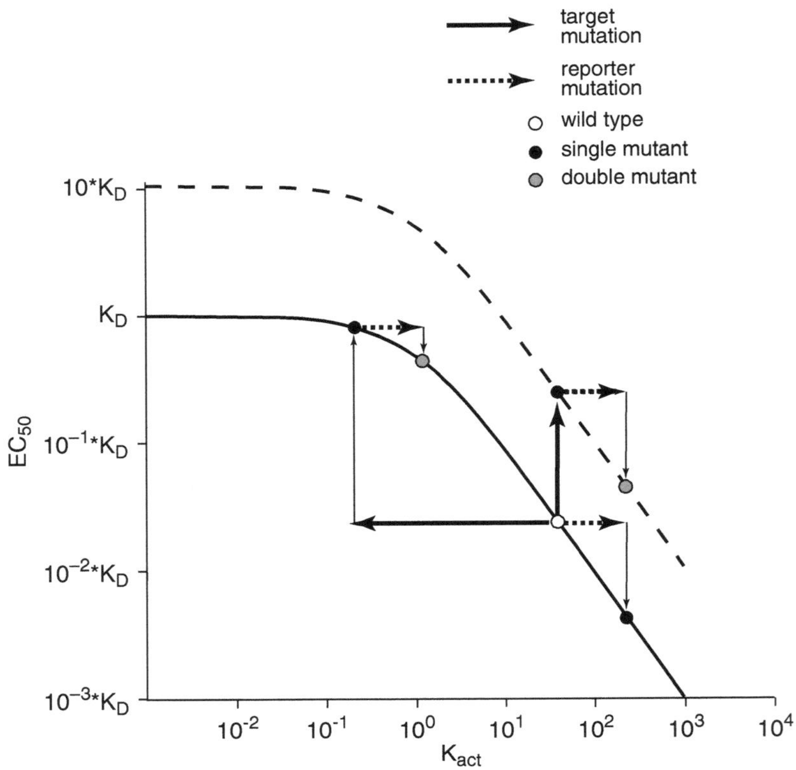

Fig. 2. Plot of EC_{50} versus K_{act}. (*Solid line*) Relationship between EC_{50} and K_{act} for a simple allosteric model (Eq. 1). (*Dotted line*) Plot of the same model, but with K_D tenfold larger. For both plots, in the high slope region changes to K_{act} produce significant changes in EC_{50} (*thin arrows*), as shown when the reporter mutation is made to the wild-type receptor. However, for allosteric receptors with K_{act} in the plateau region, such as wild-type receptors with high K_{act} into which a significant loss-of-function target mutation has been introduced, a much smaller shift in EC_{50} is seen for equivalent shifts in K_{act}.

Note that more complex kinetic models retain the essential features of this scheme (11). As noted above, the reporter mutation is chosen to be one that impacts only K_{act}. A mutation that is physically quite remote to the binding site would be appropriate, although detailed analysis of its impact on function by, for example, single-channel analysis is desirable (12–14). When plotted on a log scale, the relationship between EC_{50} and functional coupling, as judged by the equilibrium constant for receptor activation, K_{act}, is linear with high slope over a considerable range. However, as shown in Fig. 2, EC_{50} plateaus at small values of K_{act}. Since most allosteric receptors are efficiently activated by one or more agonists, the activation of the wild-type receptor will typically lie in the high slope region of the plot. By design, the gain-of-function reporter mutation solely impacts K_{act}, causing a consistent rightward shift along the *x*-axis.

Now consider a target mutation that produces a significant loss-of-function. This can impact the curve of Fig. 2 two ways. If the increase in EC_{50} is a consequence of an increase in K_D, the entire curve shifts upward (dashed line). This keeps EC_{50} in the high slope region, and so the reporter has the same fold shift in EC_{50} (thin arrows) for both the wild-type and altered target receptors. Necessarily, mutant cycle analysis will produce an $\Omega = \sim 1$. Alternatively, a loss-of-function target mutation could act by shifting K_{act} to the left (thick arrow). If the target mutation substantially inhibits allosteric function, it will produce a significant shift that will move EC_{50} into the plateau region (thin arrow). Now the effect of the reporter mutation on EC_{50} will be diminished relative to wild type and $\Omega \neq 1$. Thus, in favorable circumstances, a simple mutant cycle analysis using macroscopic measurements can provide insights into the detailed mechanism of receptor function.

Our discussion, and the protocols described below, have emphasized receptors that are activated by binding a small molecule. However, the essential feature of allosteric receptors is embedded in Eq. 1 and Fig. 2, and so any system with appropriate wild-type activation kinetics and for which a significant gain-of-function reporter can be identified will be amenable to ELFCAR analysis.

2. Materials

Unless otherwise stated, water is Millipore-filtered water with resistivity ≥18 MΩ cm. 0.2 μm filters are Nalgene (Thermo Fisher Scientific, Inc., Wilmington, DE).

2.1. Polymerase Chain Reaction-Based Site-Directed Mutagenesis, DNA Amplification, and mRNA Transcription

1. DNA Amplification: Super Competent Top 10 *Escherichia coli* cells.
2. DNA isolation: QIAprep Miniprep kit (Quiagen, Valencia, CA).
3. DNA linearization: *Not*I enzyme and 10× buffer (New England Biolabs, Ipsich, MA).
4. Transcription: mMESSAGE mMACHINE T7 kit (Applied Biosystems, Foster City, CA).
5. RNA cleanup: RNeasy Mini kit (Quiagen, Valencia, CA).
6. Gel running buffer: dilute 1 part 50× Tris/acetic acid/EDTA (TAE; Bio-Rad Laboratories, Hercules, CA) with 49 parts water.
7. 1% Agarose gel: dissolve 0.60 g agarose (Invitrogen, Carlsbad, CA) in 60 ml of 1× TAE with heat (e.g., microwave for 45–60 s). Prepare as needed; pour gel into mold with desired number of lanes and store at 4°C for 1–12 h until use.

2.2. Expression of Ligand-Gated Ion Channels in *Xenopus laevis* Oocytes

1. ND96 solution: 96 mM NaCl, 2 mM KCl, 1 mM $MgCl_2$, 1.8 mM $CaCl_2$, 5 mM HEPES, pH to 7.5 with NaOH. Filter with 0.2 μm filter and store at 4°C (see Note 1).
2. Oocyte incubation media: Add 138 mg sodium pyruvate (to 2.5 mM) and 60 mg theophylline (to 0.67 mM) to 500 ml of ND96 solution; filter with 0.2 μm filter. Add 500 μl of 50 mg/ml gentamicin solution (to ~0.11 mM). Add 1–2% horse serum and store for <10 days at 4°C (see Note 2).
3. Oocyte viewing station: Use a standard dissecting microscope with ~20× eyepieces and adjustable ~0.5×–4× objective (such as Leica S6E, Spectra Services, Ontario, NY) with a flexible light source that has multiple lighting levels (such as model 3100 Dolan-Jenner Industries, Inc., Lawrence, MA).
4. Injection needles: Pull injection needles from 8 in. glass capillary with outer diameter = 1.14 mm, inner diameter = 0.53 mm (Drummond Scientific, Broomall, PA) in groups of ~10 on a Sutter P-97 horizontal pipette puller (Sutter Instrument Co., Novato, CA).
5. Oocyte injection: In addition to the oocyte viewing station, install an x, y, z manipulator on a stable base (Nanoject Support Base, Drummond Scientific, Broomall, PA). Use a 10 μl microdispenser (Drummond Digital Microdispenser) to control the volume of injected solutions.

2.3. Whole-Cell Recording with OpusXpress

1. 1.00 M Stock solutions of agonists: for example, acetylcholine (ACh) chloride (Sigma-Aldrich/RBI) in water and immediately store at −80°C in aliquots of 100–500 μl (see Note 3).
2. Ca^{2+}-free ND96 solution: 96 mM NaCl, 2 mM KCl, 1 mM $MgCl_2$, 5 mM HEPES, pH to 7.5 with NaOH. Filter with 0.2 μm filter and store at 4°C (see Note 2).
3. Recording drug solutions: thaw then dilute previously made aliquots of 1.00 M agonist into Ca^{2+}-free ND96 solution within 12 h prior to recording. Approximately ten doses over ~3 orders of magnitude of concentration centered on the anticipated EC_{50} value is a good starting point (see Note 4).
4. Recording electrodes: Use borosilicate glass with length 15 cm, outer diameter = 1.5 mm, and inner diameter = 1.17 mm (Sutter Instrument Co., Novato, CA). Pull >30 electrodes on horizontal pipette puller to a resistance of 0.3–3 MΩ for current headstages and 0.3–10 MΩ for voltage headstages.
5. Electrophysiological recording: OpusXpress 6000 A (Molecular Devices Axon Instruments, Sunnyvale, CA).

2.4. Data Analysis

1. Clampex software (Axon Instruments, Union City, CA).

2. Software for graphing and fitting dose–response curves, e.g., KaleidaGraph, Origin, or SigmaPlot.

3. Methods

The purpose of this protocol is to use a reporter mutation to determine the functional role of distant targets by mutant cycle analysis. The first step is to identify a gain-of-function reporter. The technique then proceeds as is outlined in Fig. 3. While ELFCAR is broadly applicable to allosteric receptors, the protocol detailed here is specific for elucidating the functional role of a number of amino acids in the nicotinic acetylcholine receptor (nAChR), the prototypical ligand-gated ion channel (LGIC) of

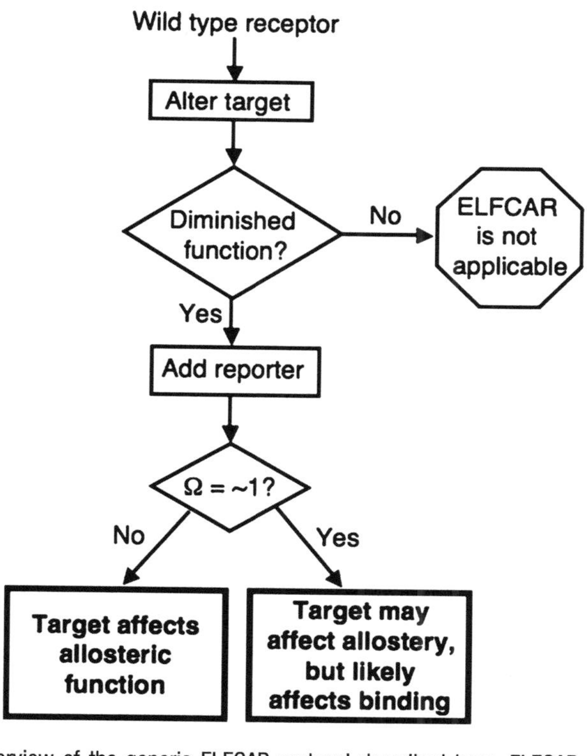

Fig. 3. Overview of the generic ELFCAR protocol described here. ELFCAR begins by altering targets and identifying those that are significantly deleterious to function. These targets are then combined with a well-characterized gain-of-function reporter mutation. Mutant cycle analysis is performed on four total allosteric receptors: by obtaining functional data for wild type as well as mutated receptors (separately and in tandem) and a value for Ω is calculated (Fig. 1a). For altered targets that produce a loss-of-function, if $\Omega \neq 1$, the altered target primarily impacts allosteric function. If $\Omega = \sim 1$, the altered target presumably primarily affects binding.

the Cys-loop receptor superfamily (15–17). The fetal mouse muscle receptor studied here is a heteropentamer composed of 2α, 1β, 1γ, and 1δ subunits. Activation of the nAChR is initiated by the binding of ligand, such as ACh, in the extracellular domain. This induces a conformational change in which the channel pore gates from a closed, nonconducting form, to an open, conducting, form. Receptor activation is readily measured in a whole-cell voltage clamp assay over a range of concentrations and a dose–response relation is plotted and fit to the Hill equation, Eq. 2, where n_H is the Hill coefficient, I_{max} is maximal response, and EC_{50} is the concentration of agonist needed to elicit half maximal response.

$$\frac{I}{I_{max}} = \frac{1}{(1 + (EC_{50}/[A])^{n_H})}. \tag{2}$$

For ELFCAR, an EC_{50} value is obtained for each of the four receptor sets – wild type, target, reporter, and target + reporter – in order to perform the mutant cycle analysis calculation of the Ω value (12–14). Details of the general ELFCAR procedure including (1) making the mutated allosteric receptors, (2) expressing them in a heterologous system, (3) obtaining functional data for the allosteric receptors, and (4) analyzing these data to calculate Ω, are described below. Once all necessary mutants are made, steps 2–4 can be performed rapidly (<1 week) and in parallel.

3.1. Polymerase Chain Reaction-Based Site-Directed Mutagenesis, DNA Amplification, and mRNA Transcription

Using the following protocol, perform site-directed mutagenesis for each of the mutations needed – reporter and target(s), if applicable. For all molecular biology reactions, combine reagents in appropriate sized Eppendorf tube (>3× reaction volume), then mix by pipetting for 30 s, vortexing for 3 s, then microcentrifuging for 1–3 s.

1. Design mutagenic primers of 25–45 bases with 50–55% GC content and $T_m \geq 78°C$; terminate with one or more C/G at 3′ end and check for primer dimers. For each mutation of interest, order forward and reverse (complementary sequence) primers.
2. Perform polymerase chain reaction (PCR) using Stratagene QuickChange method. For each mutation, a 50 μl reaction mix contains: 5 μl 10× polymerase buffer, 1 μl circular dsDNA template (5–50 ng), 1.25 μl of both mutagenic primers (~125 ng each), 1 μl 25 mM dNTP mix, 39.5 μl RNAse-free water, and 1 μl Pfu Turbo Hotstart DNA polymerase. Use a 10 min extension time in each thermocycle and modify annealing temperatures as required for successful incorporation of the mutation.

3. Once PCR is complete, remove 10 μl of the PCR mixture for gel screen.
4. Perform 1.5–3 h of digestion by *Dpn*I of the remaining PCR mixture to eliminate methylated template DNA from the PCR product, leaving the desired mutated DNA.
5. Perform a gel screen (1% agarose gel) of the PCR by loading 10 μl of each PCR mixture (from step 3) to each well and the corresponding circular DNA (control) in neighboring lanes; run for ~1 h. Develop gel to visualize with ethidium bromide (0.1% in gel running buffer) for 15 min and imagegel.
6. Amplify the PCR products by electroporation of Super Competent Top 10 *E. coli* cells at 1,800 V followed by ~12 h of growth on agar/Luria broth/ampicillin plates in an incubator at 37°C.
7. Select single colonies and amplify each separately in 4 ml liquid Luria broth + 4 μl ampicillin with orbital shaking at 37°C for ~12 h.
8. Using a QIAPrep Miniprep kit, isolate the DNA from each bacterial sample separately as follows. Pellet DNA from step 7 and resuspend in 250 μl chilled buffer P1 in a 1.5 ml Eppendorf tube. Add 250 μl buffer P2 and mix by inverting 6×, then add 350 μl buffer N3 and mix immediately by inverting six times. Microcentrifuge for 10 min at 12,000×*g*, pipette supernatant to QIAprep spin column, centrifuge 1 min and discard flow-through. Wash QIAprep spin column with 0.5 ml buffer PB, centrifuge for 1 min and discard flow-through; repeat rinse with 0.75 ml Buffer PE. Microcentrifuge an additional 1 min to remove residual buffer. Elute DNA into a clean 1.5 ml Eppendorf tube by placing 30–50 μl water on the center of the column, letting stand for 1 min, and centrifuging for 1 min (see Note 5).
9. Determine DNA concentration.
10. Sequence DNA to verify the successful incorporation of the mutation at the selected site.
11. Linearize nAChR subunit DNA in expression vector pAMV by incubating 50 μg (~50–100 μl) with 20 μl 10× buffer, 20 μl *Not*I, and RNAse-free water to a final reaction volume of 200 μl for 8–16 h at 37°C. Verify complete linearization by running a 1% agarose gel in running buffer of ~2 μl reaction mixture against 2 μl of circular DNA.
12. Transcribe linearized DNA(s) using the T7 mMessage mMachine system individually for each subunit required for the receptor to be expressed; for the fetal mouse muscle nAChR these are α, β or βL9'S – where L9'S is the reporter mutation –

γ, and δ (see Note 6). Perform a typical transcription by combining the following (in this order): 6 μl 10× transcription buffer, 30 μl 2× NTP mix, 8 μl RNAse-free water, 10 μl (~10 μg) linearized plasmid DNA, 6 μl T7 enzyme mix, and incubating for 3–4 h at 37°C.

13. Stop transcription by adding 3 μl DNAse and incubating at 37°C for 0.5–1 h.
14. To cleanup mRNA using the RNeasy Mini kit, start by adjusting sample volume to 100 μl by adding 37 μl RNase-free water. Add 350 μl buffer RLT (with 1% β-mercaptoethanol previously added to buffer RLT), vortex 5 s, and microcentrifuge 10 s. Add 250 μl 100% ethanol, mix with gentle pipetting, then apply the 700 μl sample to an RNeasy column placed in a 2 ml collection tube and microcentrifuge for 15 s at 12,000×*g*. Discard flow-through and place column in a new 2 ml collection tube. Wash column with 500 μl diluted buffer RPE (4:1, 100% ethanol: RPE concentrate), with 15 s microcentrifuging. Discard flow-through and repeat with 500 μl diluted buffer RPE and 2 min microcentrifuging, then place column in a new 2 ml collection tube and microcentrifuge for 1 min to ensure dryness and minimize ethanol carryover. Transfer column to a new 1.5 ml Eppendorf tube, pipette 30–60 μl RNase-free water directly onto the silica-gel membrane of the column and elute with 1 min microcentrifuging. Reapply eluent to silica-gel of column and re-elute to increase concentration (see Note 5).
15. Determine mRNA concentration.

3.2. Expression of Ligand-Gated Ion Channels in X. laevis *Oocytes (18)*

1. For the wild-type or mutant receptors, prepare mRNA solution for injection by combining subunits in a 2:1:1:1 ratio ($\alpha:\beta:\delta:\gamma$, by mass) with a total mRNA concentration of ~1–2 ng/nl (see Note 7).
2. With freshly flame-sterilized Inox forceps, break an injection needle to ~10–20 μm tip diameter (at injection station). Back fill completely with mineral oil, load onto a microdispenser, and place in injection manipulator. Place mRNA solution, typically 1–3 μl, on a clean coverslip and draw mRNA solution into injection needle (see Note 8).
3. Inject defolliculated stage V–VI *X. laevis* oocytes by pressing the injection needle through the cell membrane until the membrane bounces back, then expelling ~50 nl of the combined mRNA solution into the cell.
4. Change the oocyte incubation media 30–60 min after injection, and incubate injected oocytes at 16 ± 2°C for 12–36 h with orbital shaking in 35 mm dishes (see Note 9).

3.3. Whole-Cell Recording with OpusXpress

Perform semiautomated whole-cell voltage clamp recording of *X. laevis* oocytes expressing LGICs using the OpusXpress 6000 A (see Note 10).

1. Fill previously pulled electrodes with 3.0 M KCl and place in electrode holders, then place one holder with electrode on each of the 16 headstages. Lower electrodes into Ca^{2+}-free ND96 solution in chambers and check resistances: 0.3–10 M ohm for voltage electrodes, 0.3–3 M ohm for current electrodes (see Note 11).
2. Fill 96-well drug plates with increasing concentrations of drug solutions, starting with 0.
3. Place one oocyte in each chamber and impale oocytes individually. Common recording parameters are: capacitance neutralization (0.0), output gain (1), ultra-high DC gain (on), and initial recording gain (1,500). Optimize recording gain by increasing or decreasing gain individually on each impaled oocyte to achieve maximally square responses to test pulse with minimal ringing (usually 3,000–6,000 for healthy oocytes).
4. Voltage clamp each oocyte at a holding potential between −40 mV and −80 mV (see Note 12).
5. Perform a typical run by applying each concentration of drug solution for 15 s followed by a 130 s wash with Ca^{2+}-free ND96 solution between each applied drug concentration. During drug application, superfuse oocytes with a flow rate of 4 ml/min; during wash, use a flow rate of 3 ml/min. Sample data at 125 Hz and filter at 50 Hz (see Note 13).
6. Start run and obtain dose–response data for ≥6 concentrations of agonist and for ≥5 oocytes with sufficient expression for each mutant (see Note 14). It is advisable to apply at least two control doses (Ca^{2+}-free ND96 with no added agonist) during each run to verify nonresponse as well as to differentiate from leak current and fluidics effects. For each mutant, recordings should be performed on oocytes from ≥3 different donor *X. laevis* frogs.

3.4. Data Analysis

1. Open each recorded data file in Clampex and determine the response at each concentration as the largest current deflection from average baseline.
2. Plot each oocyte's normalized current response and corresponding drug concentration (see Note 15).
3. For remaining cells, average each concentration's normalized response and plot the average dose–response relation for all oocytes with error bars (standard error).
4. Fit whole-cell dose–response relations to Hill equation (Eq. 2); record results.

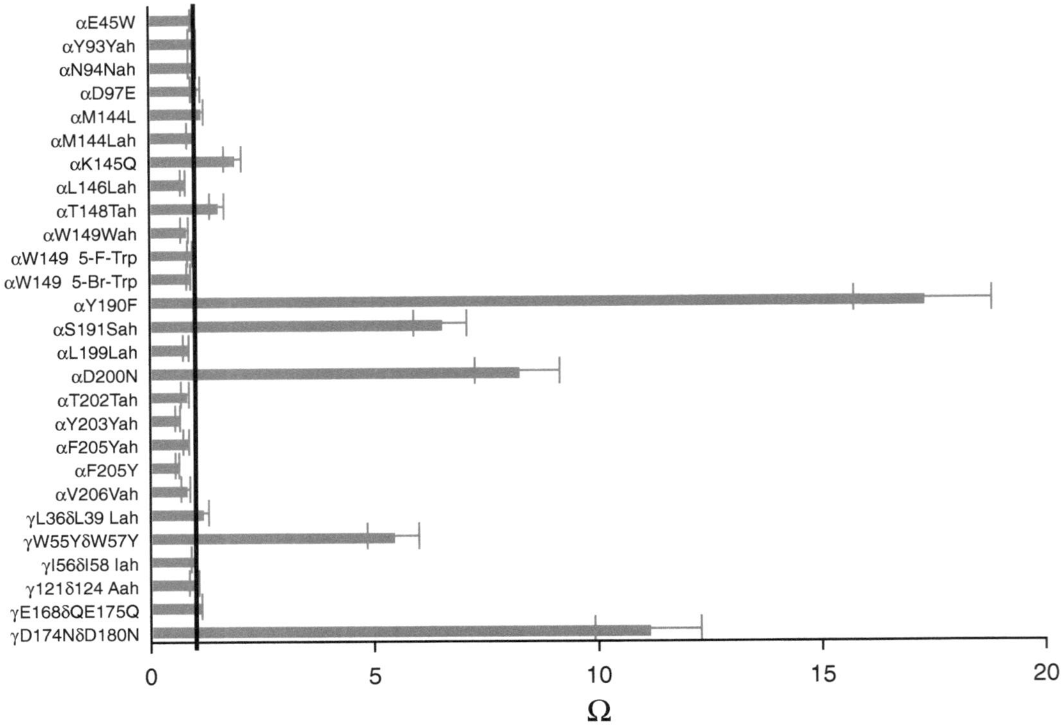

Fig. 4. Bar chart of omega values that ELFCAR produces. *Gray bars* represent Ω for 27 altered targets in the nAChR extracellular domain with βL9′S reporter. Note, βL9′S is a well-characterized gain-of-function mutation in the channel pore known primarily to impact channel activation (K_{act}) (12–14). Five altered targets show an Ω value that deviates significantly from 1 (*black line*) and are thereby assigned by ELFCAR to primarily impact allosteric function. Of the other 22, those altered targets that produce a significant loss-of-function (data not shown) are assigned as primarily impacting binding. For example, 5-F-Trp (5-fluorotryptophan) at αW149 produces an EC_{50} that is four times higher than wild type, but with $\Omega = 0.9$, so ELFCAR assigns this altered target to impact binding. 5-Br-Trp is 5-bromotryptophan. For the α-hydroxy acids (Yah, Nah, Lah, Wah, Vah, Iah, and Aah), a three-letter abbreviation is used: the one-letter code for the parent amino acid is followed by -ah; thus Yah is the α-hydroxy acid of tyrosine.

5. Using the EC_{50} values for each of the four components of ELFCAR, calculate Ω (Fig. 1a). An example of the results produced for 27 different target mutations is shown in Fig. 4.
6. As shown in Fig. 3, targets that significantly diminish receptor function and with calculated $\Omega \neq 1$ are assigned as playing a role in the allosteric function of the receptor. Targets with $\Omega = \sim 1$ and significant loss-of-function are assigned as primarily affecting binding. For further studies, see Notes 16 and 17.

4. Notes

1. It is often convenient to prepare a concentrated (20×) stock solution of ND96 and Ca^{2+}-free ND96 in order to quickly

prepare the large volume of buffers used in semiautomated electrophysiology. The 1× and 20× stock, ND96 and Ca^{2+}-free ND96 should be stored with Parafilm around the cap.

2. Osmolarity of the incubation media should be 190–230 osmol/L. It is especially important that Parafilm be used to seal the cap of the bottle containing the oocyte incubation media to reduce solvent evaporation and prevent contamination.
3. Aliquot size should be chosen to ensure ≤3 freeze–thaw cycles. For agonists that are less stable, use individual stock solution aliquots for each fresh batch of drug solutions.
4. Use four doses per order of magnitude if there is reason to suspect multiple populations of receptors in order to more definitively detect and characterize a possible biphasic dose–response behavior. To determine whether or not a dose–response relation has multiple components, as opposed to having a low n_H, use the *F*-test to determine whether one component (monophasic) or two components (biphasic) fits the dose–response data better.
5. Ensure that all water with DNA or RNA is recovered from the spin column by drawing a line on the Eppendorf tube at a level corresponding to the volume of water used in the final DNA or RNA elution step. Spin additional times if necessary to recover this total volume from the column.
6. Our α-subunit cDNAs encode an HA epitope in the M3-M4 cytoplasmic loop for Western blot studies (not described). The HA epitope provides an independent way of determining if a mutation(s) giving no current is not functional or is not being assembled, transported, and inserted into the cell membrane. Control experiments show that this epitope does not significantly alter EC_{50}.
7. Set aside bench space that is dedicated to mRNA work only. Mix the mRNA in this area.
8. Once the mRNA solution is placed on the coverslip, and especially under the light source, work quickly since these small volumes evaporate, and therefore change concentration, quickly.
9. We find that the 35 mm dishes from CELLSTAR (Grenier Bio-One) extend oocyte viability and reduce oocyte adhesion to the dish. After injection check oocytes at 30–60 min, then continue every 12–24 h. Store 10–40 oocytes per 35 mm dish. Discard oocytes that are damaged or have begun the maturation process (expanding circle of degenerating membrane); also change media. Do this as often as is necessary to keep the media clear. If oocytes aren't living as long as your

experiments require, change the dishes each time oocyte media is changed.

10. Due to heavy use (>80 h/week), we have implemented weekly and monthly maintenance procedures of the OpusXpress.

    Weekly: Clean electrode holders by completely disassembling and rinsing in water with stirring for 30–60 min (do not rinse the gold pin, threaded Teflon collar, or electrode wire with nylon sleeve and silicone seal); dry each piece thoroughly and reassemble. Remove debris from solution filters by sonicating them in water for 10 min. Inspect headstages for corrosion and inspect chambers and rinse individually by pipetting water to remove debris. Wipe the liquid handler nozzles and headstages with a moistened Kimwipe and dry.

    Monthly: In addition to the weekly maintenance, replace all peristaltic tubing on pumps A, B, and the aspirator pump. Prime the new peristaltic tubing by rinsing with 70% isopropanol for 10 min, followed by water for 10 min. In each case, use a flow rate of ≥1 ml/min for pump A and B and ≥4 ml/min for the aspirator pump. Spray the headstage tracks with Inox lubricant and apply fluorocarbon gel to the driving assembly shafts.

    Implement similar maintenance for electrophysiology rigs used to record on one oocyte at a time. Especially important are rinsing perfusion lines and chamber after each use with water, regular maintenance of electronics, and frequent checking of the wiring/cables for salt buildup, fraying, and grounding.

11. Use these electrodes for repeated experiments as long as their resistances are still in range, they do not show drift, and they are free of significant oocyte membrane; usually replace electrodes every 1–3 days of heavy use (≥15 h/day).

12. Larger holding potentials increase I_{max}. This can be used to increase the signal-to-noise ratio if it is too low. However, higher holding potentials often decrease oocyte viability through an entire run. Make I_{max} comparisons (see Note 17a) at the same holding potentials, and after similar incubation times with parallel treatment of oocytes.

13. As in most experiments on LGICs, information from macroscopic experiments is usually limited by the speed of the solution change. We have estimated solution exchange times of ~1–2 s for these flow application rates on the OpusXpress. These are much slower than nAChR kinetics, but faster than some receptors (i.e., ρ1 $GABA_A$) (19). In the rare cases where activation/deactivation kinetics can be resolved by the OpusXpress, one must first optimize electronic filter

characteristics. Use an analog-to-digital sample rate >2.5× the analog filter's cutoff frequency, to avoid aliasing.

14. For the wild-type nAChR, define functional mutants (sufficient expression) as having I_{max} of ≥~100 nA. This produces a signal-to-noise ratio of >~3–5. For these channels with single-channel conductance ($g_{channel}$) of ~40 pS (with Na^+ as the permeant ion), 100 nA at a potential of −60 mV corresponds to ~40,000 receptors on the ooycte's surface ($g_{oocyte} = I/V = 100$ nA/0.06 V $= 1.7 \times 10^{-6}$ S; $g_{oocyte}/g_{channel} =$ ~40,000). Define receptors with a lower $g_{channel}$ as functional for I_{max} values proportionally lower than 100 nA, provided that I_{max} can be determined, given the signal-to-noise ratio of the current responses.
15. Verify that there was no response to control doses (no drug). It is common to have a larger standard error for the regions of the dose–response curve that are near EC_{50}.
16. If the target significantly diminishes function, but $\Omega = \sim 1$, more detailed information regarding binding can be obtained by systematic structural and/or electronic perturbations of an amino acid side chain using site-directed and unnatural amino acid mutagenesis (20, 21).
17. If significant coupling is observed ($\Omega \neq 1$), several experimental strategies that complement, but are redundant to, ELFCAR can be pursued. Many of these can also be used in other allosteric receptors.
    (a) Increase of I_{max} (maximal function) by the reporter in the background of the target (versus target alone) (10).
    (b) Increase of P_{open} (maximal function) by a reporter in the background of the target as measured by single-channel recording (10, 22).
    (c) More detailed single-channel kinetic analyses such as the REFER technique for the measurement of Φ-values to determine the relative timing of movement of residues during allosteric transitions (23, 24). In this way, targets can be identified as moving early or late in the activation process.
    (d) Block of recovery of I_{max} (maximal function) of a partial agonist. In our studies, the relative efficacy of succinylcholine (SuCh) to ACh (I_{max}(SuCh)/I_{max}(ACh)) generally recovered from ~4% to 70–100% from the wild-type receptor to noncoupled perturbations ($\Omega = \sim 1$). Such recovery of I_{max} was not seen for coupled mutations ($\Omega \neq 1$) (10).
    (e) Other reporter mutations that couple to the target. In the case of the nAChR, we found that several reporter

mutations gave similar coupling results to the βL9'S reporter. All of the alternative reporter mutations that we tested (γL9'S, δL9'S, αL9'S, αL13'S, and αL16'S) were in the channel pore and also exhibited gain-of-function behavior on their own (10).

In particular, we have shown that steps a, b, and d are readily implemented as coarse-grained verification of the ELFCAR result. Note that combining ELFCAR with steps a, b, and d can theoretically allow for identification of functional coupling in an allosteric receptor when a traditional functional metric, such as whole-cell dose–response relation, may have concluded that no functional change occurred when the target was altered based on an observed small shift or no shift in EC_{50}. However, if the gain-of-function reporter introduces significant increases in I_{max} (a) or P_{open} (b) in the background of the altered target, the target may still play a significant role in allosteric function. This is especially true if the reporter's gain-of-function is larger than the target's loss-of-function and/or if the wild-type allosteric receptor has a K_{act} value that places the system between the negative slope and plateau regions.

Acknowledgments

We thank Kristin Rule Gleitsman for essential discussions throughout the development of ELFCAR. This work was supported by the National Institutes of Health (NS 34407; NS 11756). J.A.P.S. and S.F. were partially supported by National Research Service Awards from the NIH.

References

1. Changeux, J. P., and Edelstein, S. J. (2005) Allosteric mechanisms of signal transduction, *Science* **308**, 1424–1428.
2. Fenton, A. W. (2008) Allostery: an illustrated definition for the 'second secret of life', *Trends Biochem Sci* **33**, 420–425.
3. Colquhoun, D. (1998) Binding, gating, affinity and efficacy: the interpretation of structure-activity relationships for agonists and of the effects of mutating receptors, *Br J Pharmacol* **125**, 924–947.
4. Carter, P. J., Winter, G., Wilkinson, A. J., and Fersht, A. R. (1984) The use of double mutants to detect structural changes in the active site of the tyrosyl-tRNA synthetase (Bacillus stearothermophilus), *Cell* **38**, 835–840.
5. Kash, T. L., Jenkins, A., Kelley, J. C., Trudell, J. R., and Harrison, N. L. (2003) Coupling of agonist binding to channel gating in the GABA(A) receptor, *Nature* **421**, 272–275.
6. Gleitsman, K. R., Kedrowski, S. M., Lester, H. A., and Dougherty, D. A. (2008) An intersubunit hydrogen bond in the nicotinic acetylcholine receptor that contributes to channel gating, *J Biol Chem* **283**, 35638–35643.
7. Price, K. L., Millen, K. S., and Lummis, S. C. (2007) Transducing agonist binding to channel gating involves different interactions in

5-HT3 and GABAC receptors, *J Biol Chem* **282**, 25623–25630.
8. Venkatachalan, S. P., and Czajkowski, C. (2008) A conserved salt bridge critical for GABA(A) receptor function and loop C dynamics, *Proc Natl Acad Sci U S A* **105**, 13604–13609.
9. Alexiev, U., Mollaaghababa, R., Khorana, H. G., and Heyn, M. P. (2000) Evidence for long range allosteric interactions between the extracellular and cytoplasmic parts of bacteriorhodopsin from the mutant R82A and its second site revertant R82A/G231C, *J Biol Chem* **275**, 13431–13440.
10. Gleitsman, K. R., Shanata, J. A., Frazier, S. J., Lester, H. A., and Dougherty, D. A. (2009) Long-range coupling in an allosteric receptor revealed by mutant cycle analysis, *Biophys J* **96**, 3168–3178.
11. Kalbaugh, T. L., VanDongen, H. M., and VanDongen, A. M. (2004) Ligand-binding residues integrate affinity and efficacy in the NMDA receptor, *Mol Pharmacol* **66**, 209–219.
12. Filatov, G. N., and White, M. M. (1995) The role of conserved leucines in the M2 domain of the acetylcholine receptor in channel gating, *Mol Pharmacol* **48**, 379–384.
13. Kearney, P. C., Zhang, H., Zhong, W., Dougherty, D. A., and Lester, H. A. (1996) Determinants of nicotinic receptor gating in natural and unnatural side chain structures at the M2 9′ position, *Neuron* **17**, 1221–1229.
14. Labarca, C., Nowak, M. W., Zhang, H., Tang, L., Deshpande, P., and Lester, H. A. (1995) Channel gating governed symmetrically by conserved leucine residues in the M2 domain of nicotinic receptors, *Nature* **376**, 514–516.
15. Connolly, C. N., and Wafford, K. A. (2004) The Cys-loop superfamily of ligand-gated ion channels: the impact of receptor structure on function, *Biochemical Society Transactions* **32**, 529–534.
16. Dougherty, D. A. (2008) Cys-loop neuroreceptors: structure to the rescue?, *Chem Rev* **108**, 1642–1653.
17. Sine, S. M., and Engel, A. G. (2006) Recent advances in Cys-loop receptor structure and function, *Nature* **440**, 448–455.
18. Dascal, N., and Lotan, I. (1992) Expression of exogenous ion channels and neurotransmitter receptors in RNA-injected *Xenopus* oocytes, In *Methods in Molecular Biology* (Longstaff, A., and Revest, P., Eds.), pp 205–225, Humana Press, Totowa, NJ.
19. Wang, J., Lester, H. A., and Dougherty, D. A. (2007) Establishing an ion pair interaction in the homomeric rho1 gamma-aminobutyric acid type A receptor that contributes to the gating pathway, *J Biol Chem* **282**, 26210–26216.
20. Nowak, M. W., Gallivan, J. P., Silverman, S. K., Labarca, C. G., Dougherty, D. A., and Lester, H. A. (1998) In vivo incorporation of unnatural amino acids into ion channels in Xenopus oocyte expression system, *Methods in Enzymology* **293**, 504–529.
21. Dougherty, D. A. (2008) Physical organic chemistry on the brain, *J Org Chem* **73**, 3667–3673.
22. Mortensen, M., and Smart, T. G. (2007) Single-channel recording of ligand-gated ion channels, *Nat Protoc* **2**, 2826–2841.
23. Chakrapani, S., Bailey, T. D., and Auerbach, A. (2004) Gating dynamics of the acetylcholine receptor extracellular domain, *J Gen Physiol* **123**, 341–356.
24. Grosman, C., Zhou, M., and Auerbach, A. (2000) Mapping the conformational wave of acetylcholine receptor channel gating, *Nature* **403**, 773–776.

Part II

Monitoring Allosteric Conformational Changes

Chapter 7

A Review of Methods Used for Identifying Structural Changes in a Large Protein Complex

Owen W. Nadeau and Gerald M. Carlson

Abstract

This chapter explores the structural responses of a massive, hetero-oligomeric protein complex to a single allosteric activator as probed by a wide range of chemical, biochemical, and biophysical approaches. Some of the approaches used are amenable only to large protein targets, whereas others push the limits of their utility. Some of the techniques focus on individual subunits, or portions thereof, while others examine the complex as a whole. Despite the absence of crystallographic data for the complex, the diverse techniques identify and implicate a small region of its catalytic subunit as the master allosteric activation switch for the entire complex.

Key words: Protein structure, Hetero-oligomeric protein complexes, Chemical crosslinking, Proteolysis, Electron microscopy, Chemical footprinting, Small-angle X-ray scattering, Allosteric switch

1. Introduction

The large protein complex that is featured in this chapter as a model system for the detection of conformational changes caused by an allosteric activator is skeletal muscle phosphorylase kinase (PhK), the first protein kinase to be purified and characterized (1). It remains one of the most complex enzymes known, being composed of 16 subunits $(\alpha\beta\gamma\delta)_4$, and having a total mass of 1.3×10^6 Da (2). The γ subunit is catalytic, but accounts for only 13.7% of PhK's mass. The remaining subunits are regulatory and impose quaternary constraint on the activity of the catalytic subunit in PhK's nonactivated basal state. A wide variety of allosteric activators act upon the regulatory subunits of PhK, and these include ADP, GDP, glycogen, acidic phospholipids, certain gangliosides, and phosphorylation of the α and β subunits; however, the most fundamental allosteric activator is Ca^{2+}, because the enhanced activity induced by all other activators is completely Ca^{2+}-dependent. It appears that there is a fundamental tier of Ca^{2+}-dependent conformational changes leading to activation, with additional

Aron W. Fenton (ed.), *Allostery: Methods and Protocols*, Methods in Molecular Biology, vol. 796,
DOI 10.1007/978-1-61779-334-9_7,

conformational changes induced by the other activators being layered upon this fundamental tier to cause as much as a 100-fold cumulative activation at physiological pH. Consequently, we have sought to elucidate the Ca^{2+}-dependent conformational changes in the PhK complex as the first step in understanding its overall allosteric activation. Although the large mass of the complex limits some of the physicochemical approaches that can be used, in contrast, it opens the door to a large number of informative approaches not practical with smaller enzymes.

The primary target for Ca^{2+} ions in the PhK complex is undoubtedly the δ subunit, which is calmodulin (CaM), although a recent sequence analysis suggests the presence of calcineurin B-like domains in the α and β subunits that may also be potentially capable of binding Ca^{2+} (3). The behavior of the CaM subunit of PhK has long been known to be unusual in two ways. First, it remains a tightly bound subunit of the complex even in the complete absence of Ca^{2+}; and second, it has been reported to bind only three Ca^{2+} ions, despite its four EF-hands (4). Results from a variety of techniques are consistent with the catalytic γ subunit being a major interaction partner for the CaM (δ) subunit within the PhK complex: an isolated γδ dimer retains Ca^{2+}-dependent kinase activity (5); the carboxy-terminal regulatory domain of γ (γCRD) beyond its catalytic domain contains two distinct regions with high affinity for CaM (6), and removal of these regions from the isolated γ subunit results in Ca^{2+}-independent kinase activity (7); and a γδ conjugate can be formed by chemical crosslinking of the $(\alpha\beta\gamma\delta)_4$ complex (8, 9). A variety of chemical, biochemical, and biophysical approaches have been utilized to determine the effects of Ca^{2+} on PhK's structure and properties.

2. Effects of Ca^{2+} on Individual Subunits in the PhK Complex

Many of the approaches used have sought to identify specific subunits of PhK that are structurally affected, either directly or indirectly, by the binding of Ca^{2+}. As will be described, each of its four different subunits has been shown to be affected. Yet, one of the most informative and specific techniques for detecting conformational changes, namely, hydrogen–deuterium exchange, has yet to be evaluated with PhK $\pm$ Ca^{2+}, because of the very large size of the complex.

2.1. Single-Pair Fluorescence Resonance Energy Transfer

Fluorescence resonance energy transfer (FRET) is an extremely sensitive method for assessing conformational dynamics in proteins, and in the case of PhK, it was used to determine the conformations of its δ (CaM) subunit in the absence and presence of Ca^{2+} (10). A double mutant of CaM was first derivatized with a fluorescent donor–acceptor pair at opposite ends of the molecule. A small amount of this derivatized CaM was then exchanged for wild-type CaM in the $(\alpha\beta\gamma\delta)_4$ PhK complex by incubation with low concentrations of urea (0.5 M) in the absence of Ca^{2+}, following which, the PhK was repurified by size-exclusion HPLC. Because of the theoretical possibility of different PhK complexes containing different numbers of derivatized CaM subunits (1–4), single molecule fluorescence spectroscopy was employed to assess the conformational substates of the PhKδ subunit, with the distribution of FRET distances representing compact to extended conformations. This approach revealed that the conformational substates of CaM bound as the δ subunit of the PhK complex compared to free CaM were greatly skewed toward a higher percentage of extended conformations in the complex (40% vs. 26%). The binding of Ca^{2+} to PhKδ further increased the percentage of extended conformers to 45%, which again was markedly different than the conformational substate distributions of free Ca^{2+} CaM. These results indicate that the other subunits of PhK exert a dramatic effect on the conformation of CaM bound as PhKδ.

Intrinsic to the above interpretation that populations of bound CaM molecules with the donor–acceptor pair at different apparent distances represent different conformational substates of the δ subunits is the assumption that the conformational substates of the δ subunit are in dynamic equilibrium, as is observed with free CaM. The large size and consequent slow diffusion rate of PhK allowed evidence to be gained in support of this assumption. Trajectory studies over time, some for up to 4 ms, followed single molecules of PhK having bound derivatized CaM that was observed to repeatedly jump from one conformational substate to another (10). Thus, the small δ subunits of PhK exercise considerable conformational plasticity, even when loaded with Ca^{2+} and interacting with the other subunits within the complex.

2.2. Partial Proteolysis

Limited proteolysis followed by analysis on SDS–PAGE gels is a simple, yet effective, method for detecting ligand-induced conformational changes in proteins (see Note 1). It is especially useful for large, hetero-oligomeric complexes, because specific subunits affected by a ligand can usually be detected. Using a wide variety of 11 different proteases, the α subunit within the nonactivated $(\alpha\beta\gamma\delta)_4$ PhK complex was observed to generally be the most rapidly degraded, followed by the β subunit, with the γ and finally δ subunits being relatively resistant to hydrolysis (11). Utilizing partial proteolysis as a conformational probe successfully detected activated conformers of PhK produced by a number of ligands and phosphorylation. In the case of the allosteric activator Ca^{2+}, the β subunit was protected from hydrolysis by trypsin (11), whereas the rate of hydrolysis of the α subunit by chymotrypsin was stimulated threefold (12). These results indicate that conformational changes induced in the δ subunit by the binding of Ca^{2+} are transmitted to both of PhK's large regulatory α and β subunits.

2.3. Immunochemical Analysis

Conformation-sensitive antibodies represent another highly sensitive tool that is unaffected by the size of the protein target, and as above, it allows specific subunits in a complex to be targeted. Based upon their earlier work using overlapping synthetic peptides corresponding to the γCRD to identify two noncontiguous CaM-binding domains within that region of γ (6), Blumenthal and colleagues generated monospecific polyclonal antibodies against four synthetic peptides that covered the entire length of the γCRD. Using ELISA, these antibodies were then screened as conformational probes of the regulatory region of the γ subunit within the $(\alpha\beta\gamma\delta)_4$ complex (13). Ca^{2+} was reported to exert its greatest effect on the antibodies against regions nearest to the catalytic domain of γ, with their binding being enhanced, suggesting the activator causes increased exposure of the epitopes in question. Comparing the effects of these antibodies on Ca^{2+}-dependent activity vs. their dependence on Ca^{2+} for binding allowed insightful inferences to be made regarding the functional roles of these regions of the γCRD on the allosteric activation by Ca^{2+}.

Monoclonal antibodies specific for the individual subunits of PhK have also proved to be sensitive conformational probes for the effects of a variety of activators, but none of them detected structural changes caused by Ca^{2+} (14, 15). These antibodies were, however, valuable in mapping the relative locations of epitopes of PhK's individual subunits by immunoelectron microscopy (immunoEM), which is shown for a single αβγδ protomer in Fig. 1. Note that the catalytic γ subunit is surrounded by portions of its three regulatory subunits (14–17).

2.4. Chemical Footprinting

Chemical modification of proteins has been used for many years to assess the functional roles and solvent accessibility of specific amino acids. The latter, often referred to as chemical footprinting, is especially useful for detecting conformational changes in oligomeric complexes at the subunit level by measuring differences in the extent of their modification in the absence and presence of an allosteric ligand (see Note 2). The most useful chemical footprinting reagents contain functional groups that allow the reagent to be readily detected and that react preferentially with specific amino acid side chains (Table 1). Using ^3H-iodoacetate (a reagent that preferentially targets protein thiols) and formaldehyde (a compound that *bis*-methylates lysine ε-amines in the presence of $NaCNBH_3$), Ca^{2+} was found to increase labeling of the catalytic γ subunit in the PhK complex by both reagents (18), consistent with the immunochemical results described above.

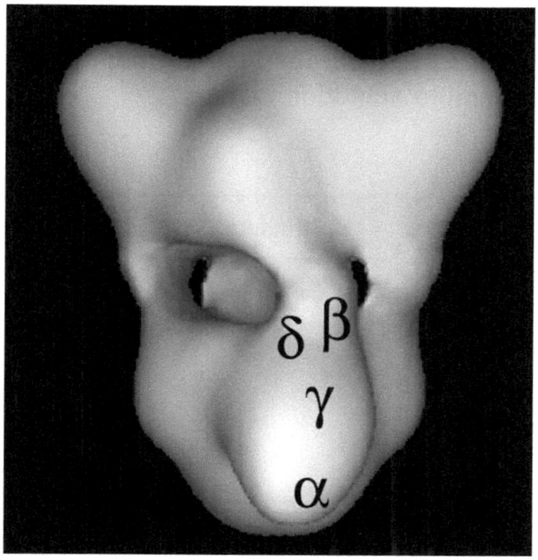

Fig. 1. Three-dimensional structure of PhK at 25 Å resolution reconstructed from ~5,000 images of frozen hydrated particles in cryoEM (16). Each lobe is an $(\alpha\beta\gamma\delta)_2$ octamer and four bridges separate the two lobes. The positions of specific regions of the four subunits of a single $\alpha\beta\gamma\delta$ protomer are denoted. The regions of the α, β, and γ subunits represent epitopes recognized by monoclonal antibodies that were localized by immunoEM of negatively stained complexes (14, 15), with those locations then transferred to the cryoEM structure. The δ subunit was directly visualized by scanning transmission EM of PhK in which a fraction of its δ subunits had been exchanged with mutant calmodulin derivatized with Nanogold (17). Its location was also then transferred to the cryoEM model.

2.5. Chemical Crosslinking

Crosslinking is a technique used to covalently link two or more chemical functional groups, in this case amino acid side chains, and it can detect conformational changes in proteins and within protein complexes (19). As an extension of general protein chemical modification, chemical crosslinking is an empirical technique that is governed by the same variables described for chemical footprinting; however, because crosslinking utilizes reagents containing more than one reactive functional group (Table 1), the number of potential products formed is far greater than those generated using simple monofunctional reagents. As depicted in Fig. 2, crosslinking is an ongoing process in which coupling reactions between two proteins may progress from a single heteromeric conjugate to large polymeric arrays of indeterminable stoichiometry. The specificity of crosslinking is directly dependent on the selectivity of the crosslinking reagent used and stringent control of the crosslinking reaction, both of which are determined empirically by screening the protein target with a variety of crosslinkers under a range of conditions (19). For an oligomeric complex, the overall goal of such a screening is to limit the formation

Table 1
A partial list of chemical functional groups that are targeted preferentially by reactive amino acid side chains

Functional group		Targeting nucleophile	Optimal pH	Amino acid target
Name	Structure			
N-Hydroxysuccinimide ester	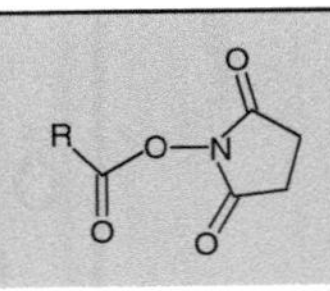	Amine	>7	Lysine Protein N-termini
Imido ester	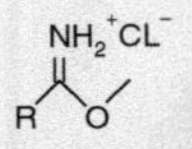	Amine	8–9	Lysine Protein N-termini
Maleimide	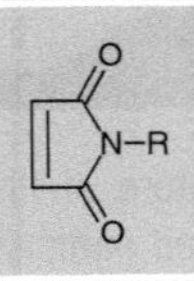	Sulfhydryl	6.5–7.5	Cysteine
Cyclohexanedione	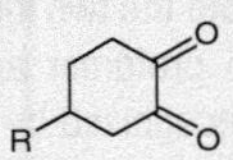	Guanidine	>7	Arginine
Phenyl azide	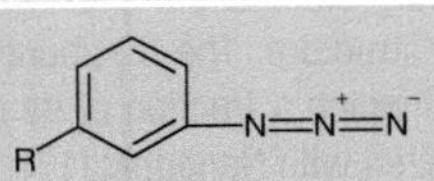	Amine Phenol Double bond	7	Cysteine Arginine Lysine Histidine Tyrosine Phenylalanine Tryptophan

of crosslinked conjugates to a few readily identifiable species, so that each is formed in sufficient quantity for analysis of its subunit composition by apparent mass (SDS–PAGE), cross-reaction against subunit-specific antibodies (Western blots), or its primary structure (gas phase sequencing, MS). Although crosslinking can occur within a given subunit of a protein complex, we refer in this section only to crosslinking between subunits of that complex (see Note 3). For in-depth reviews on chemical crosslinking, see the books of Hermanson (20) and Wong (21).

An allosteric ligand may alter the extent of formation of a given conjugate, or produce a different subunit crosslinking pattern, or both. In the case of PhK, Ca^{2+} was shown to increase the formation of $\alpha\gamma\gamma$ and $\alpha\delta$ conjugates that were initially observed after crosslinking the native complex with phenylene dimaleimide (PDM) (18). Similarly, a Ca^{2+}-dependent increase in formation of a $\beta\delta$

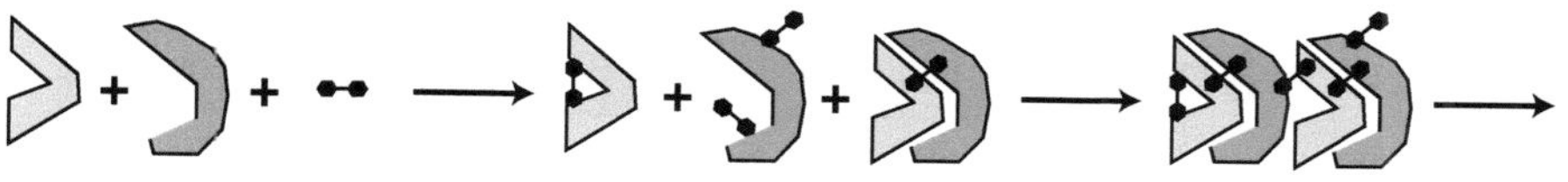

Fig. 2. Protein crosslinking. Protein side chains react with bifunctional reagents to form mono-derivatized and crosslinked forms of interacting proteins. Subclasses within each form simultaneously occur through different combinations of mono-derivatization and crosslinking. In the continuous process of crosslinking, conjugates of increasing size may also be formed, progressing from dimers, tetramers, etc. to large extensively crosslinked polymers.

conjugate by the crosslinker *m*-maleimidobenzoyl-*N*-hydroxysuccinimide ester was observed; however, Ca^{2+} promoted crosslinking of different regions of the δ subunit to β in that conjugate (18). A novel αγδδ conjugate was observed when PhK was crosslinked by PDM only in the presence of Ca^{2+} (18). Formation of a γδ conjugate by the zero-length crosslinker *N*-ethoxycarbonyl-2-ethoxy-1,2-dihydroquinoline (EEDQ) was inhibited by Ca^{2+} (9), and a Ca^{2+}-dependent linkage between the γ and α subunits was revealed when PhK was crosslinked with *N*-5-azido-2-nitrobenzoyl-oxysuccinimide (22), 4-phenyl-1,2,4-triazoline-3,5-dione (23), or formaldehyde (12). Moreover, nearly all the other known allosteric activators of PhK besides Ca^{2+} increased crosslinking between the α and γ subunits in the PhK complex (24), suggesting structural similarities among the conformers induced by these activators and that enhanced αγ crosslinking is a marker for activated forms of PhK (24). Evidence for the latter came from showing that the complex remained partially activated after crosslinking in the presence of activators, even after they were subsequently removed (18, 22, 24), suggesting that structural features of the activated conformers were trapped by crosslinking.

Recent advances in MS and search engine technologies have considerably strengthened the utility of the crosslinking approach by enhancing the identification and characterization of crosslinked peptides from digests of nanomolar amounts of protein conjugates (ref. 25 and references therein). Although Fig. 2 indicates products from the crosslinking of proteins to already be complex, the number of potential products generated after digesting conjugates of large proteins is immense, and such products are essentially impossible to characterize in their entirety without computational assistance. We have recently employed search engine technologies developed by our laboratory (26) and by Yu et al. (27) in combination with FT-ICR MS to determine specific crosslinked residues in βγ (28) and γδ (9) conjugates formed in the PhK complex. In each case, the crosslink to γ was within its CRD, with the γδ EEDQ zero-length conjugate being of special interest, because the residue of δ (CaM) crosslinked to a γ-lysyl group was an aspartyl residue in its third EF hand, indicating that within the PhK complex that Asp, normally a Ca^{2+} ligand in CaM, is in a salt-bridge with the Lys, thus

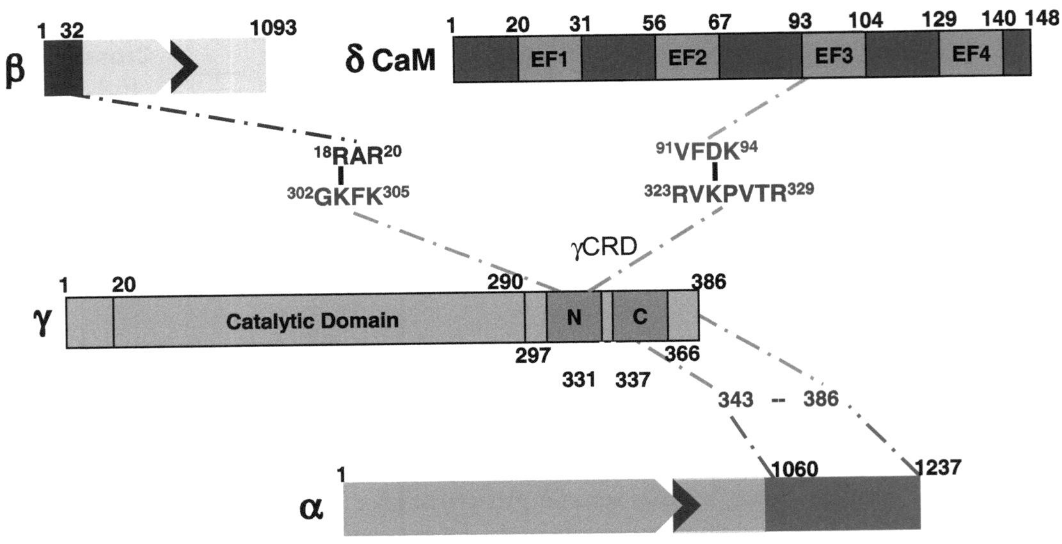

Fig. 3. Linkage map of PhK's regulatory subunit interactions with the γCRD determined by chemical crosslinking.

explaining the binding of only three Ca^{2+} ions by PhKδ (4). These crosslinking studies provided the first direct evidence for linkages in the PhK complex between functional domains of its regulatory α, β, and δ subunits and the regulatory domain of its catalytic γ subunit (Fig. 3).

3. Effects of Ca^{2+} on the Overall Structure of the PhK Complex

The experimental approaches described above indicate that the structure of every individual subunit of the PhK complex is altered by the binding of Ca^{2+} to its δ subunit. A number of biophysical techniques corroborate the notion that Ca^{2+} does indeed induce widespread structural changes throughout the large PhK complex, involving alterations in its secondary, tertiary, and quaternary structure.

3.1. Conformational Changes Observed Through Spectroscopy

Tertiary structure was examined through second derivative UV absorption spectroscopy to monitor changes in the microenvironments of PhK's 428 Phe, 460 Tyr, and 124 Trp residues upon the binding of Ca^{2+}. The second derivative spectrum for the nonactivated enzyme showed six negative peaks having the following residue assignments: Phe (~245 and 251 nm), Tyr (~260 and 270 nm), overlapping Tyr/Trp signal (~276 nm), and Trp (~285 nm) (29). The only peak shifts observed to occur upon the binding of Ca^{2+} were for those assigned to Tyr residues, which were shifted to a longer wavelength, suggesting that those residues then occupied a more apolar environment (30).

The conclusions from second derivative UV absorption spectroscopy were consistent with those from intrinsic fluorescence spectroscopy. Although the intrinsic fluorescence emission peaks of Tyr and Trp residues in proteins overlap, these residues can be distinguished through the synchronous fluorescence technique (31, 32), which utilizes simultaneous scanning by the excitation and emission monochromators. Again, the binding of Ca^{2+}, especially at pH values >7.0, showed burial of Tyr residues (29, 33). Given that the α and β subunits of PhK contain 82% of its Tyr residues, the spectral changes described most likely reflect structural changes in those subunits.

Circular dichroism was used to investigate changes in the secondary structure of PhK caused by the binding of Ca^{2+}. The spectrum showed two minima near 208 and 220 nm, which is characteristic of proteins containing significant α-helical content. At pH values >7.0, the addition of Ca^{2+} shifted the second peak to 223 nm and increased its intensity, whereas the ellipticity at 208 nm was reduced without a change in peak position (29). Because the ratio of negative ellipticity at 222–208 nm is an index of the extent of interaction between α-helices and β-sheets (34, 35), the binding of Ca^{2+} can be considered to decrease the interactions between these structural elements in the PhK complex.

Examination of secondary structure by Fourier transform infrared (FTIR) spectroscopy also showed Ca^{2+}-induced spectral changes, especially at 1,652 cm^{-1}, a frequency near the band normally assigned to α-helices (36). The second derivative spectrum of the nonactivated conformer of PhK in this region suggested the presence of simple α-helices, whereas the Ca^{2+}-activated conformer showed a new shoulder at 1,647 cm^{-1} (37). Second derivative minima near 1,645 ± 4 cm^{-1} are typically assigned to unordered structure (36) or distorted helices (38). Thus, the binding of Ca^{2+} by PhK may decrease the amount of highly ordered α-helices within the complex.

For all of the spectral techniques in this section but FTIR, it was possible to vary the temperature, and without exception, the spectral signals were maintained to a significantly higher temperature with the nonactivated, Ca^{2+}-free conformer (37). Thus, the secondary and tertiary structures were maintained to higher temperatures. Upon binding Ca^{2+}, PhK becomes increasingly labile to thermal perturbation, suggesting that activation by Ca^{2+} is associated with a less stable conformation of the complex.

3.2. Conformational Changes Detected by Zeta Potential Analysis

The zeta potential reflects the colloid shear surface to solvent interface environment (Stern Layer), a relatively ordered layer of solvent molecules that diffuses as part of the colloid particle. A change in electrostatic surface potential as a result of alterations in the interactions of solvent-accessible charged residues will alter the Stern Layer of a protein, and consequently, its zeta potential

(39). Thus, zeta potential values provide an estimate of effective surface charge. At neutral pH, the nonactivated conformer of PhK has a zeta potential of −33 mV; however, upon binding Ca^{2+}, its zeta potential becomes dramatically less negative, to between −22 (36) and −13 mV (29). That this change in zeta potential is associated with PhK's activation, rather than just the binding of Ca^{2+} ions, is confirmed by the observation that increasing the pH to >7, which also activates the enzyme through deprotonation of residues on its regulatory α and/or β subunits, causes a similar change in zeta potential (29). Consequently, activation of PhK by either mechanism is associated with either the burial or neutralization of acidic surface side chains or the increased exposure of basic residues. Inasmuch, however, as the activity at elevated pH remains Ca^{2+}-dependent, there must be additional Ca^{2+}-dependent conformational changes besides those associated with just the change in zeta potential. These have, in fact, been detected as an increase in the ratio of ellipticity at 222 nm vs. 208 nm (29), as described in the previous section; however, the increase in this ratio occurs only in conformers of the PhK complex having less negative zeta potential values. The pH- and Ca^{2+}-dependent conformers of PhK that have been observed, depicted in Fig. 4, are inter-dependent, and multiple, independent approaches, many described herein, corroborate the conformational transitions shown. The conformer with the greatest catalytic activity is the Ca^{2+}-bound form at high pH.

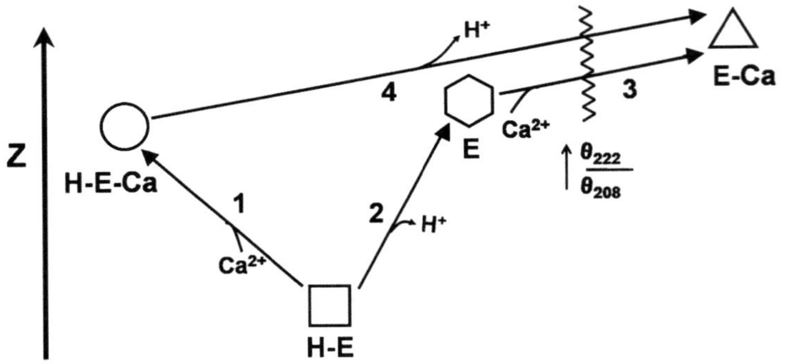

Fig. 4. Schematic depicting the effects of Ca^{2+} and pH on the structure of PhK as defined through circular dichroism and zeta potential. The four conformers, two at pH 6.8 (H-E and H-E-Ca) and two at pH 8.2 (E and E-Ca), are vertically arranged according to the values of their zeta potential, with the Ca^{2+}-free form at pH 6.8 having the most negative zeta potential and negligible catalytic activity. The *numbers* represent conformational transitions between forms, and the *waved line* intersecting conformational transitions 3 and 4 represents the occurrence of a greatly increased $\theta_{222}/\theta_{208}$ ratio. The Ca^{2+}-bound conformer at pH 8.2 has by far the greatest catalytic activity. Copyright 2008 Wiley: used with permission from Liu, Weiya; Priddy, Timothy S.; Carlson, Gerald M., Physiochemical changes in phosphorylase kinase associated with its activation, *Protein Science*, John Wiley and Sons.

3.3. Conformational Changes Detected by Electron Microscopy

Electron microscopy with single-particle reconstruction provides a means for determining low-to-medium resolution three-dimensional structures of large protein complexes that are refractory to crystallization, such as PhK. In the single molecule approach, images of proteins are obtained from either molecules stained on carbon grids (negative staining) or frozen in vitreous ice (cryoEM), generating fields of the particles, which are considered as rigid body copies in different orientations (40). Each image is assigned directional (Euler) angles on the basis of an initial model, which may be deduced (41) or determined experimentally by methods such as the random conical tilt (42). Fourier transforms of the images are then aligned, averaged, and combined using best fit Euler angles from the model in three-dimensional Fourier space, followed by back transformation to generate a real-space reconstruction. The process is then repeated using the new reconstruction as a starting model and continued iteratively to further refine the structure. Reconstructions of the nonactivated and Ca^{2+}-activated conformers of PhK were carried out with images of negatively stained molecules of the kinase using a simple spheres starting model (41), based on a previously deduced structure of the kinase (43). To account for potential differences in negative staining and enzyme preparations, reconstructions from three different PhK preparations were carried out $\pm Ca^{2+}$. Because the differences among the reconstructions for each conformer proved to be minor, all the particles from each kinase preparation were combined and averaged for each of the conformers. Comparison of the two averaged structures showed that Ca^{2+} promoted significant redistribution of density in both the lobe and bridge regions of the complex without significantly altering its maximum dimension (260 Å). Major structural changes in the lobes occurred in regions corresponding to the approximate positions mapped for the α, γ, and δ subunits in the complex (Fig. 1), corroborating Ca^{2+}-dependent structural changes detected for these subunits by crosslinking (12, 18, 22, 23). Ca^{2+} also perturbed the structure of the bridges, which is consistent with the position of the δ subunits proximal to the bridges and with the Ca^{2+}-dependent changes in crosslinking observed between the δ and β subunits, which are thought to compose the bridges (44).

3.4. Conformational Changes Detected by Small-Angle X-Ray Scattering

The overall shape and dimensions of proteins can be determined from their small-angle X-ray scattering (SAXS) patterns. A useful conformational parameter derived from SAXS measurements is $P(r)$, which is the probable distribution of distances between scattering centers in a molecule and is obtained by calculating the inverse Fourier transform of the scattering data. $P(r)$ reflects the overall asymmetry and domain structure of the scattering particle (45). SAXS complements the EM approach, in that it provides molecular weight, volume, radius of gyration (R_g: root mean square distance of the molecule's parts from its center of gravity), and the maximum dimension of a molecule. In turn, EM provides moderate resolution structural information that can be used to model SAXS data and augment structural details for potential shape changes between conformers of allosteric proteins. Utilizing this modeling approach, simple geometric shapes were used to simulate the bridges and lobes of PhK in analyzing its SAXS data collected ± Ca^{2+} (46). As was observed by EM, the binding of Ca^{2+} caused a redistribution of density in the lobe and bridge regions of PhK without significantly changing its R_g or maximum dimension (270 Å).

4. Evidence for the γCRD Being the Master Allosteric Activation Switch for the PhK Complex

Utilization of the various techniques reviewed in this chapter showed that the binding of Ca^{2+} ions to the δ subunit of the $(\alpha\beta\gamma\delta)_4$ PhK complex causes conformational changes in all four of its subunits. Specific regions affected include the regulatory domain of the γ subunit, γCRD (shown by immunochemistry), and the C-terminal region of the α subunit (shown by partial proteolysis at Phe-1014), itself a regulatory region of that subunit having multiple phosphorylation sites (2). In addition to perturbations by Ca^{2+} of the secondary and tertiary structures of PhK's subunits, it also altered the quaternary interactions of all four subunits as probed by chemical crosslinking. Of particular note was the doubling by Ca^{2+} of αγ conjugate formation by the very short crosslinker formaldehyde, with the crosslinked region of α shown biochemically to be within its C-terminal regulatory domain (12). The interaction of this region of α with γ is further supported by two-hybrid screening results, which show an interaction between the C-terminal portion of γCRD and the C terminus of α (12). Zero-length crosslinking to form a γδ conjugate between the third EF hand of δ and the γCRD was also influenced by Ca^{2+}; but unlike αγ formation by formaldehyde, γδ formation was halved (9), instead of doubled. That the α and δ subunits both interact with regions within the γCRD suggests a Ca^{2+}-sensitive network of quaternary interactions involving these three subunits that is mediated by the γCRD. In this network, the Ca^{2+}-induced conformational changes in δ observed by single-pair FRET (spFRET) would perturb γ–δ interactions and concomitantly the γ–α interactions. The locations of the subunits (δ near the bridges and the C-terminal region of α near the lobe tips) considered together with the Ca^{2+}-induced structural changes observed in these regions by EM provide additional support for the existence of this Ca^{2+}-sensitive $\alpha \leftrightarrow \gamma \leftrightarrow \delta$ communication network, which stretches from near the center of the PhK complex to the tips of its lobes, a distance of ~84 Å. A high sequence similarity between regions of γCRD and the inhibitory region of troponin I containing the Ca^{2+}-dependent allosteric activation switch for the actin ↔ troponin I ↔ troponin C system has been discussed (6, 9, 47), and of course, troponin C is a homolog of CaM, PhK's δ subunit. These similarities, coupled with the direct evidence described herein, support the hypothesis that the γCRD is likewise the Ca^{2+}-dependent allosteric activation switch for the PhK complex. Not only its proximal interactions with the δ subunit and the regulatory C-terminal region of the α subunit are Ca^{2+}-dependent, but chemical crosslinking has also shown it to be situated relatively near the N-terminal regulatory region of the β subunit that is phosphorylated by cAMP-dependent protein kinase (2, 28). The interactions of γCRD with PhK's three regulatory subunits, depicted in Fig. 3, would seem to require that it be conformationally dynamic in order to function as an allosteric activation switch, which is consistent with the experimental data. It might also be noted that the γCRD is very basic with a *p*I of 10.0 (9); thus, increased exposure of this domain upon activation could explain PhK's significantly less negative zeta potential upon its binding of Ca^{2+} ions.

5. Notes

1. When evaluating partial proteolysis as a conformational probe, a wide variety of proteases should be screened, as in ref. 11, and of course, this should be done under the conditions that the enzyme in question demonstrates its allosteric response. For analysis of a response to an allosteric effector, the proteolysis should still be in its linear phase, so that either inhibition or stimulation can be readily observed. Also, it is important to test the same protease against several other protein substrates to ensure that any observed effect of a ligand is due to its action on the protein of interest, and not on the protease being used. Finally, with the widespread availability of mass spectrometry to readily analyze protein fragments, this technique is becoming even more useful.
2. Chemical footprinting and all other protein chemical modification methods are empirical techniques that are influenced by a number of variables, including the concentrations of reactants, time and temperature of the reaction, and the pH value (19). The variability inherent in these techniques

derives primarily from the protein itself, specifically the numerous potentially reactive amino acid side chains that may be targeted selectively by different classes of reagents. With only a few exceptions, the amino acids that typically undergo chemical modification (tyrosine, lysine, cysteine, histidine, aspartic acid, and glutamic acid) contain ionizable nucleophilic side chains. The microenvironment (location in the protein) of a side chain functional group directly influences its accessibility, polarity, and noncovalent interactions with other residues or solvent. These and other factors ultimately lead to differences in reactivity for a given functional group that can occur even on the surface of a protein, thus causing difficulty in selectively targeting a particular side chain with a single reagent. To optimize selectivity and extent of modification of a protein, it is important to screen multiple reagents having different reactive functional groups (Table 1), water solubility, reporter groups, etc., while varying the time, pH, and concentration of the reactants. As with partial proteolysis, screens should be performed under conditions in which the target protein responds to its cognate allosteric ligand and undergoes chemical modification linearly with time.

3. A necessary precaution in determining whether a given crosslinking reagent is a suitable conformational probe is to verify that a covalent conjugate of interest is derived from crosslinking within a complex as opposed to between complexes. This verification can usually be achieved by size-exclusion chromatography (SEC) over a matrix capable of separating the native complex from its dimers and higher order multimers (22). Following crosslinking, the reaction is quenched (using an excess of a general nucleophilic reagent such as mercaptoethanol or lysine) and applied to SEC. The protein fractions that coelute with the native complex are collected and analyzed by SDS–PAGE. Any conjugates detected in these fractions represent intersubunit crosslinking within the complex.

References

1. Krebs, E. G., Love, D. S., Bratvold, G. E., Trayser, K. A., Meyer, W. L., and Fischer, E. H. (1964) Purification and properties of rabbit skeletal muscle phosphorylase b kinase, *Biochemistry* **3**, 1022–1033.
2. Brushia, R. J., and Walsh, D. A. (1999) Phosphorylase kinase: The complexity of its regulation is reflected in the complexity of its structure, *Front Biosci* **4**, D618-641.
3. Carriere, C., Mornon, J. P., Venien-Bryan, C., Boisset, N., and Callebaut, I. (2008) Calcineurin b-like domains in the large regulatory alpha/beta subunits of phosphorylase kinase, *Proteins* **71**, 1597–1606.
4. Burger, D., Cox, J. A., Fischer, E. H., and Stein, E. A. (1982) The activation of rabbit skeletal muscle phosphorylase kinase requires the binding of 3 Ca^{2+} per delta subunit, *Biochem Biophys Res Commun* **105**, 632–638.
5. Chan, K. F., and Graves, D. J. (1982) Rabbit skeletal muscle phosphorylase kinase. Catalytic

and regulatory properties of the active αγδ and γδ complexes, *J Biol Chem* **257**, 5948–5955.

6. Dasgupta, M., Honeycutt, T., and Blumenthal, D. K. (1989) The gamma-subunit of skeletal muscle phosphorylase kinase contains two noncontiguous domains that act in concert to bind calmodulin, *J Biol Chem* **264**, 17156–17163.
7. Harris, W. R., Malencik, D. A., Johnson, C. M., Carr, S. A., Roberts, G. D., Byles, C. A., Anderson, S. R., Heilmeyer, L. M., Jr., Fischer, E. H., and Crabb, J. W. (1990) Purification and characterization of catalytic fragments of phosphorylase kinase gamma subunit missing a calmodulin-binding domain, *J Biol Chem* **265**, 11740–11745.
8. Picton, C., Klee, C. B., and Cohen, P. (1980) Phosphorylase kinase from rabbit skeletal muscle: Identification of the calmodulin-binding subunits, *Eur J Biochem* **111**, 553–561.
9. Jeyasingham, M. D., Artigues, A., Nadeau, O. W., and Carlson, G. M. (2008) Structural evidence for co-evolution of the regulation of contraction and energy production in skeletal muscle, *J Mol Biol* **377**, 623–629.
10. Priddy, T. S., Price, E. S., Johnson, C. K., and Carlson, G. M. (2007) Single molecule analyses of the conformational substates of calmodulin bound to the phosphorylase kinase complex, *Protein Sci* **16**, 1017–1023.
11. Trempe, M. R., and Carlson, G. M. (1987) Phosphorylase kinase conformers. Detection by proteases, *J Biol Chem* **262**, 4333–4340.
12. Rice, N. A., Nadeau, O. W., Yang, Q., and Carlson, G. M. (2002) The calmodulin-binding domain of the catalytic γ subunit of phosphorylase kinase interacts with its inhibitory α subunit: Evidence for a Ca^{2+} sensitive network of quaternary interactions, *J Biol Chem* **277**, 14681–14687.
13. Wangsgard, W. P., Dasgupta, M., and Blumenthal, D. K. (1997) Antipeptide antibodies as probes of subunit-dependent structural changes in the regulatory domain of the gamma-subunit of phosphorylase kinase, *Biochem Biophys Res Commun* **230**, 179–183.
14. Wilkinson, D. A., Marion, T. N., Tillman, D. M., Norcum, M. T., Hainfeld, J. F., Seyer, J. M., and Carlson, G. M. (1994) An epitope proximal to the carboxyl terminus of the α-subunit is located near the lobe tips of the phosphorylase kinase hexadecamer, *J Mol Biol* **235**, 974–982.
15. Wilkinson, D. A., Norcum, M. T., Fizgerald, T. J., Marion, T. N., Tillman, D. M., and Carlson, G. M. (1997) Proximal regions of the catalytic γ and regulatory β subunits on the interior lobe face of phosphorylase kinase are structurally coupled to each other and with enzyme activation, *J Mol Biol* **265**, 319–329.
16. Nadeau, O. W., Gogol, E. P., and Carlson, G. M. (2005) Cryoelectron microscopy reveals new features in the three-dimensional structure of phosphorylase kinase, *Protein Sci* **14**, 914–920.
17. Traxler, K. W., Norcum, M. T., Hainfeld, J. F., and Carlson, G. M. (2001) Direct visualization of the calmodulin subunit of phosphorylase kinase via electron microscopy following subunit exchange, *J Struct Biol* **135**, 231–238.
18. Nadeau, O. W., Sacks, D. B., and Carlson, G. M. (1997) The structural effects of endogenous and exogenous Ca^{2+}/calmodulin on phosphorylase kinase, *J Biol Chem* **272**, 26202–26209.
19. Nadeau, O. W., and Carlson, G. M. (2005) Protein interactions captured by chemical cross-linking, In *Protein–protein interactions: A molecular cloning manual* (Golemis, E., and Adams, P. D., Eds.) 2nd ed., pp 105–127, Cold Spring Harbor Laboratory Press, Cold Spring Harbor, New York.
20. Hermanson, G. T. (2008) Bioconjugate techniques, 2nd ed., Elsevier Academic Press, Amsterdam; Boston.
21. Wong, S. S. (1993) Chemistry of protein conjugation and cross-linking, CRC Press, Boca Raton, FL.
22. Nadeau, O. W., Traxler, K. W., Fee L. R., Baldwin, B. A. and Carlson, G. M (1999) Activators of phosphorylase kinase alter the cross-linking of its catalytic subunit to the C-terminal one-sixth of its regulatory α subunit, *Biochemistry* **38**, 2551–2559.
23. Ayers, N. A., Nadeau, O. W., Read, M. W., Ray, P., and Carlson, G. M. (1998) Effector-sensitive cross-linking of phosphorylase b kinase by the novel cross-linker 4-phenyl-1,2,4-triazoline-3,5-dione, *Biochem J* **331 (Pt 1)**, 137–141.
24. Nadeau, O. W., Sacks, D. B., and Carlson, G. M. (1997) Differential affinity cross-linking of phosphorylase kinase conformers by the geometric isomers of phenylenedimaleimide, *J Biol Chem* **272**, 26196–26201.
25. Chu, F., Baker, P. R., Burlingame, A. L., and Chalkley, R. J. (2010) Finding chimeras: A bioinformatics strategy for identification of cross-linked peptides, *Mol Cell Proteomics* **9**, 25–31.
26. Nadeau, O. W., Wyckoff, G. J., Paschall, J. E., Artigues, A., Sage, J., Villar, M. T., and Carlson, G. M. (2008) Crosssearch, a user-friendly search engine for detecting chemically cross-

linked peptides in conjugated proteins, *Mol Cell Proteomics* 7, 739–749.

27. Yu, E. T., Hawkins, A., Kuntz, I. D., Rahn, L. A., Rothfuss, A., Sale, K., Young, M. M., Yang, C. L., Pancerella, C. M., and Fabris, D. (2008) The collaboratory for MS3D: A new cyberinfrastructure for the structural elucidation of biological macromolecules and their assemblies using mass spectrometry-based approaches, *Journal of Proteome Research* 7, 4848–4857.
28. Nadeau, O. W., Anderson, D. W., Yang, Q., Artigues, A., Paschall, J. E., Wyckoff, G. J., McClintock, J. L., and Carlson, G. M. (2007) Evidence for the location of the allosteric activation switch in the multisubunit phosphorylase kinase complex from mass spectrometric identification of chemically crosslinked peptides, *J Mol Biol* **365**, 1429–1445.
29. Liu, W., Priddy, T. S. and Carlson, G. M. (2008) Physiochemical changes in phosphorylase kinase associated with its activation, *Protein Sci.* **17**, 2111–2119.
30. Demchenko, A. P. (1986) Derivative spectroscopy of aromatic amino acids and proteins, In *Ultraviolet spectroscopy of proteins* Rev. and enl. translation of the Russian ed., pp 121–135, Springer-Verlag, Berlin; New York.
31. Lloyd, J. B. F. (1971) Synchronised excitation of fluorescence emission spectra, *Nature physical sciences* **231**, 64–65.
32. Miller, J. N. (1979) Recent advances in molecular luminescence analysis *Proceedings of the Analytical Division of the Chemical Society* **16**, 203–208.
33. Cui, F. L., Wang, J. L., Cui, Y. R., and Li, J. P. (2006) Fluorescent investigation of the interactions between n-(p-chlorophenyl)-n′-(1-naphthyl) thiourea and serum albumin: Synchronous fluorescence determination of serum albumin, *Anal Chim Acta* **571**, 175–183.
34. Chothia, C., Levitt, M., and Richardson, D. (1977) Structure of proteins: Packing of alpha-helices and pleated sheets, *Proc Natl Acad Sci USA* **74**, 4130–4134.
35. Manavalan, P., and Johnson, W. C. (1983) Sensitivity of circular dichroism to protein tertiary structure class, *Nature* **305**, 831–832.
36. Susi, H., and Byler, D. M. (1986) Resolution-enhanced fourier transform infrared spectroscopy of enzymes, *Methods Enzymol* **130**, 290–311.
37. Priddy, T. S., Middaugh, C. R., and Carlson, G. M. (2007) Electrostatic changes in phosphorylase kinase induced by its obligatory allosteric activator Ca^{2+}, *Protein Sci* **16**, 517–527.
38. Trewhella, J., Liddle, W. K., Heidorn, D. B., and Strynadka, N. (1989) Calmodulin and troponin c structures studied by fourier transform infrared spectroscopy: Effects of Ca^{2+} and Mg^{2+} binding, *Biochemistry* **28**, 1294–1301.
39. McNeil-Watson, F., Tscharnuter, W., and Miller, J. (1998) A new instrument for the measurement of very small electrophoretic mobilities using phase analysis light scattering (pals), *Colloids and Surfaces A: Physicochemical and Engineering Aspects* **140**, 53–57.
40. Frank, J. (2002) Single-particle imaging of macromolecules by cryo-electron microscopy, *Annu Rev Biophys Biomol Struct* **31**, 303–319.
41. Nadeau, O. W., Carlson, G. M. and Gogol, E. P. (2002) A Ca^{2+}-dependent global conformational change in the 3D structure of phosphorylase kinase obtained from electron microscopy, *Structure* **10**, 23–32.
42. Radermacher, M., Wagenknecht, T., Verschoor, A., and Frank, J. (1987) Three-dimensional reconstruction from a single-exposure, random conical tilt series applied to the 50s ribosomal subunit of *Escherichia coli*, *J Microsc* **146**, 113–136.
43. Norcum, M. T., Wilkinson, D. A., Carlson, M. C., Hainfeld, J. F., and Carlson, G. M. (1994) Structure of phosphorylase kinase. A three-dimensional model derived from stained and unstained electron micrographs, *J Mol Biol* **241**, 94–102.
44. Trempe, M. R., Carlson, G. M., Hainfeld, J. F., Furcinitti, P. S., and Wall, J. S. (1986) Analyses of phosphorylase kinase by transmission and scanning transmission electron microscopy, *J Biol Chem* **261**, 2882–2889.
45. Trewhella, J. (1997) Insights into biomolecular function from small-angle scattering, *Curr Opin Struct Biol* 7, 702–708.
46. Priddy, T. S., MacDonald, B. A., Heller, W. T., Nadeau, O. W., Trewhella, J., and Carlson, G. M. (2005) Ca^{2+}-induced structural changes in phosphorylase kinase detected by small-angle x-ray scattering, *Protein Sci* **14**, 1039–1048.
47. Paudel, H. K., and Carlson, G. M. (1990) Functional and structural similarities between the inhibitory region of troponin I coded by exon VII and the calmodulin-binding regulatory region of the catalytic subunit of phosphorylase kinase, *Proc Natl Acad Sci USA* **87**, 7285–7289.

Chapter 8

Allosteric Mechanisms of G Protein-Coupled Receptor Signaling: A Structural Perspective

Tarjani M. Thaker, Ali I. Kaya, Anita M. Preininger, Heidi E. Hamm, and T.M. Iverson

Abstract

G protein-Coupled Receptors (GPCRs) use a complex series of intramolecular conformational changes to couple agonist binding to the binding and activation of cognate heterotrimeric G protein (Gαβγ). The mechanisms underlying this long-range activation have been identified using a variety of biochemical and structural approaches and have primarily used visual signal transduction via the GPCR rhodopsin and cognate heterotrimeric G protein transducin (G_t) as a model system. In this chapter, we review the methods that have revealed allosteric signaling through rhodopsin and transducin. These methods can be applied to a variety of GPCR-mediated signaling pathways.

Key words: G protein-coupled receptor, Heterotrimeric G proteins, Rhodopsin, Transducin, Receptor-mediated nucleotide exchange

1. Introduction

GPCRs regulate physical processes ranging from vision to olfaction to cardiac contractility to neurotransmission, yet these receptors likely act via a conserved signaling cycle (Fig. 1). In this cycle, the activation states of GPCRs and cognate G proteins are encoded into protein conformations that influence the binding of partner signaling molecules. For example, agonist binding to a GPCR induces a conformational change that forms a binding site for heterotrimeric G proteins 25 Å away (Fig. 2) (1, 2). The high affinity GPCR-G protein complex promotes exchange of GDP for GTP in the guanine nucleotide-binding pocket of the G protein α subunit (Gα) another 30 Å away from the proposed receptor-binding site (3, 4). The identity of the bound nucleotide then influences the formation of an effector-binding site, yet another 10 Å from bound GTP (Fig. 3) (5–7). The activation steps in this

Aron W. Fenton (ed.), *Allostery: Methods and Protocols*, Methods in Molecular Biology, vol. 796,
DOI 10.1007/978-1-61779-334-9_8, © Springer Science+Business Media, LLC 2012

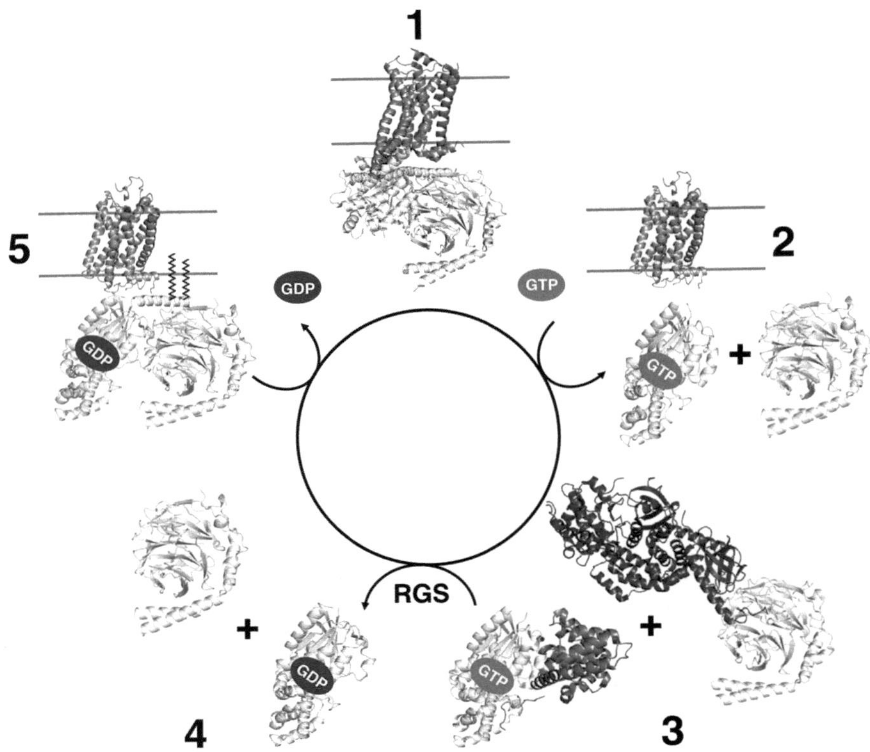

Fig. 1. Summary of the G protein signaling cycle. GPCRs signal through soluble G proteins. In *State 1*, activated receptor binds to the GDP-bound heterotrimeric G protein and promotes release of GDP from the Gα subunit. Binding of GTP to the Gα subunit results in dissociation of this high-affinity complex into GTP-bound Gα and Gβγ (*State 2*), each of which is now able to bind to downstream effectors (*State 3*) and elicit downstream responses. The Gα subunit has intrinsic GTPase activity which is enhanced by *R*egulators of *G* protein *S*ignaling (RGS). Following hydrolysis of GTP, GDP-bound Gα subunits reassociate with Gβγ subunits (*State 4*) and traffic to the membrane, where they can interact with receptors in the next signaling cycle (*State 5*).

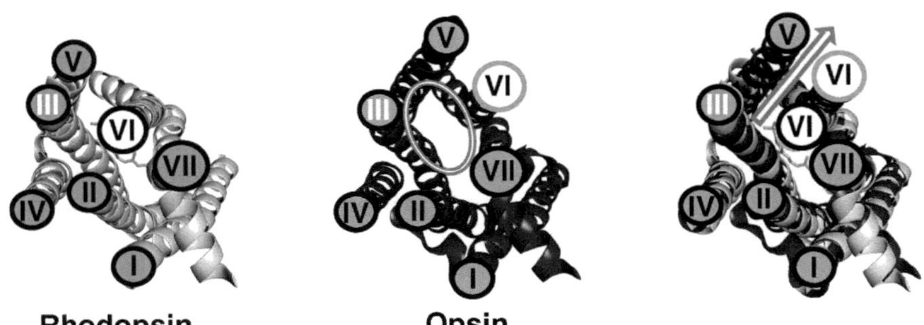

Fig. 2. Differences in conformation changes between rhodopsin and opsin. *Left panel*: Rhodopsin (PDBID 1U19; (54)) is shown in *gray*. *Middle panel*: Opsin (PDBID 3DQB; (1)) is shown in *black*. *Right panel*: An overlay of the structures of rhodopsin and opsin highlights the differences in these structures. These two structures have a 7 Å shift in the position of helix VI (*arrow*) that allows opsin to bind the C-terminus of $G\alpha_t$ in an intracellular binding pocket (1). The conformation of opsin is similar to the conformation proposed for rhodopsin upon light activation.

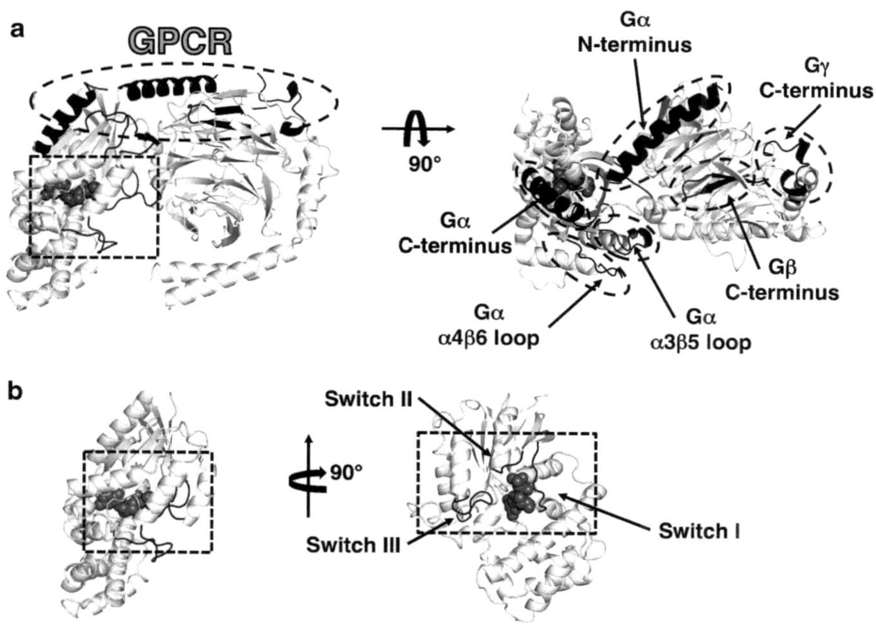

Fig. 3. The binding regions of G proteins. Heterotrimeric G proteins are composed of three subunits: the nucleotide-binding Gα and the dimeric Gβγ subunits. (**a**) In the heterotrimer (PDBID 1GOT; (5)), the receptor-binding surface (*oval*) is composed of the N-terminus, C-terminus, α4β6 loop, α3β5 loop of the Gα subunit, and the C-termini of the Gβ and Gγ subunits. The switch regions on the Gα subunit (*box*) are in conformations that do not bind to effectors. (**b**) Interactions with activated receptors catalyzes nucleotide exchange in the guanine nucleotide binding site of the Gα subunit (PDBID 1TAD; (16)), and the heterotrimer disassociates into GTP-bound Gα and Gβγ subunits. The binding of GTP changes the conformation of the switch regions (*box*) and the disassociation of the complex leaves this surface unobstructed, allowing Switch II and the surrounding region to interact with effectors.

complex series have been investigated by a number of complementary methods.

1.1. Introduction to Allostery in Receptor Activation

Each GPCR has incredible fidelity to a small set of cognate ligands that act as agonists, antagonists, or inverse agonists to promote different activation states (8). Despite this diversity, GPCRs all share a common architecture comprising seven transmembrane helices (9, 10) and several conserved sequence motifs believed to be critical for regulation of receptor activation (11–15). As a result, many of the general mechanisms of activation are assumed to be shared amongst all GPCRs.

The best-characterized GPCR, rhodopsin, is responsible for perception of light under low-light conditions. Numerous properties of this photoreceptor have made it amenable to characterization by both in vivo and in vitro methods. Rhodopsin is naturally abundant (16), easily purified, stable in a variety of detergents, and has spectroscopic properties that provide a convenient method to monitor activation. In addition, rhodopsin covalently ligates an inverse agonist, 11-*cis*-retinal, that locks it into an inactive conformation, which decreases the conformational heterogeneity. Accordingly, rhodopsin was the first GPCR to be characterized by X-ray crystallography (17), and 8 years elapsed before the structures of other GPCRs (18–22) verified the conserved arrangement of transmembrane helices in the family.

Cryoelectron microscopy (10, 23), site-directed spin labeling, and electron paramagnetic resonance spectroscopy (SDSL-EPR) (24, 25) suggested that large conformational changes accompany rhodopsin activation, the most dramatic of which is the rigid body movement of helix VI away from helix III (Fig. 2). This helix movement has long been hypothesized to be required for the formation of a G protein-binding site, and the recent crystal structures of opsin (2) and opsin in complex with a high-affinity peptide mimicking the C-terminus of the $G\alpha_t$ subunit (1) confirmed this hypothesis.

1.2. Introduction to Allostery in Heterotrimeric G Proteins

Heterotrimeric G proteins comprise Gα, Gβ, and Gγ (26). Upon activation by GPCRs, heterotrimeric G proteins disassociate into GTP-bound Gα, and Gβγ, both of which can interact with effector proteins to regulate downstream signaling (Fig. 1) (27). In humans, there are currently 18 known Gα subtypes belonging to four subfamilies, 5 Gβ isoforms, and 12 Gγ isoforms. Each has differing binding specificities for both effectors and GPCRs (28).

Gα and Gβγ can act on a diverse variety of effectors, but like their cognate GPCRs, their mechanisms of activation and deactivation are likely conserved throughout the family. Numerous X-ray crystal structures have revealed how the binding of GDP, GDP-AlF_4^-, or GTPγS to the Gα subunit influences the conformations of three nearby surface loops known as switches I–III (4, 6, 29). These switch regions form a contiguous surface for effector binding (Fig. 3b) (30).

While rhodopsin is by far the most extensively characterized GPCR, its cognate G protein G_t does not heterologously express to levels sufficient for biochemical characterization, which limits techniques that require mutagenesis. However, the large quantities of endogenously expressed transducin in bovine retina have allowed the native protein to be studied and crystallized. For techniques requiring mutagenesis, the close homolog of the $G\alpha_t$ subunit, $G\alpha_i$ (with 67% identity and 82% similarity to $G\alpha_t$) is frequently used to understand general mechanisms of heterotrimeric G protein activation (31–34). $G\alpha_i$ can be activated by rhodopsin, is easily expressed in bacteria, and can be reconstituted with Gβγ to form a fully functional system (4–6).

1.3. Introduction to Receptor-Catalyzed Nucleotide Exchange

The rate-limiting step in GPCR signaling is receptor-catalyzed GDP release from the Gα subunit of the heterotrimer. A combination of site-directed mutagenesis (36–39), peptide-mapping (3, 40–43), and chemical cross-linking (42–45) demonstrated that the α4-β6 loop, α3-β5 loop, N-terminus, and C-terminus of the Gα subunit as well as the C-termini of the Gβ and Gγ subunits are implicated in binding to rhodopsin. While these motifs are distantly arranged in the primary sequence, they form a contiguous surface on the folded protein (Fig. 3a) (46, 47).

The α5 helix, located at the C-terminus of the Gα subunit, is perhaps the best-studied region of receptor interaction and independently binds within a pocket on activated receptor (1). Multiple studies have suggested that receptor-induced rotation and translation of the α5 helix is a requirement for GDP release from Gα (26), and movement of the α5 helix dipole may be important for weakening the interactions between the bound nucleotide and Gα (43).

Much less is understood about how receptor interactions with the Gβγ subunits contribute to G protein activation. Until recently, two major hypotheses existed: the gearshift model (50) and the lever-arm model (51). Mutagenesis was used to support both hypotheses. The gearshift model proposes that the Gβ subunit rotates into the GTP-binding domain of the Gα subunit and pushes the helical domain away from the bound nucleotide. In comparison, the lever-arm model proposes that nucleotide exchange is catalyzed by induced tilt of the Gβγ subunits, which act as a lever to open the binding pocket through interactions with switch II. New insights into the mechanism of G protein activation were revealed by two recent crystal structures of activated β2 adrenergic receptor (β2AR). The first structure was that of the activated form of the receptor stabilized by llama antibodies termed nanobodies (52). The second structure was that of the activated receptor stabilized by its physiologically-relevant signaling partner, the Gasβγ heterotrimer (Gs) (53). Both structures exhibited conformational changes in transmembrane helices 5 and 6. The β2AR-Gs structure additionally confirmed the stabilizing effects of Gαβγ association on the agonist-bound state of the receptor and the presence of a translation of the a5 helix of Gα. Unanticipated observations revealed by the β2AR-Gs structure include ordering of intracellular loop 2 on β2AR, a lack of extensive contacts between the receptor and the Gβγ subunits of Gαsβγ, and a large displacement of the helical domain relative to the Ras-like GTPase domain of Gs. These data, taken together, offer a possible mechanism for receptor-catalyzed GDP release from Gαβγ that is largely consistent with the models suggested by biochemical and structural data obtained thus far.

2. Materials

2.1. Dark-Adapted Rod Outer Segment Membrane Isolation

1. 200 Bovine retinas (see Note 1).
2. Isolation Buffer: 90 mM KCl, 30 mM NaCl, 2 mM $MgCl_2$, 10 mM 3-(*N*-morpholino)propanesulfonic acid (MOPS), 0.1 mM ethylenediaminetetraacetic acid (EDTA), pH 8.0.
3. 26% Sucrose Solution: Isolation Buffer containing 26% w/v sucrose, 1 mM dithiothreitol (DTT), and 50 μM phenylmethylsulfonyl fluoride (PMSF) (see Notes 2 and 3).
4. 30% Sucrose Solution: Isolation Buffer containing 30% w/v sucrose, 1 mM DTT, and 50 μM PMSF.
5. Gradient Solution 1: 0.84 M sucrose, 10 mM MOPS, 60 mM KCl, 30 mM NaCl, and 2 mM $MgCl_2$.
6. Gradient Solution 2: 1.0 M sucrose, 10 mM MOPS, 60 mM KCl, 30 mM NaCl, and 2 mM $MgCl_2$.
7. Gradient Solution 3: 1.14 M sucrose, 10 mM MOPS, 60 mM KCl, 30 mM NaCl, and 2 mM $MgCl_2$.
8. 10 mg/mL aprotinin.
9. 10 mg/mL leupeptin.
10. 2.5 mg/mL pepstatin.

2.2. Rhodopsin–G_t Separation: ROS Membrane Urea Wash

1. Dark-adapted rod outer segment (ROS) membranes isolated from retinas.
2. EDTA Buffer: 10 mM Tris(hydroxymethyl)aminomethane (Tris–Cl) (see Note 4), 1 mM ethylenediaminetetraacetic acid (EDTA), 1 mM DTT, pH 7.5 (see Note 2).
3. Urea Buffer: 10 mM Tris–Cl, 1 mM EDTA, 1 mM DTT, 6 M urea, pH 7.5.
4. Buffer A: 10 mM MOPS pH 7.5, 2 mM $MgCl_2$, 200 mM NaCl, 1 mM DTT, 0.1 mM PMSF.

2.3. Rhodopsin Extraction from Native Membranes

1. Urea-washed, dark-adapted ROS membranes.
2. Concanavalin A (ConA) Binding Buffer: 20 mM Tris–Cl pH 7.4, 250 mM NaCl, 1 mM $MgCl_2$, 1 mM $CaCl_2$, 1 mM $MnCl_2$.
3. Rhodopsin Solubilization Buffer: 80 mM SOL-grade *n*-Dodecyl-β-D-maltoside (DDM) (Anatrace) (see Note 5) dissolved in ConA Binding Buffer (see Note 6).

2.4. Purification of Detergent-Extracted Rhodopsin by ConA Affinity Chromatography

1. ConA Sepharose column (GE Healthcare).
2. ConA Binding Buffer: 20 mM Tris–Cl pH 7.4, 250 mM NaCl, 1 mM $MgCl_2$, 1 mM $CaCl_2$, 1 mM $MnCl_2$ (see Note 4).
3. Wash Buffer 1: ConA Binding Buffer containing 0.5 mM ANAGRADE DDM (Anatrace) (see Notes 5 and 6).
4. Wash Buffer 2: 20 mM Tris–Cl pH 7.4, 100 mM NaCl, 0.5 mM ANAGRADE DDM (Anatrace), pH 7.4.
5. Elution Buffer 1: 20 mM Tris–Cl pH 7.4, 100 mM NaCl, 0.5 mM ANAGRADE DDM (Anatrace), 400 mM α-D methylglucoside (Acros Organics), pH 7.4.
6. Elution Buffer 2: 20 mM Tris–Cl pH 7.4, 100 mM NaCl, 0.5 mM ANAGRADE DDM (Anatrace), 500 mM sucrose, pH 7.4.
7. Extra Meta II Exchange Buffer: 50 mM HEPES pH 8.0, 100 mM NaCl, 1 mM $MgCl_2$, 0.5 mM DDM.

2.5. Determining the Concentration of Rhodopsin

1. Rhodopsin in native membranes or purified into detergent micelles.
2. 20 mM hexadecyl-cetyltrimethylammonium chloride (HTAC) buffer (Sigma Aldrich).

2.6. Metarhodopsin II Assay

1. 10 μM purified rhodopsin.
2. 10 μM purified transducin.
3. Metarhodopsin II (Meta II) Buffer: 50 mM HEPES pH 8.0, 100 mM NaCl, 1 mM $MgCl_2$, 1 mM DTT (see Note 2).

2.7. Crystallization of Rhodopsin

1. Pregreased 24-well Linbro plates (Hampton Research).
2. Siliconized Glass or Plastic coverslips (Hampton Research) (see Note 7).
3. Refrigerant-based propelled compressed air cleaner, such as Dust-off.

2.8. Light-Adapted ROS Membrane Isolation

1. 200 Bovine retinas (see Note 1).
2. Isolation Buffer: 90 mM KCl, 30 mM NaCl, 2 mM $MgCl_2$, 10 mM MOPS, 0.1 mM EDTA, pH 8.0.
3. 26% Sucrose Solution: Isolation Buffer containing 26% w/v sucrose, 1 mM dithiothreitol (DTT), and 50 μM phenylmethylsulfonyl fluoride (PMSF) (see Notes 2 and 3).
4. 30% Sucrose Solution: Isolation Buffer containing 30% w/v sucrose, 1 mM DTT, and 50 μM PMSF.
5. Gradient Solution 1: 0.84 M sucrose, 10 mM MOPS, 60 mM KCl, 30 mM NaCl, and 2 mM $MgCl_2$.

6. Gradient Solution 2: 1.0 M sucrose, 10 mM MOPS, 60 mM KCl, 30 mM NaCl, and 2 mM $MgCl_2$.
7. Gradient Solution 3: 1.14 M sucrose, 10 mM MOPS, 60 mM KCl, 30 mM NaCl, and 2 mM $MgCl_2$.
8. 10 mg/mL aprotinin.
9. 10 mg/mL leupeptin.
10. 2.5 mg/mL pepstatin.

2.9. Isolation and Purification of G_t

1. Isotonic Buffer: 5 mM Tris–Cl, 130 mM KCl, 0.6 mM $MgCl_2$, 1 mM EDTA, 5 mM βME or 1 mM DTT, 0.1 mM PMSF, pH 8.0 (see Notes 2–4).
2. Hypotonic Buffer: 5 mM Tris–Cl, 0.6 mM $MgCl_2$, 1 mM EDTA, 5 mM βME or 1 mM DTT, 0.1 mM PMSF, pH 8.0.
3. Hypotonic buffer + GTP: Hypotonic Buffer containing 0.1 mM guanosine-5′-triphosphate (GTP) (see Note 8).

2.10. Expression of $G\alpha_{i1}$

1. 2× YT (Yeast-Tryptone) Media: 1.6% w/v Bacto-tryptone, 10% w/v Bacto-yeast extract, 0.5% w/v NaCl, autoclaved.
2. 100 mg/mL ampicillin, filter sterilized (see Note 9).
3. Isopropyl-β-D-thiogalactopyranoside (IPTG), 1.0 M in distilled water, filter sterilized.

2.11. Purification of $G\alpha_{i1}$

1. Lysis Plus Buffer: 50 mM NaH_2PO_4, 300 mM NaCl, 5 mM β-mercaptoethanol (βME), 5 mM imidazole pH 8.0. Add 5 mM βME or 1 mM DTT, 20 μM GDP, 0.1 mM PMSF, 10 μg/mL aprotinin, 10 μg/mL leupeptin, and 2.5 μg/mL pepstatin (see Notes 2 and 3).
2. Loading Buffer: 50 mM NaH_2PO_4, 300 mM NaCl, 5 mM βME or 1 mM DTT, pH 8.0.
3. Wash buffer: 50 mM NaH_2PO_4, 300 mM NaCl, 5 mM βME or 1 mM DTT, 10 mM imidazole, pH 8.0.
4. Elution buffer: 50 mM NaH_2PO_4, 300 mM NaCl, 5 mM βME or 1 mM DTT, 100 mM imidazole, pH 8.0.
5. Dialysis Buffer: 50 mM Tris–Cl pH 8.0, 50 mM NaCl, 1 mM $MgCl_2$, 10% (v/v) glycerol (see Note 4). Add 0.1 mM PMSF, 5 mM βME or 1 mM DTT, and 20 μM GDP right before buffer is used.
6. Buffer A: 50 mM Tris–Cl pH 8.0, 50 mM NaCl, 2 mM $MgCl_2$, 5 mM βME or 1 mM DTT.
7. Buffer B: 50 mM Tris–Cl pH 8.0, 1 M NaCl, mM $MgCl_2$, 5 mM βME or 1 mM DTT.
8. Storage Buffer: 50 mM Tris-Cl pH 8.0, 50 mM NaCl, 2 mM $MgCl_2$, 5 mM βME or 1 mM DTT, 10 μM GDP.

9. Collection Buffer: 50 mM Tris–Cl pH 8.0, 50 mM NaCl, 2 mM $MgCl_2$, 5 mM βME or 1 mM DTT, 100 μM GDP.
10. $G\alpha_i$ Buffer: 50 mM Tris–Cl pH 8.0, 200 mM NaCl, 2 mM $MgCl_2$, 1 mM EDTA pH 8.0, 1 mM DTT, 20 μM GDP.
11. TALON or Nickel NTA Resin (Clontech).
12. SOURCE™ 15Q Anion Exchange resin (GE Healthcare).
13. Superdex 200 GL 10/300 Gel Filtration Column (GE Healthcare).

2.12. Intrinsic Tryptophan Fluorescence Assay

1. NaF, 500 mM in distilled water.
2. $AlCl_3$, 10 mM in distilled water.
3. Storage Buffer: 50 mM Tris–Cl pH 8.0, 50 mM NaCl, 2 mM $MgCl_2$, 5 mM βME or 1 mM DTT, 10 μM GDP (see Notes 2 and 4).

2.13. Determining the Concentration of G_t or $G\alpha_{i1}$

1. Quick Start™ Bradford Protein Assay Kit containing 1× dye reagent and seven BSA standards ranging in concentration from 0.125 to 2.0 mg/mL (BIORAD).
2. 1:10 and 1:50 dilutions of purified G_t or $G\alpha_{i1}$ sample.
3. 1 mL quartz cuvettes.

2.14. Crystallization of G_t or $G\alpha_{i1}$

1. Pregreased 24-well Linbro plates (Hampton Research).
2. Siliconized Glass or Plastic coverslips (Hampton Research) (see Note 7).
3. Refrigerant-based propelled compressed air cleaner, such as Dust-off.

2.15. Chemical Cross-Linking of Rhodopsin–G_t

1. Dark-adapted ROS membranes.
2. Buffer A: 10 mM MOPS pH 7.4, 200 mM NaCl, 1 mM $MgCl_2$, 1 mM DTT, 1 mM EDTA, 0.1 mM PMSF (see Notes 2 and 3).
3. 3,3′-dithio-bis succini-midylpropionate (DTSSP), cleavable cross-linking agent (Pierce).
4. 1 M Tris–Cl pH 7.5 (see Note 4).
5. 100 μM GTPγS.
6. 50 mM DTT.

2.16. In Vitro Rho–G_t Complex Formation

1. Purified G_t.
2. Purified rhodopsin.
3. Rho–G_t Buffer: 50 mM HEPES pH 8.0, 100 mM NaCl, 1 mM $MgCl_2$, 0.5 mM DDM, 40 μM all-*trans* retinal (see Notes 10 and 11).

4. Superdex 200 GL 10/300 Gel Filtration Column (GE Healthcare).

2.17. Receptor-Catalyzed GTPγS Exchange Assay

1. 200 nM purified G_t.
2. 100 nM purified rhodopsin.
3. G_t Activation Buffer: 10 mM MOPS, 130 mM NaCl, 1 mM $MgCl_2$, pH 7.2.
4. 10 μM GTPγS.

3. Methods

Large quantities of pure protein are essential for studying biophysical properties of GPCRs and heterotrimeric G proteins. Therefore, the identification of mechanisms of GPCR signaling has been facilitated by reproducible methods for obtaining and purifying rhodopsin, transducin, and $G\alpha_i$. Spectroscopic monitoring is commonly used to follow the activation state in these proteins, while X-ray crystal structures have provided snapshots of stable signaling intermediates.

3.1. Dark-Adapted Rod Outer Segment Membrane Isolation

1. All steps are to be performed under dim red light and on ice or at 4°C.
2. Thaw bovine retinas from 200 bovine eyes on ice and decant into a 600-mL beaker. Wash the containers with 30% Sucrose Solution, pool the wash with the thawed retinas, and bring the final volume to 200 mL with 30% Sucrose Solution.
3. Stir the retina slurry for 1 h at 4°C to separate the outer segments from the rod cell inner segments and basolateral membrane.
4. Evenly distribute approximately 30 mL of the retina slurry into eight Oak Ridge centrifuge tubes and balance with 30% Sucrose Solution. Cap and invert the tubes two to three times to evenly mix the solution. Centrifuge for 4 min at 4°C in a Sorvall SS-34 rotor at 4,000 rpm (or 2,000 RCF(max) in a similar rotor).
5. Carefully pour the supernatant into a beaker placed on ice and set aside (see Notes 12 and 17). Resuspend each pellet with a small volume of 30% Sucrose Solution. Evenly distribute the resuspended pellet into four clean Oak Ridge centrifuge tubes. Rinse the old centrifuge tubes with a small volume of 30% Sucrose Solution. Evenly distribute this wash solution into each of the new centrifuge tubes and balance with 30% Sucrose Solution. Cap and invert the tubes two to three times to evenly mix the solution. Centrifuge at 4°C for 4 min in a Sorvall SS-34 rotor at 6,000 rpm (or 4,300 RCF(max) in a similar rotor).

6. Carefully pool the supernatant with the supernatant from the previous round of centrifugation. The pellet may now be discarded. Evenly distribute the pooled supernatant into four, clean Oak Ridge centrifuge tubes and balance with Isolation Buffer. Cap and invert the tubes two to three times to evenly mix the solution. Centrifuge for 20 min at 4°C in a Sorvall SS-34 rotor at 19,000 rpm (or 43,000 RCF(max) in a similar rotor).
7. Discard the supernatant and resuspend the pellets with 1 mL of 26% Sucrose Solution for every two pellets. Pool the resuspended pellet slurry in a clean beaker placed on ice. Wash the centrifuge tubes with 1 mL of 26% Sucrose Solution for every two tubes. Pool the wash solution with the resuspended pellet slurry.
8. Prepare a sucrose gradient by layering each of the Sucrose Gradient Solutions (1–3) in six 36 mL Sorvall AH629 swinging-bucket centrifuge tubes (or similar). To prepare this gradient, first use a large needle or pipette to carefully layer 7 mL of Sucrose Solution 1 on the bottom of each centrifuge tube. Next, using the same pipette, layer 7 mL of Solution 2 by inserting the pipette tip through Solution 1 and touching the bottom of the centrifuge tube below Solution 1. Dispensing Solution 2 will cause Solution 1 to rise. Finally, pipette 7 mL of Solution 3 below Solution 2 using the same technique. Dispensing Solution 3 will cause Solutions 1 and 2 to rise. The final order of the sucrose solutions is Sucrose Solution 3 on the bottom, Sucrose Solution 2 in the middle, and Sucrose Solution 1 on the top (see Note 13).
9. Carefully layer the pellet slurry on top of each of each tube, balance with 26% Sucrose Solution, and centrifuge at 4°C for 30 min in a Sorvall AH-629 swinging-bucket rotor at 25,000 rpm (or a similar rotor at 112,000 RCF(max)).
10. Once the spin is complete, there will be a gradient of different colored layers in each of the centrifuge tubes. The bottom layer will be somewhat translucent, the next layer will be opaque, and the top layer will, again, be translucent. The orange layer in the center is the ROS membrane layer. Carefully discard the upper layer using a pipette and collect the opaque layer in the middle (see Note 14).
11. Pool the ROS membrane layers, evenly distribute into clean Oak Ridge centrifuge tubes and balance with Isolation Buffer. Cap and invert the tubes two to three times to evenly mix the solution. Centrifuge for 20 min at 4°C in a Sorvall SS-34 rotor at 19,000 rpm (or 43,000 RCF(max) in a similar rotor) to pellet the ROS membranes. During this centrifugation step, prepare 15 mL of Isolation Buffer containing 5 μg/mL pepstatin, 10 μg/mL aprotinin. and 10 μg/mL leupeptin.

12. Discard the supernatant and resuspend each pellet with 1 mL of freshly prepared Isolation buffer containing 2.5 μg/mL pepstatin, 10 μg/mL aprotinin, and 10 μg/mL leupeptin. Thoroughly homogenize the resuspended ROS membranes by pipetting up and down. Pool the ROS membranes in a 15-mL conical tube placed on ice.
13. Wash the empty centrifuge tubes with enough of the remaining freshly prepared Isolation Buffer so that the final volume of ROS membranes does not exceed 15 mL. Dark-adapted ROS membranes can be stored in tubes wrapped in aluminum foil at −80°C for long-term storage.

3.2. Rhodopsin–Transducin Separation: Dark-Adapted ROS Membrane Urea Wash

1. All steps are to be performed under dim red light and on ice or at 4°C.
2. Thaw dark-adapted ROS membranes on ice and transfer to a 40 mL glass homogenizer. Fill the homogenizer with EDTA Buffer and homogenize ROS membranes (see Note 15). Pour the homogenate into ultracentrifuge tubes and balance with EDTA buffer (see Note 16). Cap and invert the tubes two to three times to evenly mix the solution. Centrifuge for 30 min in a Ti-70 rotor at 30,000 rpm or 92,000 RCF(max) in a similar rotor.
3. Discard the supernatant. Resuspend each pellet with 1 mL EDTA Buffer, and transfer to a glass homogenizer. Rinse the tubes with a small volume of EDTA Buffer, pool with the resuspended pellets in the homogenizer, fill the homogenizer with EDTA Buffer, and homogenize (see Note 17). Evenly distribute the homogenate to new ultracentrifuge tubes and balance with EDTA Buffer. Cap and invert the tubes two to three times to evenly mix the solution. Centrifuge for 30 min in a Ti-70 rotor at 30,000 rpm (or 92,000 RCF(max) in a similar rotor).
4. Discard the supernatant and resuspend each pellet with 1 mL of Urea Buffer. Transfer the pellet resuspension to a glass homogenizer. Rinse the tubes with a small volume of Urea Buffer, pool with the resuspended pellets, fill the homogenizer with Urea Buffer, and homogenize. Evenly distribute the homogenate to clean ultracentrifuge tubes and balance with Urea Buffer. Cap and invert the tubes two to three times to evenly mix the solution. To pellet the membranes, centrifuge for 30 min in a Ti-70 rotor at 45,000 rpm (or 208,000 RCF(max) in a similar rotor).
5. Discard the supernatant and resuspend the pellets with Buffer A (see Note 17). Transfer the pellet resuspension to a glass

homogenizer. Rinse the tubes with a small volume of Buffer A, pool with the resuspended pellets, fill the homogenizer with Buffer A, and homogenize. Evenly distribute the homogenate to clean ultracentrifuge tubes and balance with Buffer A. Cap and invert each tube two to three times to evenly mix the solution. Centrifuge for 30 min in a Ti-70 rotor at 30,000 rpm (or 92,000 RCF(max) in a similar rotor).

6. Discard the supernatant and resuspend each pellet with 1 mL of Buffer A by pipetting up and down with a pipette in the centrifuge tubes. Prepare 100 μL aliquots of the urea-washed, dark-adapted ROS membranes in 650 μL tubes, wrap in foil, and store at −80°C.

3.3. Rhodopsin Extraction from Native Membranes

1. All steps are to be performed under dim red light and on ice or at 4°C.
2. Thaw three to four aliquots of urea washed, dark-adapted ROS membranes on ice and pool in a 1.5-mL tube. Use 400 μL of ConA Binding Buffer wash each tube and pool with the thawed membranes (see Note 18).
3. Centrifuge the membranes for 25 min at 4°C at top speed (20,000 RCF(max)) in a benchtop Eppendorf Centrifuge.
4. Discard the supernatant and resuspend the pellet with 400 μL Rhodopsin Solubilization Buffer (see Note 19) by pipetting up and down until the resuspension is no longer cloudy. Incubate the resuspended pellet on ice for 45 min. Mix the sample by pipetting every 15 min during the incubation.
5. After 45 min, centrifuge the sample for 25 min at 4°C at top speed (20,000 RFC(max)) in a benchtop Eppendorf Centrifuge.
6. Collect the supernatant (~500 μL) with a pipette and discard the pellet.

3.4. Purification of Detergent-Extracted Rhodopsin by ConA Affinity Chromatography

1. Assemble a light-protected "ConA Purification Hutch" in a 4°C environment to protect the purification apparatus from light. Place a clamp stand for the ConA Sepharose column, a peristaltic pump, and a tube rack inside of the hutch (see Fig. 4).
2. Attach a 30-mL syringe to the top of a 1 mL ConA column for holding buffers (see Note 20). Load 10 mL of Wash Buffer 1 to wash off the ethanol-containing storage buffer in which the column is packed. Equilibrate the column in 30 mL of Wash Buffer 1 in a continuous loop using a peristaltic pump flowing at 0.3 mL/min for at least 48 h (see Note 21).
3. Dilute ~500 μL of detergent-extracted rhodopsin (see Subheading 4) to a total volume of 10 mL with ConA Binding Buffer (see Note 22). When the volume of Wash Buffer 1 in

Fig. 4. ConA Purification Hutch. a) A schematic showing the contents of a ConA purification hutch. b) A simple cardboard box lined with aluminum foil and sealable flaps is sufficient for creating a dark-adapted environment for ConA purification of rhodopsin. *Box 1* highlights the 1 mL ConA Sepharose Column (GE Healthcare) attached to a 30-mL syringe. Both are held to a ring stand with a standard clamp. *Box 2* shows the peristaltic pump with two leads, one of which remains connected to the bottom end of the ConA Sepharose Column. Depending on the requirement of the protocol, the second lead can either rest in a collection tube (*Box 3*) or can be secured in place inside of the syringe (*Box 1*) for flowing solutions over the column in a continuous loop (i.e., while loading the column with rhodopsin or while preequilibrating the column with detergent).

the syringe is ≤1 mL, load the rhodopsin fraction. Flow the rhodopsin load fraction in a continuous loop over the column using a peristaltic pump flowing at 0.3 mL/min for 4 h (see Note 23).

4. Collect the flowthrough and all subsequent fractions for later analysis of the purification by SDS-PAGE (see Fig. 5 for expected results). When the volume of loaded protein in the syringe is ≤1 mL, load 10 mL of Wash Buffer 1 and collect the flowthrough as Wash 1.

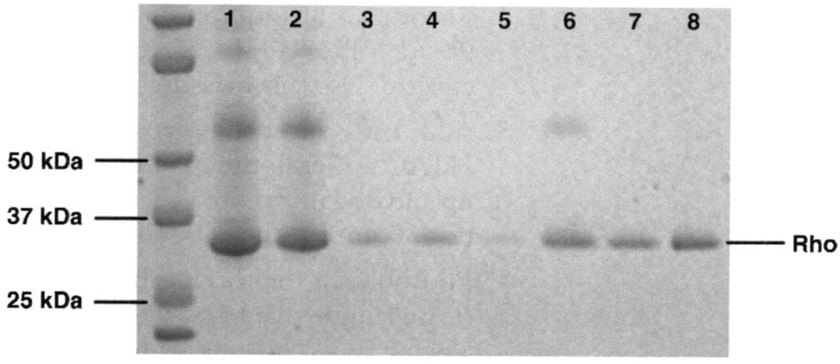

Fig. 5. ConA purification fractions analyzed by SDS-PAGE. Fractions from a ConA purification of dark-adapted rhodopsin are shown on a 10 to 12% gradient polyacrylamide gel (Invitrogen). *Lanes 1 and 2*: Load Fraction, *Lane 3*: Flowthrough, *Lane 4*: Wash 1, *Lane 5*: Wash 2, *Lane 6*: Elution 1, *Lane 7*: Elution 2 , *Lane 8*: pooled Elution 1 and 2. Rhodopsin runs at an apparent molecular weight of ~34 kDa.

5. When the volume of loaded Wash Buffer 1 in the syringe is ≤1 mL, load 5 mL of Wash Buffer 2 and collect the flowthrough as Wash 2.
6. When the volume of loaded Wash Buffer 2 in the syringe is ≤1 mL, load 50 mL of Elution Buffer 1 and collect the flowthrough as Elution 1. This fraction contains ConA-purified rhodopsin.
7. When the volume of loaded Elution Buffer 1 in the syringe is ≤1 mL, load 5 mL of Elution Buffer 2 and collect the flowthrough as Elution 2. This fraction should also contain ConA-purified rhodopsin and can be pooled with Elution 1.
8. Concentrate the pooled Elution fractions in a 30-kDa molecular weight cutoff Amicon concentrator wrapped in aluminum foil. Centrifuge the concentrator in a benchtop centrifuge at 4°C at 1,600 RCF(max) in 20 min intervals until the total volume is ~2 mL (see Note 24).
9. Using the same concentrator, buffer exchange the concentrated, ConA-purified rhodopsin by diluting to 50 mL in Extra Meta II Exchange Buffer and serial centrifugation at 4°C and 1,600 RCF(max) in 20 min intervals. Concentrate the sample to a volume of ~200–500 μL and measure the concentration using the rhodopsin concentration assay (see Subheading 5).

3.5. Determining the Concentration of Rhodopsin

1. The concentration of solubilized rhodopsin is determined by measuring its absorbance at 500 nm and 650 nm both before and after photobleaching in the presence of 20 mM fluorescence enhancing detergent, HTAC.

2. Record an absorption spectrum from 350 to 650 nm for 96 μL of 20 mM HTAC buffer added to an appropriate cuvette. Use this measurement to establish a baseline.
3. Add 4 μL of a dark-adapted rhodopsin sample to 96 μL of HTAC buffer in the cuvette and mix thoroughly by pipetting up and down. Record a second absorption spectrum from 350 to 650 nm for the dark-adapted sample.
4. Photobleach the rhodopsin-containing sample in the cuvette by pulsing with a burst of light from a camera flash, such as a Vivitar 283. Thoroughly mix the sample by pipetting up and down in the cuvette, being careful to avoid generating air bubbles, and pulse the sample with light a second time (see Notes 25 and 27). Mix the sample again and record a third spectrum from 350 to 650 nm for the light-adapted sample.
5. Calculate the difference in the absorbance between 650 and 500 nm for both the dark-adapted and light-adapted readings using Eq. 1:

$$\Delta A_{500-650} = A_{500} - A_{650} \tag{1}$$

6. Calculate the difference in this value between dark-adapted and light-adapted rhodopsin using Eq. 2.

$$\begin{aligned} A_{\text{dark-light}} = {} & \Delta A_{500-650}\ (\text{dark - adapted}) \\ & - \Delta A_{500-650}\ (\text{light - adapted}) \end{aligned} \tag{2}$$

7. Use this value to then calculate the concentration with Beer's Law (Eq. 3):

$$\begin{aligned} A_{\text{dark-light}} &= \varepsilon b C \\ C &= A_{\text{dark-light}}/(\varepsilon \times b) \end{aligned} \tag{3}$$

   where A corresponds to the absorbance value, ε corresponds to the molar extinction coefficient for rhodopsin, b corresponds to the path length (in cm), and C corresponds to concentration (in units of molarity (mol/L)). The molar extinction coefficient is 42,000 $M^{-1}cm^{-1}$ for rhodopsin in ROS membranes and 40,600 $M^{-1}cm^{-1}$ for detergent-extracted rhodopsin. Use the latter value to determine the concentration of purified rhodopsin (see Note 26).
8. See Fig. 6 for expected results.

3.6. Metarhodopsin II Assay

The metarhodopsin II assay can be used to determined the stability of light-activated rhodopsin, meta II, in the conditions in which it has been purified. Chimeric, heterotrimeric G protein ($G\alpha_i\beta\gamma$) can be substituted for G_t to characterize the contribution of mutations on protein complex stability.

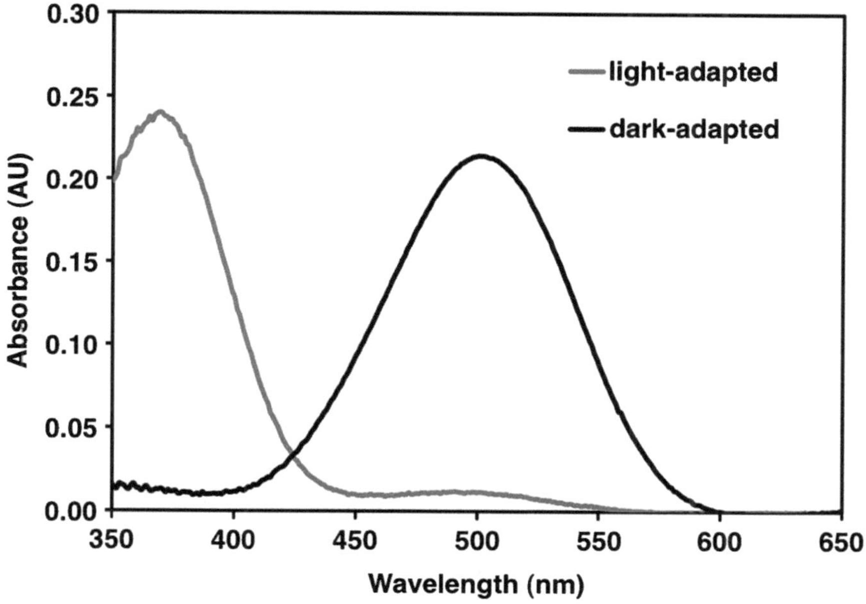

Fig. 6. Solubilized rhodopsin absorption spectrum. The dark-adapted spectrum is shown in *black*, while the light-adapted spectrum is shown in *light gray*. The concentration of rhodopsin is calculated as a function of the absorbance difference between dark-adapted and light-adapted rhodopsin at 500 nm (see Eqs. 1–3). Upon light activation, there is an observable shift in the absorbance maxima to 380 nm, the wavelength characteristic of metarhodopsin II.

1. Prepare a 10 μM sample of purified, dark-adapted rhodopsin (see Subheadings 4 and 5) in Meta II Buffer and mix with 10 μM of purified G_t (see Subheadings 9 and 13) in Meta II Buffer.
2. Incubate the sample on ice for 15 min in the dark or under dim red light. Pipette the incubated sample into an appropriate cuvette, place in a spectrophotometer, and collect an absorption spectrum from 350 to 650 nm for the dark-adapted sample.
3. Light-activate the sample by pulsing with two bursts of light from a camera flash. Wait 1 min to allow the sample to be completely activated, mix thoroughly by pipetting, and collect a second absorption spectrum from 350 to 650 nm for the light-adapted sample (see Note 27).
4. To determine the signal contributed by meta II, calculate the difference in the absorbance between 380 and 440 nm for both the dark-adapted and light-adapted readings using Eq. 4.

$$\Delta A_{380-440} = A_{380} - A_{440} \tag{4}$$

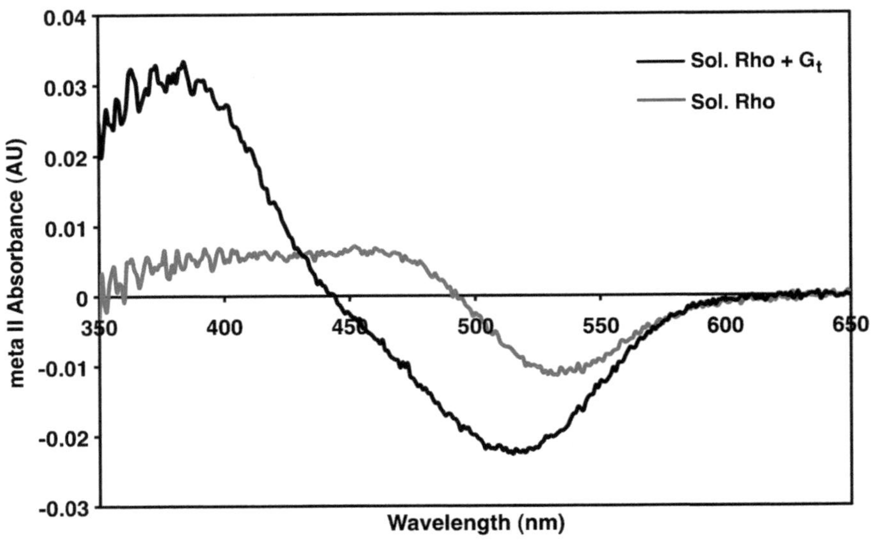

Fig. 7. Metarhodopsin II absorption spectrum. The absorbance spectra of detergent-solubilized rhodopsin (*gray*) and detergent-solubilized rhodopsin in the presence of G_t (*black*) after light activation. In the absence of GTP, G_t significantly increases the meta II signal (380 nm). This likely reflects the stabilization of the activated form of receptor by the heterotrimeric G protein.

5. Then determine the meta II signal by calculating the difference in $\Delta A_{380\text{–}440}$ between dark-adapted and light-adapted rhodopsin using Eq. 5.

$$A_{\text{light}-\text{dark}} = \Delta A_{380-440}(\text{light - adapted}) - \Delta A_{380-440}(\text{dark - adapted}) \quad (5)$$

6. See Fig. 7 for expected results.

3.7. Crystallization of Rhodopsin

While multiple methods of crystallization exist (sitting drop, hanging drop, batch, etc.), the hanging drop vapor diffusion method is the most commonly used (Fig. 8). In this method, a closed system is designed in which a droplet of purified protein mixed with reservoir solution, usually in a 50:50 ratio, is allowed to equilibrate with a solution contained in a reservoir below. As water evaporates from the droplet hanging above the reservoir solution, the solubility of the purified protein decreases, the concentration of precipitant increases, and nucleation of crystals occurs. The system is then maintained at equilibrium until crystallization is complete.

1. Filter all reagents to be used in the crystallization screen and equilibrate at the crystallization temperature (see Notes 28 and 29).
2. Mix 1 mL of reservoir solution in the bottom of each well of a pregreased 24-well Linbro tray. Crystallization conditions for rhodopsin are given in Table 1. Place the tray on a rotating table or pipette reservoir solution up and down to mix the reagents.

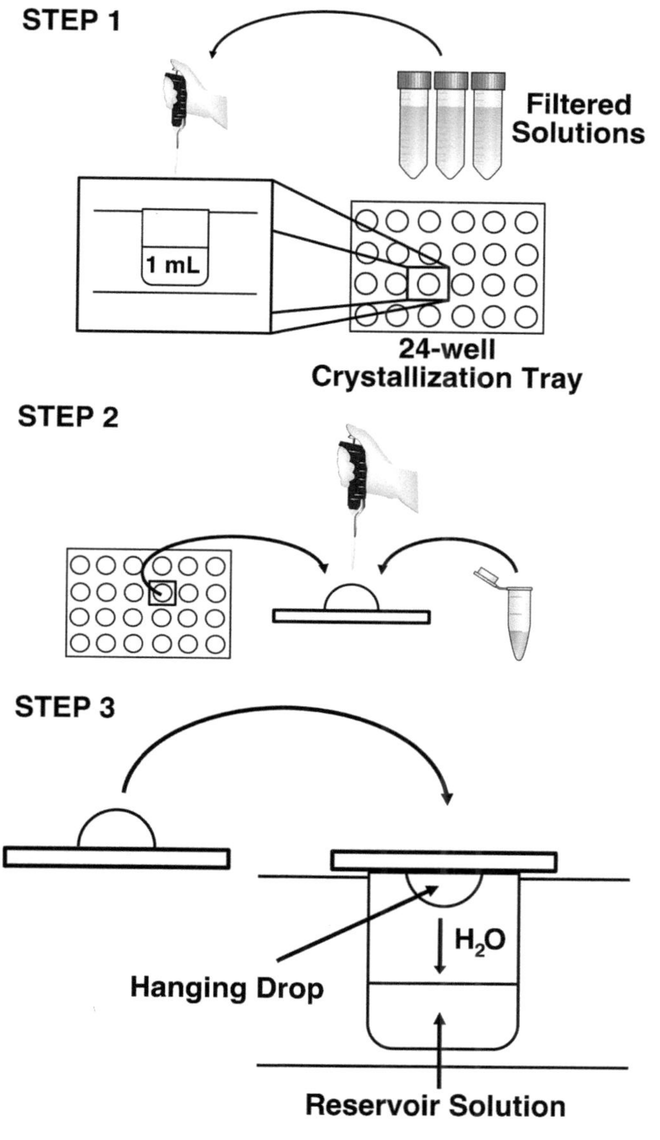

Fig. 8. Hanging drop vapor diffusion. *Step 1*: Prepare the reservoir solution by mixing appropriate volumes of filtered solutions in a single well of a 24-well crystallization tray. The final volume of reservoir solution is typically 1 mL. *Step 2*: Pipette equal volumes of the purified protein and the reservoir solution onto a siliconized coverslip. There is no need to mix the two components. A common starting volume for the hanging drop is 2 μL (1 μL protein + 1 μL of reservoir solution), but can range from <1 μL to 10 μL. *Step 3*: Carefully invert the coverslip without disturbing the droplet and place over the corresponding well. Gently press down on the coverslip to ensure an air-tight seal between the coverslip and the well.

Table 1
Crystallization of Rhodopsin

	Rhodopsin (17)	Rhodopsin (54)	Squid Rhodopsin (20)	Meta II (55)	Meta II (55)	Opsin (2)	Opsin (1)
PDB ID	1F88	1U19	2Z73	3PXO	3PQR	3CAP	3DQB
Space Group	$P4_1$	$P4_1$	$P6_2$	H32	H32	H3	H32
Resolution	2.8Å	2.2Å	2.5Å	3.0Å	2.85Å	2.9Å	3.2Å
Method	Hanging Drop	Hanging Drop	Sitting Drop	Hanging Drop	Hanging Drop	Hanging Drop	Hanging Drop
Protein Concentration	10 mg/mL	6-8 mg/mL	8 mg/mL	5 mg/mL	5 mg/mL	5 mg/mL	7-10 mg/mL
Drop Size	9-11 μL	4-10 μL	10 μL	4 μL	4 μL	4 μL	4 μL
Source	Bovine ROS	Bovine ROS	Microvillar membranes of photoreceptor rhabdoms from squid retina	Bovine ROS	Bovine ROS	Bovine ROS	Bovine ROS
Protein Buffer	MES or sodium acetate pH 6.3-6.4, 5-7 mM βME, 90-120 mM $Zn(OAc)_2$, 0.50-0.65% heptane-1,2,3-triol, 2.2:1 ratio of nonyl glucoside: rhodopsin	6-12 mM βME, 0.1-0.5% heptyl-thioglucoside, 0.5-0.7 M ammonium sulfate	30 mM MES, pH 6.4, 30 mM $Zn(OAc)_2$, ~100 mM β-D-octyl-glucopyranoside	1% β-D-octyl-glucopyranoside, 0.02% *n*-dodecyl- β-D-maltopyranoside	1% β-D-octyl-glucopyranoside, 0.02% *n*-dodecyl-β-D-maltopyranoside, GαCT2 peptide	1% β-D-octyl-glucopyranoside	1% β-D-octyl-glucopyranoside, GαCT peptide (4:1 ratio of protein:peptide)
Reservoir Solution	MES-buffered 3.0-3.4 M ammonium sulfate pH 6.0-6.1	20-30 mM MES pH 5.9-6.1, 2.5-3.0 M ammonium sulfate	3.2 M ammonium sulfate, 32 mM MES pH 6.0-6.7, 38 mM EDTA, 10 mM βME	3.0-3.4 M $(NH_4)_2SO_4$, 0.1 M sodium acetate pH 5.0-5.8	3.0-3.4 M $(NH_4)_2SO_4$, 0.1 M sodium acetate pH 5.0-5.8	2.8-3.2 M ammonium sulfate, 0.1 M MES or sodium acetate pH 5.6	3.2 M ammonium sulfate, 0.1 M MES or sodium citrate pH 6.0
Drop Size	3-5°C	10°C	4°C	4°C	4°C	4°C	4°C

3. Remove dust from siliconized coverslips using an air duster. Pipette 1–2 μL of purified protein onto the coverslip. Pipette an equal volume of reservoir solution on top of the protein drop (see Note 30). It is not necessary to mix this drop. Carefully invert the coverslip and place over the well from which the reservoir solution was taken. Gently push down on the coverslip to make sure the well is tightly sealed. There should be no gaps between the coverslip and silicone grease.
4. Allow the crystallization reactions to equilibrate without disruption in a vibration-free location at the appropriate temperature for at least 24 h before monitoring (see Notes 31–33).
5. Adjust crystallization conditions based on the results of preliminary screens (see Notes 34 and 35).

3.8. Light-Adapted ROS Membrane Isolation

1. All steps should be performed under ambient light and on ice or at 4°C.
2. Thaw bovine retina preparations from 200 eyeballs on ice and decant into a 600 mL beaker. Wash the containers with 30% Sucrose Solution, pool the wash with the thawed retinas, and bring the final volume to 200 mL with 30% Sucrose Solution.
3. After diluting the retinal slurry to 200 mL with 30% Sucrose Solution, photobleach the retinal slurry without stirring for 30 min to allow for the formation of the high affinity Rhodopsin–G_t complex. After 30 min, continue to photobleach the retinal slurry for an additional 1 h with stirring to separate the outer segments from the rod cell inner segments and basolateral membrane.
4. Repeat steps 4–13 as given in Subheading 1. It is not necessary to protect the purified membranes from the light.

3.9. Isolation and Purification of Transducin

1. All steps should be performed at 4°C or on ice.
2. Thaw light-adapted ROS membranes (see Subheading 8) and pour into a 40 mL glass homogenizer. Rinse the frozen ROS membrane container with Isotonic Buffer. Pool the wash solution with the thawed ROS membranes, fill the homogenizer with Isotonic Buffer, and homogenize.
3. Evenly distribute homogenate into ultracentrifuge tubes and balance with Isotonic Buffer (see Note 16). Cap and invert the tubes two to three times to evenly mix the solution. Centrifuge for 20 min in a Ti-70 rotor at 30,000 rpm (or 92,000 RCF(max) in a similar rotor).
4. Discard the supernatant and resuspend the pellets with a small volume of Isotonic Buffer. Transfer the resuspended pellets to the glass homogenizer. Rinse each centrifuge tube with Isotonic Buffer. Pool the wash solution with resuspended pellets, fill the homogenizer with Isotonic Buffer, and homogenize.

5. Repeat steps 3 and 4 twice, followed by step 3 a third time for a total of four washes with Isotonic Buffer.
6. After the fourth Isotonic Buffer wash, discard the supernatant and resuspend the pellets with a small volume of Hypotonic buffer. Transfer the resuspended pellets to a glass homogenizer. Rinse each centrifuge tube with Hypotonic Buffer, pool the wash solution with resuspended pellets, fill the homogenizer with Hypotonic Buffer, and homogenize.
7. Evenly distribute homogenate into ultracentrifuge tubes and balance with Hypotonic Buffer. Cap and invert the tubes two to three times to evenly mix the solution. Centrifuge for 20 min in a Ti-70 rotor at 30,000 rpm (or 92,000 RCF (max) in a similar rotor).
8. Repeat steps 6 and 7 for a total of two Hypotonic Buffer washes.
9. Prepare 400 mL of fresh Hypotonic Buffer containing 0.1 mM GTP (see Note 8). After the second Hypotonic Buffer wash, resuspend each pellet with 25 mL of Hypotonic Buffer containing 0.1 mM GTP. Transfer pellet resuspension to a glass homogenizer. Rinse each tube with a small-volume of the Hypotonic Buffer containing 0.1 mM GTP and homogenize.
10. Evenly distribute homogenate into ultracentrifuge tubes and balance with Hypotonic Buffer containing 0.1 mM GTP. Cap and invert the tubes two to three times to evenly mix the solution. Centrifuge for 45 min in a Ti-70 rotor at 30,000 rpm (or 92,000 RCF(max) in a similar rotor).
11. Decant the supernatant into a clean container and store on ice. This supernatant contains G_t. Repeat steps 9 and 10 for a total of two washes with Hypotonic Buffer containing 0.1 mM GTP.
12. Pool the supernatant with the supernatant from the previous round of washing with Hypotonic Buffer containing 0.1 mM GTP. The pellet can now be discarded.
13. Concentrate G_t-containing supernatant to 1 mL. Particulate contaminants and any remaining membranes may be removed by spinning at top speed in a benchtop Eppendorf centrifuge at 4°C for 30 min (optional).
14. If the purified G_t is to be stored for use at a later time, add glycerol to a final concentration of 10% and freeze the purified protein at −20°C for short-term storage, or −80°C for long-term storage (see Note 36).
15. See Fig. 9 for expected results of an SDS-PAGE analysis of G_t purification fractions.

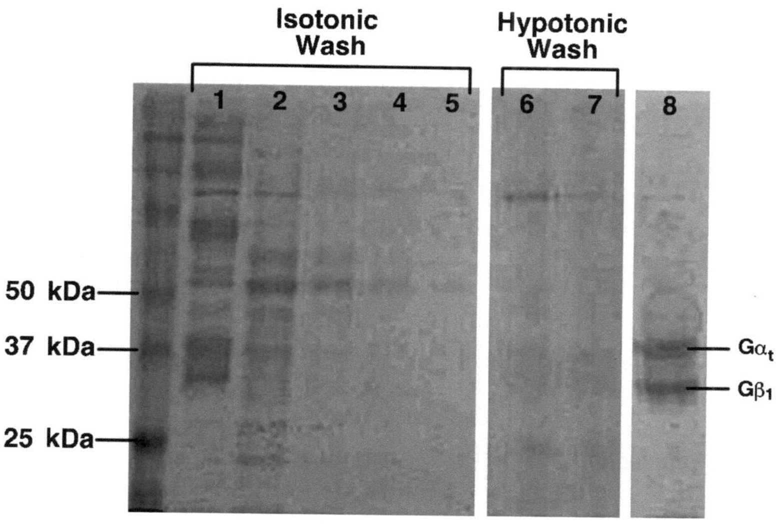

Fig. 9. Purification of G_t. Fractions from a typical purification of G_t are shown on a 10% polyacrylamide gel. *Lane 1*: urea-washed, light-adapted ROS membranes, *Lane 2*: Isotonic Wash 1, *Lane 3*: Isotonic Wash 2, *Lane 4*: Isotonic Wash 3, *Lane 5*: Isotonic Wash 4, *Lane 6*: Hypotonic Wash 1, *Lane 7*: Hypotonic Wash 2, *Lane 8*: Concentrated, buffer-exchanged G_t. Gα runs on the 10% gel as a 38 kDa protein, while Gβ runs as a 36 kDa protein. Gγ is not observed on the gel.

3.10. Expression of $G\alpha_{i1}$

1. $G\alpha_i$ subunits are usually expressed with hexahistidine-tags and from expression vectors with ampicillin-resistance. If your mutant is expressed using an alternative antibiotic resistance marker, use the appropriate antibiotic.
2. Transform BL21 cells with plasmid DNA encoding Gα and plate onto a luria-broth (LB) agar plate containing 100 μg/mL ampicillin. Incubate the plate overnight at 37°C to allow colonies to grow (see Note 37).
3. The day before you begin cell culture, prepare eight 2 L baffled flasks containing 500 mL 2× YT media and one 500-mL baffled flask containing 150 mL of 2× YT media. Autoclave and cool the media to room temperature.
4. In the morning, prepare a starter culture by pipetting 2 mL 2× YT medium out of the 150 mL flask into a 14-mL cell culture tube and supplementing with 100 μg/mL ampicillin. Inoculate this starter culture with a single colony from the overnight LB agar plate, by transferring on a sterile pipette tip or toothpick (see Note 38). Incubate this culture for 8 h at 37°C while shaking.
5. After 8 h, supplement the 500 mL baffled flask containing 150 mL 2× YT media with 100 μg/mL ampicillin and inoculate with your 2 mL starter culture. Incubate this culture overnight, at room temperature (22–25°C), while shaking.

6. On the following morning, supplement each 2-L baffled flask containing 500 mL 2× YT media with 100 μg/mL ampicillin and inoculate with 10 mL of the 150 mL 2× YT overnight culture. Incubate the cultures at room temperature while shaking until optical density at 600 nm (OD_{600}) reaches 0.3–0.4 (see Note 39).
7. Once the OD_{600} reaches the appropriate level, induce protein expression by adding 30 μM IPTG to each flask. Incubate the cultures at room temperature while shaking for 16–18 h (see Note 40).
8. After 18 h, stop the cell growth and harvest cells for purification (see Note 41).

3.11. Purification of Gα_{i1}

1. Distribute overnight cultures (see Subheading 10) into 1-L centrifuge bottles, balance each bottle, and pellet the cells by centrifuging for 15 min at 4°C at 5,000 rpm in Sorvall F9S rotor (or 4,500 RCF(max)).
2. Discard the supernatant and resuspend the cell pellets with 25 mL of Lysis Plus Buffer on ice. Pool the cell resuspension in a sonication beaker on ice and bring the total volume to 100 mL with Lysis Plus Buffer.
3. Lyse the cell resuspension by sonicating at 50% amplitude in 20 s intervals for a total of 4 min. Rest the cells for 40 s between each sonication cycle. Supplement the cell lysate with 0.1 mM PMSF after each minute of sonication. Add 0.1 mM GDP to the lysate, transfer to ultracentrifuge tubes (see Note 16), and balance the tubes with Lysis Plus Buffer. Cap and invert the tubes two to three times to mix the solution. Centrifuge at 4°C for 1 h in a Ti-70 rotor at 50,000 rpm (or 257,000 RCF(max) in a similar rotor). During this time, begin to equilibrate the resin for affinity chromatography (step 5).
4. Collect the supernatant and discard the pellet. The supernatant is your cleared lysate. Supplement the lysate with 0.1 mM PMSF and store on ice if the resin for affinity chromatography is not fully prepared.
5. \Equilibrate 5 mL of Ni-NTA resin or Co^{2+}-TALON resin (10 mL of a 50% slurry) per 4 L of cell culture in Lysis Plus Buffer. To do this, add 5 mL slurry to each of two 50-mL conical tubes, fill each tube to 50 mL with Lysis Plus Buffer, and mix by gently inverting (see Note 42). Pellet the resin by centrifugation at 4°C and 700 RCF(max) for 2 min, discard supernatant, and resuspend each resin pellet with 25 mL (10 × column volume) of Lysis Plus Buffer. Repeat centrifugation and resuspension steps three to five times (see Note 43).

6. Once equilibration of the resin is nearly complete, add DNase and RNase to the cleared lysate and incubate at room temperature for ~5–10 min to digest genomic DNA and decrease the viscosity of your lysate. Filter the lysate through a 0.45-μm filter.
7. Pellet the resin by centrifugation, discard the supernatant, and mix each 2.5 mL aliquot of equilibrated resin with 50 mL lysate in conical tubes for 1 h at 4°C while rotating.
8. Pour each lysate–resin mixture into a separate 20-mL gravity flow column and collect the lysate flowthrough (see Note 44).
9. Wash the tubes containing the filtered lysate–resin mixture with 10 mL of Loading Buffer to recover any remaining protein or resin and add to the gravity flow columns.
10. Add 25 mL (10 × column volume) of Loading Buffer to each column and collect the flowthrough as Wash fraction 1 in a new collection tube.
11. Add 25 mL of Wash Buffer (10 × column volume) to each column and collect the flowthrough as a Wash fraction 2 in a new collection tube.
12. Add 5 mL of Elution Buffer (10 × column volume) to each column and collect the flowthrough as an Elution Fraction 1 in a new collection tube.
13. Pool the elution fractions, transfer to 10-kDa molecular weight cutoff dialysis tubing, and dialyze the sample in 1.8 L of Dialysis Buffer overnight at 4°C with stirring.
14. On the following morning, collected the dialyzed protein sample, and concentrate to 5 mL in a 10-kDa molecular weight cutoff Amicon concentrator. Pass the concentrated, dialyzed protein sample through a 0.22-μm Whatman filter.
15. Connect a column packed with 5 mL of SOURCE™ 15Q resin (GE Healthcare) to an HPLC or FPLC system. Place one buffer inlet into Buffer A, and the other inlet into Buffer B. Wash the attached column with 15 mL (3 × column volume) of Buffer B.
16. Following Buffer B, wash the column with 30 mL (6 × column volume) of Buffer A. Load the filtered protein sample onto a 5 mL sample loop and flow Buffer A over the column at 1.0 mL/min until the entire volume of protein sample has been loaded. While the sample is loading, add 300 μL of Collection Buffer to prechilled test tubes (see Note 45).
17. Flow a linear gradient of Buffer B at 1.0 mL/min so that the final concentration of the buffer flowing over the column at the end of 30 min is 20% Buffer B and 80% Buffer A. Once the gradient reaches 20% Buffer B, flow Buffer B in a linear gradient at 1.0 mL/min so that the final concentration at the end of 10 min is 100% Buffer B and 0% Buffer A (see Note 46).

Collect 500 μL fractions in the prechilled test tubes containing 300 μL of Collection Buffer until the gradient is complete.

18. Use the intrinsic tryptophan fluorescence assay to test the activity of each fraction (see Subheading 12). Pool fractions with a minimum of 40% increase in intrinsic tryptophan activation by AlF_4^- and analyze purity by SDS-PAGE (see Note 47).
19. Subject pooled fractions to Bradford Assay (see Subheading 13) to determine the concentration of the pooled protein. If necessary, concentrate the pooled fractions to 2 mL using a 10-kDa molecular weight cutoff Amicon concentrator. Purified protein can be stored at −80°C as 0.5 mg protein aliquots in 10% glycerol or immediately purified by gel filtration chromatography.
20. For gel filtration chromatography, equilibrate a Superdex 200 10/300 GL gel filtration column in $G\alpha_i$ Buffer (see Notes 48 and 49).
21. Filter the protein sample using a 0.22-μm SpinX filter and load onto your column (see Note 50).
22. Collect 500 μL fractions corresponding to purified Gα, which is expected to elute 15.5 mL after sample injection onto a 24 mL Supderdex 200 10/300 GL gel filtration column (see Note 51). Analyze the fractions for purity by SDS-PAGE (see Note 52) and test for activity using the intrinsic tryptophan fluorescence assay (see Subheading 12). Pure, active protein fractions can then be pooled, concentrated, and frozen at −80°C for long-term storage.
23. See Fig. 10 for expected results of an SDS-PAGE analysis of Gα purification fractions.

3.12. Intrinsic Tryptophan Fluorescence Assay

1. Set excitation wavelength to 280 nm and emission wavelength to 340 nm on a spectrofluorometer.
2. Add 900 μL of Storage Buffer to a quartz cuvette and zero the spectrofluorometer.
3. Add 20 μL of purified protein (~200 nM) to 900 μL of Storage Buffer in the cuvette, mix thoroughly, and record basal fluorescence.
4. Add AlF_4^- by mixing 20 μL of 500 mM NaF (10 mM final concentration) and 8 μL of 10 mM $AlCl_3$ (80 μM final concentration) simultaneously in the cuvette containing protein in Storage Buffer (see Note 53). Mix thoroughly by pipetting and record fluorescence emission at 340 nm.
5. Calculate the percentage increase in fluorescence between the two readings using the following formula:

$$\%\text{increase} = \frac{I(\text{protein} + \text{GDP} \cdot \text{AIF}_4^-) - I(\text{protein})}{I(\text{protein})} \times 100 \quad (6)$$

6. See Fig. 11 for expected results.

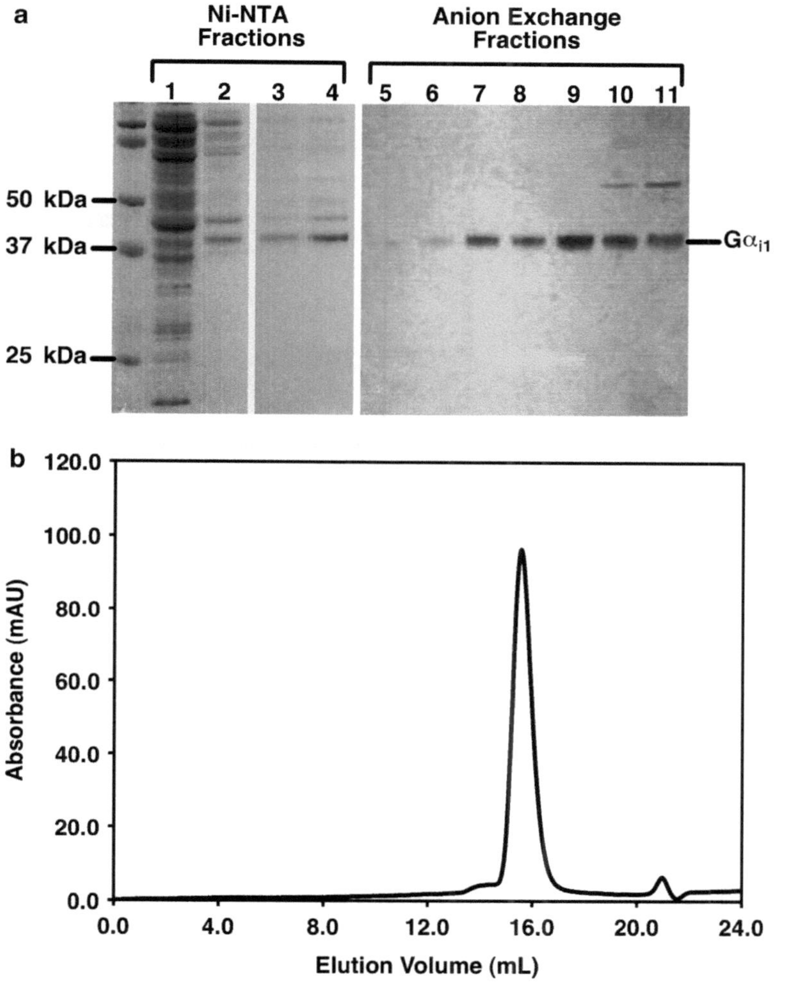

Fig. 10. Purification of $G\alpha_{i1}$. The results from a typical purification of wild-type $G\alpha_{i1}$ expressed in BL21-GOLD Codon Plus are shown here. (**a**) SDS-PAGE of fractions from affinity chromatography and anion exchange chromatography separated on a 4–12% polyacrylamide gel (Invitrogen). *Lane 1*: Wash Fraction 1, *Lane 2*: Wash Fraction 2, *Lane 3*: Elution Fraction 1, *Lane 4*: Dialyzed protein. *Lanes 5–11:* Fractions from the anion exchange purification that were shown to have a 40% increase in their intrinsic tryptophan fluorescence signal. Protein fractions shown in *lanes 7–10* were pooled and concentrated for further purification by gel filtration chromatography. (**b**) Typical chromatogram of wild-type $G\alpha_{i1}$ (100 μL of 1.3 mg/mL) separated on a Superdex 200 10/300 GL gel filtration column equilibrated in 50 mM Tris–Cl pH 8.0, 200 mM NaCl, 2 mM $MgCl_2$, 1 mM EDTA, 1 mM DTT, and 20 μM GDP. Purified $G\alpha_{i1}$ elutes as a well-defined Gaussian peak at 15.5 mL.

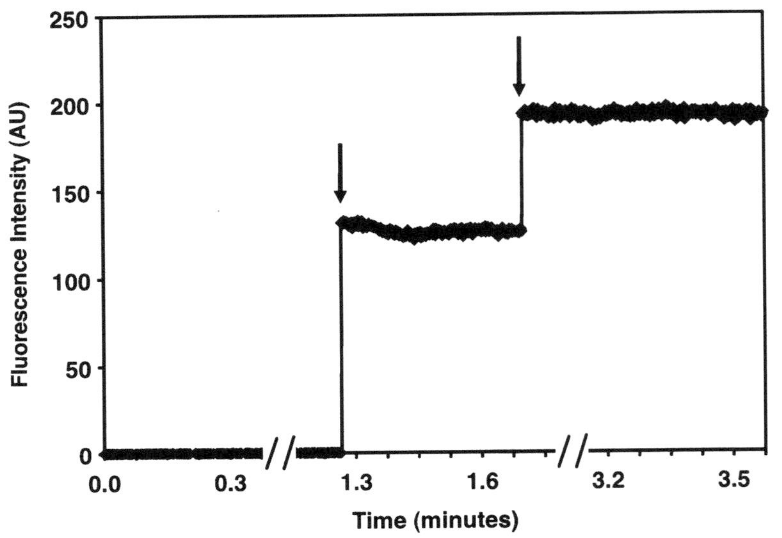

Fig. 11. Intrinsic tryptophan fluorescence of heterologously expressed Gα subunits. The spectrum shows the fluorescence signal from ~200 nM of purified wild-type $G\alpha_{i1}$ in a buffer containing 50 mM Tris–Cl pH 8.0, 50 mM NaCl, 2 mM $MgCl_2$, 1 mM DTT, and 10 μM GDP before and after the addition of AlF_4^-. Purified protein added to Storage Buffer (50 mM Tris–Cl pH 8.0, 50 mM NaCl, 2 mM $MgCl_2$, 1 mM DTT, and 10 μM GDP) results in basal fluorescence (*arrow at 1.3 min*). AlF_4^- is added by bringing the concentration of NaF in the cuvette to 10 mM and $AlCl_3$ to 50 μM. Binding of AlF_4^- to GDP-Gα results in an increase in the fluorescence from Trp^{211} (*arrow at 2 min*). The increase in fluorescence is calculated as the percentage difference in the average fluorescence signal of GDP- versus GDP-AlF4⁻-bound $G\alpha_{i1}$.

3.13. Determining the Concentration of G_t or $G\alpha_{i1}$

1. Warm 1× Quick Start ™ Bradford dye to room temperature (see Notes 54 and 55).
2. Add 20 μL of water or buffer for a blank standard, each BSA standard, and 1:10 or 1:50 dilutions of purified, soluble G_t or $G\alpha_{i1}$ into 1 mL cuvettes (see Notes 56 and 57).
3. Add 1 mL of room temperature 1× Quick Start™ dye to each cuvette. Mix each sample by pipetting and incubate at room temperature for at least 5 min, but no more than 1 h (see Note 58).
4. Measure the absorbance of each sample at 595 nm. Use the blank sample prepared with water or buffer to zero the instrument before recording the absorbance for the standards and purified $G\alpha_{i1}$ or G_t.
5. Use a graphing program to plot the absorbance values and to fit the data to a line with the following equation:

$$\begin{aligned} A_{595} &= mC + b \\ C &= (A_{595} - b)/m \end{aligned} \quad (7)$$

where A_{595} corresponds to the absorbance at a wavelength of 595 nm for each standard, C corresponds to the concentration of each known standard, b is the y-intercept of the line fit to the data collected, and m is the slope of the line.

6. After determining the equation of the line that best fits the absorbance data collected for the known standards, use the recorded A_{595} collected for the protein sample of unknown concentration to determine its concentration (C).

3.14. Crystallization of Gα Proteins

Table 2 summarizes the conditions used to determine the structures of a subset of Gα proteins known to interact with rhodopsin. These can be used to guide your own vapor diffusion experiment using Subheading 7.

3.15. Chemical Cross-Linking of Rhodopsin to G_t

1. Determine the concentration of rhodopsin in dark-adapted ROS membranes that are not urea-washed and resuspend to 10 μM in Buffer A (see Note 59).
2. Light-activate the ROS-G_t sample by incubating on ice for 5 min under ambient light.
3. Add a tenfold molar excess of DTSSP to the light-activated ROS sample and incubate for 30 min on ice.
4. Add 1 M Tris–Cl pH 7.5 to a final concentration of 50 mM and incubate for 15 min at room temperature to terminate the cross-linking reaction.
5. Centrifuge the cross-linked sample at 100,000 RCF(max) for 15 min at 4°C, discard the supernatant, and resuspend the pellet with Buffer A. Repeat membrane pellet wash three times. This removes G_t that is not associated with rhodopsin.
6. After the third wash, resuspend the membrane pellet with Buffer A containing 100 μM GTPγS. Incubate the resuspended membranes for 5 min at 4°C.
7. After 5 min, centrifuge the sample at 100,000 RCF(max) for 15 min at 4°C. Separate the supernatant from the pellet and analyze both by SDS-PAGE (see Fig. 12 for expected results).

3.16. In Vitro Rho–G_t Complex Formation

1. Mix stoichiometric amounts of dark-adapted, ConA-purified rhodopsin with purified G_t and allow the sample to mix for 1 h at 4°C in the dark.
2. Light-activate rhodopsin and induce complex formation by pulsing the sample with bursts of light from a camera flash. Mix the sample by pipetting up and down and expose to a second burst of light from a camera flash. Incubate the light-activated sample for 20 min on ice under ambient light.
3. Filter the sample through a 0.22-μm filter and load the activated complex on a Superdex 200 10/300 GL gel filtration column equilibrated with at least ten column volumes of Rho–G_t Buffer (see Note 60).
4. Monitor the absorbance of eluted species at 280 nm (total protein), 380 nm (meta II), and 440 nm (meta I) to quantify

Table 2
Crystallization of G protein -subunits

	Gα_t-GDP (5)	Gα_t-GDP-AlF$_4^-$ (6)	Gα_t- GTPγS (4)	Gα_{i1}-GDP-AlF$_4^-$ (31)	Gα_{i1}- GTPγS (31)
PDB ID	1TAG	1TAD	1TND	1GFI	1GIL
Space Group	I222	P2$_1$	P2$_1$	P3$_2$21	P3$_2$21
Resolution	1.8 Å	1.7 Å	2.2 Å	2.2 Å	2.3 Å
Method	Hanging Drop	Batch	Hanging Drop	Sitting Drop	Sitting Drop
Protein Concentration	13.3 mg/mL	20 mg/mL	20-25 mg/mL	25-30 mg/mL	25-30 mg/mL
Drop Size	3-5 μL	20 μL	3-5 μL	20 μL	20 μL
Source	Bovine ROS	Bovine ROS	Bovine ROS	Heterologous expression in E. coli	Heterologous expression in E. coli
Protein Buffer	10 mM Tris-Cl pH 7.5, 0.5 mM $MgCl_2$, 50 μM $AlCl_3$, 15 mM NaF, 10% glycerol, 100-200 mM NaCl, 0.1% βME	10 mM Tris-Cl pH 7.5, 0.5 mM $MgCl_2$, 50 μM $AlCl_3$, 15 mM NaF, 10% glycerol, 100-200 mM NaCl, 0.1% βME	5 mM Tris-Cl pH 7.5, 10 mM $MgCl_2$, 25 mM βME, 40% glycerol	300 μM $AlCl_3$, 5 mM NaF, 5 mM $MgCl_2$, 10 mM DTT, 200 mM sodium acetate pH 6.0	300 μM GTPγS, 5 mM NaF, 5 mM $MgCl_2$, 10 mM DTT, 200 mM sodium acetate pH 6.0
Reservoir Solution	9% PEG 8000, 10% glycerol, 50 mM MES pH 7.5, 200 mM $MgCl_2$, 0.1% βME	20% PEG 8000, 200 mM $CaCl_2$, 100 mM sodium cacodylate pH 6.0, 0.2% βME, 5 mM $MgCl_2$, 100 μM $AlCl_3$, 30 mM NaF, 40% glycerol	5% PEG 8000, 250 mM sodium cacodylate pH 6.0, 350 mM $CaCl_2$, 25 mM βME, 30% glycerol	1.8 -1.9 M ammonium sulfite	1.8 -1.9 M ammonium sulfite
Crystallization Temperature	4°C	4°C	-12.5°C	21°C	21°C

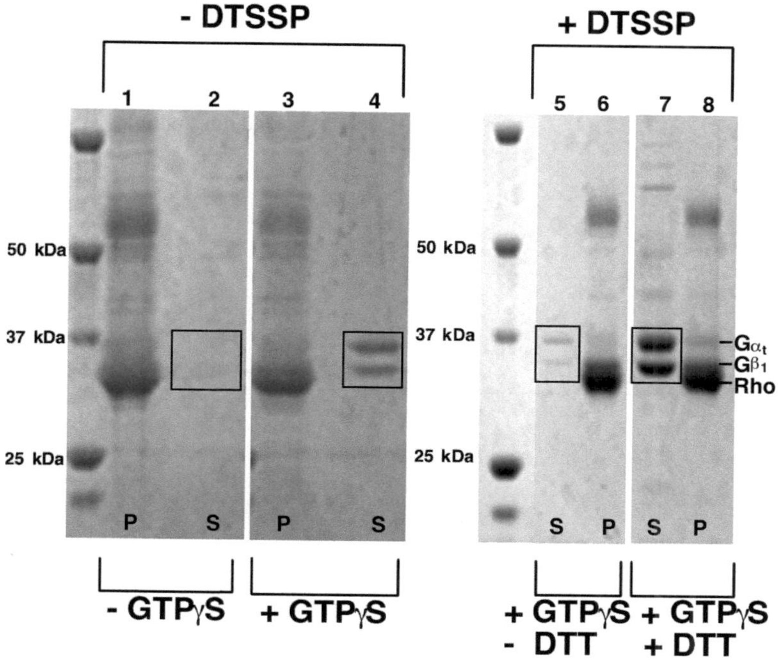

Fig. 12. Chemical cross-linking of Rhodopsin–G_t. Dark-adapted ROS membranes were resuspended to 10 μM total protein and photoactivated to form the rhodopsin–G_t complex in the presence and absence of the cross-linking agent DTSSP. Samples were centrifuged to separate the membrane fraction from the soluble fraction and each analyzed by SDS-PAGE on a 4–12% gradient gel (Invitrogen). The cross-linked sample does not migrate into the gel, and the formation of the cross-link is verified by monitoring disappearance of the $G\alpha_t$ and $G\beta_{t1}$ bands. *Lanes 1–4* contain samples that were prepared without the cross-linker DTSSP. *Lanes 5 and 6* contain samples that were incubated with DTSSP. *Lanes 7 and 8* contain samples that were incubated with DTSSP, but were cleaved with DTT. Lanes marked with S represent the supernatant fractions. Lanes marked with P represent the pellet fraction. *Lane 1*: dark-adapted ROS membranes. The bands corresponding to $G\alpha_t$ and $G\beta_{t1}$ are present but are obscured by other proteins. *Lane 2*: supernatant after three washes in Buffer A. $G\alpha_t$ and $G\beta_{t1}$ are not released into the supernatent as they are in a high-affinity complex with rhodopsin. Their expected position is noted in *Box 1*. *Lane 3*: ROS-membranes after three washes in Buffer A containing 100 μM GTPγS. These are now depleted of $G\alpha_t$ and $G\beta_{t1}$. *Lane 4*: supernatant after three washes in Buffer A containing 100 μM GTPγS linking should not be in bold. The addition of GTPγS to light-activated ROS disassociates the rhodopsin–G_t complex. In the absence of a cross-linking agent, G_{t1} is released into the supernatent, and $G\alpha_t$ and $G\beta_t$ are observed on the gel (*Box 2*) *Lane 5*: supernatant from a DTSSP-treated sample after three washes in Buffer A containing 100 μM GTPγS. The covalent cross-link prevents disassociation of the complex after GTPγS, and not much G_t is released (*Box 3*). *Lane 6*: DTSSP-treated ROS membranes after three washes in Buffer A containing 100 μM GTPγS. *Lane 7*: supernatant of a DTSSP-treated sample after incubation with 50 mM DTT and three washes in Buffer A containing 100 μM GTPγS. As in *lane 4*, G_t is released from the complex upon treatment with GTPγS, and robust bands corresponding to $G\alpha_t$ and $G\beta_t$ are observed on the gel (*Box 4*). *Lane 8*: DTSSP-treated ROS membranes after incubation with 50 mM DTT and three washes in Buffer A containing 100 μM GTPγS.

the completeness of rhodopsin activation and to separate heterogeneous species (see Fig. 13a for expected results).

5. Collect 500 μL fractions corresponding to purified Rhodopsin–G_t complex. The 2:1 complex is expected to elute at 11.5 mL and the 1:1 complex is expected to elute at 12.6 mL on a Superdex 200 10/300 GL column equilibrated in Rho–G_t Buffer (see Note 61). Analyze the fractions by SDS-PAGE (see Fig. 13b for expected results) to verify the purity and presence of all components of the complex (see

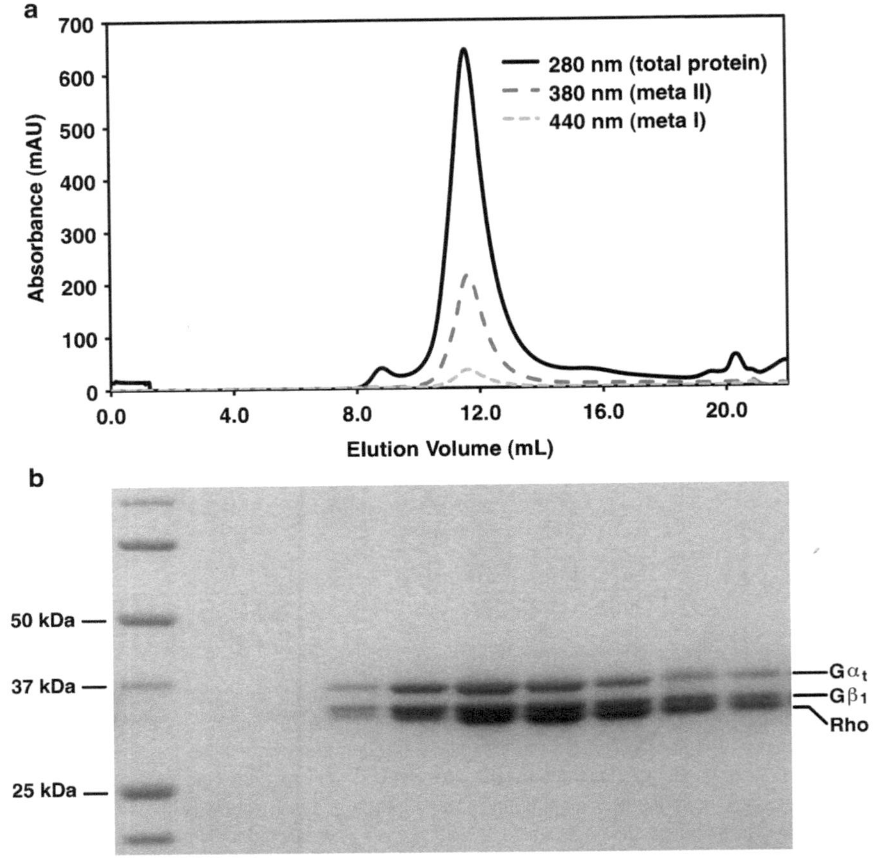

Fig. 13. In vitro formation of the rhodopsin–G_t complex. Purified G_t and purified rhodopsin in 0.5 mM DDM were mixed in equimolar amounts. The sample was light activated, incubated under ambient light for 20 min, and analyzed by gel filtration chromatography on a Superdex 200 10/300 GL size-exclusion column equilibrated in 50 mM HEPES pH 8.0, 100 mM NaCl, 1 mM $MgCl_2$, 1 mM EDTA, and 0.5 mM DDM. (**a**) The rhodopsin:G_t complex elutes as a homogeneous, Gaussian peak at 11.5 mL. (**b**) Fractions collected across the peak in (**a**) analyzed on a 4–12% gradient polyacrylamide gel (Invitrogen). Rhodopsin, $G\alpha_t$, and $G\beta_1$ are observed.

Notes 52 and 62). Analyze the activity of the complex using the meta II assay (see Subheading 6) or the GTPγS exchange assay (see Subheading 17).

3.17. Receptor-Catalyzed GTPγS Exchange Assay

The rate of receptor-catalyzed nucleotide exchange can be measured by monitoring the increase in intrinsic tryptophan fluorescence in heterotrimeric G proteins in the presence of light-activated rhodopsin. This assay can be applied to G_t or recombinant $G\alpha_i$ reconstituted with $G\beta\gamma$ (see Note 63).

1. Mix 200 nM G_t or 200 nM $G\alpha_i$ mixed with 200 nM $G\beta\gamma$ in a cuvette containing Activation Buffer (see Note 64).
2. Set the excitation wavelength to 280 nm and emission wavelength to 340 nm.

a

	Initial Crystallization Condition:	Optimization Screen:
Buffer	50 mM Tris-Cl pH 7.5	50 mM Tris-Cl pH 7.5
Salt	200 mM NaCl	200 - 275 mM NaCl
Precipitant	25% PEG 2000	25-35% PEG 2000

b

	1	2	3	4	5	6
A	50 mM Tris-Cl pH 7.5, 200 mM NaCl, 25% PEG 2000	50 mM Tris-Cl pH 7.5, 200 mM NaCl, 27% PEG 2000	50 mM Tris-Cl pH 7.5, 200 mM NaCl, 29% PEG 2000	50 mM Tris-Cl pH 7.5, 200 mM NaCl, 31% PEG 2000	50 mM Tris-Cl pH 7.5, 200 mM NaCl, 33% PEG 2000	50 mM Tris-Cl pH 7.5, 200 mM NaCl, 35% PEG 2000
B	50 mM Tris-Cl pH 7.5, 225 mM NaCl, 25% PEG 2000	50 mM Tris-Cl pH 7.5, 225 mM NaCl, 27% PEG 2000	50 mM Tris-Cl pH 7.5, 225 mM NaCl, 29% PEG 2000	50 mM Tris-Cl pH 7.5, 225 mM NaCl, 31% PEG 2000	50 mM Tris-Cl pH 7.5, 225 mM NaCl, 33% PEG 2000	50 mM Tris-Cl pH 7.5, 225 mM NaCl, 35% PEG 2000
C	50 mM Tris-Cl pH 7.5, 250 mM NaCl, 25% PEG 2000	50 mM Tris-Cl pH 7.5, 250 mM NaCl, 27% PEG 2000	50 mM Tris-Cl pH 7.5, 250 mM NaCl, 29% PEG 2000	50 mM Tris-Cl pH 7.5, 250 mM NaCl, 31% PEG 2000	50 mM Tris-Cl pH 7.5, 250 mM NaCl, 33% PEG 2000	50 mM Tris-Cl pH 7.5, 250 mM NaCl, 35% PEG 2000
D	50 mM Tris-Cl pH 7.5, 275 mM NaCl, 25% PEG 2000	50 mM Tris-Cl pH 7.5, 275 mM NaCl, 27% PEG 2000	50 mM Tris-Cl pH 7.5, 275 mM NaCl, 29% PEG 2000	50 mM Tris-Cl pH 7.5, 275 mM NaCl, 31% PEG 2000	50 mM Tris-Cl pH 7.5, 275 mM NaCl, 33% PEG 2000	50 mM Tris-Cl pH 7.5, 275 mM NaCl, 35% PEG 2000

Fig. 14. Example optimization of a crystallization condition. Small variations in stock solutions between individuals and lab-to-lab variations in temperature, and techniques often require that crystallization conditions undergo optimization if the crystals are to be replicated from a published protocol. Crystallization conditions are usually optimized by varying the concentration or identity of each crystallization component in a stepwise manner. (**a**) Example crystallization conditions for a hypothetical protein. (**b**) Example of a 24-well screen optimizing the crystallization conditions in (**a**). These conditions sample concentrations of reagents comprising the reservoir solution above and below those of the initial crystallization condition.

3. Add 10 μM GTPγS and 100 nM purified rhodopsin, mix thoroughly by pipetting, and collect the emission spectrum at 340 nm for 40 min at 21°C.
4. Normalize the data to the baseline (0%) and the fluorescence maximum (100%).
5. Fit the normalized data to an exponential association curve to determine the receptor-catalyzed nucleotide exchange rate (see Fig. 14 for expected results).

4. Notes

1. Bovine eyes can be obtained from your local slaughterhouse.
2. All solutions that include volatile reagents (DTT or βME) or protease inhibitors with short half-lives (PMSF) should be prepared fresh each time the solution is made.
3. PMSF rapidly degrades in aqueous solutions where its half-life is approximately 110 min (56). Prepare stock solutions of this protease inhibitor in organic solvents such as acetone or isopropanol.
4. The pH of all buffers containing Tris–Cl listed in protocols throughout this chapter should be adjusted with HCl.
5. Many detergents are available in two grades. SOL-grade detergents are less pure and less expensive. These can be used for solubilization of membrane proteins but may be too heterogeneous to support crystallization. ANAGRADE detergents are of higher purity and should be used for techniques requiring homogeneous solutions, such as X-ray crystallography.
6. All buffers used for chromatographic techniques should be filtered to remove large particles and dust. Buffers containing detergents should be filtered in sufficiently large bottles (use a 1-L bottle for filtering 500 mL of buffer) so that the bubbles generated during filtering will not be suctioned into the vacuum pump. If you are using Millipore Centricon® filters, add DDM after filtering. DDM will concentrate in Millipore Centricon® filters, which can affect the final concentration of detergent in your buffer.
7. It is important to use high-quality, siliconized or plastic coverslips versus nonsiliconized glass coverslips. While more expensive, the hydrophobic surface of siliconized coverslips promotes protein droplets more conducive to crystallization.
8. GTP can be hydrolyzed in aqueous solutions. Prepare this buffer right before use and add GTP from a fresh stock.
9. Ampicillin is usually prepared as a 1,000× (100 mg/mL) stock solution that can withstand long-term storage at −20°C. The stock solution can be made in either a water solution or a 70% ethanol solution. The latter will not freeze at −20°C, thus decreasing the wait time for thawing the reagent during experiments.
10. The rhodopsin-G_t complex retinal is light sensitive. Protect the gel filtration column from ambient light by performing gel filtration experiment under dim red light or by wrapping the column in aluminum foil.
11. The half-life of a rhodopsin–G_t complex in the presence of detergent micelles is approximately 30 min. Supplementing

the gel filtration buffer or eluted protein fractions with all-*trans* retinal can improve the half-life of the complex (57).

12. If the volume of the retinal slurry exceeds the volume held by the maximum number of centrifuge tubes that will fit in your rotor, leave the pellet intact, layer remaining slurry on top of the pellet, and repeat the centrifugation step until all of the retinal slurry is centrifuged as described. Pool the fresh supernatant with supernatant from the previous round of centrifugation.
13. Sucrose Solution 1 is the least dense, while Sucrose Solution 3 is the densest. To ensure that you maintain a distinct interface, be slow to dispense each sucrose solution into each of the centrifuge tubes and slow to remove the pipette tip from the centrifuge tube once the solution has been dispensed.
14. The sucrose layer below the ROS membrane layer and the pellet at the bottom of the centrifuge tube contain contaminants. To avoid these contaminants, remove the top layer by pipetting slowly and without disrupting the gradient.
15. Homogenize all resuspended fractions by inserting and removing the plunger into the glass homogenizer at least ten times. This will ensure adequate separation of the ROS membranes from other retinal components.
16. Extra caution is required when balancing ultracentrifuge tubes. The weight of each tube should be within 0.1 g of each other.
17. The pellet in this step of the purification contains the ROS membranes. It is very soft and may decant with the supernatant. Instead of decanting, pipette the supernatant off of the pellet to control the amount of pellet loss.
18. Detergent in this step of the membrane wash will extract rhodopsin from the membrane pellet and result in a loss of protein. It is important to use a solution without any detergent when washing this membrane pellet.
19. The volume of Solubilization Buffer should be proportional to the initial volume of urea-washed ROS membranes used in your ConA purification so that the final concentration of DDM after receptor solubilization is less than 40 mM.
20. Be careful to avoid getting air on the column, which can damage the resin and prevent the elution buffer from accessing the air pockets (thus decreasing yield). Attach the syringe to the column using a "wet-connect" technique by adding buffer to an adaptor connected to the top of the column and filling the syringe tube with a small volume of buffer once it is connected to the column before the column plug on the bottom end is removed.
21. It is important to equilibrate the ConA resin for at least 2 days in a buffer containing 0.5 mM (or more) DDM before it is

used for the first time (SOL-GRADE detergent is acceptable for this). Most chromatography columns contain hydrophobic binding sites that will strip the detergent micelles from protein if they are not filled. This results in aggregation of membrane proteins and large decreases in yield.

22. The detergent concentration in the solubilized rhodopsin sample needs to be sufficiently decreased by dilution with ConA Binding Buffer. While the effects of high detergent concentrations on ConA affinity chromatography are unknown, it is possible that the high concentration of DDM may interfere with rhodopsin–ConA interactions.
23. The rhodopsin-containing load fraction needs to be looped over the ConA column during the loading step because the kinetics of binding are slow.
24. ConA-purified rhodopsin samples contain the sugars α-D methylglucoside and sucrose, which can stick to concentrator filters. During concentration, wash the sides of the filter with the concentrated protein sample before refilling with unconcentrated, purified protein to minimize the amount of protein stuck to the filter.
25. While the sample is being light-activated with bursts of bright light, remember to close your eyes to preserve night-vision.
26. The concentration measured in this experiment is for 4 μL of protein diluted to 100 μL (1:25 dilution). Be sure to account for such dilution factors when calculating the concentration. Multiply the concentration obtained using Eq. 3 by 25 to obtain the actual rhodopsin concentration in your undiluted sample.
27. This photoactivation technique activates only 10% of the total rhodopsin population in your sample.
28. Filter all buffers and protein samples before using in the vapor diffusion experiment to remove dust particles.
29. The temperature of the reagents should match that of the crystallization system. Drastic changes in temperature often interfere with crystallization.
30. Be careful to avoid air bubbles during pipetting. Watching your pipette as solutions are dispensed and stopping before air is dispensed can prevent bubble formation in your protein drop. If you do introduce an air bubble into the crystallization reaction, these can be removed by touching the droplet gently with a clean, dry pipette tip.
31. Crystallization can be temperature sensitive. If you do not have a dedicated crystallization incubator, place the crystallization trays into a room with low temperature variation and away from windows, as sunlight can cause dramatic local temperature changes.
32. Physical vibrations, such as those from air compressors, can impair crystallization. Store crystallization trays in a room that

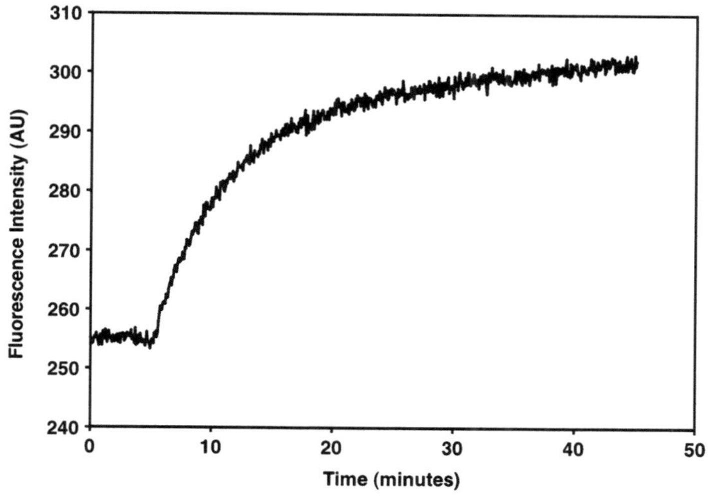

Fig. 15. Receptor-catalyzed GTPγS exchange. A fluorescence spectrum obtained for 200 nM G_t activated by 100 nM purified rhodopsin in the presence of 10 μM GTPγS. Excitation is at 280 nm, and emission is at 340 nm. The baseline is collected from 0 to 10 min (fluorescence in the absence of GTPγS). Upon addition 10 μM GTPγS, the fluorescence signal increases until G_t is maximally activated (~30 min).

lacks signification vibrations. Avoid disrupting trays for at least 1 day after having set them up.

33. Because rhodopsin undergoes a conformational change upon activation, dark-adapted rhodopsin crystals must be examined under dim red light, either by using a red filter fitted to a light microscope or with an infrared camera.

34. The crystallization condition for obtaining the highest quality crystals can be optimized by slightly varying the components in the condition. This includes testing alternative precipitants, alternative salts, buffer pH, temperature, concentrations of each chemical, protein concentrations, hanging drop size, and the method of crystallization used. For example, in the hanging drop vapor diffusion example given in Fig. 15a, the buffer identity and concentration are held constant, but salt and precipitant concentrations are varied. Condition optimization is often performed in 24-well crystallization trays with each well containing a unique crystallization condition. Figure 15b is an example of a optimization screen.

35. While the majority of protein crystals exhibit the best diffraction when they are large, the best crystals of rhodopsin were very small and needle-shaped. The diffraction from rhodopsin crystals is inconsistent, with only a small percentage of optically equivalent crystals exhibiting reasonable diffraction.

36. If G_t is to be separated into GDP-bound $G\alpha_t$ and $G\beta\gamma$, concentrate the G_t-containing supernatant to ~7–8 mL. Add $MgSO_4$ to a final concentration of 40 mM and purify the protein sample on a Blue Sepharose gel filtration column

(GE Healthcare). Collect the fractions corresponding to Gα and Gβγ separately, and analyze their purity by SDS-PAGE.

37. It is preferable to use cells that have been transformed with fresh plasmid DNA (less than 3 months old) immediately before cell expression. Both glycerol stocks of transformed cells and older DNA (even stored at −20°C) commonly result in lower protein yields for unknown reasons.
38. Wrap parafilm around the bacterial plate, to keep moisture and air out, and store at 4°C. Colonies from this plate can be used for protein growths for up to 1 week. After 1 week, yields may be compromised.
39. Most cell cultures of Gα proteins expressed in BL-21 *E. coli* will reach an OD_{600} of 0.3–0.4 within 2–3 h postinoculation when grown at room temperature; however, this is dependent on the Gα mutant and expression system used. Monitor the absorbance carefully, as late induction can affect the yield.
40. Induction for longer than 18 h decreases the yield.
41. Cell pellets can be stored at −80°C. Freezing and thawing the pellets has been shown to actually improve the protein yield by contributing to cell lysis.
42. When resuspending the resin, gently invert the bottle until all of the resin is in solution. Shaking can cause the resin to break apart into small pieces, called fines, which can clog fritted filters and interfere with affinity chromatography. To remove fines, remix the resin in the desired buffer and allow it to settle for ~5 min. This is enough time to allow the beads to settle, but the small particles remain in solution. Decant the resin supernatant, add fresh buffer, and remix by inverting. Repeat the process until the resin settles within 5 min and leaves behind a clear supernatant.
43. Commercially available resin is usually stored in ethanol, which will precipitate the protein if it is not completely exchanged prior to use. Perform three to five successive washes of the resin slurry using ten column volumes of Loading Buffer per wash or by continuous flow of Loading Buffer over resin in a gravity flow column to remove the ethanol.
44. While G proteins are stable at room temperature, it is beneficial to perform all steps of purification at 4°C.
45. If anion exchange is performed at room temperature, be sure to collect protein fractions on ice.
46. A slow, low salt buffer gradient followed by a fast, high-salt buffer gradient is needed to separate Gα proteins from contaminants by anion exchange chromatography.
47. Purified Gα is best resolved on 10% polyacrylamide gels.

48. Filter all buffers and protein samples prior to gel filtration chromatography as large particles and aggregates can irreversibly damage both the column and the instrument. Small volumes of buffer or protein can be spin-filtered on a Spin-X column, while large volumes can be vacuum-filtered.
49. $G\alpha$ subunits are more stable in Tris–Cl buffers than HEPES buffers. The inclusion of a guanine nucleotide is critical for the stability of the $G\alpha$ subunit, and GDP is most frequently used.
50. For achieving the best resolution, use a sample volume of 500 μL or less for gel filtration purification of proteins on a 24 mL Superdex 200 10/300 GL gel filtration column.
51. $G\alpha_i$ purified on a column equilibrated in 50 mM Tris–Cl pH 7.5 buffer containing 200 mM NaCl will elute at 15.5 mL. This is not necessarily the case for $G\alpha_i$ purified in other buffers. If using other buffers, the gel filtration column should be calibrated in that buffer using standards.
52. The G_t heterotrimer and $G\alpha_i$ are both best resolved on a 10% polyacrylamide gel. Of the three G_t subunits, $G\gamma$ is the smallest (approximately 8 kDa) and is not observed on 10% acrylamide gels.
53. AlF_4^- is highly unstable in solution for extended periods of time and must be formed by mixing $AlCl_3$ and NaF immediately before intrinsic tryptophan fluorescence is measured.
54. In the case of limiting quantities of protein, this assay can also be performed in a 96-well plate using 5 μL of protein and 250 μL of 1× Quick Start™ Bradford dye.
55. Mix the dye thoroughly by inverting the bottle multiple times before use so that all samples develop evenly.
56. Use a clean pipette tip for each sample to avoid cross-contamination, as this will affect the fit of a regression line to your data.
57. In order to determine the appropriate dilution of your protein of unknown concentration, try loading different dilutions of the purified sample on a gel and analyze by SDS-PAGE. Compare the intensity of each dilution to that of a protein sample of known concentration. You will want to pick a dilution that falls into the range of the protein standard concentrations.
58. The Bradford dye is light sensitive. Keep the samples protected from ambient light during the incubation step.
59. If substituting buffers, do not use Tris–Cl or any buffer that contains a primary amine, as these will interfere with the cross-linking reaction.
60. For the reasons discussed in Note 21, it is important to equilibrate the gel filtration column in buffer containing

detergent for a minimum of 48 h prior to the first use. To save money on detergent, a fixed volume of buffer may be looped over the column.

61. The given elution volumes are for rhodopsin–G_t complexes purified on a Superdex 200 10/300 GL gel filtration column equilibrated in Rho–G_t Buffer. This is not necessarily the volume at which the complex will elute if purified in different buffers or different detergents.
62. Many membrane proteins will aggregate upon boiling in SDS-PAGE buffer. Room temperature incubation is sufficient.
63. To reconstitute G_t using recombinant Gα and G$\beta\gamma$, first mix purified proteins in equimolar amounts and incubate on ice for 10 min prior to addition to the sample cuvette containing buffer.
64. The Activation Buffer should not contain GDP. The presence of free GDP in solution will inhibit GTPγS binding.

References

1. Scheerer, P., Park, J. H., Hildebrand, P. W., Kim, Y. J., Krauss, N., Choe, H. W., Hofmann, K. P., and Ernst, O. P. (2008) Crystal structure of opsin in its G-protein-interacting conformation, *Nature 455*, 497–502.
2. Park, J. H., Scheerer, P., Hofmann, K. P., Choe, H. W., and Ernst, O. P. (2008) Crystal structure of the ligand-free G-protein-coupled receptor opsin, *Nature 454*, 183–187.
3. Hamm, H. E., Deretic, D., Arendt, A., Hargrave, P. A., Koenig, B., and Hofmann, K. P. (1988) Site of G protein binding to rhodopsin mapped with synthetic peptides from the alpha subunit, *Science 241*, 832–835.
4. Noel, J. P., Hamm, H. E., and Sigler, P. B. (1993) The 2.2 A crystal structure of transducin-alpha complexed with GTP gamma S, *Nature 366*, 654–663.
5. Lambright, D. G., Sondek, J., Bohm, A., Skiba, N. P., Hamm, H. E., and Sigler, P. B. (1996) The 2.0 angstrom crystal structure of a heterotrimeric G protein, *Nature 379*, 311–319.
6. Sondek, J., Lambright, D. G., Noel, J. P., Hamm, H. E., and Sigler, P. B. (1994) GTPase mechanism of Gproteins from the 1.7-A crystal structure of transducin alpha-GDP-AIF-4, *Nature 372*, 276–279.
7. Van Eps, N., Oldham, W. M., Hamm, H. E., and Hubbell, W. L. (2006) Structural and dynamical changes in an alpha-subunit of a heterotrimeric G protein along the activation pathway, *Proc Natl Acad Sci U S A 103*, 16194–16199.
8. Millar, R. P., and Newton, C. L. (2010) The year in G protein-coupled receptor research, *Mol Endocrinol 24*, 261–274.
9. Hargrave, P. A., McDowell, J. H., Curtis, D. R., Wang, J. K., Juszczak, E., Fong, S. L., Rao, J. K., and Argos, P. (1983) The structure of bovine rhodopsin, *Biophys Struct Mech 9*, 235–244.
10. Schertler, G. F., Villa, C., and Henderson, R. (1993) Projection structure of rhodopsin, *Nature 362*, 770–772.
11. Sakmar, T. P., Franke, R. R., and Khorana, H. G. (1991) The role of the retinylidene Schiff base counterion in rhodopsin in determining wavelength absorbance and Schiff base pKa, *Proc Natl Acad Sci U S A 88*, 3079–3083.
12. Karnik, S. S., Sakmar, T. P., Chen, H. B., and Khorana, H. G. (1988) Cysteine residues 110 and 187 are essential for the formation of correct structure in bovine rhodopsin, *Proc Natl Acad Sci U S A 85*, 8459–8463.
13. Ballesteros, J. A., Jensen, A. D., Liapakis, G., Rasmussen, S. G., Shi, L., Gether, U., and Javitch, J. A. (2001) Activation of the beta 2-adrenergic receptor involves disruption of an ionic lock between the cytoplasmic ends of transmembrane segments 3 and 6, *J Biol Chem 276*, 29171–29177.
14. Shi, L., Liapakis, G., Xu, R., Guarnieri, F., Ballesteros, J. A., and Javitch, J. A. (2002) Beta2 adrenergic receptor activation. Modulation of the proline kink in transmembrane 6 by

a rotamer toggle switch, *J Biol Chem 277*, 40989–40996.

15. Mirzadegan, T., Benko, G., Filipek, S., and Palczewski, K. (2003) Sequence analyses of G-protein-coupled receptors: similarities to rhodopsin, *Biochemistry 42*, 2759–2767.
16. Papermaster, D. S., and Dreyer, W. J. (1974) Rhodopsin content in the outer segment membranes of bovine and frog retinal rods, *Biochemistry 13*, 2438–2444.
17. Palczewski, K., Kumasaka, T., Hori, T., Behnke, C. A., Motoshima, H., Fox, B. A., Le Trong, I., Teller, D. C., Okada, T., Stenkamp, R. E., Yamamoto, M., and Miyano, M. (2000) Crystal structure of rhodopsin: A G protein-coupled receptor, *Science 289*, 739–745.
18. Cherezov, V., Rosenbaum, D. M., Hanson, M. A., Rasmussen, S. G. F., Thian, F. S., Kobilka, T. S., Choi, H. J., Kuhn, P., Weis, W. I., Kobilka, B. K., and Stevens, R. C. (2007) High-resolution crystal structure of an engineered human beta(2)-adrenergic G protein-coupled receptor, *Science 318*, 1258–1265.
19. Jaakola, V. P., Griffith, M. T., Hanson, M. A., Cherezov, V., Chien, E. Y. T., Lane, J. R., IJzerman, A. P., and Stevens, R. C. (2008) The 2.6 Angstrom Crystal Structure of a Human A(2A) Adenosine Receptor Bound to an Antagonist, *Science 322*, 1211–1217.
20. Murakami, M., and Kouyama, T. (2008) Crystal structure of squid rhodopsin, *Nature 453*, 363–367.
21. Warne, T., Serrano-Vega, M. J., Baker, J. G., Moukhametzianov, R., Edwards, P. C., Henderson, R., Leslie, A. G. W., Tate, C. G., and Schertler, G. F. X. (2008) Structure of a beta (1)-adrenergic G-protein-coupled receptor, *Nature 454*, 486-U482.
22. Wu, B., Chien, E. Y., Mol, C. D., Fenalti, G., Liu, W., Katritch, V., Abagyan, R., Brooun, A., Wells, P., Bi, F. C., Hamel, D. J., Kuhn, P., Handel, T. M., Cherezov, V., and Stevens, R. C. (2010) Structures of the CXCR4 chemokine GPCR with small-molecule and cyclic peptide antagonists, *Science 330*, 1066–1071.
23. Unger, V. M., and Schertler, G. F. (1995) Low resolution structure of bovine rhodopsin determined by electron cryo-microscopy, *Biophys J 68*, 1776–1786.
24. Farrens, D. L., Altenbach, C., Yang, K., Hubbell, W. L., and Khorana, H. G. (1996) Requirement of rigid-body motion of transmembrane helices for light activation of rhodopsin, *Science 274*, 768–770.
25. Altenbach, C., Kusnetzow, A. K., Ernst, O. P., Hofmann, K. P., and Hubbell, W. L. (2008) High-resolution distance mapping in rhodopsin reveals the pattern of helix movement due to activation, *Proc Natl Acad Sci U S A 105*, 7439–7444.
26. Fung, B. K., Hurley, J. B., and Stryer, L. (1981) Flow of information in the light-triggered cyclic nucleotide cascade of vision, *Proc Natl Acad Sci U S A 78*, 152–156.
27. Gilman, A. G. (1987) G-Proteins - Transducers of Receptor-Generated Signals, *Annual Review of Biochemistry 56*, 615–649.
28. Downes, G. B., and Gautam, N. (1999) The G protein subunit gene families, *Genomics 62*, 544–552.
29. Oldham, W. M., Van Eps, N., Preininger, A. M., Hubbell, W. L., and Hamm, H. E. (2006) Mechanism of the receptor-catalyzed activation of heterotrimeric G proteins, *Nature Structural & Molecular Biology 13*, 772–777.
30. Tesmer, J. J., Sunahara, R. K., Gilman, A. G., and Sprang, S. R. (1997) Crystal structure of the catalytic domains of adenylyl cyclase in a complex with Gsalpha.GTPgammaS, *Science 278*, 1907–1916.
31. Coleman, D. E., Berghuis, A. M., Lee, E., Linder, M. E., Gilman, A. G., and Sprang, S. R. (1994) Structures of active conformations of Gi alpha 1 and the mechanism of GTP hydrolysis, *Science 265*, 1405–1412.
32. Mixon, M. B., Lee, E., Coleman, D. E., Berghuis, A. M., Gilman, A. G., and Sprang, S. R. (1995) Tertiary and quaternary structural changes in Gi alpha 1 induced by GTP hydrolysis, *Science 270*, 954–960.
33. Wall, M. A., Coleman, D. E., Lee, E., Iniguez-Lluhi, J. A., Posner, B. A., Gilman, A. G., and Sprang, S. R. (1995) The structure of the G protein heterotrimer Gi alpha 1 beta 1 gamma 2, *Cell 83*, 1047–1058.
34. Coleman, D. E., and Sprang, S. R. (1998) Crystal structures of the G protein Gi alpha 1 complexed with GDP and Mg2+: a crystallographic titration experiment, *Biochemistry 37*, 14376–14385.
35. Denker, B. M., Schmidt, C. J., and Neer, E. J. (1992) Promotion of the GTP-liganded state of the Go alpha protein by deletion of the C terminus, *J Biol Chem 267*, 9998–10002.
36. Denker, B. M., Boutin, P. M., and Neer, E. J. (1995) Interactions between the amino- and carboxyl-terminal regions of G alpha subunits: analysis of mutated G alpha o/G alpha i2 chimeras, *Biochemistry 34*, 5544–5553.
37. Grishina, G., and Berlot, C. H. (2000) A surface-exposed region of G(salpha) in which substitutions decrease receptor-mediated activation and increase receptor affinity, *Mol Pharmacol 57*, 1081–1092.

38. Preininger, A. M., Parello, J., Meier, S. M., Liao, G., and Hamm, H. E. (2008) Receptor-mediated changes at the myristoylated amino terminus of Galpha(il) proteins, *Biochemistry 47*, 10281–10293.
39. Kisselev, O. G., Ermolaeva, M. V., and Gautam, N. (1994) A farnesylated domain in the G protein gamma subunit is a specific determinant of receptor coupling, *J Biol Chem 269*, 21399–21402.
40. Martin, E. L., Rens-Domiano, S., Schatz, P. J., and Hamm, H. E. (1996) Potent peptide analogues of a G protein receptor-binding region obtained with a combinatorial library, *J Biol Chem 271*, 361–366.
41. Rasenick, M. M., Watanabe, M., Lazarevic, M. B., Hatta, S., and Hamm, H. E. (1994) Synthetic peptides as probes for G protein function. Carboxyl-terminal G alpha s peptides mimic Gs and evoke high affinity agonist binding to beta-adrenergic receptors, *J Biol Chem 269*, 21519–21525.
42. Taylor, J. M., Jacob-Mosier, G. G., Lawton, R. G., Remmers, A. E., and Neubig, R. R. (1994) Binding of an alpha 2 adrenergic receptor third intracellular loop peptide to G beta and the amino terminus of G alpha, *J Biol Chem 269*, 27618–27624.
43. Taylor, J. M., Jacob-Mosier, G. G., Lawton, R. G., VanDort, M., and Neubig, R. R. (1996) Receptor and membrane interaction sites on Gbeta. A receptor-derived peptide binds to the carboxyl terminus, *J Biol Chem 271*, 3336–3339.
44. Itoh, Y., Cai, K., and Khorana, H. G. (2001) Mapping of contact sites in complex formation between light-activated rhodopsin and transducin by covalent crosslinking: use of a chemically preactivated reagent, *Proc Natl Acad Sci U S A 98*, 4883–4887.
45. Cai, K., Itoh, Y., and Khorana, H. G. (2001) Mapping of contact sites in complex formation between transducin and light-activated rhodopsin by covalent crosslinking: use of a photoactivatable reagent, *Proc Natl Acad Sci U S A 98*, 4877–4882.
46. Lichtarge, O., Bourne, H. R., and Cohen, F. E. (1996) Evolutionarily conserved Galphabetagamma binding surfaces support a model of the G protein-receptor complex, *Proc Natl Acad Sci U S A 93*, 7507–7511.
47. Onrust, R., Herzmark, P., Chi, P., Garcia, P. D., Lichtarge, O., Kingsley, C., and Bourne, H. R. (1997) Receptor and betagamma binding sites in the alpha subunit of the retinal G protein transducin, *Science 275*, 381–384.
48. Marin, E. P., Krishna, A. G., and Sakmar, T. P. (2002) Disruption of the alpha5 helix of transducin impairs rhodopsin-catalyzed nucleotide exchange, *Biochemistry 41*, 6988–6994.
49. Preininger, A. M., Funk, M. A., Oldham, W. M., Meier, S. M., Johnston, C. A., Adhikary, S., Kimple, A. J., Siderovski, D. P., Hamm, H. E., and Iverson, T. M. (2009) Helix Dipole Movement and Conformational Variability Contribute to Allosteric GDP Release in G alpha(i) Subunits, *Biochemistry 48*, 2630–2642.
50. Cherfils, J., and Chabre, M. (2003) Activation of G-protein G alpha subunits by receptors through G alpha-G beta and G alpha-G gamma interactions, *Trends in Biochemical Sciences 28*, 13–17.
51. Rondard, P., Liu, J., Huang, S., Malhaire, F., Vol, C., Pinault, A., Labesse, G., and Pin, J. P. (2006) Coupling of agonist binding to effector domain activation in metabotropic glutamate-like receptors, *J Biol Chem 281*, 24653–24661.
52. Rasmussen, S. G., Choi, H. J., Fung, J. J., Pardon, E., Casarosa, P., Chae, P. S., Devree, B. T., Rosenbaum, D. M., Thian, F. S., Kobilka, T. S., Schnapp, A., Konetzki, I., Sunahara, R. K., Gellman, S. H., Pautsch, A., Steyaert, J., Weis, W. I., and Kobilka, B. K. (2011) Structure of a nanobody-stabilized active state of the beta(2) adrenoceptor, *Nature 469*, 175–180.
53. Rasmussen, S. G., Devree, B. T., Zou, Y., Kruse, A. C., Chung, K. Y., Kobilka, T. S., Thian, F. S., Chae, P. S., Pardon, E., Calinski, D., Mathiesen, J. M., Shah, S. T., Lyons, J. A., Caffrey, M., Gellman, S. H., Steyaert, J., Skiniotis, G., Weis, W. I., Sunahara, R. K., and Kobilka, B. K. (2011) Crystal structure of the beta(2) adrenergic receptor-Gs protein complex, *Nature*.
54. Okada, T., Sugihara, M., Bondar, A. N., Elstner, M., Entel, P., and Buss, V. (2004) The retinal conformation and its environment in rhodopsin in light of a new 2.2 A crystal structure, *J Mol Biol 342*, 571–583.
55. Choe, H. W., Kim, Y. J., Park, J. H., Morizumi, T., Pai, E. F., Krauss, N., Hofmann, K. P., Scheerer, P., and Ernst, O. P. (2011) Crystal structure of metarhodopsin II, *Nature 471*, 651–655.
56. James, G. T. (1978) Inactivation of the protease inhibitor phenylmethylsulfonyl fluoride in buffers, *Anal Biochem 86*, 574–579.
57. Kaya, A. I., Thaker, T. M., Preininger, A. M., Iverson, T. M., and Hamm, H. E. (2011) Coupling efficiency of rhodopsin and transducin in bicelles, *Biochemistry 50*, 3193–3203.

Chapter 9

Dynamic Light Scattering to Study Allosteric Regulation

Aaron L. Lucius, P. Keith Veronese, and Ryan P. Stafford

Abstract

The *Escherichia coli* ClpA protein, like many AAA+ motor proteins, is allosterically regulated by nucleotide binding. We have combined analytical ultracentrifugation approaches with dynamic light scattering (DLS) to examine the self-association properties and the allosteric linkage of assembly to nucleotide binding. Here we present a protocol for the rapid and precise determination of the diffusion coefficient using DLS measurements in a model-independent fashion. When combined with analytical ultracentrifugation experiments, such an approach can yield a more complete understanding of the hydrodynamic and thermodynamic properties of the system.

Key words: Dynamic light scattering, Diffusion coefficient, Assembly, Allostery, Polysteric linkage

1. Introduction

Escherichia coli ClpA couples ATP binding and hydrolysis to enzyme catalyzed polypeptide translocation and protein unfolding. Its primary role in the cell is to associate with the serine protease, ClpP, and use its polypeptide translocation activity to feed a newly unfolded protein to ClpP for protein degradation (1–5). In addition to this activity, ClpA uses its protein unfolding activity to remodel protein complexes and is therefore considered a molecular chaperone (6–8).

ClpA, from sequence analysis, has been found to be a member of the *A*TPases *A*ssociated with various cellular *A*ctivities (AAA+) family of proteins (9). AAA+ proteins are a large superfamily of proteins that include a number of molecular chaperones, some DNA helicases and cellular cargo transporters such as Dynein (9, 10). One commonality among many AAA+ family members is that they require nucleotide binding to stabilize or drive assembly of the oligomeric form that is active in binding and catalysis, usually a hexamer (11, 12). This linkage of the assembly state to

Aron W. Fenton (ed.), *Allostery: Methods and Protocols*, Methods in Molecular Biology, vol. 796,
DOI 10.1007/978-1-61779-334-9_9, © Springer Science+Business Media, LLC 2012

the nucleotide ligation state is termed polysteric linkage (13). Polysteric linkage is a form of allostery characterized by the binding of a ligand that induces changes in the aggregation state of the macromolecule. One can imagine this as a classical allosteric interaction where the binding of ligand *x*, in this case nucleotide, induces binding of ligand *y*, in this case additional protomers of the protein.

The *E. coli* ClpA protein is a complex allosteric machine that is regulated by polysteric linkage. The enzyme exists in a mixture of monomers and tetramers at 25°C (14). This mixture of oligomers is inactive in polypeptide binding in the absence of nucleoside triphosphate (15). Upon nucleoside triphosphate binding, ClpA assembles into a hexameric ring with polypeptide binding activity (Veronese, manuscript in preparation) (15). This interaction shows nucleoside triphosphate to be an allosteric effector that, upon binding, induces ClpA to associate into a hexameric ring. In addition to its role in driving assembly, ATP binding and hydrolysis is coupled to repeated cycles of polypeptide translocation (15). Thus, there exists a complex network of allosteric interactions that involve binding of two different ligands and ligand-linked association of the macromolecule. This level of complexity is confounded by the fact that ClpA contains two nucleotide binding and hydrolysis sites per protomer (1).

We have initiated a detailed examination of the allosteric interactions between nucleotide binding and ClpA association (Veronese, manuscript in preparation). To this end, we have employed a combination of sedimentation velocity, sedimentation equilibrium, and dynamic light scattering. Analytical ultracentrifugation (AUC) techniques, such as sedimentation velocity and sedimentation equilibrium have been reviewed extensively, including in this issue, and will not be the focus of this review. Instead, the focus of this review is on the application and incorporation of DLS techniques to a complete examination of the hydrodynamic and thermodynamic properties of ClpA.

2. Materials

2.1. Preparation and Dialysis of ClpA

1. Overexpress and purify ClpA protein as described (14).
2. Buffer H; 25 mM HEPES (pH 7.5 at 25°C), 300 mM NaCl, 10% Glycerol, 2 mM 2-mercaptoethanol, 10 mM $MgCl_2$, twice distilled H_2O.
3. Storage Buffer; 25 mM HEPES (pH 7.5 at 25°C), 300 mM NaCl, 50% Glycerol, 2 mM 2-mercaptoethanol, 10 mM $MgCl_2$, twice distilled H_2O.

4. Twice distilled water produced using the Purelab Ultra Genetic system (Siemens Water Technologies, Warrendale, PA).
5. 50,000 MWCO Spectra/Por dialysis membrane (Spectrum Labs, Rancho Dominguez, CA).
6. Microcentrifuge.
7. 1.5 or 2.0 mL microcentrifuge tubes.
8. Cary 100 Bio UV-Vis spectrophotometer (Varian, Inc., Palo Alto, CA) or equivalent.

2.2. Instrument Preparation

1. HPLC grade toluene.
2. Anotop 25 inorganic membrane filter: 0.2 μm/25 mm and 0.02 μm/25 mm (Whatman/GE Healthcare Piscataway, NJ).
3. 50 mL glass syringe.

2.3. Data Collection

1. Borosilicate glass culture tubes (Fisherbrand, Fairlawn, NJ).
2. Ribbed plugs (12 × 75) for borosilicate glass culture tubes (Fisherbrand, Fairlawn, NJ) (see Note 2).
3. BD 1 mL tuberculin slip tip syringe (BD, Franklin Lakes, NJ).
4. Kimwipe delicate task wipers (Kimberly-Clark, Irving, TX).
5. 0.22 μm/4 mm Millex-GV syringe driven filter unit (Millipore, Billerica, MA).
6. Minicentrifuge.
7. Means of collecting sample and buffer density, viscosity, and partial specific volume. This is done either experimentally or computationally, i.e., SEDNTERP (John Philo, Alliance Protein Laboratories).
8. Water bath or other means of controlling sample and toluene bath temperature.
9. Data acquisition software is also needed. In our experiments, we use the ALV-5000/E software packaged with the ALV Dynamic light scattering/Static light scattering Goniometer System (ALV-GbmH, Langen, Germany).

3. Methods

Recently, there has been a resurgence in the application of sedimentation velocity experiments to examine macromolecular assembly (16–21). From such experiments, one gains hydrodynamic information about the macromolecule of interest. Specifically, the Svedberg coefficient or sedimentation coefficient, *s*, is determined and is defined by Eq. 1.

$$s = \frac{M(1 - \bar{v}\rho)}{Nf} = \frac{MD(1 - \bar{v}\rho)}{RT}, \tag{1}$$

where M is the molecular weight, $\bar{v}$ is the partial specific volume of the macromolecule, ρ is the density of the buffer, N is Avogadro's number, f is the frictional coefficient, D is the diffusion coefficient, R is the ideal gas constant, and T is absolute temperature. The two forms of Eq. 1 are related by Eq. 2,

$$D = \frac{k_b T}{f}, \tag{2}$$

where k_b is Boltzman's constant.

According to Eq. 1, the sedimentation coefficient is the product of the molecular weight, M, and the diffusion coefficient, D. Thus, the sedimentation coefficient is influenced by both size and shape of the macromolecule.

If an independent determination of the diffusion coefficient is possible, then the diffusion coefficient can be combined with the sedimentation coefficient to determine the molecular weight (22). The overarching benefit of such a strategy is that both sedimentation velocity and DLS experiments are relatively fast and easy to perform. In contrast, the molecular weight and self-association parameters are often determined from the application of sedimentation equilibrium experiments. We prefer to incorporate results from sedimentation equilibrium experiments whenever possible. However, a problem that is occasionally encountered with sedimentation equilibrium is that the macromolecule being sedimented is not amenable to the several days of centrifugation that is often required for the system to achieve sedimentation equilibrium. For example, ClpA in the presence of ATPγS does not achieve sedimentation equilibrium at 25°C, most likely due to the slow hydrolysis of ATPγS (Veronese, manuscript in preparation). However, ATPγS binding is required to drive the formation of hexamers, and other non-hydrolysable nucleotide analogs are poor substitutes. On the other hand, DLS experiments can be performed in ~1 h and sedimentation velocity experiments can often be performed in ~2–3 h, timescales much shorter than a typical sedimentation equilibrium experiment.

By combining sedimentation velocity and dynamic light scattering approaches, ClpA can be easily examined in the presence of ATPγS without concern for slow hydrolysis. Moreover, the approach outlined here could serve as an alternative strategy when sedimentation equilibrium experiments are not tractable. Equally, it can complement other approaches such as direct fitting of reaction boundaries or analysis of weight average sedimentation coefficients (23, 24). With that, even if sedimentation equilibrium experiments are possible, the addition of DLS measurements still serves as a complementary hydrodynamic approach.

3.1. Preparation of ClpA Samples

1. Dialyze samples of ClpA stored at −80°C in storage buffer against Buffer H. After 6 h of dialysis, the old dialysate is removed and fresh Buffer H is added. Dialysis is allowed to continue for at least another 6 h.
2. Collect dialysate and save for individual sample dilutions.
3. Transfer dialyzed ClpA sample to a 1.5 mL microcentrifuge tube and centrifuged in a 4°C environment at 13,000 rpm (17,000 × *g*) for 15 min using a Fisher Scientific accuSpin microcentrifuge or equivalent to sediment any possible precipitants (see Note 9).
4. After centrifugation, carefully draw away the protein supernatant from the pellet and place in a fresh microcentrifuge tube. The microcentrifuge tube containing the pellet is discarded.
5. Determine the concentration of the protein using a Cary 100 Bio UV-Vis spectrophotometer. The extinction coefficient of ClpA was determined to be $(3.1 \pm 0.2) \times 10^4\ \text{M}^{-1}\ \text{cm}^{-1}$ for the monomer in Buffer H at 280 nm (17, 25).
6. ClpA in Buffer H should be kept at 4°C until it is needed for experiments. ClpA exhibits full polypeptide translocation activity under these conditions for up to 1 week (15).

3.2. Instrument Preparation

1. Coarse filter toluene into a glass flask using a Whatman 0.2 μm filter. Approximately 50 mL of toluene is required to fill the bath in our instrument (see Note 5).
2. The coarse filtered toluene is then fine filtered directly into the toluene bath using a Whatman 0.02 μm filter. If excessive contaminants are still seen in the incident beam, the filtered toluene is drawn out of the toluene bath and filtered again using the same process outlined above (see Notes 3 and 5).
3. Turn on water bath and set to 25°C or desired temperature at least 1 h prior to data acquisition.

3.3. Obtaining Raw Data

1. All measurements are obtained using an ALV DLS/SLS Goniometer System using a 22 mW HeNe-Laser monitoring at 632.8 nm (JDS Uniphase, Manteca, CA), an ALV-125 goniometer, and an ALV-5000E auto-correlator card (ALV-GbmH, Langen Germany). The instrument is schematized in Fig. 1 (see Notes 4 and 7).
2. The temperature of the toluene bath is maintained at 25°C using a water bath that is in thermal contact with the toluene bath of the ALV DLS/SLS Goniometer System. The temperature of the toluene is continuously monitored with a digital probe and the data is transferred to the computer.

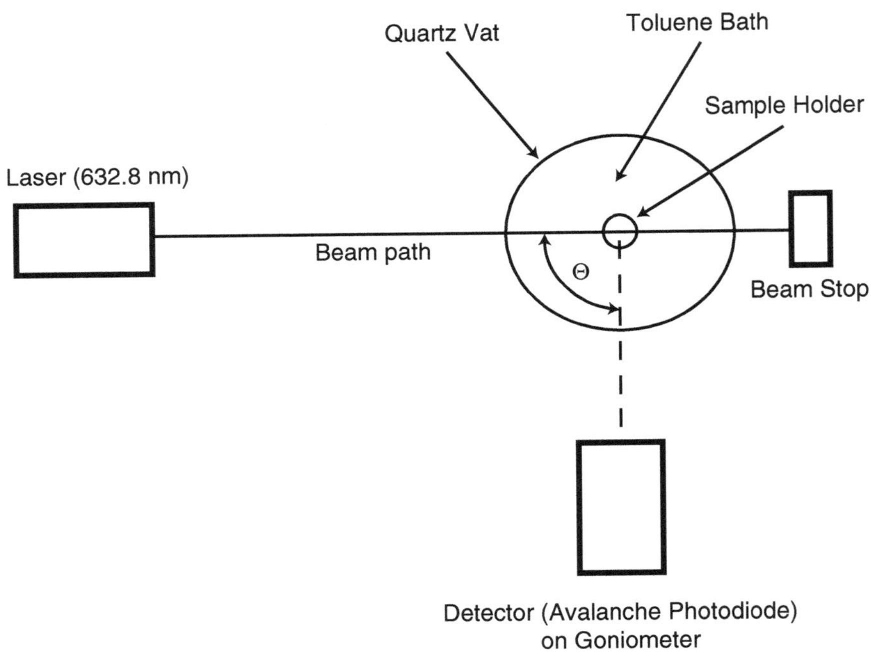

Fig. 1. Schematic representation of the ALV light scattering instrument.

3. Dilute ClpA protein stock sample to individual experimental samples with a concentration of 3, 6, 9, 12, 15, and 18 μM using the collected dialysate (see Note 1).
4. Place diluted ClpA samples in a 1.5 or 2.0-mL microcentrifuge tube and centrifuge in a minicentrifuge (6,600 rpm/ 2,200 × *g*) for 20–30 s (see Note 8).
5. Choose borosilicate culture tubes that have a minimal amount of inherent imperfections as determined by inspection (see Note 6).
6. One milliliter of each of the protein samples is filtered directly into a borosilicate culture tube using a disposable syringe and a 0.22 μm/4-mm syringe filter (see Notes 9–11).
7. Wipe borosilicate culture tubes with a Kimwipe to remove any residual smudges or fingerprints before being placed into the toluene bath.
8. Allow every protein sample to temperature equilibrate in the toluene bath for 10 min prior to use (see Note 12).
9. Collect autocorrelation functions at angles of 60, 90, and 120° for each sample.
10. Collect autocorrelation functions every 3 s. Twenty autocorrelation functions for a given angle are collected and averaged (see Note 13).
11. Do not reuse borosilicate culture tubes.

3.4. Data Analysis

1. For each concentration of ClpA, the three autocorrelation functions collected at different angles are globally fit to Eq. 3 (see Notes 13–16).
2. Global nonlinear least squares (NLLS) analysis is accomplished using Conlin (14, 26) and Eq. 3. Each autocorrelation function is assigned an amplitude, A_i, and the diffusion constant, D_i, was a global parameter, i.e., the same for all angles.

$$g_2(\tau) = (g_1(\tau))^2 + c, \tag{3}$$

where c is a baseline offset and $g_1(\tau)$ is given by Eq. 4.

$$g_1(\tau) = \sum_i A_i e^{\Gamma_i \tau}, \tag{4}$$

where "i" is the number of diffusing species, A_i is the amplitude of the i-th species and Γ_i for the i-th species is given by Eq. 5.

$$\Gamma_i = D_i q^2, \tag{5}$$

where D_i is the diffusion coefficient for the i-th species and q is the scattering vector given by Eq. 6.

$$q = \frac{4 n_o \pi}{\lambda} \sin\left(\frac{\theta}{2}\right), \tag{6}$$

where n_o is the refractive index of the solution, λ is the wavelength of the laser, and θ is the angle of the detector relative to the incident beam.

As previously noted, at a given concentration of ClpA, normalized autocorrelation functions are collected at 60, 90, and 120°. The set of three angles are globally analyzed using Eq. 3, with $i = 2$, D_i as a global parameter, i.e., the same for each trace, and θ is 60, 90, or 120°. A representative set of three auto correlation functions collected at three different angles are shown in Fig. 2a with the associated global fit. From this method of analysis, a Z-average diffusion coefficient is obtained (27). All diffusion coefficients reported should be corrected to 20°C and infinite dilution in water, i.e., $D_{20,\mathrm{w}}$ (28, 29). The Z-average diffusion coefficient, $D_{z,20,\mathrm{w}}$, as a function of the free ClpA concentration is then fit to Eq. 7 (27).

$$D_{z,20,\mathrm{w}} = \frac{(M_\mathrm{m}) D_{\mathrm{m},20,\mathrm{w}} + 4(M_\mathrm{t}) D_{\mathrm{t},20,\mathrm{w}} L_{40} [\mathrm{ClpA}]^3_{\mathrm{free}}}{(M_\mathrm{m}) + 4(M_\mathrm{t}) L_{40} [\mathrm{ClpA}]^3_{\mathrm{free}}}, \tag{7}$$

where $D_{z,20,\mathrm{w}}$ is the Z-average diffusion coefficient, L_{40} is as described above, $D_{\mathrm{m},20,\mathrm{w}}$ and $D_{\mathrm{t},20,\mathrm{w}}$ are the monomer and tetramer diffusion coefficients, respectively, and M_m and M_t are the monomer and tetramer molecular weights, respectively.

The Z-average diffusion coefficient as a function of total ClpA concentration, shown in Fig. 2b, was subjected to NLLS analysis

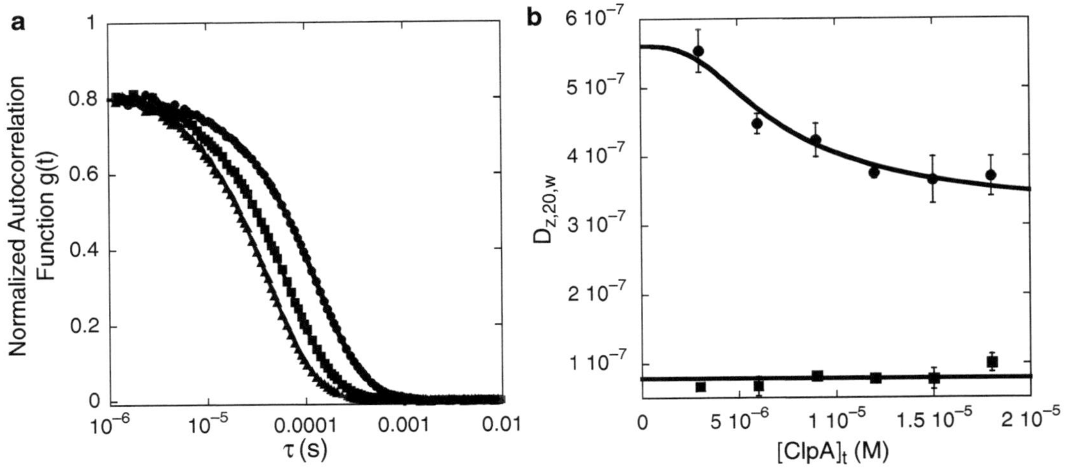

Fig. 2. Dynamic light scattering data. (**a**) Example of normalized autocorrelation functions for ClpA in Buffer H collected at 60° (*circles*), 90° (*squares*), and 120° (*triangles*) *solid lines* are the result of a global NLLS fit using Eq. 3 with $i = 2$. (**b**) Plot of the two *Z*-average diffusion coefficients from the NLLS analysis in *panel* (**a**) as a function of total [ClpA]. The *solid black squares* show a diffusion coefficient that is independent of [ClpA] with the *solid line* representing the average. The *solid black circles* represent the *Z*-average diffusion coefficient that depends upon [ClpA]. The *solid line* represents the NLLS fit of the data to Eq. 7. Figure reproduced from Veronese et al. with permission (14).

by implicitly solving Eq. 7 and the conservation of mass equation given by Eq. 8.

$$[\text{ClpA}]_\text{t} = [\text{ClpA}]_\text{free} + 4L_{40}[\text{ClpA}]^4_\text{free}. \tag{8}$$

4. Notes

1. Concentrations of protein should be chosen so as to span as wide a concentration range as possible to fully populate the largest assembly state. The thermodynamic driving force for self assembly is the free monomer concentration; thus, the concentration of the protein samples bears on what assembly states are observed.
2. Care should be taken when attaching the ribbed plugs to the top of the glass sample tubes. If the cap is forced down, the sample can spill or be agitated to produce bubbles, or worse, the fragile tube can easily shatter.
3. The intensity of the scattered light increases with increasing size of the particle. Therefore, dust and bubbles are the bane of the light-scattering experimentalist. Because of this, we must first ensure that there are no particulates in the index-matching solution, i.e., the toluene bath, see Fig. 1. This is accomplished by first filtering the toluene using a 0.2-μm

filter and then filtering again with a 0.02-μm filter (see Subheadings 2.2 and 3.2).

4. Because the intensity of the scattered light increases with increasing size of the particle it is necessary to optimize the laser power before starting the experiment. The ALV instrument splits the beam so that part of the beam is detected by a diode and the remainder goes through the sample. Therefore, the laser intensity can be monitored at the diode, where the intensity can be modulated with a built in filter. The laser intensity is kept below 200 kHz. High quality data are acquired between 50 and 190 kHz.
5. Toluene is used to fill the quartz vat as its refractive index matches that of the quartz (see Fig. 1). Thus, there is no bending of light as the beam passes through the quartz vat. Often, after filling the toluene bath and directly observing the laser beam passing through the toluene, one will see frequent sparkles in the beam. This is due to large particles that have settled on the bottom of the bath being disturbed during the filling process. Because of this, after filling the bath with fresh toluene, experiments should not be performed for at least 24 h in order to give time for dust and other large particles to settle. If after 24 h, the sparkles in the beam do not diminish, the toluene should be removed with a glass pipette and replaced with newly filtered toluene. If the instrument has not been used in some time, this process may need to be repeated several times.
6. Considerable time should be taken to select borosilicate culture tubes that have a minimal number of observable flaws that potentially bend or scatter light. These culture tubes are often packaged tightly, 250 to a package, causing a number of scratches on a large population of tubes.
7. Prior to data acquisition, turn on the laser that accompanies the dynamic light-scattering instrument. It is advised to leave the laser powered on the entire week in which one is acquiring data. Multiple on/off cycles decrease the lifespan of the laser (manufacturer recommendation).
8. Once the laser is powered on, observe the laser's path through the toluene bath. If the beam exhibits few disturbances, the previously filtered toluene bath is more than likely able to be re-used.
9. Removal of bubbles, particulates, etc. from the sample is essential. This is accomplished in two steps. First, the sample is centrifuged in order to pellet any precipitants that formed during dialysis (see Subheading 3.1). Second, immediately before the sample is used in an experiment, it is pushed through a 0.2-μm syringe filter directly into the

borosilicate glass tube that will be used for measurements (see Subheading 3.1). It is essential during this step that bubbles are not produced in the sample. Thus, the sample must be filtered by pushing the plunger of the syringe very slowly. Lastly, the sample should be dribbled down the side of the borosilicate culture tube in order to prevent bubble formation.

10. It is common for proteins to stick to syringe filters. It is also common for biological macromolecules to stick to the surfaces of glass. To control this, one should check the concentration before and after a light-scattering experiment by absorbance methods. Because light scattering is a nondestructive technique, the concentration should be the same before and after experiments. If the concentration decreases, significant amounts of protein may be sticking to the filter or other surfaces. In the case of ClpA, we have not observed significant loss of material after light-scattering experiments.
11. Preferably, each protein concentration is made directly into microcentrifuge tubes and then filtered directly into the borosilicate tube. In principle, measurements can be made on the highest concentration sample and then the next lower concentration sample can be prepared by serial dilution. However, upon diluting and mixing, the sample will need to be refiltered as it is difficult to remove all of the micro bubbles that form during the dilution step. Thus, in the serial dilution strategy, the probability of losing protein due to sticking to the filter is much greater.
12. Once the protein sample is adequately filtered, the borosilicate tube containing the sample is gently placed in the toluene bath. The sample is allowed to temperature equilibrate for 10 min. The 10-min wait time has been determined empirically for ClpA and longer wait times may be required depending on the temperature difference between the sample and the toluene bath. Two issues are addressed by allowing for temperature equilibration of the sample. First, a temperature gradient will produce Schlieren lines, which is a refractive index change. This will cause different sections of the solution to scatter light differently. Second, whatever reactions are taking place in solution must have sufficient time to achieve thermodynamic equilibrium at the given temperature.
13. At a given angle, autocorrelation functions are collected every 3 s within a 1 min timeframe. This results in 20 different curves. The 20 curves are then averaged. After the extensive filtering of the protein sample, if any particles remain, their appearance in the beam should be infrequent. By averaging 20 time courses, the infrequent appearance of spikes will be averaged out.

14. The time courses collected for the three angles are subjected to global nonlinear least squares analysis using Eq. 3. The benefit of this strategy is that there are two independent variables, the autocorrelation time, τ, and the angle, θ. By globally fitting the data as a function of both τ and θ, better constraints on the diffusion coefficient, D, are obtained.
15. To make your diffusion coefficient data more accessible and comparable to sedimentation velocity values for further study and literature comparison, convert diffusion coefficient data to a value corresponding to 20°C and infinite dilution in water, i.e., $D_{20,w}$.
16. Although sedimentation equilibrium experiments yield molecular weight estimates for protein assembly states, we have had success in estimating the molecular weight of the mixture of monomeric and tetrameric ClpA algebraically using a combination of dynamic light scattering and sedimentation velocity experiments. From the dynamic light scattering data shown in Fig. 1, a $D_{20,w}$ value of $(5.6 \pm 0.3) \times 10^{-7}$ cm^2/s for ClpA monomer and $(2.9 \pm 0.3) \times 10^{-7}$ cm^2/s for a tetramer of ClpA were obtained. When combined with the corresponding monomer and tetramer values of $s_{20,w}$ obtained from NLLS fitting (4.50 ± 0.1) S and (9.8 ± 0.3) S, respectively, Eq. 1 yields an estimated molecular weight of 86,500 ± 5,000 Da for the monomer and 360,000 ± 40,000 Da for the tetramer. These values are very close to the calculated molecular weight of ClpA monomer from sequence, 84,200 Da, and within error for that of a tetramer of ClpA (336,800 Da).

Acknowledgments

We thank Wilhelm Peters for alignment of the laser and pointing out many of the pitfalls discussed here. This work was supported by NSF grant MCB-0843746 to ALL and the University of Alabama Department of Chemistry Start-up Funds.

References

1. Singh, S. K., and Maurizi, M. R. (1994) Mutational analysis demonstrates different functional roles for the two ATP-binding sites in ClpAP protease from Escherichia coli, *J Biol Chem 269*, 29537–29545.
2. Reid, B. G., Fenton, W. A., Horwich, A. L., and Weber-Ban, E. U. (2001) ClpA mediates directional translocation of substrate proteins into the ClpP protease, *Proc Natl Acad Sci USA 98*, 3768–3772.
3. Katayama, Y., Gottesman, S., Pumphrey, J., Rudikoff, S., Clark, W. P., and Maurizi, M. R. (1988) The two-component, ATP-dependent Clp protease of Escherichia coli. Purification, cloning, and mutational

analysis of the ATP-binding component, *J Biol Chem 263*, 15226–15236.

4. Hwang, B. J., Woo, K. M., Goldberg, A. L., and Chung, C. H. (1988) Protease Ti, a new ATP-dependent protease in Escherichia coli, contains protein-activated ATPase and proteolytic functions in distinct subunits, *J Biol Chem 263*, 8727–8734.
5. Weber-Ban, E. U., Reid, B. G., Miranker, A. D., and Horwich, A. L. (1999) Global unfolding of a substrate protein by the Hsp100 chaperone ClpA, *Nature 401*, 90–93.
6. Levchenko, I., Luo, L., and Baker, T. A. (1995) Disassembly of the Mu transposase tetramer by the ClpX chaperone, *Genes & development 9*, 2399–2408.
7. Wickner, S., Gottesman, S., Skowyra, D., Hoskins, J., McKenney, K., and Maurizi, M. R. (1994) A molecular chaperone, ClpA, functions like DnaK and DnaJ, *Proc Natl Acad Sci USA 91*, 12218–12222.
8. Hoskins, J. R., Pak, M., Maurizi, M. R., and Wickner, S. (1998) The role of the ClpA chaperone in proteolysis by ClpAP, *Proceedings of the National Academy of Sciences of the United States of America 95*, 12135–12140.
9. Tucker, P. A., and Sallai, L. (2007) The AAA+ superfamily-a myriad of motions, *Current opinion in structural biology 17*, 641–652.
10. Neuwald, A. F., Aravind, L., Spouge, J. L., and Koonin, E. V. (1999) AAA+: A class of chaperone-like ATPases associated with the assembly, operation, and disassembly of protein complexes, *Genome research 9*, 27–43.
11. Walker, J. E., Saraste, M., Runswick, M. J., and Gay, N. J. (1982) Distantly related sequences in the alpha- and beta-subunits of ATP synthase, myosin, kinases and other ATP-requiring enzymes and a common nucleotide binding fold, *The EMBO journal 1*, 945–951.
12. Gottesman, S., Clark, W. P., and Maurizi, M. R. (1990) The ATP-dependent Clp protease of Escherichia coli. Sequence of clpA and identification of a Clp-specific substrate, *J Biol Chem 265*, 7886–7893.
13. Colosimo, A., Brunori, M., and Wyman, J. (1976) Polysteric linkage, *J Mol Biol 100*, 47–57.
14. Veronese, P. K., Stafford, R. P., and Lucius, A. L. (2009) The Escherichia coli ClpA Molecular Chaperone Self-Assembles into Tetramers, *Biochemistry 48*, 9221–9233.
15. Rajendar, B., and Lucius, A. L. (2010) Molecular mechanism of polypeptide translocation catalyzed by the Escherichia coli ClpA protein translocase, *J Mol Biol 399*, 665–679.
16. Schuck, P. (2003) On the analysis of protein self-association by sedimentation velocity analytical ultracentrifugation, *Analytical biochemistry 320*, 104–124.
17. Veronese, P. K., Stafford, R. P., and Lucius, A. L. (2009) The Escherichia coli ClpA molecular chaperone self-assembles into tetramers, *Biochemistry 48*, 9221–9233.
18. Laue, T. (2004) Analytical ultracentrifugation: a powerful 'new' technology in drug discovery, *Drug Discovery Today: Technologies 1*, 309–315.
19. Cole, J. L., and Hansen, J. C. (1999) Analytical Ultracentrifugation as a Contemporary Biomolecular Research Tool. *Journal of Biomolecular Techniques 10*, 163–176.
20. Ralston, G. (1993) *Introduction to Analytical Ultracentrifugation*, Beckman Coulter Inc., Fullerton, CA.
21. McRorie, D. K., and Voelker, P. J. (1993) *Self-Associating Systems in the Analytical Ultracentrifuge*, Beckman Coulter Inc., Fullerton, CA.
22. Tinoco, I., Sauer, K., and Wang, J. C. (1995) *Physical chemistry : principles and applications in biological sciences*, 3rd ed., Prentice Hall, Englewood Cliffs, NJ.
23. Correia, J. J. (2000) Analysis of weight average sedimentation velocity data, *Methods Enzymol 321*, 81–100.
24. Correia, J. J., and Stafford, W. F. (2009) Extracting equilibrium constants from kinetically limited reacting systems, *Methods Enzymol 455*, 419–446.
25. Lohman, T. M., Chao, K., Green, J. M., Sage, S., and Runyon, G. T. (1989) Large-scale purification and characterization of the Escherichia coli rep gene product, *J Biol Chem 264*, 10139–10147.
26. Williams, D. J., and Hall, K. B. (2000) Monte Carlo applications to thermal and chemical denaturation experiments of nucleic acids and proteins, *Methods Enzymol 321*, 330–352.
27. Dennis, E. K. (1972) Analysis of Macromolecular Polydispersity in Intensity Correlation Spectroscopy: The Method of Cumulants, *The Journal of Chemical Physics 57*, 4814–4820.
28. Tanford, C. (1961) *Physical chemistry of macromolecules*, Wiley, New York,.
29. Cantor, C. R., and Schimmel, P. R. (1980) *Techniques for the study of biological structure and function*, W. H. Freeman, San Francisco.

Chapter 10

Dissecting the Linkage Between Transcription Factor Self-Assembly and Site-Specific DNA Binding: The Role of the Analytical Ultracentrifuge

Amie D. Moody, James P. Robblee, and David L. Bain

Abstract

A long-standing goal of biomedical research has been to determine the quantitative mechanisms responsible for gene regulation and transcriptional activation. These events occur through numerous protein–protein and protein–DNA interactions, many of which are allosterically coupled. For systems where highly purified protein is available, analytical ultracentrifugation provides a means to study these linked reactions, allosteric or otherwise. Sedimentation velocity is an ultracentrifugation technique that provides rigorous insight into protein self-association, homogeneity, and gross structure. Because self-association is often in dynamic equilibrium with other reactions such as DNA binding, an explicit and independent analysis of each interaction is critical to revealing mechanism. This chapter details a protocol for using sedimentation velocity to dissect the linkage between transcription factor self-association and site-specific DNA binding.

Key words: Analytical ultracentrifugation, Sedimentation velocity, Allostery, Linkage, Steroid receptors, Protein–DNA interactions

1. Introduction

Many proteins responsible for biological regulation carry out their functions via allosteric communication. That is, binding of one ligand is linked or coupled to binding of another at a topographically distinct site, resulting in change of functional outcome. This communication allows regulatory proteins to constantly sense their environment and respond accordingly. For proteins involved in gene regulation, allostery can manifest itself in multiple forms, such as coupling DNA binding with cooperativity (1, 2). However, not all such interactions are necessarily allosteric in nature, despite the reactions being linked. Perhaps the simplest example

Aron W. Fenton (ed.), *Allostery: Methods and Protocols*, Methods in Molecular Biology, vol. 796,
DOI 10.1007/978-1-61779-334-9_10,

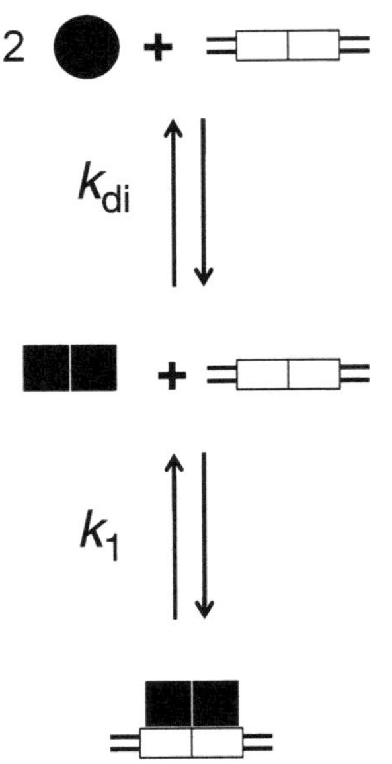

Fig. 1. Linkage of protein self-assembly to DNA binding. The equilibrium between monomers (*black circles*) and dimers (two adjoined *black squares*) is described by the dimerization constant, k_{di}. Dimer binding to DNA (*two lines with open white rectangle* as the binding site) is described by the intrinsic binding affinity, k_1.

of a non-allosteric linkage is that between protein self-association (e.g., dimerization) and site-specific DNA binding. Simple as this might be, understanding this linkage is critical to deciphering more complex reactions such as cooperative assembly at multi-site promoter sequences (3, 4).

The steroid receptor family of ligand-activated transcription factors represents a classic example of linked reactions (5, 6). Dogma states that steroid receptors form stable dimers in the absence of DNA, then bind to their palindromic response elements. This linked pathway is described schematically in Fig. 1, where k_{di} and k_1 represent dimerization and DNA binding, respectively. Typically measured in most DNA binding assays, however, is the apparent dissociation constant, K_{app}, as determined from half-saturation of the DNA. Unfortunately, this parameter is a nonlinear composite of the two microscopic parameters and has little molecular meaning. Specific examples are presented below.

Shown in Fig. 2 are DNA-binding isotherms for the steroid-activated transcription factors estrogen receptor (ER), progesterone receptor (PR), and glucocorticoid receptor (GR). By visual inspection, the apparent binding affinities (K_{app}) for ER, PR, and GR are 9.7, 28, and 93 nM, respectively. These binding affinities arise from fitting each data set (dotted line) to the

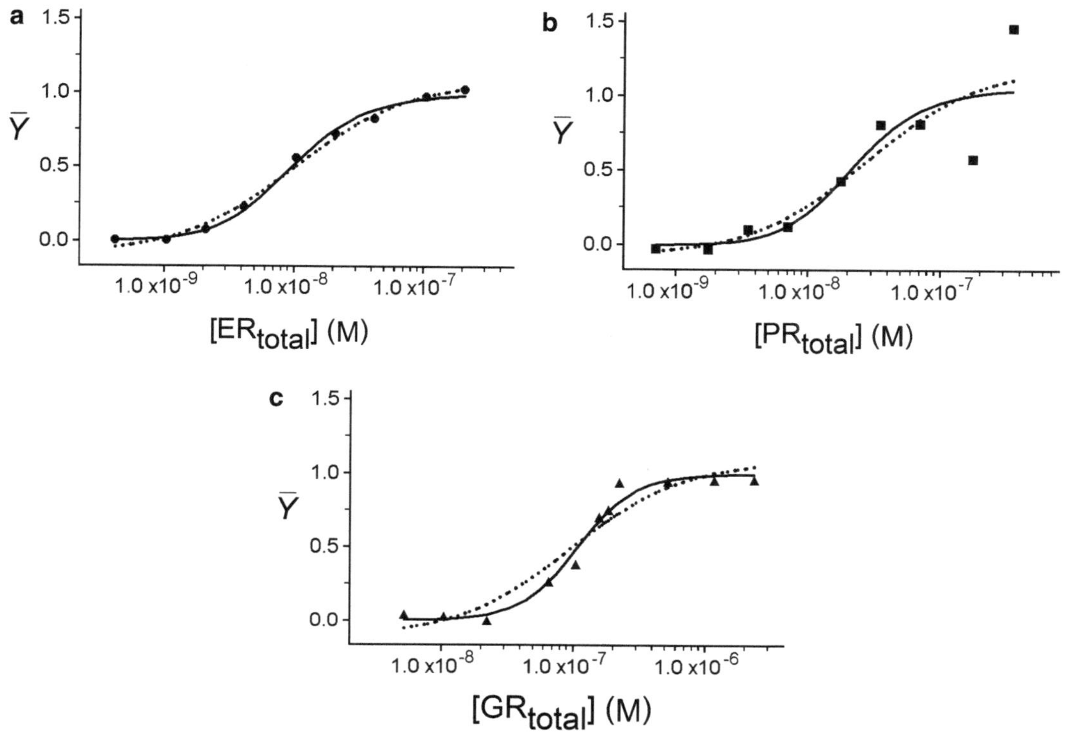

Fig. 2. Binding isotherms for full-length, human steroid receptor assembly at a palindromic response element in the presence of cognate hormone. ER binding is shown in *Panel A* (*circles*), PR binding is shown in *Panel B* (*squares*), and GR binding is shown in *Panel C* (*triangles*). The *dotted lines* represent the best fits using Eq. 1, and the *solid lines* represent the best fits using Eq. 2. Other than hormone type, experiments were carried out under identical conditions.

familiar Langmuir binding equation describing a bimolecular interaction:

$$\overline{\mathrm{Y}} = \frac{K_a \cdot x_{\mathrm{tot}}}{1 + K_a \cdot x_{\mathrm{tot}}} \tag{1}$$

where $\overline{\mathrm{Y}}$ is the fraction of DNA bound by protein, K_a is the apparent association constant, and x_{tot} is the total protein concentration. Setting the fractional saturation equal to 0.5 results in x_{tot} equaling the inverse of K_a – the dissociation constant, K_d (7). However, since the reactions in question are not bimolecular, the K_d has no physical meaning and is thus "apparent." Based on the model in Fig. 1, the equation that correctly describes each binding curve is (8):

$$\overline{\mathrm{Y}} = \frac{k_{\mathrm{di}} k_1 \cdot x_{\mathrm{m}}^2}{1 + k_{\mathrm{di}} k_1 \cdot x_{\mathrm{m}}^2} \tag{2}$$

where the association constants are as defined above and x_{m} is the free monomer concentration rather than the total protein concentration. However, it is impossible to successfully fit Eq. 2 to the data because the two parameters, k_1 and k_{di} are highly correlated – any change in the estimate of one parameter is equally compensated by a change in the other. The only way one can determine the intrinsic DNA-binding affinity from the data in Fig. 2 is to first independently determine the dimerization constant, calculate the free monomer concentration, then fix (lock) k_{di} in Eq. 2. As we will show later in this chapter, upon estimating the dimerization constant for each receptor, the intrinsic affinity for each receptor–DNA interaction (k_1) bears no clear relationship to its apparent affinity. This is so because of the differing extent of linkage for each receptor.

The only remaining issue is how to experimentally determine the dimerization constant. Historically, a number of techniques have been used to assess self-association energetics including fluorescence anisotropy (9, 10), size-exclusion chromatography (11) and sedimentation equilibrium (12). However, with the advent of new algorithms and faster computing power, sedimentation velocity (SV) has some attractive properties, particularly for researchers interested in a quick diagnosis of the quality and characteristics of their protein sample. Experiments can be carried out in a day; small amounts of material are required; and the results offer insight into protein structural homogeneity not normally available from conventional approaches. Finally, for those interested in pursuing more long-term investigations using sedimentation equilibrium, for example, the results of SV allow one to interpret the equilibrium results with confidence (13). The following protocol describes how to carry out SV experiments for self-associating transcription factors, and the implementation of the results to obtain molecular insight into interacting systems.

2. Materials

2.1. Protein

1. At least 1 mg of highly purified protein (>95%) at a concentration of at least 1 mg/ml. Purity should be determined by densitometric analysis of an overloaded Coomassie Blue stained gel. Shown in Fig. 3 is a representative example of highly purified ER as used in these experiments. Concentration should be determined by absorption spectroscopy using the protein's extinction coefficient. Colorimetric techniques such as the Bradford assay should be avoided due to their imprecision.
2. The protein should be dialyzed into a buffer that allows for high solubility, long-term stability, and low background absorbance (see Note 1).
3. Dialysis cups (Thermo Scientific, cat #69570) and accompanying dialysis raft (Thermo Scientific, cat # 69588).

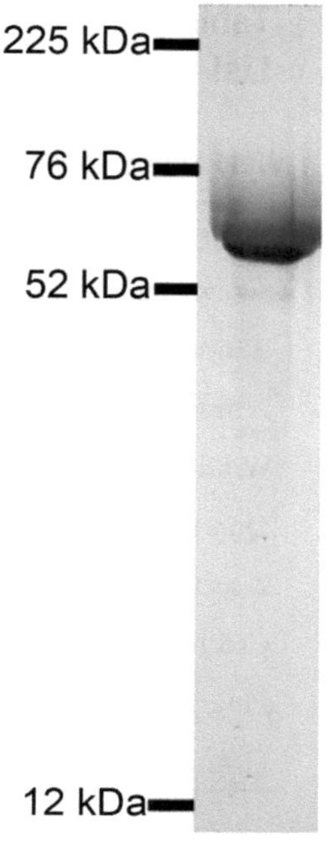

Fig. 3. Purification of human full-length ER. FLAG-tagged receptor (67 kDa) was expressed in Sf9 insect cells, and purified using affinity and size exclusion chromatography followed by Q-Sepharose resin. 14 μg of purified ER was fractionated by SDS-PAGE and visualized with Coomassie Blue stain.

2.2. Ultracentrifugation Cells

1. Sedimentation velocity cells (two-sector centerpiece) (see Table 1).
2. Cell alignment tool (Beckman Coulter, cat # 362340).
3. Spincote (Beckman Coulter, cat # 306812).
4. Torque stand assembly (Beckman Coulter, cat # 361318).
5. Gel-loading tips (Life Science Products, cat # LSRD05).

2.3. Analytical Ultracentrifuge

1. Beckman Coulter XLA analytical ultracentrifuge equipped with monochrometer, personal computer, and ProteomeLab (Beckman Coulter) computer program (see Note 2).
2. An-60 or An-50 titanium rotor.

2.4. Analysis Programs

An in-depth discussion of data analysis is beyond the scope of this chapter. However, there are resources available in the literature (14), and several state-of-the-art analysis programs are listed in Note 3.

Table 1
List of components and part numbers for SV cell

Velocity cell parts	Beckman part number
Screw ring (1)	301922
Screw-ring washer (1)	362328
Window holder (2)	305037
Window gasket (2)	327021 (100/package)
Window liner (2)	362329
Quartz Window (2)	301730
2-sector centerpiece (1)	306493
Cell housing (1)	305552
Plug gaskets (2)	327022 (100/package)
Housing plugs (2)	362327 (6/package)

3. Methods

For most analytical ultracentrifugation (AUC) techniques, it is imperative that the protein of interest be highly pure (>95%). Since the experimental signal in AUC is typically based on absorption optics, using pure protein ensures that all data collected arise from the desired sample and not from contaminating material. It is also desirable to cover as wide a protein concentration range as possible, ideally a tenfold span using three different protein concentrations. The following methods describe the experimental set-up and data collection for sedimentation velocity.

3.1. Prepare the Samples for Sedimentation Velocity

1. Since sedimentation data is collected by subtracting the absorbance of the reference sector from the absorbance of the sample sector, the buffer used for protein suspension and dilution must be exactly matched to the reference buffer. This is best achieved by extensive dialysis. Prepare 1 L of the experimental buffer.
2. Each SV sample requires 400 μl. Dilute the stock protein down to the desired experimental concentrations using the experimental buffer. Using a benchtop spectrophotometer, ensure that the final protein concentration will generate a strong absorbance signal at the desired wavelength. Keep in

mind that the pathlength of the centerpiece is 1.2 cm. Calculate the desired concentration accordingly.

3. Split the sample into dialysis cups, as per the manufacturer's instructions. Place prepared dialysis cups into dialysis raft available from manufacturer.
4. Place dialysis raft into 500 ml of the experimental buffer and dialyze for at least 8 h. Exchange the experimental buffer with the second 500 ml. Allow this second dialysis to proceed for at least another 4 h.
5. Remove samples from dialysis cups and pool the samples. Determine the protein concentration via UV absorbance to ensure no sample loss or dilution during dialysis.
6. Save several milliliters of the second dialysis buffer for use in loading into the reference sector of the ultracentrifugation cells.

3.2. Assemble the Sedimentation Velocity Cells

Turn on the AUC and set to the appropriate experimental temperature. Place the monochrometer into the AUC and initiate a vacuum. If performing the experiment at 4°C, allow the ultracentrifuge at least an hour to reach temperature. Likewise, place the rotor in a cold room to equilibrate to temperature. These steps will reduce the amount of time it will take the AUC to re-equilibrate to the experimental temperature after loading the cells into the instrument.

1. Due to the expensive and delicate nature of the cell components, great care in handling is paramount. Check that the centerpiece and windows are thoroughly clean. Any visible blemishes (streaks, scratches, dust, etc.) on the windows will detract from the quality of the data.
2. Apply a thin layer of Spincote to the screw ring. There should be no visible Spincote around the edges; remove any excess. Set screw ring aside.
3. Carefully inspect the window gaskets and liners for obvious signs of wear, such as a visible groove or cracks. Any ill-fitting liners or gaskets could result in broken windows. In the bottom of each window holder, place a single window gasket. Then place a window liner in each holder such that the apex of the liner is aligned with the notch of the window holder.
4. After ensuring that the windows are clean, place a dust-free window into each window holder from step 3. Make sure the scribed hash mark on the window is face up in the holder and that it is perfectly aligned with the notch in the window holder. Carefully place one assembled window holder into the cell housing; this will be the bottom of the cell. The notch on the window holder will fit over the spine running

down the one side of the cell housing. Do not force the holder into the bottom of the cell! Ensure that the face of the window is oriented toward the center of the cell housing, not the bottom of the cell.

5. Place the centerpiece into the cell housing. Once again, the notch on the centerpiece must align perfectly with the spine of the cell housing. When orienting the centerpiece, place it into the cell housing so that the numbers under the load holes are readable (not upside down). This is to ensure that the reference and sample sectors are always the same. Do not use excessive force to place the centerpiece into the housing.
6. Repeat steps 4 and **5** with the second window and window holder. Place the assembled window holder into the cell housing, making sure that the window face is toward the centerpiece, not toward the top of the cell.
7. Set the screw ring gasket on top of the second window holder, and hand-tighten the screw ring. Carefully put the assembled cell into the torque stand. The final desired torque is 130 in·lbs. Slowly tighten to a torque between 75 and 100 in·lbs to ensure that the ring and cell housing are threaded properly. Slowly loosen the screw ring all the way. Tighten again up to a torque between 100 and 125 in·lbs and then loosen completely. All motions should be slow and smooth. If there are any sharp movements/jumps, make sure that the cell is firmly clamped into the torque stand and that the screw ring has not become cross-threaded. Finally, torque to 130 in·lbs. Remove the cell from the torque stand and observe the septum of the centerpiece. It should be solid black with very sharply defined edges of the sectors against the window. Look for any scratches that span the septum, since these could lead to leaks between the two sectors resulting in mixing of reference and sample.
8. Repeat steps 1–7 for all cells to be used in the experiment (one cell for each protein concentration). It is not recommended to use more than three cells in a SV experiment when using absorbance optics since collection of absorbance scans is slow compared to interference optics. A visual approach to assembling AUC cells is available online (15).

3.3. Loading Velocity Cells and Carrying Out an Experiment

1. The AUC is calibrated to subtract reference absorbance from sample absorbance – if the reference and sample are loaded in the wrong sectors, then the resultant data will be inverted. With this in mind, when loading the cell lay it on its side so that the loading holes are on top and the screw ring (top of the cell) is facing you.

2. Taking a P-200 pipetter, load 200 μl of experimental buffer (the dialysis buffer) into the left sector two times for a total of 400 μl. When ejecting sample into the sector, make sure the pipet tip does not come into contact with the loading hole. If liquid is left there, this can lead to leaking of the sector due to capillary action. Before removing the pipet tip, ensure that all buffer is ejected and does not remain on the tip.
3. The technique for loading the sample into the right (sample) sector is the same as for the left (reference) sector, including all precautionary measures. However, in this case, only 390 μl of sample is loaded. This creates a slight volume mismatch between sectors and therefore allows for a reading of the buffer absorbance and identification of the sample meniscus.
4. With both sectors loaded, place a single housing plug gasket onto the top of each load hole. Use a pipet tip to press each gasket flush with the centerpiece. Then screw in the housing plugs. Tighten by hand, but do not strip the screws.
5. When loading the cells into the rotor, the counterbalance cell is always in the last hole (hole 4 in the 4-hole rotor or hole 8 in the 8-hole rotor). This is essential for calibrating the AUC. Using a benchtop balance and the weights supplied by Beckman Coulter, ensure that the counterbalance cell has the same mass as the cell that will be placed opposite of it in the rotor (i.e., the cell for either hole 2 in the 4-hole rotor, or hole 4 in the 8-hole rotor). The AUC is designed to tolerate up to 0.5 g difference between the two; if using three sample cells, the remaining two cells must also be within 0.5 g of each other.
6. Load the cells into their respective holes in the rotor. Make sure the cells are evenly distributed around the rotor. When placing the cells in the rotor, the load holes of the cell must face the interior of the rotor and the screw ring must face up. With cells properly placed into the rotor, carefully inspect the bottom of the rotor. Do not turn the rotor over or the cells and counterbalance could fall out! Instead lift up the rotor and look up at the bottom. Use the aligning tool to align the small notches on the bottom of the cells with those on the rotor. (The hash marks are not on the screw ring, hence it will be obvious if the cell was inserted upside down).
7. Release the vacuum from the AUC and remove the monochrometer. Place the rotor onto the AUC spindle, being careful to not bump any of the sensors on the floor of the chamber. Replace the monochrometer and close the AUC. Turn the vacuum back on. Let the AUC reach

experimental temperature. Once at temperature, wait at least 1 h for the monochrometer, rotor and samples to fully equilibrate to temperature.

8. After an hour, turn on the work station PC attached to the AUC. Open the AUC software ProteomeLab. Wait several seconds for the software to connect to the AUC.
9. A text box will open entitled "NONAMExx.scn". This box contains all necessary parameters that need to be set before starting an AUC run. This can be saved as a unique file if desired.
10. Choose which rotor will be used in the experiment (4 or 8 hole).
11. Before running the velocity experiment at full speed, collect several low velocity scans of each cell to look for obvious leaks and to ensure adequate signal at the desired wavelength. In "XL Settings" box, input 3,000 rpm under "Speed" and the desired temperature (°C) in "Temp."
12. For each cell being run: (a) select "velocity", (b) check "Absorbance" box, (c) enter desired wavelength in "W1", and (d) enter sample information in "Comment." Click on the "Details" button. There are options for collecting data at a second and third wavelength – do not do this for velocity experiments. Ensure that the radial step size is 0.003 and the number of replicates is 1. Also make sure that "continuous" mode is selected, as well as "2" for "centerpiece" (as in two-sector centerpiece). Press "OK."
13. Press "Method" to enter the method for collecting the 3,000 rpm scans. "Delayed start" and "Time between scans" should both have zeros in the boxes. For "Number of scans," enter at least "2". This means 2 scans at 3,000 rpm will be collected for each cell. More scans are generally not necessary. Press "OK" to leave menu.
14. Press "Options" to go to the options menu. Make sure that "Stop XL after last scan" is unchecked, and that the number of scans being collected per cell (two in the above example) is entered into "Overlay last xx scans."
15. Press "Start method scan" to collect 3,000 rpm scans. Once the new windows (one for each cell) have opened on the screen with the completed scans check to make sure the absorbance is between 0.1 and 1.0. After the second scan is collected, make sure that the peaks signifying the reference and sample meniscus (Fig. 4) do not change position between scans. A shift to the right of either (or both) meniscus indicates a decrease in volume of the sector due

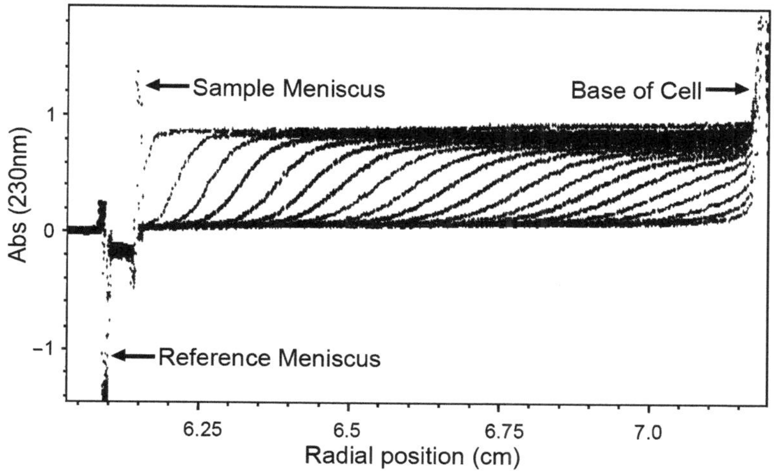

Fig. 4. SV data collected at 230 nm for human ER. The data are shown as absorbance vs. radial position of the sample. Only every fifth scan is shown for clarity.

to cell leakage. If there are changes in meniscus position, IMMEDIATELY stop the run and remove the cell(s).

16. If the initial scans at 3,000 rpm look acceptable, then stop the rotor. This can be accomplished by pressing the stop rotor button (red circular arrow with a cross-hashmark through it) on the centrifuge or by pressing "Speed," entering in "0," and pressing "Enter." Wait for the AUC speed to read "0." Reset the time stamp on the AUC by pressing "Time," entering "0," and pressing "Enter."
17. To set up the velocity experiment, most settings will stay the same. The first that needs to be changed is the desired rotor speed. The optimal experimental rotor speed can be estimated using simulation software available in UltraScan ((16); see Note 3). Replace 3,000 rpm in the "XL settings" box with the experimental speed. Go back to the "Method" menu. Change the "Number of scan" box to a number between 300 and 500. For most systems, this will be excessive, but can easily be reduced in subsequent experiments once the appropriate scan number is determined. Press "OK." A standard example is to run at 50,000 rpm and collect a very high number of scans per cell, such as 300. For proteins as small as 10 kDa, this speed should be sufficient for complete sedimentation to the base of the cell. Go to the "Options" menu. Check the box for "Stop XL after last scan" and change the number in "Overlay last xx scans" to "10."
18. Press "Start method scan." It is best to monitor the acceleration of the AUC to speed and observe at least the first several scans of each cell. If there is a problem, such

as a leak that was not observed at the low velocity scan, the AUC can be stopped before any damage has occurred. In this case, the cells must be disassembled, cleaned, reassembled, and reloaded.

19. The velocity run is finished when the absorbance across the cell is zero and there is no more movement of the boundaries. Figure 4 shows a complete SV run at a single concentration.
20. If the run is complete as determined in step 19, stop scan collection and AUC if necessary. Otherwise, release vacuum, remove the monochromoter, and then the rotor.
21. Copy experimental files onto a removable medium such as a flash drive. It is not recommended to analyze data on the same PC used to collect data.

3.4. Sample Recovery and Cleaning Centrifugation Cells

1. If desired, most of the protein sample can be recovered at the conclusion of the experiment. Invert the cell several times to resuspend any material sedimented to the base of the cell. Unscrew the plug and remove the plug gasket from the sample sector; remove as much of the sample as possible with a P-200 gel loading tip.
2. Upon removing sample, disassemble the cell in the reverse order that it was assembled.
3. Thoroughly rinse each piece under hot water.
4. Window liners, gaskets and housings should be rinsed thoroughly with diH_2O. Set out to dry on a Kimwipe.
5. For the cell housing and screw ring, wash the threads thoroughly with a detergent such as Contrad. This is essential to remove all Spincote residue. Wash the detergent off completely with hot water and rinse cell housing and screw ring with diH_2O.
6. Take the centerpiece and each window, rinse with hot water and let soak in dilute detergent. Do not let the centerpiece soak for more than 30 min as this may cause it to swell.
7. Wash all detergent off of the windows and centerpiece with hot water after soaking. Rinse with diH_2O and let air dry.
8. Store all components once dry. The centerpiece can be stored in the cell housing. Individually wrap each window in a Kimwipe to prevent scratches and dust accumulation.

3.5. Data Analysis and Interpretation

There are numerous programs available for SV data analysis, many of which are free (see Note 3). Each program offers its own functions and advantages. For example, Sednterp is useful for calculating the buffer density and partial specific volume at experimental temperature. Other programs such as Sedfit (17),

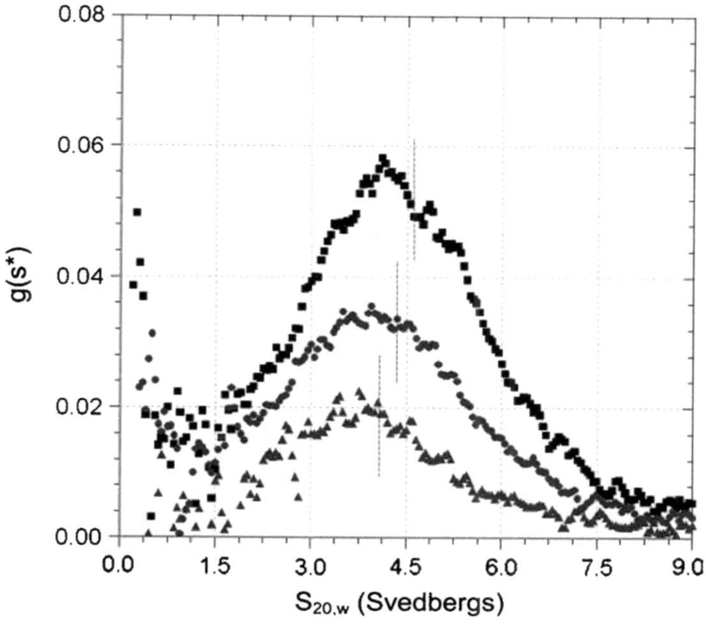

Fig. 5. Concentration dependence PR self-association using DCDT+. Initial loading concentrations were 0.5 (*triangles*), 1.0 (*circles*), and 1.5 μM (*squares*). *g*(*s**) distributions were determined by analysis of successive scans taken at each loading concentration. *Vertical dashed lines* indicate the weight-average sedimentation coefficient. Reprinted with permission from (13). Copyright 2006 American Chemical Society.

DCDT+ (18), Svedberg (19), SedAnal (20), and Ultrascan are used to find the sedimentation coefficient distributions from the raw data. From these distributions, a sedimentation coefficient (s) is resolved for each sedimenting species. This is useful because the average sedimentation coefficient is directly related to the average molecular weight (M) as seen below:

$$s_{20,\mathrm{w}} = \frac{M(1 - \bar{\upsilon} \cdot \rho)}{N_{\mathrm{A}} \cdot f} \tag{3}$$

where $s_{20,\mathrm{w}}$ is the sedimentation coefficient corrected to 20°C and water, $\bar{\upsilon}$ is the partial specific volume of the protein, ρ is the density of water at 20°C, N_{A} is Avogadro's number, and f is the frictional coefficient of the resolved species.

When analyzing SV data, concentration dependent increases in the sedimentation coefficient, due to the increase in the average value of M, are indicative of a dynamic self-association reaction. Analysis of PR SV data provides an example of this concentration dependence in $s_{20,\mathrm{w}}$. Figure 5 shows data collected at three concentrations (0.5, 1.0, and 1.5 μM). These data are from a model independent $g(s^*)$ analysis using DCDT+ that generates a sedimentation coefficient distribution (x-axis). It can be seen

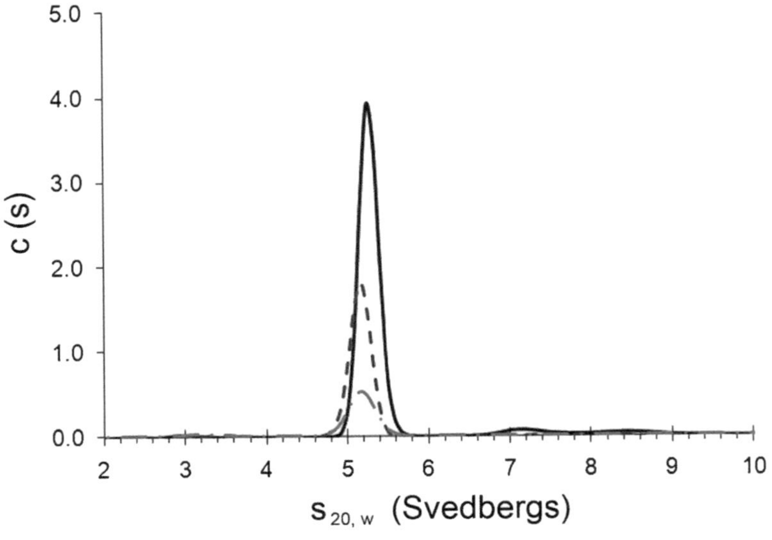

Fig. 6. Analysis of ER SV data showing a single, non-interacting species. Loading concentrations for these experiments were 0.3 (*dash-dot line*), 1.0 (*dashed line*), and 2.0 μM (*solid black line*). The resolved $s_{20,w}$ values were 5.2, 5.2, and 5.3 s respectively.

that there is a broad distribution of *s*-values (1.5–7.5 s) possibly suggestive of multiple species in solution. More importantly, the weight-average sedimentation coefficient (vertical black line) increases with increasing PR concentration, indicating that PR self-associates in the low micromolar range. Consistent with this interpretation, globally fitting the SV data to a monomer-dimer model using Sedphat (21) yielded a k_{di} for PR of 2 μM (13).

In contrast to PR, ER shows very different self-association properties. Figure 6 shows the sedimentation distribution using a different analysis software package, Sedfit. In contrast to the $g(s^*)$ distribution, the $c(s)$ analysis directly fits the data to the Lamm equation in a model-dependent manner. These sedimentation coefficient distributions cover a very narrow range of *s*-values (4.8–5.8 s) suggestive of a limited number of species. Additionally, individual analysis of each concentration shows a predominant peak with sedimentation coefficients of 5.2, 5.2, and 5.3 s at 0.3, 1.0, and 2.0 μM, respectively. Analysis of the size distribution ($c(M)$) (22) shows that the molecular weight of this species is 110 kDa, which is consistent with a dimer of ER. Since the *s*-value of the peak that represents ER dimer changes little over the observed concentrations, there must be no ER monomer-dimer equilibrium over this concentration range. Therefore, the dimerization constant must be at least tenfold less than the lowest protein concentration (i.e., $k_{di} \leq 0.030$ μM). Finally, studies of GR also result in a single peak (data not shown) at

Table 2
Resolved apparent and intrinsic binding affinities for ER, PR and GR

Steroid receptor	Apparent DNA binding affinity (K_{app}, nM)[a]	Dimerization constant (k_{di}, μM)[b]	Intrinsic DNA binding affinity (k_1, nM)[c]
ER	9.7	≤0.030	≥1.5
PR	28	1.1	0.4
GR	93	≥10	≤1.1

[a]Binding isotherm data were fit using Eq. 1
[b]Values obtained by sedimentation velocity analysis
[c]Binding isotherm data were fit using Eq. 2. The limits for ER and GR were obtained by using the limits of their respective dimerization constants

1 μM. However, analysis in Sedfit resolves an $s_{20,w}$ of 3.9 s and a molecular weight of only 78 kDa. Since the molecular weight of GR is 85 kDa, this indicates that the majority of GR exists as a single monomeric species. As with determining limits for ER, this means that the dimerization constant for GR is 10 μM or greater.

Now that the self-association reactions of PR, ER, and GR have been independently assessed using SV, their intrinsic DNA-binding affinities can be estimated using Eq. 2. Each data set in Fig. 2 was refit (solid lines) and the results are shown in Table 2. Although k_1 cannot be precisely determined for ER and GR, it is still possible to show the affect that dimerization has on resolution of k_1 by using the limits of k_{di} determined in the SV experiments. For example, the apparent binding constants (K_{app}) resolved by fitting to Eq. 1 (or by visual inspection of the data) would lead one to conclude that the affinities for each receptor span a tenfold range from roughly 10–100 nM. As can be seen when the data is fit to the correct model (Eq. 2), the overall conclusion is quite different. With the imposed limits on ER and GR dimerization, the intrinsic binding affinity for each receptor is approximately 1 nM; it is the dimerization constant that varies over several orders of magnitude. There is no way to reach this conclusion without performing a dissection of the linked dimerization reaction. This type of analysis becomes increasingly important with more complex systems, such as the interactions of steroid receptors with multisite natural promoters.

4. Notes

1. When monitoring samples at 230 nm, special attention must be paid to the buffer components. Phosphate is the preferred buffer for carrying out experiments at this wavelength because it has little to no absorbance. However, other buffers such as Tris–HCL or HEPES can still be used. Reducing agents are the most significant contributors to signal at 230 nm. β-mercaptoethanol (β-ME) is not recommended because it has too strong of an absorbance at this wavelength. On the other hand, a low concentration of DTT is acceptable, although caution must be exercised when using DTT at 280 nm because oxidized DTT has an absorbance peak at this wavelength. In any case, before using a buffer in an AUC experiment, it is a recommended to check the absorbance of the buffer against water to make sure the O.D. is less than 0.2 at the experimental wavelength.
2. Although this protocol focuses on absorbance optics, other options available are interference (Beckman Coulter XLI) and fluorescence (Beckman Coulter XLF). Any AUC that is equipped with interference optics is also capable of doing absorbance experiments, but fluorescence experiments must be performed on a specially equipped instrument. Using either of these optics involves modifications to the method that are outside the scope of this protocol, but are adaptable from the literature (23).
3. As mentioned in the text, there are numerous sources for data analysis software. Two of those mentioned are Sedfit and Sedphat, both of which are available for free download from: http://www.analyticalultracentrifugation.com/default.htm. The website also includes detailed tutorials for loading and analyzing data. Sedfit uses numerical solutions to the Lamm equation for direct fitting of the data while taking into account the effects of diffusion of the sample during sedimentation. It is useful for analysis of individual cells in an SV experiment to get information regarding molecular weights and evidence for self-association reactions occurring across several concentrations. Sedphat is used to simultaneously analyze data collected from multiple concentrations of sample in an SV experiment. This global analysis results in resolution of a dimerization constant. Other software for download includes Svedberg, DCDT+ and Sednterp, all available from: http://www.jphilo.mailway.com/default.htm, and SedAnal from: http://www.sedanal.bbri.org/. DCDT+ and SedAnal fit SV data using the time derivative method that does not

account for diffusion in the data. Svedberg provides another method of directly fitting the data and is particularly good at providing information on minor species. While not used for data analysis, Sednterp is very useful for calculating buffer properties and the temperature corrected partial specific volume of the protein. Finally, there is a suite of programs available called Ultrascan II. It is available for free download from: http://www.ultrascan.uthscsa.edu/. This suite provides analysis software for both SV and SE data, and offers several methods and models for these various analyses.

Acknowledgments

The authors would like to thank Michael T. Miura and Keith D. Connaghan. This work was supported by NIH grant DK061933 and the Avon Foundation for Women.

References

1. Heneghan, A. F., Connaghan-Jones, K. D., Miura, M. T., and Bain, D. L. (2007) Coactivator assembly at the promoter: efficient recruitment of SRC2 is coupled to cooperative DNA binding by the progesterone receptor. *Biochemistry* ***46***, 11023–32.
2. Connaghan, K. D., Heneghan, A. F., Miura, M. T., and Bain, D. L. (2010) Na(+) and K(+) allosterically regulate cooperative DNA binding by the human progesterone receptor. *Biochemistry* ***49***, 422–31.
3. Connaghan-Jones, K. D., Heneghan, A. F., Miura, M. T., and Bain, D. L. (2008) Thermodynamic dissection of progesterone receptor interactions at the mouse mammary tumor virus promoter: monomer binding and strong cooperativity dominate the assembly reaction. *J Mol Biol* ***377***, 1144–60.
4. Ackers, G. K., Johnson, A. D., and Shea, M. A. (1982) Quantitative model for gene regulation by lambda phage repressor. *Proc Natl Acad Sci U S A* ***79***, 1129–33.
5. Bain, D. L., Heneghan, A. F., Connaghan-Jones, K. D., and Miura, M. T. (2007) Nuclear receptor structure: implications for function. *Annu Rev Physiol* ***69***, 201–20.
6. Tsai, M. J., and O'Malley, B. W. (1994) Molecular mechanisms of action of steroid/thyroid receptor superfamily members. *Annu Rev Biochem* ***63***, 451–86.
7. Wyman, J., Gill, S. J. (1990) *Binding and linkage: functional chemistry of biological macromolecules*, University Science Books, Mill Valley, CA.
8. Heneghan, A. F., Connaghan-Jones, K. D., Miura, M. T., and Bain, D. L. (2006) Cooperative DNA binding by the B-isoform of human progesterone receptor: thermodynamic analysis reveals strongly favorable and unfavorable contributions to assembly. *Biochemistry* ***45***, 3285–96.
9. Heyduk, T., Ma, Y., Tang, H., and Ebright, R. H. (1996) Fluorescence anisotropy: rapid, quantitative assay for protein-DNA and protein-protein interaction. *Methods Enzymol* ***274***, 492–503.
10. Jameson, D. M., and Sawyer, W. H. (1995) Fluorescence anisotropy applied to biomolecular interactions. *Methods Enzymol* ***246***, 283–300.
11. Beckett, D., Koblan, K. S., and Ackers, G. K. (1991) Quantitative study of protein association at picomolar concentrations: the lambda phage cI repressor. *Anal Biochem* ***196***, 69–75.
12. Laue, T. M. (1995) Sedimentation equilibrium as thermodynamic tool. *Methods Enzymol* ***259***, 427–52.
13. Connaghan-Jones, K. D., Heneghan, A. F., Miura, M. T., and Bain, D. L. (2006) Hydrodynamic analysis of the human progesterone receptor A-isoform reveals that self-association

occurs in the micromolar range. *Biochemistry* *45*, 12090–9.

14. Correia, J. J. Analysis of tubulin oligomers by analytical ultracentrifugation. *Methods Cell Biol* *95*, 275–88.
15. Balbo, A., Zhao, H., Brown, P. H., and Schuck, P. (2009) Assembly, loading, and alignment of an analytical ultracentrifuge sample cell. *J Vis Exp*.
16. Demeler, B., Brookes, E., Wang, R., Schirf, V., and Kim, C. A. (2010) Characterization of Reversible Associations by Sedimentation Velocity with UltraScan. *Macromol Biosci* *10*, 775–82.
17. Schuck, P., and Rossmanith, P. (2000) Determination of the sedimentation coefficient distribution by least-squares boundary modeling. *Biopolymers* *54*, 328–41.
18. Philo, J. S. (2000) A method for directly fitting the time derivative of sedimentation velocity data and an alternative algorithm for calculating sedimentation coefficient distribution functions. *Anal Biochem* *279*, 151–63.
19. Philo, J. S. (1994) Measuring sedimentation, diffusion, and molecular weights of small moleules by direct fitting of sedimentation velocity concentration profiles, in *Modern analytical ultracentrifugation* (Schuster, T. M., and Laue, T. M., Eds.) pp 156–170, Birkhauser, Boston.
20. Stafford, W. F. (2000) Analysis of reversibly interacting macromolecular systems by time derivative sedimentation velocity. *Methods Enzymol* *323*, 302–25.
21. Schuck, P. (2003) On the analysis of protein self-association by sedimentation velocity analytical ultracentrifugation. *Anal Biochem* *320*, 104–24.
22. Schuck, P. (2000) Size-distribution analysis of macromolecules by sedimentation velocity ultracentrifugation and lamm equation modeling. *Biophys J* *78*, 1606–19.
23. MacGregor, I. K., Anderson, A. L., and Laue, T. M. (2004) Fluorescence detection for the XLI analytical ultracentrifuge. *Biophys Chem* *108*, 165–85.

Chapter 11

Fluorescence Correlation Spectroscopy and Allostery: The Case of GroEL

Gabriel A. Frank, Amnon Horovitz, and Gilad Haran

Abstract

Fluorescence correlation spectroscopy (FCS) is an experimental technique in which the equilibrium fluctuations of the fluorescent signal of molecules diffusing through a focused laser beam are measured. An autocorrelation analysis of these fluctuations provides information on dynamic processes, such as allosteric transitions, that the molecules undergo provided that they are fast relative to the diffusion time through the beam. In cases when the dynamics are slow relative to the diffusion time through the beam, FCS curves can yield information about the number of conformational states and their relative populations. Hence, FCS can be used to investigate allosteric systems with either slow or fast dynamics but the type of information gained in these two situations is different.

Here, the utility of the FCS technique is exemplified in the case of the single-ring version of the *Escherichia coli* molecular chaperone GroEL that interconverts with relatively slow dynamics between two allosteric states: a *T* state with low affinity for ATP and an *R* state with high affinity for ATP. Thermodynamic analysis suggests that the *T*-state population should become negligible with increasing ATP concentrations, in conflict with the requirement for conformation cycling, which is essential for the operation of molecular machines. Surprisingly, FCS experiments showed that, even at ATP saturation, ~50% of the molecules still populate the *T* state at any instance of time, indicating constant out-of-equilibrium cycling between *T* and *R*.

Key words: GroEL, Chaperonins, Fluorescence correlation spectroscopy, Allostery, Cooperativity, Singular-value decomposition, Oxazine dyes, Electron transfer-based quenching

1. Introduction

Fluorescence correlation spectroscopy (FCS) is an experimental technique first introduced by Magde et al. (1), in which the equilibrium fluctuations of the fluorescent signal of molecules diffusing through a focused laser beam are measured and analysed (2). An autocorrelation analysis of the fluctuations yields information on the time-scales of the different dynamic processes the system undergoes. FCS is most commonly used to study diffusion, which usually occurs on a time-scale of a few hundreds of

Aron W. Fenton (ed.), *Allostery: Methods and Protocols*, Methods in Molecular Biology, vol. 796,
DOI 10.1007/978-1-61779-334-9_11,

microseconds, but it can also report on faster dynamics, such as transitions between various photophysical or conformational states of a biomolecule. In the case of allosteric transitions, the characteristic time-scales are often relatively slow since they involve the correlated displacement of a large number of atoms and/or conformational changes with high energetic barriers. In cases when the conformational dynamics are slow relative to the diffusion time through the beam, FCS curves can be analysed as linear combinations of the contributions from different (allosteric) states of the system, thereby yielding information about the number of states and their relative populations. Hence, the FCS technique can be used to investigate allosteric systems when their dynamics are either slow or fast relative to their diffusion time through the beam but the type of information gained in these two situations is different. Here, the utility of the FCS technique is exemplified by the case of the single-ring version of the *Escherichia coli* molecular chaperone GroEL (SR1), which has a diffusion time of ~1 ms, which is shorter than the time-scale for inter-conversion between its allosteric states (3). As will be seen below, the main advantage of FCS over other methods that can be used to study the population of allosteric states is that no prior knowledge about the signals associated with the different states is required.

GroEL is a striking allosteric system that consists of two heptameric rings, stacked back-to-back, with a cavity at each end (4) in which protein folding can take place in a protective environment (5, 6). It functions in conjunction with a heptameric ring-shaped co-chaperonin, GroES, that caps the cavity of the so-called *cis* ring (7), thereby triggering dissociation of bound protein substrates into the cavity. GroEL undergoes allosteric transitions that are promoted by ATP binding. These transitions occur with positive cooperativity within rings, and negative cooperativity between rings (8). A nested allosteric model that accounts for these findings was put forward (8) in which, in accordance with the Monod–Wyman–Changeux (MWC) representation (9), each ring is in equilibrium between two states: a tense (T) state, with low affinity for ATP and high affinity for non-folded protein substrates, and a relaxed (R) state with high affinity for ATP and low affinity for non-folded protein substrates (10, 11). In the presence of increasing concentrations of ATP, the GroEL double-ring switches in a sequential manner from the TT state (both rings are in the T state) via the TR state to the RR state in accordance with Koshland–Némethy–Filmer (KNF) allosteric model (12). Binding of GroES to the ATP-bound R state stabilizes an R' state that is structurally distinct from the R state (7).

A prediction of the nested model is that the MWC representation can be used to describe the allosteric behavior of a single-ring version of GroEL (SR1). According to this model, the fraction of single-rings in the R state, f_R, is given by:

$$f_R = \frac{(1 + K_R[S])^N}{L(1 + K_T[S])^N + (1 + K_R[S])^N}, \quad (1)$$

where $[S]$ is the substrate (ATP) concentration, K_R and K_T are the association constants of ATP for the T and R states and L is the allosteric constant for the equilibrium between the unligated T and R states ($L = [T]/[R]$). Inspection of Eq. 1 shows that at high substrate concentrations the value of f_R should approach one since it is usually the case that $L(K_T/K_R)^N \ll 1$. In the case of GroEL, for example, this analysis implies that high ATP concentrations would fully shift the equilibrium of GroEL molecules to the R conformation (or the R' conformation in the presence of GroES). However, since a protein substrate prefers binding to the T form, this analysis suggests that ATP saturation would impede the folding activity of the chaperonin. Given this conundrum, we decided to determine the relative populations of molecules in the T and R states using FCS as described below.

In brief, it was first necessary to label the SR1 molecules with a probe that reports on the allosteric state of the molecule. We designed a switch-like probe based on a fluorophore, the oxazine dye Atto 655 (Atto-tec), which can be quenched by photo-induced electron transfer from an intrinsic tryptophan residue upon contact (13, 14). The switch is "off" (i.e., weakly fluorescent) in the T state and "on" (highly fluorescent) in the R or R' states. The labeled SR1 was then subjected to FCS experiments in which the fluctuations of the fluorescent signal of molecules diffusing through a focused laser beam were measured and analysed (15). Using singular-value decomposition (SVD), it was possible to describe the FCS curves as linear combinations of the contributions of the different allosteric states of the SR1 molecules, thereby providing information on their relative populations at different ATP concentrations. The analysis revealed that a significant T-state population remains even at saturating ATP concentrations. This type of analysis can be readily adopted to determine the populations of different states in other allosteric systems.

2. Methods

2.1. Rational Design of a Conformationally Sensitive Optical Switch

The design procedure described here is based on the static quenching of fluorescence resulting from the interaction of tryptophan and certain fluorophores such as oxazine derivatives, rhodamines, fluorescein (16), Alexa 488, Alexa 555, and Alexa 594 (17). The fluorophore and the tryptophan need to be positioned so that they will interact in one conformation of the protein but not in the other, thereby forming a conformationally sensitive optical switch. The oxazine derivative Atto 655 undergoes the strongest tryptophan-mediated quenching among the fluorophores listed above. The interaction between the tryptophan and the fluorophore is essentially short-range on the order of several angstroms (18). The formation and breakdown of the ground-state complex formed between the fluorophore and the tryptophan can be observed as a fast exponentially decaying process in the fluorescence autocorrelation curve (see below). The creation of the switch will usually involve replacing a carefully selected pair of amino acids with a cysteine and a tryptophan. If the protein contains reactive (e.g., solvent-exposed) cysteines that are not part of the switch then they should be replaced with, for example, serine or alanine. The desired purified protein is then labeled with the fluorescent dye. The fact that the design of the switch requires only one chemical modification simplifies the labeling and purification procedures involved with formation of the switch. This feature of the design becomes crucial for homo-oligomers where control over the relative position of several chemical modifications is difficult.

2.2. Protein Labeling

A stock solution of the labeling reagent [e.g., Atto655-maleimide (ATTO-TEC, Germany)] is prepared by dissolving some or all of the dry material in the vial in dimethylsulfoxide (DMSO) (e.g., 0.1 mg in 60 μL). Alkylation of cysteine by maleimide and maleimide hydrolysis are competing reactions. In order to minimize hydrolysis, care must be taken to keep the labeling reagent in dry conditions. When kept in ambient atmosphere, DMSO contains water from atmospheric humidity. Consequently, hydrolysis will start immediately after dissolving the labeling reagent in DMSO. It is, therefore, important to prepare fresh labeling stock solution before every labeling reaction. In order to minimize hydrolysis, the preparation of the stock solution should be the last step before the start of the labeling reaction. Certain fluorophores tend to form complexes when dissolved at high concentrations. The concentration of the stock solution of the labeling reagent should, therefore, be determined by absorbance measurements of a series of different dilutions in buffer. A linear relationship between the absorbance and the dilution factor will indicate that complexes did not form. The molar extinction coefficient of Atto655 at 663 nm is ~ 125,000 $(\text{cm}\cdot\text{M})^{-1}$. The labeling reaction begins by adding the protein solution (e.g., 0.5 mL of 100 μM SR1 in a buffer that does not contain functional groups that can interact with maleimides or cysteines) to the stock labeling solution (e.g., 25 μL) and mixing them thoroughly without producing foam. The labeling reaction mixture is incubated at room temperature for 2 h in the dark and then quenched by adding

2.5 μL of β-mercaptoethanol diluted tenfold in buffer. The protein and dye concentrations used in the labeling reaction depend on the desired labeling ratio that in the case of SR1 was ~0.7 fluorophores per seven subunits (15).

2.3. Removal of the Unreacted Dye

It is essential to separate the labeled protein from the unreacted dye. This can be done by loading the quenched reaction mixture on a PD-10 desalting column (GE Healthcare) equilibrated with a buffer in which the protein of interest will be stored [e.g., 50 mM Tris–HCl (pH 7.5) containing 10 mM $MgCl_2$, 10 mM KCl, and 1 mM DTT (buffer A) in the case of SR1]. Sample loading and elution are carried out according to the manufacturer's instructions. These steps should then be repeated using a second PD-10 column. The eluted protein is then diluted to 20 mL with buffer A and transferred to a centrifugation concentration device with a molecular weight cut-off smaller than that of the protein of interest, for further removal of any traces of free dye. The protein is concentrated to 0.5 mL and this procedure is then repeated six times.

2.4. Assessment of the Labeling Ratio

The labeling ratio (LR) can be assessed by comparing the absorbance spectrum of the labeled protein (after removal of the unreacted dye) with the absorbance spectra of the dye and the unlabeled protein. The interactions of the dye and the tryptophan residue(s) with each other and with other groups or the solvent might affect their spectra. Hence, spectrophotometric measurements should be carried out on the unfolded state, which can be generated by adding, for example, 10 μL of the concentrated labeled protein to 490 μL of a buffered 6 M urea solution. The unfolding reaction should then be allowed to proceed for 3 h at room temperature in the dark. The dye solution should be prepared by adding 5 μL of the stock dye solution to 45 μL β-mercaptoethanol diluted tenfold in buffer and then transferring 10 μL of this solution to 490 μL of buffered urea solution. The absorbance measurements should be done over a wavelength range that spans the absorbance peaks of the dye and the protein, e.g., 260–740 nm, and be carried out using the buffered urea solution as a blank. The ratio of the molar absorptivities of the dye, η_{dye}, at the wavelengths corresponding to maximum absorption of the protein is then calculated. In the case of Atto655, for example, $\eta_{dye} = \varepsilon^{dye}_{663nm}/\varepsilon^{dye}_{280nm} \approx 12.5$. Using this ratio, the contribution of the dye to the labeled protein's absorption at 280 nm can be readily calculated ($A^{dye}_{280} = A^{dye}_{\lambda dye}/\eta_{dye}$). The labeling ratio is then determined using the following equation:

$$LR = \frac{A_{\lambda dye}/\varepsilon^{dye}_{\lambda dye}}{(A_{280} - A^{dye}_{280})/\varepsilon^{protein}_{\lambda 280}}, \quad (2)$$

where $A_{\lambda dye}$ and A_{280} are the total absorption of the labeled protein at the wavelengths corresponding to maximum absorption by the dye and at 280 nm, respectively, and $\varepsilon^{dye}_{\lambda dye}$ and $\varepsilon^{protein}_{\lambda 280}$ are the molar extinction coefficients of the dye at the wavelength corresponding to the maximum absorption of the dye and of the protein at 280 nm, respectively. The molar extinction coefficient of the dye at the wavelength corresponding to its maximum absorption can be found in handbooks or obtained from the manufacturer's data. The molar extinction coefficient of the protein at 280 nm can be calculated based on the sequence (19). In the case of a homo-oligomer when labeling of one monomer is not affected by the labeling of other monomers, the fraction of multiply-labeled proteins, f_{mult}, can be predicted using the binomial distribution, as follows:

$$f_{\mathrm{mult}} = 1 - \left(1 - \frac{\mathrm{LR}}{n}\right)^{n} - \mathrm{LR}\left(1 - \frac{\mathrm{LR}}{n}\right)^{n-1}, \qquad (3)$$

where f_{mult} is the fraction of multiply-labeled oligomers, n is the number of subunits and LR, as before, is the labeling ratio.

2.5. Preparation of Flow Cells for FCS Experiments

18 × 18 mm and 24 × 24 mm #1.5 cover slips (Paul Marienfeld GmbH & Co, Germany) should be cleaned by sonication in 10% HF, followed by careful rinsing with double-distilled water, and drying. Next, ~3 × 22 mm spacers should be prepared using parafilm. Place on each 24 × 24 cover slip two parallel parafilm spacers ~17 mm apart, cover them with the 18 × 18 mm cover slip and then press gently. Allow the two sides of the flow cells to stick to the spacers by baking them for 10 min at 110°C. The flow cells should be kept in a clean vessel until use.

2.6. Description of the Fluorescence Correlation Spectrometer

The signal picked up by a fluorescence correlation spectrometer is a result of the fluctuations around equilibrium of the system under study (Fig. 1a). The amplitude of these fluctuations scales with $N^{-0.5}$, where N is the number of molecules. Consequently, the auto-correlation function calculated from the fluorescence signal scales with $1/N$. FCS measurements are, therefore, conducted while having a small number of molecules in the observation volume (typically 0.5–10 on average), which can be readily achieved if this volume is of the order of 1 fL. Such a small observation volume is naturally realized in a microscopy system based on confocal optics. While commercial FCS instruments now abound, we chose to work with a home-built system, designed for our needs.

Figure 1b is a schematic drawing of our FCS system. The excitation laser beam (continuous arrows going right) is expanded by a telescope (TL) and sent through the tilted glass (TG). The laser is reflected by the dichroic mirror (DM) into the infinity-corrected water objective (OB), which focuses the laser into the

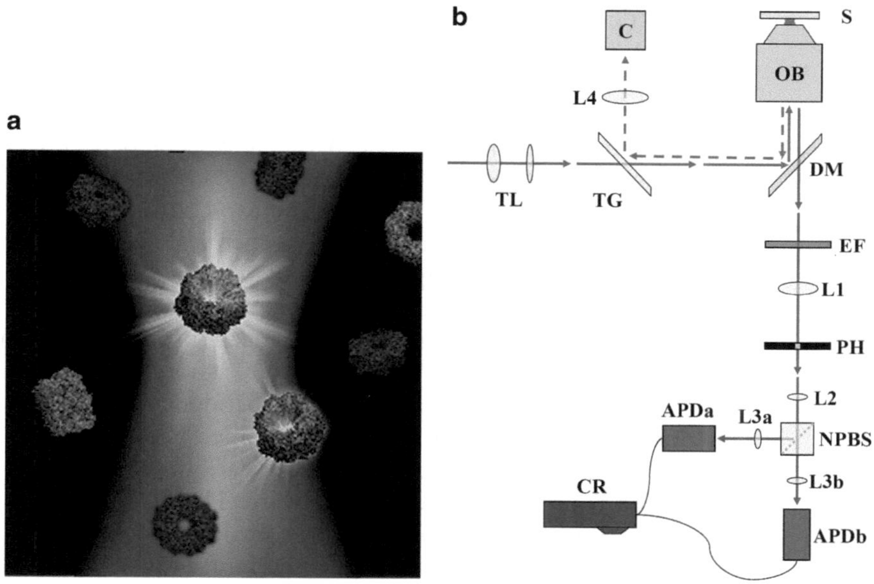

Fig. 1. (**a**) When labeled molecules of SR1 pass through a focused laser beam they emit fluorescence. More fluorescence is emitted by molecules in the *R* state (*light gray*) than by molecules in the *T* state (*dark gray*). The overall fluorescence intensity fluctuates due to the diffusion of the molecules through the beam. A correlation analysis picks up these fluctuations in the FCS experiment. (**b**) A schematic drawing of our home-built FCS setup. The telescope TL expands the excitation beam in order to fill the back aperture of the objective leading to as small as possible diffraction limited spot. Further reduction of the observation volume is achieved by passing the emission through the spatial filter formed by the lenses L1, L2, and the pinhole PH. The emission is divided by a non-polarizing beam splitter NPBS into two APD. The resulting stream of digital signals resulting from light impinging on the detector is fed into a correlator.

sample (S). The OB collects the emission and transmits it as a collimated beam. The emission (arrows going down) passes through the DM and the emission filters (EF) into a spatial filter consisting of: the lens L1 which focuses the emission on the pinhole (PH) and the lens L2 which collimates the emission once again. A non-polarizing-beam-splitter (NPBS) splits the emitted light into two equal intensity channels. Lenses on each channel (L3a and L3b) focus the light on the avalanche photo diodes (APDa and APDb). The correlator (CR) records the digital signals resulting from the arrival of photons to the APDs. The OB collects the reflection of the laser from the glass surface of the sample and the reflected laser light is transmitted to the DM (broken arrows). The DM reflects the laser back to the TG. The TG reflects a small fraction of the laser to the lens (L4) which focuses the laser on the camera (C), thereby providing a means for monitoring the position of the sample relative to the laser focus.

2.7. Sample Preparation and Data Collection

The surface of the flow cell should be blocked with BSA before the protein sample is introduced into it. Load the flow cell with 200 μL of 2 mg/mL BSA solution, removing excess liquid by gentle vacuum suction. Incubate in closed humid vessel for ~30 min, then wash the flow cell with 600 μL of buffer A. Load 250 μL of the desired fluorescently labeled protein solution into the flow cell. FCS measurements are typically conducted with samples containing 1–10 nM of fluorescent particles. Measurement time of a single sample can take between 30 min to several hours. Consequently, it is important to control substrate concentrations in samples that undergo a chemical reaction (like nucleotide hydrolysis) during the measurement.

In order to start the measurements, place the flow cell in the FCS system and set the laser focus deep into the sample (typically 20 μm). Perform multiple (at least 30) short ~60 s FCS measurements. Perform three or four independent repeats of the above process, using a fresh sample each time. These separate data sets will be used for parameter error estimation. Observe each data set using your favorite plotting software. The FCS curves of each set should overlap each other well. A small number of curves in each data set might be found to be aberrant and should be removed. Use the remaining curves to calculate an average curve, as well as to obtain the variance at each time point.

3. Data Analysis

3.1. Singular-Value Decomposition

Singular-value decomposition (SVD) is a method that enables the representation of complex data sets as a linear combination of the smallest required number of "singular vectors". The analysis also provides the weights of these vectors in terms of "singular values". A detailed description of SVD with references to the mathematical literature can be found in the book Numerical Recipes (20). Briefly, suppose one has n vectors in a certain vector space (each vector is a data set), SVD constructs out of these vectors a basis set of n orthonormal vectors (singular vectors). An $n \times n$ diagonal matrix of weights, the so-called singular values, is also obtained in the analysis. The singular values are ranked by their magnitude. When analyzing experimental data, it often happens that only m of these weights ($m < n$) have a significant value, implying that the underlying phenomenon is controlled by a linear combination of a small number of processes. In such cases, it should be possible to reconstruct the original data using only those data sets that correspond to the m non-zero singular values.

To perform SVD analysis on a series of (averaged) FCS curves measured as a function of solution conditions, first arrange them as vectors in one matrix. Perform SVD analysis on this matrix [using, e.g., the command SVD in Matlab (Mathworks)], to obtain the matrix of singular values, the matrix of singular vectors and a matrix of weights. Check whether the first few singular values are dominant, implying that the whole data set can be spanned by the corresponding singular vectors. The following discussion assumes that there are two dominant singular values, as was the case in the SVD analysis of ATP-dependent FCS curves of SR1 molecules. The discussion can be readily generalized to the case of a larger number of dominant singular values. It can be directly verified that two singular vectors span the whole data set by reconstructing each FCS curve using these two vectors (consult the Help section in Matlab or the chapter on SVD in Numerical Recipes (20) in order to acquaint yourself with the vector algebra required for this reconstruction). A successful reconstruction leads to the conclusion that the conformational space of the studied molecules comprises only two states with respect to the observable of the experiment (e.g., the fluorescence of the optical switch).

3.2. Physical Interpretation of SVD Results

To cast the SVD results in physically meaningful terms, select vectors which correspond to stable states, or pure species of the studied molecule, for example, the FCS vectors of the T and R' states in the case of GroEL. The SVD weights can be used to express all FCS curves as linear combinations of the curves corresponding to the pure species of the system. To see how this is done, we first need to understand the relationship between a linear combination of correlation functions and the populations of the contributing species. The derivation below is written for the case of two pure species, but can be extended if more than two states are involved.

Consider a solution of two uncorrelated species, with n_1 and n_2 molecules in the sampling volume, and quantum efficiencies Q_1 and Q_2, respectively. The correlation function of the solution can be written in terms of the correlation functions of the two pure species in the following way:

$$G(t) = \frac{{Q_1}^2 {n_1}^2}{(Q_1 n_1 + Q_2 n_2)^2} G_1(t) + \frac{{Q_2}^2 {n_2}^2}{(Q_1 n_1 + Q_2 n_2)^2} G_2(t). \quad (4)$$

$G_1(t)$ and $G_2(t)$ are inversely proportional to n_1 and n_2, respectively, that is, the number of molecules of each species in the mixture. In reality the experiment is performed over a range of conditions, and we therefore have series of functions $G_i(t)$. The indices 1 and 2 are reserved for the pure species functions.

To proceed, one needs to get rid of this dependency. To do that, one first needs to fit $G_1(t)$ and $G_2(t)$ to the standard functional form of fluorescence correlation functions:

$$G_{1(2)}(t) = \frac{1}{N_{1(2)}} \left[1 + \sum_{\mathrm{j}} K_{\mathrm{j}} \mathrm{e}^{-t/\tau_{\mathrm{j}}} \right] \left[1 + \frac{\tau}{\tau_{\mathrm{D}}} \right]^{-1} \left[1 + \frac{\tau}{\omega^2 \tau_{\mathrm{D}}} \right]^{-\frac{1}{2}}. \quad (5)$$

Here, $N_{1(2)}$ is the number of molecules of the pure species 1 or 2 in the sampling volume. The terms in the first square parentheses in this equation represent correlation loss due to all intramolecular dynamic processes, described by a series of exponential functions with amplitudes K_j and lifetimes τ_j. The terms in the other two square parentheses represent correlation loss due to diffusion, with τ_D being the diffusion time through the laser beam and ω the aspect ratio of the beam at the focus (2). The parameters $N_{1,2}$ are obtained as fit parameters that can be used to multiply the functions $G_{1(2)}(t)$ to obtain new functions $g_{1(2)}(t) = N_{1(2)}G_{1(2)}(t)$. All other correlation functions are then normalized so that their long-time parts, which are only due to diffusion, match those of the two pure correlation functions. This normalization effectively leads to a new set of correlation functions, which have the following form:

$$g_i(t) = a_1^i g_1(t) + a_2^i g_2(t) = \frac{Q_1{}^2 n_1^i}{Q_1{}^2 n_1^i + Q_2{}^2 n_2^i} g_1(t) + \frac{Q_2{}^2 n_2^i}{Q_1{}^2 n_1^i + Q_2{}^2 n_2^i} g_2(t). \tag{6}$$

Since each of the functions $g_i(t)$ can be written in terms of the two singular vectors obtained in the SVD analysis, the weights a_1^i and a_2^i can be readily calculated from the SVD weights. With these new weights in hand, one can calculate the population fractions $f_{1(2)}^i = (n_{1(2)}^i)/(n_1^i + n_2^i)$ by using the ratio of quantum efficiencies, Q_1/Q_2. This ratio can be obtained from experiments on the pure species, using standard methods (note that the absolute quantum yields are not required, only their ratio!).

4. Demonstration of the Utility of the FCS Method in the Case of GroEL

We performed a series of FCS experiments on SR1 molecules over a broad range of ATP concentrations (15). SVD analysis was carried out on the measured FCS curves. In the SR1 experiments, the ratio of the quantum efficiencies of the R and T states is between 2 and 3 (14), and it can be used to obtain the population fractions of the T and R states. In Fig. 2, we show the population fractions calculated using a quantum efficiencies ratio of 2.5. By definition, at 0 M ATP the whole population is in the T state (the fraction of the T state is 1 and the fraction of the R state is 0) whereas in the presence of saturating ATP concentrations and GroES, the whole population is in the R (R') state (the fraction of the T state is 0 and the fraction of the R state is 1). It may be seen that in between these two extremes, the fraction of the T state

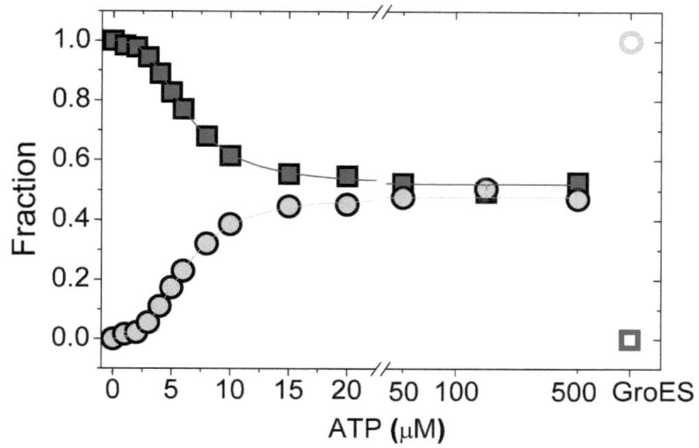

Fig. 2. The fractions of the *T* and *R* states (in *filled squares* and *circles*, respectively), as obtained from SVD analysis of curves measured at varying ATP concentrations, using a quantum efficiencies ratio of 2.5. At ATP saturation, the fraction of the R state is not more than 50%. The empty symbols designate the fractions in the presence of GroES. The full lines are the fits of the data to a Hill-like function.

decreases while the fraction of the *R* state increases in a sigmoidal manner. Fitting the data in Fig. 2 to the Hill equation yields a Hill coefficient of 2.7 ± 0.1 (15). Surprisingly, at ATP saturation (and in the absence of GroES) the fraction of the *T* state does not reach 0, and the fraction of the *R* state does not reach 1. Thus, under these conditions, *only about half of the molecules populate the R state, while the other half reside in the T state.* This result is in contrast to notions based on thermodynamic analysis of the allosteric transitions in GroEL (see Eq. 1 and discussion thereof) and suggests that ATP saturation drives GroEL strongly out-of-equilibrium. It shows the strength of the method in enumerating stable states and identifying their populations even when standard equilibrium assumptions break down.

References

1. Magde D, Elson EL, Webb WW (1974) Fluorescence correlation spectroscopy II. An Experimental realization. Biopolymers 13:29–61.
2. Krichevsky O, Bonnet G (2002) Fluorescence correlation spectroscopy: the technique and its applications. Rep Prog Phys 65:251–297.
3. Amir A, Horovitz A (2004) Kinetic analysis of ATP-dependent inter-ring communication in GroEL. J Mol Biol 338:979–988.
4. Braig K, Otwinowski Z, Hegde R. Boisvert DC, Joachimiak A, Horwich AL, Sigler PB The crystal structure of the bacterial chaperonin GroEL at 2.8 Å. (1994) Nature 371:578–586.
5. Horovitz A, Willison KR (2005) Allosteric regulation of chaperonins. Curr Opin Struct Biol 15:646–651.
6. Horwich AL, Fenton WA (2009) Chaperonin-mediated protein folding: using a central cavity to kinetically assist polypeptide chain folding. Q Rev Biophys 42:83–116.
7. Xu Z, Horwich AL, Sigler PB (1997) The crystal structure of the asymmetric GroEL-GroES-$(ADP)_7$ chaperonin complex. Nature 388:741–750.
8. Yifrach O, Horovitz A (1995) Nested cooperativity in the ATPase activity of the

oligomeric chaperonin GroEL. Biochemistry 34:5303–5308.

9. Monod J, Wyman J, Changeux JP (1965) On the nature of allosteric transitions: a plausible model. J Mol Biol 12:88–118.
10. Staniforth RA, Burston SG, Atkinson T, Clarke AR (1994) Affinity of chaperonin-60 for a protein substrate and its modulation by nucleotides and chaperonin-10. Biochem J 300:651–658.
11. Yifrach O, Horovitz A (1996) Allosteric control by ATP of non-folded protein binding to GroEL. J Mol Biol 255:356–361.
12. Koshland DE Jr., Némethy G, Filmer D (1966) Comparison of experimental binding data and theoretical models in proteins containing subunits. Biochemistry 5:365–385.
13. Neuweiler H, Schulz A, Böhmer M, Enderlein J, Sauer M (2003) Measurement of submicrosecond intramolecular contact formation in peptides at the single-molecule level. J Am Chem Soc 125:5324–5330.
14. Frank GA, Kipnis Y, Smolensky E, Daube SS, Horovitz A, Haran G (2008) Design of an optical switch for studying conformational dynamics in individual molecules of GroEL. Bioconjug Chem 19:1339–1341.
15. Frank GA, Goomanovsky M, Davidi A, Ziv G, Horovitz A, Haran G (2010) Out-of-equilibrium conformational cycling of GroEL at saturating ATP concentrations. Proc Natl Acad Sci USA 107:6270–6274.
16. Götz M, Hess S, Beste G, Skerra A, Michel-Beyerle ME (2002) Ultrafast electron transfer in the complex between fluorescein and a cognate engineered lipocalin protein, a so-called anticalin. *Biochemistry* 41:4156–4164.
17. Chen H, Ahsan SS, Santiago-Berrios MB, Abruña HD, Webb WW (2010) Mechanisms of quenching of Alexa fluorophores by natural amino acids. J Am Chem Soc 132: 7244–7245.
18. Vaiana AC, Neuweiler H, Schulz A, Wolfrum J, Sauer M, Smith JC (2003) Fluorescence quenching of dyes by tryptophan: interactions at atomic detail from combination of experiment and computer simulation. J Am Chem Soc 125:14564–14572.
19. Gill SC, von Hippel PH (1989) Calculation of protein extinction coefficients from amino acid sequence data. Anal Biochem 182:319–326.
20. Press WH, Teukolsky SA, Vetterling WT, Flannery BP (2003) *Numerical Recipes, the Art of Scientific Computing* (Cambridge University Press, Cambridge, UK).

Chapter 12

The Morpheein Model of Allostery: Evaluating Proteins as Potential Morpheeins

Eileen K. Jaffe and Sarah H. Lawrence

Abstract

An equilibrium mixture of alternate quaternary structure assemblies can form a basis for allostery. The morpheein model of allostery is a concerted dissociative model that describes an equilibrium of alternate quaternary structure assemblies whose architectures are dictated by alternate conformations in the dissociated state. Kinetic and biophysical anomalies that suggest that the morpheein model of allostery applies for a given protein of interest are briefly described. Two methods are presented for evaluating proteins as potential morpheeins. One is a subunit interchange method that uses chromatography, dialysis, and mass spectroscopy to monitor changes in multimer composition. The other is a two-dimensional native gel electrophoresis method to monitor ligand-induced changes in an equilibrium of alternate multimeric assemblies.

Key words: Allostery, Morpheein, Quaternary structure equilibrium, 2D native PAGE, Subunit interchange

1. Introduction

1.1. The Morpheein Model of Allosteric Regulation

The morpheein model of allosteric regulation is a concerted dissociative model wherein the allosteric protein exists in a dynamic and reversible equilibrium of alternate oligomers, the interconversion of which requires a conformational change of composite subunits in the dissociated state prior to reassembly to an alternate oligomer (Fig. 1a) (1). In this model, one conformer of the protein allows assembly to one particular multimer, while an alternate conformer of the protein allows assembly to a structurally and functionally distinct alternate multimer. Small molecule binding that stabilizes one multimer over the other multimer will shift the quaternary structure equilibrium, altering the protein's function toward that of the targeted multimer, thus

Aron W. Fenton (ed.), *Allostery: Methods and Protocols*, Methods in Molecular Biology, vol. 796,
DOI 10.1007/978-1-61779-334-9_12, © Springer Science+Business Media, LLC 2012

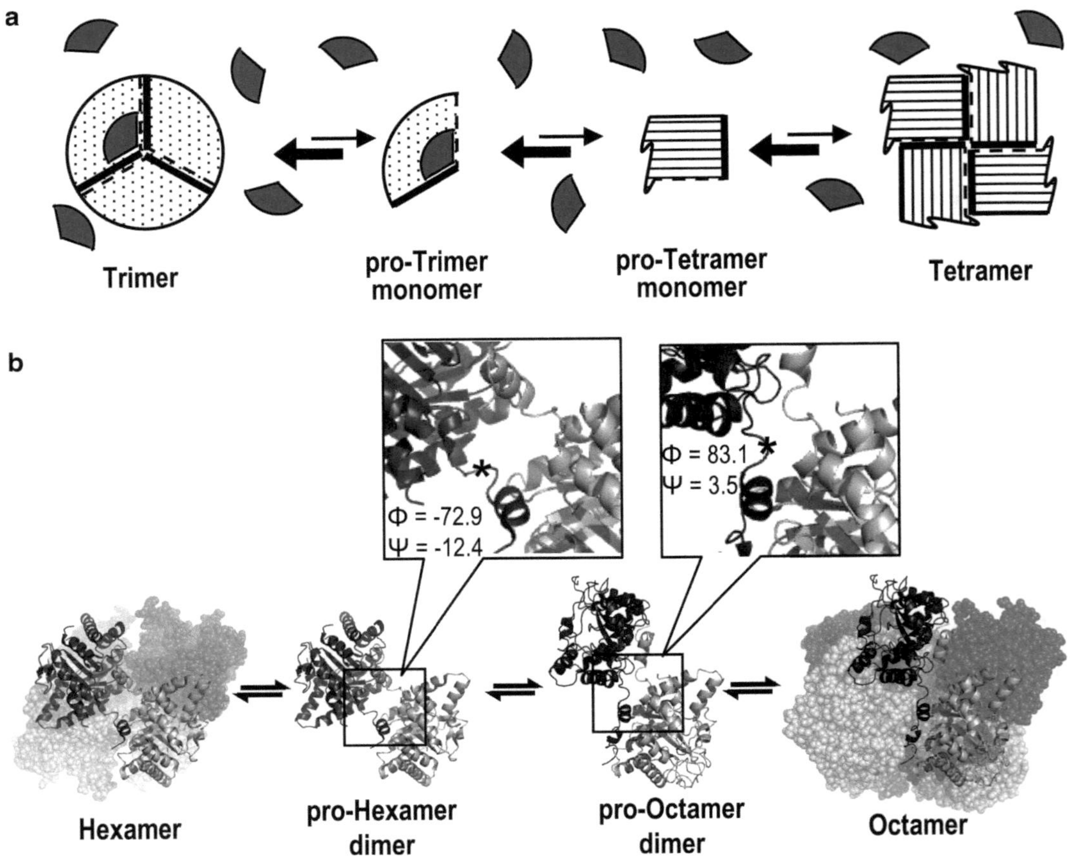

Fig. 1. The morpheein model of allosteric regulation illustrated with a schematic example and with *Hs*PBGS. *Panel a* depicts a homo-oligomeric protein for which the available functionally distinct multimeric states are a trimer and a tetramer, and two conformations of a monomer, one that can trimerize and one that can tetramerize. The "rule of engagement" for multimeric assembly is the association of a *thick line* with a *dashed line*. An allosteric regulator molecule (depicted as a *grey wedge*) has the appropriate geometry to bind only to the pro-trimer and trimer forms and shift the equilibrium in that direction, thus allosterically stabilizing the trimer and promoting its function. *Panel b* illustrates the dynamic equilibria between the quaternary structure assemblies of *Hs*PBGS. The two predominant forms in solution are the high activity octamer (PDB code 1E51) and the inactive hexamer (PDB code 1PV8); interconversion between these forms requires dissociation to a dimer, which can take on alternate conformations (4, 7). The interchange between dimer conformations involves alterations in a small number of Φ/Ψ angles that change the orientation of the αβ-barrel domains of each subunit with respect to each other. One significant change in Φ/Ψ angles is at residue 23 (*asterisk*), highlighted in the magnified panels.

providing a mechanism for allostery. In this model, the interconversion between conformers is fully reversible and does not require core structure refolding such as is seen in amyloidogenic proteins.

Proteins that can function in this way are called morpheeins. At present, the best-characterized morpheein is the enzyme porphobilinogen synthase (PBGS, EC 4.2.1.24), which catalyzes the first common step in tetrapyrrole biosynthesis. For PBGS, the functional distinction between alternate oligomers is the level of enzymatic activity; one oligomer is the "on" state and the other is the "off" state. This chapter covers two biophysical methods that can be used to address whether or not a particular protein functions as a morpheein; PBGS is used as an illustrative example. These methods are applicable to soluble proteins that migrate in native PAGE.

1.2. Atypical Data Suggestive of Morpheeins

In general, one is taught to think about homo-oligomeric protein structure and function in terms of a single multimer or derivatives thereof (e.g., a tetramer that dissociates into its component dimers or monomers; a decamer that dissociates into its component pentamers, dimers, or monomers). This was the structural framework used to interpret data on PBGS for several decades. Most of these data were consistent with the idea that PBGS formed an octamer, as was our own laboratory data and emerging crystal structures (e.g., (2)). These octamers were considered to dissociate into their component tetramers, dimers, or monomers. However, select publications using a variety of biochemical and or biophysical methods suggested the possibility of a hexameric assembly.

Our subsequent detailed investigation into the root of these anomalous findings led to the realization that PBGS from many species can exist as an equilibrium of high activity octamers and low activity hexamers whose interconversion requires dissociation and occurs at the level of a conformationally flexible dimer (Fig. 1b) (3–6). For human PBGS (*Hs*PBGS), the mole fraction of the dimers is ~0.5%, and the position of the equilibrium is pH dependent (7). At neutral pH, the protein is predominantly octameric and the octamer exhibits a high V_{max} and a low K_m. At basic pH, the equilibrium shifts toward the hexamer, which exhibits a low V_{max} and a high K_m. At basic pH, where there is a mix of octamers and hexamers, an experimental determination of V_{max} and K_m does not fit well to a single Michaelis–Menten equation. The data may conform to a fit reflecting negative cooperativity; however, the best fit (and the one that correctly reflects the behavior of PBGS in solution) is to a double hyperbolic equation reflecting a mixture of the two kinetically distinct (and slowly interconverting) multimeric states as we have described previously (8). Furthermore, small molecules have been demonstrated to bind to the protein in such a way as to shift this quaternary structure equilibrium, thus effecting function and establishing the equilibrium as a basis for allostery (4, 6, 9–11).

We have published extensively on physical and kinetic analyses of the PBGS morpheein equilibrium and do not reiterate these findings here (3, 7, 8, 12). There are many kinetic and biophysical anomalies in the literature that suggest that various proteins may be morpheeins. While the purpose of this chapter is not to provide a comprehensive discussion of such anomalies, the following list provides a brief description of some characteristics that might suggest that a particular homo-oligomeric protein may use the morpheein model of allosteric regulation.

1.2.1. Biophysical Indications of Alternate Quaternary Structure Stoichiometries

There are numerous examples of proteins for which different quaternary structure stoichiometries have been reported based on the use of different biophysical techniques for analysis (e.g., analytical ultracentrifugation, size exclusion chromatography, X-ray crystallography). Historically, such data is typically reconciled by presuming that one oligomer is "correct" and the others are artifacts. However, as part of its normal function, a morpheein can experience different quaternary assemblies under different solution conditions, or in the presence of different ligands (Fig. 2a). Alternatively, a protein that is deemed homogeneous by SDS PAGE may be seen to separate into multiple forms using other methods such as native PAGE (Fig. 2b), or by various chromatographic methods (ion exchange, size exclusion, or hydrophobic). These alternate forms are often interpreted as being chemically distinct (e.g., having a covalent modification, or different disulfide linkages), but for morpheeins this interpretation is not necessarily valid. It is important to note that the use of tags for affinity purification will prevent the chromatographic identification of alternate morpheein forms that might otherwise separate. It is our experience that the addition of tags can also alter the quaternary structure equilibrium (13).

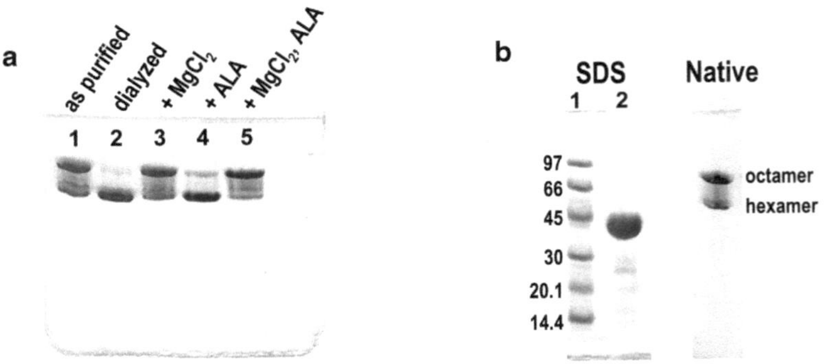

Fig. 2. Electrophoretic analysis of PBGS from *Pisum sativum* (greeen pea) (*Ps*PBGS). (**a**) Native PAGE analysis of *Ps*PBGS (1 mg/mL) at low ionic strength. *Lane 1* – as purified; *lane 2* – dialyzed *vs.* 10 mM BTP-HCL pH 8.5; *lane 3* – the sample shown in *lane 2* with 10 mM $MgCl_2$ added in the protein dilution step; *lane 4* – the sample shown in *lane 2* with 1 mM ALA added in the protein dilution step; *lane 5* – the sample shown in *lane 2* with 10 mM $MgCl_2$ and 1 mM ALA added in the protein dilution step. (**b**) SDS and native PAGE analysis of purified *Ps*PBGS. SDS *lane 1* – molecular weight markers; SDS *lane 2* – *Ps*PBGS (3 mg/mL) as purified (100 mM Tris–HCl, pH 8.5, 10 mM $MgCl_2$); Native – *Ps*PBGS (1 mg/mL) as purified.

1.2.2. Kinetic Indications of Multiple Protein Forms

In the event that a component of the assay mix induces a dissociative quaternary structure transition from a low activity form to a high activity form, this can manifest as kinetic hysteresis, first defined by Frieden (14). An example of such a transition is seen for GDP-mannose dehydrogenase, which is a putative morpheein (1, 15). A dependence of the kinetic behavior of a protein on the order of addition of reaction components also suggests multiple protein forms may be in a metastable equilibrium (15, 16). A protein concentration dependence to the specific activity of an enzyme can occur when a high activity oligomer of one stoichiometry is in equilibrium with a low activity oligomer of a different stoichiometry, and the protein concentration used in the assay is in the range of the dissociation constant for one or another oligomer (e.g., (17)).

1.2.3. Moonlighting Functions

Many proteins are being reported to have alternate and unrelated moonlighting functions (18). An outstanding example of a homo-oligomeric moonlighting protein is glyceraldehyde-3-phosphate dehydrogenase (GAPDH), which has been reported to have many alternate functions (e.g., (19–21)). Alternate GAPDH quaternary structure assemblies have been observed (22); while GAPDH functions as a homo-tetramer in glycolysis, it is possible that some of the moonlighting functions may arise from different oligomeric states. PBGS itself has two reported moonlighting functions, which have not yet been correlated with its alternate assemblies (23, 24).

1.2.4. Inconsistent Subunit–Subunit Interactions in Different Protein Crystal Structures

It is sometimes found that multi-domain proteins do not yield to protein crystal structure determination, while it is possible to obtain crystal structures from constructs containing a subset of the domains found in the mature protein. Subunit interactions apparent from these crystal structures are sometimes inconsistent with each other, though these discrepancies are rarely discussed in the literature. Although these inconsistencies may be attributed to crystal contact forces, it is possible that each is correct but represents a different multimeric form of the protein.

2. Experiment 1: Establishing Subunit Disproportionation During Quaternary Structure Equilibration

2.1. Introduction to Subunit Disproportionation

A defining characteristic of the morpheein model of allostery is an obligatory dissociation to a lower order oligomer (or monomer) as part of oligomeric interconversion (Fig. 1). This is distinct from a mechanism in which an octamer would simply lose one dimer to become a hexamer. In the case of PBGS, the transition through a dimeric intermediate implies that a given dimer can exchange between various PBGS hexamers and octamers throughout its "life" in a cell. The suite of techniques described below (which includes chromatography, dialysis, and mass spectroscopy) demonstrates a migration of subunits between the higher order oligomers.

Any one of the aforementioned experimental anomalies can suggest that a particular protein may function as a morpheein, but none establish the dissociative allosteric mechanism illustrated in Fig. 1. Here we describe two techniques, one of which established the dissociative mechanism of the PBGS quaternary structure equilibrium, and the other of which illustrates how one can monitor changes in quaternary structure as a result of allosteric modulators.

Proteins that participate in a physiologically significant morpheein equilibrium must, by definition, sample the alternate assemblies under physiological conditions. Perturbations such as ligand binding, pH or ionic strength changes, or point mutations to a protein can alter the energy landscape to favor formation of one oligomer over another. A key tool in the disproportionation experiment is a mutant protein that alters the position of the quaternary structure equilibrium. Many single amino acid changes to *Hs*PBGS alter the quaternary structure equilibrium (3, 9, 12). One naturally occurring mutation produces a variant, F12L, for which the equilibrium is dramatically shifted toward the low activity hexamer. This distinction between F12L and WT *Hs*PBGS, for which the equilibrium favors the high activity octamer, makes F12L an invaluable experimental tool. The F12L mutation, where the site of mutation is distant from the enzyme active site yet the impact on catalysis is dramatic, can be considered within a frequently observed class of mutations that exert long-range effects. Many proteins have been characterized with mutations that have long-range effects, some of which may reflect alterations to an as-yet-uncharacterized equilibrium of quaternary structure assemblies. We propose that these mutants can serve the role that F12L serves in the described experiment.

2.2. Materials Used to Establish Subunit Disproportionation

1. WT+F12L hetero-oligomeric *Hs*PBGS: A pET17b-derived plasmid coding for both WT and F12 *Hs*PBGS was heterologously expressed in *E. coli,* and purified as described previously (3, 4). Briefly, the purification involved ammonium sulfate fractionation followed by hydrophobic (phenyl Sepharose), anion exchange (Q-Sepharose), and size exclusion (Sephacryl S-300) chromatography. The purified protein was stored at −80°C in buffer comprising 0.1 M potassium phosphate, pH 7, 10 mM β-mercaptoethanol, and 10 μM $ZnCl_2$.
2. Analytical chromatography resin: Mono-Q anion exchange resin purchased from GE Healthcare equilibrated in 30 mM potassium phosphate, pH 7.0, 10 mM 2-mercaptoethanol, 10 μM $ZnCl_2$.

3. Analytical chromatography buffer: 30 mM potassium phosphate, pH 7.0, 10 mM 2-mercaptoethanol, 10 μM $ZnCl_2$ with a gradient of 0.02–1.0 M KCl.
4. Trypsin: Sequencing grade, purchased from Promega.
5. Equilibrium dialysis buffer: 0.1 M bis tris-propane hydrochloride (BTP-HCL) prepared in the absence or presence of the *Hs*PBGS substrate, 5-aminolevulinic acid (ALA) (purchased as the hydrochloride from Sigma), and adjusted to pH 7.0 with HCl.
6. PAGE reagents: All SDS and native PAGE analyses were performed on a GE PhastSystem using 12.5% polyacrylamide gels with native (880 mM L-alanine, 250 mM Tris-acetate [pH 8.8], made of 3% agarose IEF) or SDS (0.2 M tricine, 0.2 M Tris-acetate, 0.55% SDS [pH 8.1], made of 2% agarose IEF) buffer strips purchased from GE Healthcare.

2.3. Methods Used to Establish Subunit Disproportionation

2.3.1. Initial Isolation of WT+F12L *Hs*PBGS Hetero-Hexamers and Hetero-Octamers During Purification

Paratially purified WT+F12L hetero-oligomeric *Hs*PBGS was subjected to anion exchange chromatography on Mono-Q resin. The column was equilibrated with analytical chromatography buffer (30 mM potassium phosphate, pH 7.0, 10 mM β-mercaptoethanol, 10 μM $ZnCl_2$), and a gradient of 0.02–1.0 M KCl was used to resolve the mixed oligomers into their components. The surface charge difference between hexameric and octameric *Hs*PBGS is sufficient to achieve baseline separation on a Mono-Q column (3). The WT+F12L oligomers eluted from the Mono-Q column in two baseline-separated pools: Pool I eluted at a position consistent with that of homo-hexameric F12L; Pool II eluted at a position consistent with that of homo-octameric WT. Native PAGE analysis of the resolved pools confirmed that Pool I contained hexameric *Hs*PBGS and Pool II contained octameric *Hs*PBGS.

2.3.2. Analysis of the Subunit Composition of Isolated WT+F12L *Hs*PBGS Hetero-Hexamers and Hetero-Octamers

The pools containing hetero-hexameric or hetero-octameric *Hs*PBGS were each concentrated to 1 mg/mL. The concentrated pools were dialyzed against 300 mL of 2 mM BTP-HCl buffer at pH 7.0 for 3 h to remove the phosphate from the Mono-Q buffer. Samples were subject to overnight trypsin digestion using a 1:20 (w/w) trypsin:protein ratio. The tryptic peptide mixtures were spotted on a gold plate with cyano-4-hydroxycinnamic acid matrix on top of that. The mass spectral data were collected using Reflex IV matrix-assisted laser desorption/ionization time-of-flight mass spectrometer (Bruker Inc.). Mass spectrometry of both pools, focusing on the N-terminal peptide of the tryptic digest, confirmed that each pool comprised hetero-oligomers (4). The hetero-hexamers (Pool I) contained ~70% Leu12 chains, while the hetero-octamers (Pool II) contained ~70% Phe12 chains.

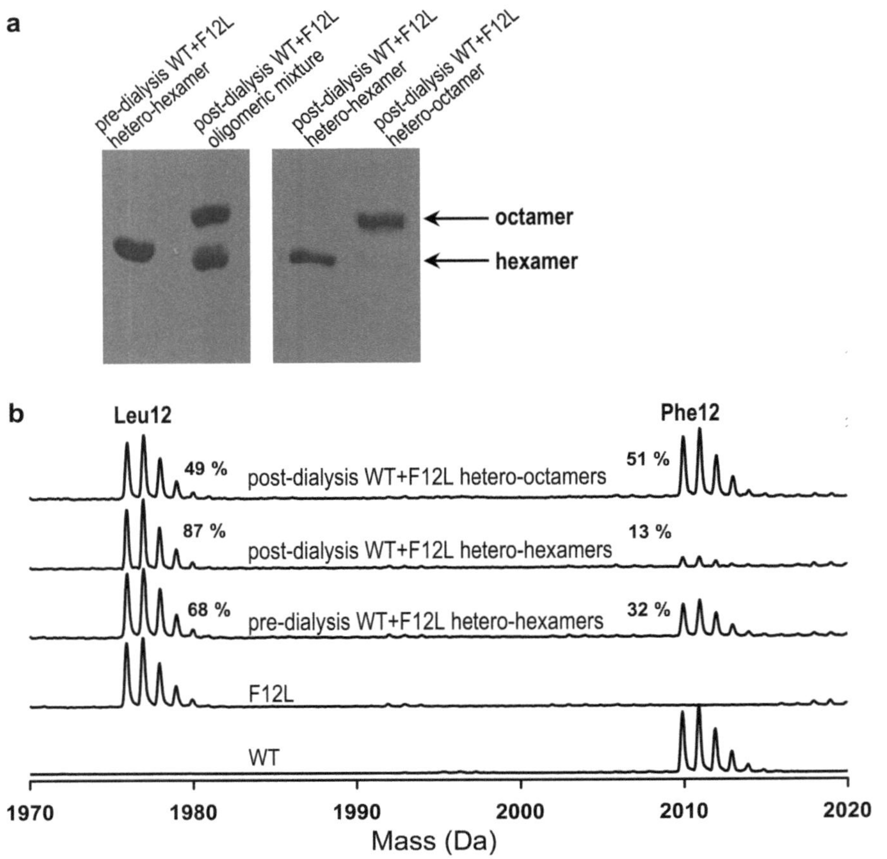

Fig. 3. Analysis of the disproportionation of subunits between *Hs*PBGS oligomers. (**a**) Native PAGE analysis of WT+F12L *Hs*PBGS oligomers before and after dialysis under turnover conditions demonstrates that substrate turnover facilitates oligomeric interconversion. Following dialysis to yield mixed hetero-oligomers, the hetero-hexamers were resolved from the hetero-octamers by anion exchange chromatography. (**b**) Mass spectroscopy focusing on the N-terminal tryptic peptides of *Hs*PBGS oligomers before and after substrate turnover conditions demonstrates subunit exchange during oligomeric interconversion.

2.3.3. Analysis of the Stability of WT+F12L **Hs**PBGS Hetero-Hexamers and Hetero-Octamers

The isolated hetero-oligomers appeared stable during storage. However, a time-dependent activation of the hetero-hexamer (kinetic hysteresis) was observed during activity assay, suggesting that it could rearrange to the octamer during turnover. To address the dissociative mechanism of this phenomenon, hetero-hexamers were subjected to dialysis in the absence or presence of the substrate, ALA. The dialysis of *Hs*PBGS proceeded for 48 h at 37°C, followed by analysis of the oligomeric composition by native PAGE. As shown previously, the hetero-hexamers remain hexameric in the absence of substrate, while the presence of substrate readily induces the formation of octamers (4). A general application of this method would use a ligand that has been demonstrated to shift the quaternary structure equilibrium of the protein being studied.

2.3.4. Demonstration of Subunit Disproportionation Between HsPBGS Hetero-Oligomers

During dialysis in the presence of substrate, WT+F12L hetero-hexamer is seen to convert to hetero-octamer (4). At approximately 50% conversion of hetero-hexamer to hetero-octamer the protein was subjected to anion exchange chromatography to resolve and isolate the resultant oligomers, as shown by native PAGE (Fig. 3a). Mass spectrometry of each oligomer (focusing again on the N-terminal tryptic peptide) revealed a dramatic disproportionation of chains between the oligomers (Fig. 3b). The pre-dialysis hetero-hexameric WT+F12L contained Leu12 and Phe12 chains in a ratio of 68:32. Following dialysis and chromatographic separation of the oligomers, the remaining hexamers contained Leu12 and Phe12 chains in a ratio of 87:13, while the newly assembled octamers contained Leu12 and Phe12 chains in a ration of 49:51. The observed disproportionation of chains is consistent with the oligomeric equilibria favoring the octamer for the WT, and the hexamer for F12L. The observed disproportionation of subunit chains definitively establishes a dissociative mechanism for oligomer interconversion, but does not define the quaternary structure of the dissociated form. If the oligomer did not dissociate during the transition, the ratio of chains in the oligomers would remain unchanged.

3. Experiment 2: Two-Dimensional Native PAGE Analysis of the Impact of Allosteric Modulators on Oligomeric State

3.1. Allosteric Modulators that Shift an Equilibrium of Oligomeric States

As discussed above, the equilibrium between the alternate oligomers is freely reversible, and a protein that functions as a morpheein samples the available oligomers under physiological conditions. The impact of an allosteric ligand on a morpheein derives from ligand binding to one of the oligomeric forms in a manner that stabilizes it, thereby drawing the quaternary structure equilibrium toward that oligomer. An allosteric inhibitor (morphlock-1) for the *Pisum sativum* (green pea) PBGS (*Ps*PBGS) has been characterized and found to inhibit *Ps*PBGS by binding to and stabilizing the low activity hexamer, drawing the equilibrium toward that form (10). The equilibrium perturbing mechanism whereby the allosteric inhibitor binds to the hexamer is fundamentally different from a ligand induced "switch" mechanism where the ligand would bind to the octamer and induce a "switch" to the hexamer.

3.2. Introduction of Two-Dimensional Native PAGE to Assess Oligomeric State Changes

The quaternary structure equilibrium of PBGS is sufficiently dynamic that oligomeric interconversion can occur within the matrix of a native polyacrylamide gel (25). This phenomenon allowed us to monitor the impact of substrate or allosteric modifiers using a two-dimensional (2D) native PAGE technique. As detailed below for *Ps*PBGS, the protein is resolved into its composite oligomers in the first dimension, the gel is subsequently incubated in a solution of the oligomer-perturbing ligand, and the results of the incubation are analyzed by a second dimension native PAGE separation. The method described below has been optimized for the GE PhastSystem, whose cassette-style open-faced plastic-backed gels facilitate manipulations of the gels between dimensions. The method could be modified to run in a standard vertical slab gel apparatus for the first dimension, followed by a second dimension on a flat-bed apparatus.

3.3. Materials Used for Two-Dimensional Native PAGE

1. *Pisum sativum* PBGS (*Ps*PBGS) expressed and purified as previously described, stored at −80°C in buffer comprising 100 mM Tris–HCl pH 8.5, 10 mM $MgCl_2$ (17).
2. Dilution buffer: For *Ps*PBGS, dilution of the protein to 1 mg/mL in 100 mM BTP-HCl at pH 8.5 yields a mixture of octamers and hexamers.
3. PAGE reagents: All native PAGE analyses were performed on a GE PhastSystem using 12.5% polyacrylamide gels with native (880 mM L-alanine, 250 mM Tris-acetate [pH 8.8], made of 3% agarose IEF) (0.2 M tricine, 0.2 M Tris-acetate, 0.55% SDS [pH 8.1], made of 2% agarose IEF) buffer strips purchased from GE.
4. Gel incubation buffer: 100 mM BTP-HCl, pH 8.5. This buffer was also supplemented with 10 mM $MgCl_2$ and 10 mM ALA to create assay turnover conditions for *Ps*PBGS, or with 1 mM morphlock-1 (with a final DMSO concentration of 20%) to induce allosteric inhibition by stabilizing the inactive *Ps*PBGS hexamer.
5. Activity assay buffer: 100 mM BTP-HCl, 10 mM ALA, 10 mM $MgCl_2$, pH 8.5.
6. STOP reagent: 8 mL of 0.1 M BTP-HCl, pH 8.5 and 8 mL of 20% trichloroacetic acid.
7. Developing reagent: 4 mL of 0.1 M BTP-HCl, pH 8.5 and 8 mL of modified Ehrlich's reagent (20 g/L *p*-dimethylaminobenzaldehyde in 80% glacial acetic acid and 20% concentrated perchloric acid).

3.4. Methods for Two-Dimensional Native PAGE

3.4.1. Preparation of the Protein for Electrophoresis

The starting sample conditions will vary for different proteins, and for the nature of the oligomeric change being monitored. Generally the protein of interest should first be incubated in or dialyzed against a buffer in which the oligomeric equilibrium is pliable, and which yields an appropriate oligomeric distribution for the experiment. For example, when analyzing a hexamer-stabilizing inhibitor of *Ps*PBGS, a starting sample that is predominantly octameric will yield the most dramatic results. Prior to the two-dimensional native PAGE experiment, appropriate initial conditions can be optimized by incubating the protein under various conditions (e.g., different pH, buffers, ligands), and examining the resultant oligomeric composition by one-dimensional native PAGE (as shown for *Ps*PBGS in Fig. 2a).

Prior to electrophoresis, *Ps*PBGS was diluted to 1 mg/mL in dilution buffer (100 mM BTP-HCl, pH 8.5). These conditions were chosen to yield a mixture of octamers and hexamers in the first dimension to illustrate the different effects of substrate and allosteric inhibitor on the protein in the second dimension. One-dimensional native PAGE (Fig. 4a) confirms the desired oligomeric composition of the starting sample. The position of the *Ps*PBGS quaternary structure equilibrium is influenced by ionic strength (higher ionic strength favors the octamer); protein concentration (higher protein concentration favors the octamer); the presence of an allosteric magnesium ion, which stabilizes the octamer by binding to an octamer-specific subunit–subunit interface; and the presence of substrate in conjunction with a catalytic magnesium ion, which also stabilizes the octamer (5).

3.4.2. Electrophoresis in the First Dimension

A single 1 μL *Ps*PBGS sample is applied to an 8 × 1 μL applicator comb in the 7th position from the left. It is critical to utilize the most narrow lane width possible, as wide bands will not meaningfully resolve in the second dimension. The sample should be loaded in a lane near the edge of the gel to allow room for second dimension migration, but not in the lane nearest the edge of the gel as protein in these lanes can run aberrantly. For the examples in Fig. 4b–d, electrophoresis in the first dimension was performed on a 12.5% polyacrylamide gel following a standard protocol. Different gel matrices could be selected based on the size of the oligomers of the protein of interest. The ideal matrix will cleanly resolve the alternate oligomers, while keeping both of them toward the center of the resolving gel (rather than abutting the stacking gel or the buffer front).

3.4.3. Incubation Between Dimensions of Native PAGE

When the alternate oligomers have been resolved in the first dimension of native PAGE, the discrete effects of ligand binding on the oligomeric state of each can be examined by incubating the gel in a solution of that ligand. For the examples, shown in Fig. 4b–d, the gels were incubated under three different

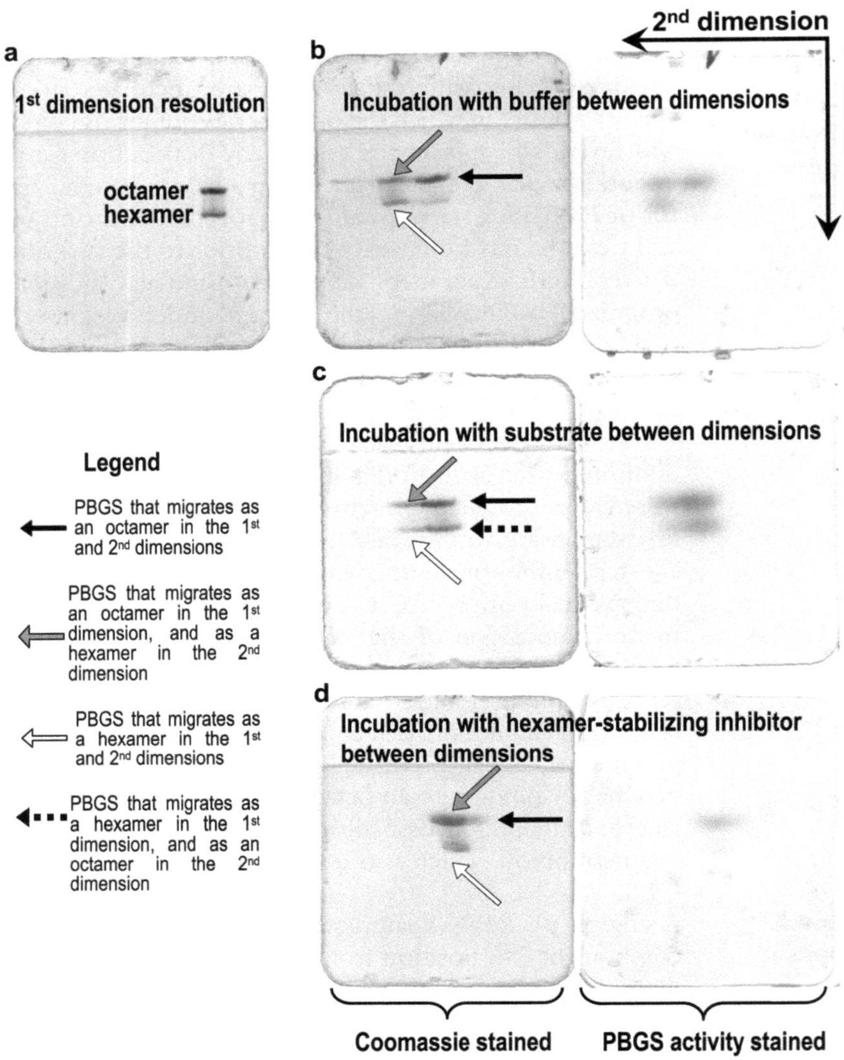

Fig. 4. The impact of substrate or allosteric inhibitor binding on the oligomeric state of PBGS, as demonstrated by 2-dimensional native PAGE. For all gels, a single 1 μL sample of 1 mg/mL *Ps*PBGS was applied in the next to the right-most sample well of a 12.5% PhastGel. (**a**) *Ps*PBGS resolves into ~70% octamer and ~30% hexamer after electrophoresis in the first dimension. The resolved gels were then incubated in following conditions for 20 min: (**b**) 0.1 M BTP-HCl, pH 8.5 at 37°C, (**c**) 0.1 M BTP-HCl, pH 8.5, 10 mM ALA, 10 mM Mg^{2+} (assay conditions) at 37°C, (**d**) 0.1 M BTP-HCl, pH 8.5, 1 mM morphlock-1, 20% DMSO. Following these incubations, the second dimension separation was carried out immediately. All gels were then stained for PBGS activity (*right panels*) followed by Coomassie staining (*left panels*). The overall retarded protein mobility observed for the second dimension separation for gel d *vs.* gels b and c is due to the presence of DMSO in the incubation buffer on the gel matrix (control not shown).

conditions in gel incubation buffer for 20 min at 37°C with gentle shaking. These were (1) a control comprising buffer (100 mM BTP-HCl, pH 8.5), (2) assay turnover conditions (100 mM BTP-HCl pH 8.5, 10 mM Mg^{2+}, and 10 mM ALA), and (3) in the presence of a hexamer stabilizing allosteric inhibitor

(100 mM BTP-HCl pH 8.5, 1 mM morphlock-1, 20% DMSO). Following the incubation, gels were rinsed briefly in distilled water prior to native PAGE separation in the second dimension.

3.4.4. Electrophoresis in the Second Dimension

The second dimension separation is performed perpendicular to the first dimension. For the GE PhastSystem, the gel is placed on the gel bed with a 90° clockwise rotation relative to the position for the first dimension. The gel should be carefully aligned to ensure that the top and bottom buffer strips contact the gel evenly, as further described by Samuel et al. (26). The duration of second dimension electrophoresis will vary among different proteins. The conditions used for the first dimension are a good starting point for optimization.

3.4.5. Visualization of Gel Bands

Any common protein gel-staining technique can be used to reveal the position of the bands following this 2D native PAGE protocol. In some cases, an activity assay can be performed prior to the protein visualization step. Analyzing the enzymatic activity associated with each provides an elegant demonstration of activation or inhibition of the protein by the ligand exposure between the two dimensions. This in-gel assay method is most suitable for enzyme assays that measure a colorimetric signal. In the illustrated case of *Ps*PBGS, the formation of the product, porphobilinogen, can be monitored by the pink color that forms by reaction with modified Ehrlich's reagent. Gels are incubated at 37°C for 20 min in 15 mL of activity assay buffer (0.1 M BTP-HCl, 10 mM 5-aminolevulinic acid, 10 mM $MgCl_2$, pH 8.5), and then transferred into 16 mL of STOP solution (8 mL of 0.1 M BTP-HCl, pH 8.5 and 8 mL of 20% trichloroacetic acid) for 1–2 min at room temperature. Gels are then transferred into 12 mL of developing reagent (4 mL of 0.1 M BTP-HCl, pH 8.5 and 8 mL of modified Ehrlich's reagent) at room temperature for 4 min during which time a pink color develops. If the assay for a protein of interest has a colorimetric signal outside of the visible range, an appropriate scanner could be used for detection.

3.4.6. Interpretation of 2D Native PAGE Results

Electrophoresis in the first dimension resolves a morpheein into its component oligomers, a roughly equal mixture of octamers and hexamers in the illustrated case of *Ps*PBGS. The incubation conditions between the first and second dimension will have different effects on the octamers and hexamers. In the control experiment shown in Fig. 4b, incubation of the gel in buffer between dimensions causes a portion of the protein that migrated as an octamer in the first dimension to migrate as a hexamer in the second dimension; the majority of the protein that migrated as a hexamer in the first dimension remains hexameric in the second dimension. Activity is observed for all bands, which could derive from two possible interpretations: (1) the octameric and hexameric *Ps*PBGS are both active forms of the protein, or (2) the inactive hexameric *Ps*PBGS is converted to active octamers in the presence of substrate during the activity assay, and the observed activity is from the newly assembled octameric *Ps*PBGS. The experiments illustrated in Fig. 4c, d confirm the latter interpretation.

In the experiment illustrated in Fig. 4c, incubation of the gel under assay turnover conditions between dimensions causes the majority of the protein in both the octameric and hexameric first dimension bands to migrate as octamer in the second dimension. This experiment conclusively demonstrates that the substrate-mediated oligomeric interconversion can occur within the native PAGE matrix. Activity is again observed for all bands; the intensity of the activity staining roughly correlates with the Coomassie staining for protein.

In the experiment illustrated in Fig. 4d, incubation of the gel in the presence of a hexamer stabilizing inhibitor (morphlock-1) between dimensions causes the majority of the protein from the first dimension octameric band to migrate as hexamer in the second dimension. All of the first dimension hexameric band remains as hexamer. Activity staining reveals that the only active component is the octamer. The presence of the hexamer-stabilizing inhibitor is sufficient to prevent the interconversion in the presence of substrate during activity staining, and demonstrates that the activity observed in the hexamer bands of the gels in Fig. 4b, c derives from *Ps*PBGS that has converted to octamers.

Acknowledgments

This work was supported by the National Institutes of Health grants R01ES003654 (to E.K.J), R56AI077577 (to E.K.J), and CA006927 (to the Institute for Cancer Research). The authors thank our colleagues Drs. Gregory Adams, Mark Andrake, Erica Golemis and George D. Markham for helpful comments.

References

1. Jaffe, E. K. (2005) Morpheeins - a new structural paradigm for allosteric regulation, *Trends in Biochemical Sciences 30*, 490–497.
2. Erskine, P. T., Senior, N., Awan, S., Lambert, R., Lewis, G., Tickle, L. J., Sarwar, M., Spencer, P., Thomas, P., Warren, M. J., Shoolingin-Jordan, P. M., Wood, S. P., and Cooper, J. B. (1997) X-ray structure of 5-aminolaevulinate dehydratase, a hybrid aldolase, *Nature Structural Biology 4*, 1025–1031.
3. Breinig, S., Kervinen, J., Stith, L., Wasson, A. S., Fairman, R., Wlodawer, A., Zdanov, A., and Jaffe, E. K. (2003) Control of tetrapyrrole biosynthesis by alternate quaternary forms of porphobilinogen synthase, *Nature Structural Biology 10*, 757–763.
4. Tang, L., Stith, L., and Jaffe, E. K. (2005) Substrate-induced interconversion of protein quaternary structure isoforms, *Journal of Biological Chemistry 280*, 15786–15793.

5. Kokona, B., Rigotti, D. J., Wasson, A. S., Lawrence, S. H., Jaffe, E. K., and Fairman, R. (2008) Probing the oligomeric assemblies of pea porphobilinogen synthase by analytical ultracentrifugation, *Biochemistry 47*, 10649–10656.
6. Ramirez, U. D., Myachina, F., Stith, L., and Jaffe, E. K. (2010) Docking to large allosteric binding sites on protein surfaces, In *Advances in Computational Biology*, Springer in book series, Advances in Experimental Medicine and Biology *680*, 481–488.
7. Selwood, T., Tang, L., Lawrence, S. H., Anokhina, Y., and Jaffe, E. K. (2008) Kinetics and thermodynamics of the interchange of the morpheein forms of human porphobilinogen synthase, *Biochemistry 47*, 3245–3257.
8. Lawrence, S. H., and Jaffe, E. K. (2008) Expanding the Concepts in Protein Structure-Function Relationships and Enzyme Kinetics: Teaching using Morpheeins, *Biochem Mol Biol Educ 36*, 274–283.
9. Tang, L., Breinig, S., Stith, L., Mischel, A., Tannir, J., Kokona, B., Fairman, R., and Jaffe, E. K. (2006) Single amino acid mutations alter the distribution of human porphobilinogen synthase quaternary structure isoforms (morpheeins), *Journal of Biological Chemistry 281*, 6682–6690.
10. Lawrence, S. H., Ramirez, U. D., Tang, L., Fazliyez, F., Kundrat, L., Markham, G. D., and Jaffe, E. K. (2008) Shape shifting leads to small-molecule allosteric drug discovery, *Chem Biol 15*, 586–596.
11. Lawrence, S. H., Ramirez, U. D., Selwood, T., Stith, L., and Jaffe, E. K. (2009) Allosteric inhibition of human porphobilinogen synthase, *J Biol Chem 284*, 35807–35817.
12. Jaffe, E. K., and Stith, L. (2007) ALAD porphyria is a conformational disease, *American Journal of Human Genetics 80*, 329–337.
13. Shanmugam, D., Wu, B., Ramirez, U., Jaffe, E. K., and Roos, D. S. (2010) Plastid-associated Porphobilinogen Synthase from Toxoplasma gondii - Kinetic and structural properties validate therapeutic potential, *Journal of Biological Chemistry 285*, 22122–22131.
14. Frieden, C. (1970) Kinetic aspects of regulation of metabolic processes. The hysteretic enzyme concept, *J Biol Chem 245*, 5788–5799.
15. Naught, L. E., Gilbert, S., Imhoff, R., Snook, C., Beamer, L., and Tipton, P. (2002) Allosterism and cooperativity in Pseudomonas aeruginosa GDP-mannose dehydrogenase, *Biochemistry 41*, 9637–9645.
16. Jaffe, E. K., Bagla, S., and Michini, P. A. (1991) Reevaluation of a Sensitive Indicator of Early Lead-Exposure - Measurement of Porphobilinogen Synthase in Blood, *Biological Trace Element Research 28*, 223–231.
17. Kervinen, J., Dunbrack, R. L., Litwin, S., Martins, J., Scarrow, R. C., Volin, M., Yeung, A. T., Yoon, E., and Jaffe, E. K. (2000) Porphobilinogen synthase from pea: Expression from an artificial gene, kinetic characterization, and novel implications for subunit interactions, *Biochemistry 39*, 9018–9029.
18. Jeffery, C. J. (2009) Moonlighting proteins–an update, *Mol Biosyst 5*, 345–350.
19. Karpel, R. L., and Burchard, A. C. (1981) A basic isozyme of yeast glyceraldehyde-3-phosphate dehydrogenase with nucleic acid helix-destabilizing activity, *Biochim Biophys Acta 654*, 256–267.
20. Meyer-Siegler, K., Mauro, D. J., Seal, G., Wurzer, J., deRiel, J. K., and Sirover, M. A. (1991) A human nuclear uracil DNA glycosylase is the 37-kDa subunit of glyceraldehyde-3-phosphate dehydrogenase, *Proc Natl Acad Sci U S A 88*, 8460–8464.
21. Hara, M. R., Cascio, M. B., and Sawa, A. (2006) GAPDH as a sensor of NO stress, *Biochim Biophys Acta 1762*, 502–509.
22. Ferreira-da-Silva, F., Pereira, P. J., Gales, L., Roessle, M., Svergun, D. I., Moradas-Ferreira, P., and Damas, A. M. (2006) The crystal and solution structures of glyceraldehyde-3-phosphate dehydrogenase reveal different quaternary structures, *J Biol Chem 281*, 33433–33440.
23. Guo, G. G., Gu, M., and Etlinger, J. D. (1994) 240-kDa proteasome inhibitor (CF-2) is identical to delta-aminolevulinic acid dehydratase, *J Biol Chem 269*, 12399–12402.
24. Gross, M., Hessefort, S., and Olin, A. (1999) Purification of a 38-kDa protein from rabbit reticulocyte lysate which promotes protein renaturation by heat shock protein 70 and its identification as delta-aminolevulinic acid dehydratase and as a putative DnaJ protein, *J Biol Chem 274*, 3125–3134.
25. Jaffe, E. K., Ali, S., Mitchell, L. W., Taylor, K. M., Volin, M., and Markham, G. D. (1995) Characterization of the Role of the Stimulatory Magnesium of Escherichia-Coli Porphobilinogen Synthase, *Biochemistry 34*, 244–251.
26. Samuel, M. (1995) "PHAST 2D," a two-dimensional electrophoretic technique on a single gel under native and denaturing conditions using pharmacia PhastSystem, *Anal Biochem 224*, 457–459.

Part III

Monitoring Allosteric Changes in Protein Dynamics/Subpopulation Distribution

Chapter 13

Combining NMR and Molecular Dynamics Studies for Insights into the Allostery of Small GTPase–Protein Interactions

Liqun Zhang, Sabine Bouguet-Bonnet, and Matthias Buck

Abstract

Combinations of experimentally derived data from nuclear magnetic resonance spectroscopy and analyses of molecular dynamics trajectories increasingly allow us to obtain a detailed description of the molecular mechanisms by which proteins function in signal transduction. This chapter provides an introduction into these two methodologies, illustrated by example of a small GTPase–effector interaction. It is increasingly becoming clear that new insights are provided by the combination of experimental and computational methods. Understanding the structural and protein dynamical contributions to allostery will be useful for the engineering of new binding interfaces and protein functions, as well as for the design/in silico screening of chemical agents that can manipulate the function of small GTPase–protein interactions in diseases such as cancer.

Key words: Protein–protein interaction, Plexin RhoGTPase binding domain, Rac1, Rnd1, GTPase–effector interaction, Nuclear magnetic resonance relaxation, Lipari–Szabo order parameters, Cross-correlation analysis

1. Introduction

1.1. Allostery Based on Changes in Dynamics

Monitoring protein dynamics and uncovering its role in protein conformation and function is a major challenge in biophysics and structural biology. All three aspects – dynamics, structure, and function – are intimately linked, at least for key regions of proteins. In an allosteric system, the binding of ligand or modification at one site of the protein causes a change in affinity or catalytic efficiency at a spatially distinct and often distant site (1). Although allostery is most often associated with multimeric proteins, long-range communication via structural and/or dynamics changes can also occur in small single-domain proteins (2). However, in the majority of cases the protein features (i.e., conformational and/or

Aron W. Fenton (ed.), *Allostery: Methods and Protocols*, Methods in Molecular Biology, vol. 796,
DOI 10.1007/978-1-61779-334-9_13,

dynamical) involved in such "communication conduits" or coupling networks are still elusive. Comparison of NMR relaxation measurements and molecular dynamics simulations of proteins in different bound and unbound states can identify the role of dynamics in these networks and thus provide insight into the mechanistic and molecular basis of the allosteric behavior.

Several models have been developed to describe allosteric mechanisms. Smock and Gierasch (3) discussed the dynamic properties in cellular signaling proteins, suggesting that proteins fluctuate among multiple states on energy landscapes that have been selected during evolution. Upstream signals remodel those landscapes and cause signaling proteins to transmit information to downstream partners. Using the landscape paradigm, fluctuations may already be sufficient in the unbound protein to sample the bound conformation (1). The bound conformation would have higher affinity for a ligand or substrate, and thus association with such a molecule would move the equilibrium to populate the bound state (so-called "equilibrium shift mechanism"). As an alternative mechanism, it has recently become clear that communication between two or more different sites in proteins can be accomplished by changes in protein dynamics alone, with no change in the average protein structure (4, 5). NMR can show, for example, that the average structure remains the same (e.g., by the lack of significant chemical shift changes), whereas the fluctuations are increased in frequency or amplitude. These dynamic changes will still have significant effect on the recognition of binding sites as they alter the reference state for entropy changes that occur upon complex formation. Below, we describe such a case from our studies on a small GTPase–effector protein interaction.

1.2. NMR as a Powerful Tool to Investigate Protein Structure and Dynamics

A number of experimental techniques are available for the detailed characterization of protein structures and structural changes, including optical spectroscopies, X-ray crystallography, and Nuclear Magnetic Resonance (NMR) spectroscopy. Initially, the main objective of studying proteins by NMR was to obtain knowledge of the architectural organization of these molecules, necessary for understanding the links between structural motifs and biological functions. During the last 30 years, theoretical, methodological, and instrumental developments of NMR have concentrated on the structure determination of proteins with increasing molecular weights. Under favorable circumstances, the possibilities offered by the labeling with nitrogen-15, carbon-13, and deuterium, as well as TROSY techniques (6) now allow the NMR structural analysis of soluble proteins with molecular weights up to and greater than 800 kDa.

NMR is unique in that it yields site-specific information on multiple timescales (Fig. 1). As a result, NMR has become a technique of choice for the study of macromolecular dynamics

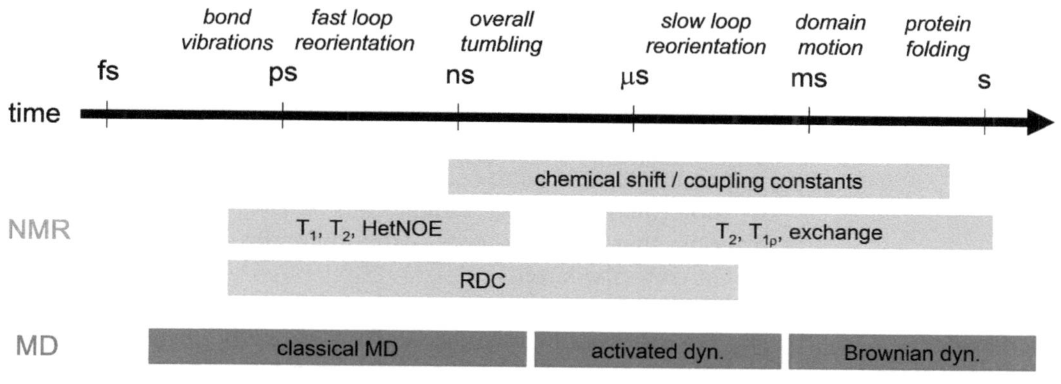

Fig. 1. NMR and MD timescales.

(7, 8). In particular, backbone relaxation measurements (^{15}N spin probes are ideal because of their uniform distribution along the protein backbone) can be used to monitor internal motions on the fast (pico to nanosecond) timescale, to indicate fluctuations on a slower (micro to millisecond) timescale and to assess the overall rotational diffusion of the molecule (nanosecond timescale). Measurement of protein relaxation has a long history. For example, Wagner and colleagues worked on the relaxation of carbon-13 nuclei at natural abundance in the Bovine Pancreatic Trypsin Inhibitor (9). Since most proteins cannot be concentrated sufficiently to allow a study using carbon-13 (or nitrogen-15) at natural abundance, techniques were then developed to label these molecules completely, partially or selectively with different isotopes. A direct consequence of these labeling techniques has been to greatly facilitate the study of main- and side-chain relaxation in proteins. Seminal investigations were carried out by Kay et al. (10), Clore et al. (11), Palmer et al. (12) and Wagner et al. (13); Buck et al. (14), first measured Arg/Asn/Gln/Trp side-chain ^{15}N relaxation and methyl ^{13}C- and ^{2}H relaxation were first monitored by Palmer et al. (15), and Nicholson et al. (16) as well as by Muhandiram et al. (17). Below we illustrate aspects of working with ^{15}N main-chain relaxation measurements, largely on the ps–ns timescale that is comparable to currently achievable molecular dynamics simulations. Other NMR techniques specifically aimed at timescales beyond the ns (global tumbling) regime of proteins are mentioned at the end of the chapter(see Subheading 3).

1.3. Studies Are Complemented by Use of Molecular Dynamics Simulations

Since the first few ps-length simulations of proteins by molecular dynamics in the late 1970s (18), molecular dynamics have revolutionized our understanding of protein structure and function. Nowadays simulations of tens, if not hundreds of ns are standard and multi-μs simulations are becoming increasingly common (19). This is exciting because many of the motions at correlation times at or faster than the global motion of the protein and that are detected experimentally by NMR relaxation, can be directly compared between calculation and experiment. Simulations can provide atom-level insights that are difficult to obtain from the experimental data. The aim of our project on small GTPase–protein interactions is to correlate the dynamic motions predicted by molecular dynamics simulations with the same timescale motions that are observed experimentally, providing greater insights into the latter.

1.4. Allostery in Small GTPases

The Ras superfamily of small GTPases is nearly ubiquitous in cell signaling, functioning as molecular switches. The structure and dynamics of several so-called switch regions within the protein largely depends on the nucleotide, GDP or GTP, which is bound. The different states (structural and dynamical) are in turn recognized by GTPase regulatory or effector proteins in signaling mechanisms (20). Trivially, the shift that occurs when GDP exchanges for GTP in the neighboring protein switch regions could be regarded as allosteric. However, we believe that the conformational behavior of these switch regions by themselves is rather mundane and that allostery in small GTPases involves longer-range communication networks in the protein. The identity of residues which allow this longer-range coupling and the mechanistic role of these residues in allostery is, therefore, of paramount interest. It has become clear, for instance, that the conformational equilibria in GTPases that are normally associated with the presence or absence of the terminal γ-phosphate of the nucleotide are sensitive to mutations not just in or close to the switch regions (21, 22), but also to sequence changes outside these regions. Specifically, an extensive mutagenesis study found that even though such residues were distant to the nucleotide, they can significantly establish the functional specificity of small GTPases (23). A statistical coupling analysis, followed by a mutagenesis study on G-proteins revealed that the nucleotide binding regions are indeed linked via a network of conserved side chains to other regions of the proteins, some far away from the nucleotide binding site (24). Thus, the GTPases may communicate signals also in an allosteric manner across the protein.

Allosteric communication in GTPases has been corroborated by a number of other studies: experimental and computational work, carried out on Ras-family proteins as they associate with lipid membranes (25) or are posttranslationally modified (26), has suggested longer-range communication utilizing specific structures outside the switch regions. Importantly, the residues involved in these communication conduits may be GTPase specific. As yet, there are only a few examples of GTPase–protein interactions which involve binding to surfaces other than the switch regions. Yet, it is thought that such regions can provide some of the specificity of the GTPase–protein interactions (27–29). In studies that have suggested allostery, evidence either derives from functional work in vivo (where the exact binding partners and their interactions are typically not identified), from sequence analyses or from investigations of the free, but modified or lipid-bound GTPases. However, the concept of allostery requires characterization also of the ternary complex (30). Although such structures are likely to be central to the function of GTPases, to the knowledge of these authors, no GTPase ternary complex structures have been determined to date.

In this article, we discuss work from our and other laboratories that concerns GTPase–protein binary complexes. The comparisons of unbound and bound states, already provides information on the allosteric nature of GTPase–protein interactions. This became apparent in our study of the plexin-B1 RhoGTPase binding domain (RBD) and its interaction with the small GTPases, Rac1 and Rnd1, which we have studied over a number of years. The binding domain has a ubiquitin-like fold (with insertion of several lengthy loops, see Fig. 2) and is part of the intracellular region of plexin-B1, part of a family of transmembrane receptors that function in axon guidance and cell migration/positional maintenance. Rac1 and Rnd1 are members of the

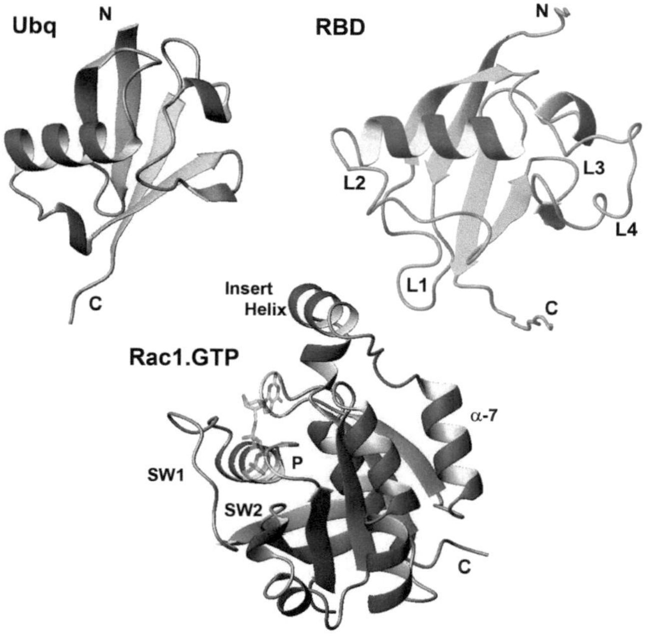

Fig. 2. Main-chain fold of ubiquitin in comparison with the ubiquitin-like RhoGTPase binding domain (RBD) of plexin-B1. The longer loops in the plexin domain are labeled, L1-4. Shown below is the RBD binding partner, Rac1, with GTP bound. Regions of interest, switch 1 and 2, the P loop, the insert helix, and helix α-7 are labeled.

Rho family of small GTPases, which bind specifically to the plexin domain and effect a conformational change that is used to activate the receptor. Our combined NMR and MD studies suggest an allosteric mechanism that is propagated through the different GTPases in the complexes with plexin and is specific to them.

2. Methods of Analysis and Examples of Results and Their Interpretation

2.1. NMR Relaxation Measurements and Analysis

Backbone motions are characterized by measuring heteronuclear relaxation parameters, which are then translated into dynamic parameters (actually a representation to describe the motional amplitudes and timescales) using models of varying complexity. However, the interpretation of relaxation parameters in terms of dynamics models is typically not unique because the number of experimentally measureable parameters is limited and is insufficient to describe physically realistic models. For instance most analyses are essentially based on three parameters (observables possibly measured at several values of the magnetic field): (1) the T_1 or longitudinal relaxation time of ^{15}N is usually measured with the inversion-recovery method, (2) the T_2 or transverse relaxation time is usually measured with the CPMG (Carr–Purcell–Meiboom–Gill) method, and (3) the $\{^1H\text{-}\}^{15}N$ NOE (Nuclear Overhauser Effect) carried out in an interleaved manner, with and without a proton saturation period applied before the start of the ^1H–^{15}N correlation experiment (31). For ^{15}N T_1 and ^{15}N T_2 measurements, several spectra are recorded corresponding to different delays to follow the recovery of the magnetization (T_1) or its disappearance in the transverse plan (T_2), see Fig. 3.

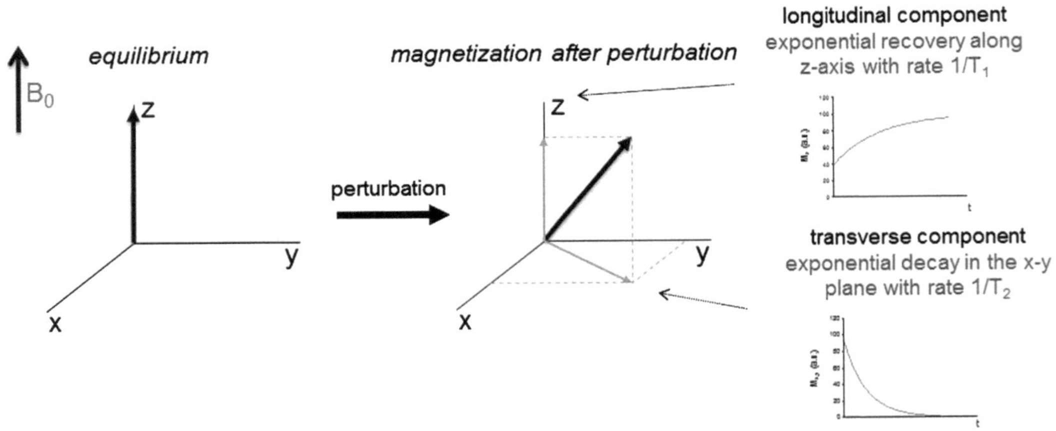

Fig. 3. Following ^{15}N relaxation by three different parameters, T_1, T_2, and *Heteronuclear* NOE. Heteronuclear NOE (not represented) is the intensity change for ^{15}N when its attached ^1H is saturated.

More elaborate NMR experiments, increasing the number of observables has also been proposed (e.g., ref. 13). Raw data are processed and the relaxation rate constants are obtained by fitting of the peak intensities (± sampled uncertainties) to a single exponential function using the nonlinear least-squares method. Experimental uncertainties are determined from baseline noise and/or duplicate measurements.

NMR parameters were measured for the plexin-RBD and Rac1 GTPase at two fields, separately and in the complex. Since the size of the protein complex is considerable (~37 kDa) and crowding of resonances in 2D spectra would have been severe, the measurements of the complex were done on separate samples. These samples had one protein ^{15}N labeled, while the other was unlabeled, with the unlabeled protein at an excess concentration to ensure that all labeled protein is in the bound state. The most widely used model to extract dynamical information from relaxation parameters is the Lipari–Szabo model (32). In this approach, the considered relaxation vector feels two molecular motions at very different timescales (Fig. 4). This model does not rely on knowledge of the orientation of the relaxation vector regarding the rotational-diffusion tensor but introduces the concept of a local director; this is called the "model-free (MF) approach". In order to describe the tumbling of the relaxation vector, one uses a so-called correlation function $C(t)$. It represents the correlation between the orientation of the relaxation vector at time 0 and its orientation at time t. MF supposes that the motions

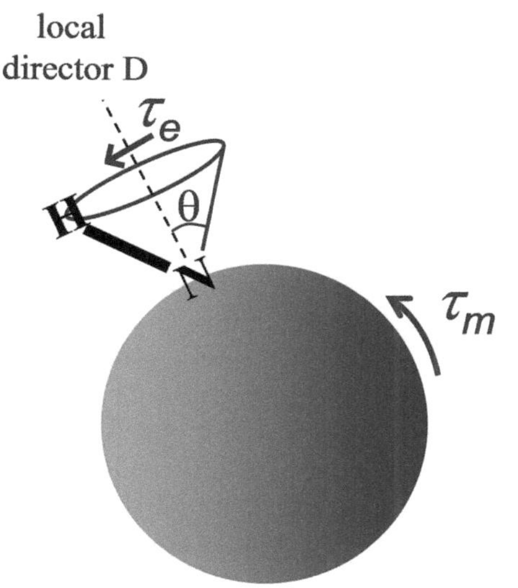

Fig. 4. Model-free representation. The local director D has the same motion as the global molecule (slow motion characterized with τ_m), while local motions occur around this director (fast motions characterized with τ_e, and restricted according to the angle θ).

can be decomposed into global and internal protein dynamics and that the two are not interdependent; so the correlation function can be decomposed into a series of independent contributions:

$$C(t) = C_o(t)C_i(t) \quad (1)$$

Here, $C_o(t)$ represents the overall rotational motion, it can be written as e^{-t/τ_m} where τ_m is the correlation time characteristic of this motion that is supposed to be isotropic. $C_i(t)$ stands for the effective internal (local) motion, it can be written as $C_i(t) = S^2 + (1 + S^2)e^{-t/\tau_e}$. Because this motion is restricted, the correlation function does not decay to zero but to the value S^2 which is called the order parameter. τ_e is the correlation time for the effective internal motion. As depicted in Fig. 4, the local director D of the MF approach moves according to the overall molecular motion (slow), while local motions (fast and small) are held around the director. S^2 is the order parameter describing the restriction of fast motions around semi-angle θ. The higher the order parameter, S^2, the more restrained the N–H bond motion is by its environment, i.e., the more tightly the proteins is packed (S^2 value of 1 means there is no internal motion). Other approaches exist, for cases where the assumption of two independent motions cannot be made (e.g., ref. 32).

Fourier transformation of the above equation leads to the spectral density function, $J(\omega_i)$, as

$$J(\omega_i) = S^2 \frac{2\tau_m}{1 + \omega_i^2 \tau_m^2} + (1 - S^2) \frac{2\tau}{1 + \omega_i^2 \tau^2} \quad (2)$$

This spectral density is sampled at different NMR frequencies ω and allows us to quantify NMR relaxation since relaxation rates (inverse of relaxation times, T_1, T_2) and heteronuclear NOE are a linear combination of spectral densities with some scaling factors (equations given elsewhere, e.g., ref. 31 and see Fig. 5). In Eq. 2, the first part on the right-hand side represents the overall motion while the second part represents the internal motion. For each kind of motions in the MF approach, the relaxation time can be calculated by fitting parameters to the correlation function of the internal motions. $\tau^{-1} = \tau_m{}^{-1} + \tau_e{}^{-1}$ with the τ_m being the correlation time for global molecular motion and the τ_e for effective internal motion. S^2 can be calculated from the average of plateau value of the correlation functions (Fig. 5). In some cases, the Lipari–Szabo model proved to be insufficient and led to an extended model by introducing two types of internal motions (33). The local motions are then characterized by two correlation times, τ_f and τ_s, and two order parameters S_f^2 and S_s^2. Other issues such as anisotropy are also omitted here for lack of space, but are extensively presented in the literature (34). Thus, the relaxation parameters can be calculated from the order parameters and global

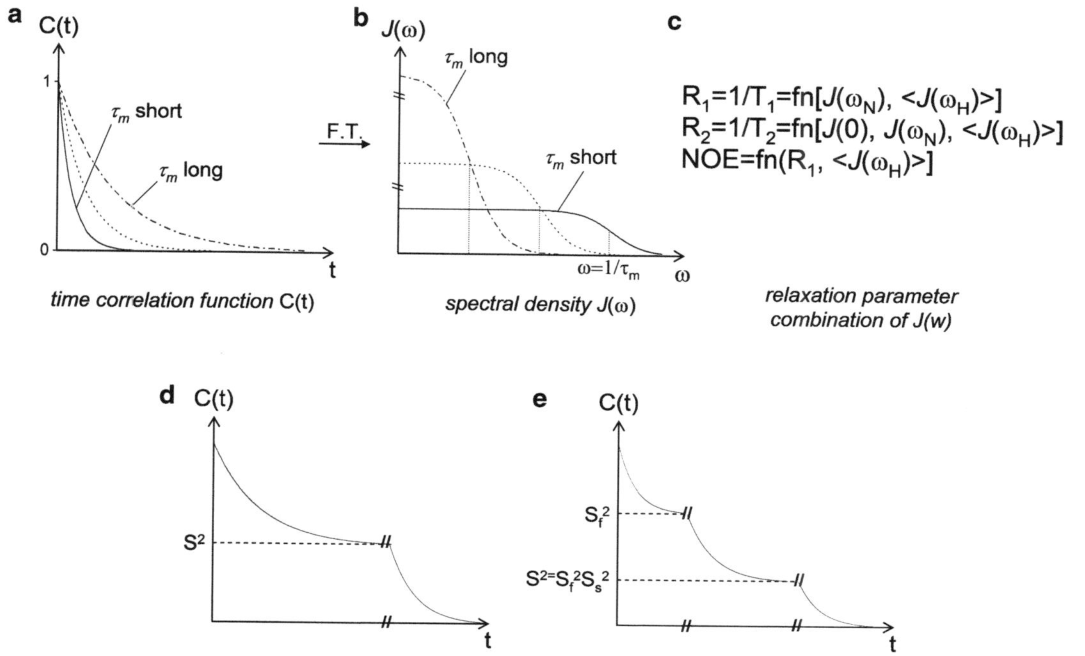

Fig. 5. Relationship between correlation function (**a**), spectral density (**b**) and NMR relaxation measurements (**c**). In the analysis of simulation data, a correlation function is Fourier Transformed to give the spectral density which then predicts NMR relaxation rates (sampling the spectral density function at certain frequencies). For the analysis of the NMR measurements, the process is the reverse. Comparisons between all three sets of parameters (**a–c**) test the models and self-consistency. Plot (**d**) shows the standard MF correlation function and (**e**) shows the more complicated model, incorporating two internal motions, sometimes used to fit experimental data.

correlation time. In fact, the latter variables in Eq. 2 above are fitted iteratively to minimize the difference between the observed and back-calculated NMR relaxation parameters.

For the MF analysis, ^{15}N experimental relaxation data are first used as input, together with the protein structures (pdb files) to fit the global motion, and then to derive the local motions (34, 35). Specifically, after residues that exhibit either significant conformational exchange (high $1/T_2$), or rapid motion on the fast timescale (low NOE factor) are excluded, the remaining T_1/T_2 ratios are used to determine the rotational diffusion tensors that describe the overall tumbling. It should be noted that the relative orientation of proteins in the complex is not required for the analysis. The structures of the proteins in their free states are used to assess the overall motion of each protein in the complex as well as issues such as anisotropy (not discussed here), assuming that the topology of each protein is not changed dramatically upon complex formation. Results reported for the plexin-B1 RBD–Rac1 system show that both proteins experience a similar axially symmetric global motion when they are part of the complex. The rotational diffusion directors for the global motion are superimposable, indicating the relative position of protein in the complex (36). The average values for the global correlation time and of the anisotropy were then used to perform the Lipari–Szabo analysis to derive values for the generalized order parameter (S^2) and the effective correlation time for internal motions (τ_{loc}) for each residue. The procedure also takes into account possible exchange contributions (R_{ex}), which are associated with fluctuations on the microsecond to millisecond timescale. Specific experiments exist that characterize more extensively motions on this timescale (see Note 1). Similar to our work on the RBD–Rac1 complex (36), other studies have used NMR relaxation to characterize protein–protein (e.g., refs. 37, 38) and protein–ligand (e.g., refs. 39, 40) interactions.

2.2. Protein Internal Dynamics Changes Upon Plexin-B1 RBD–Rac1 Complex Formation as Viewed by NMR-Derived Order Parameters

As an example of the analysis and interpretation of the experimental data, we have characterized the backbone dynamics of Rac1 (bound to nonhydrolyzable GTP analog, GMPPNP) and of plexin-B1 RBD both as free proteins and in the plexin-B1 RBD–Rac1.GMPPNP complex. Molecular correlation times, τ_m, extended Lipari–Szabo model-free order parameters, $S^2 = S_f^2 S_s^2$ and local correlation times, τ_e or τ_{loc} as well as chemical-exchange (R_{ex}) contributions, were derived using ^{15}N relaxation data acquired at two spectrometer frequencies (36). The global and internal dynamics of the two proteins change upon their association in a manner that is wide ranging and complex; specifically, changes in the dynamics are observed far from the site of binding in the Rac1 GTPase and several decreases in the extent of fluctuations appear to be compensated by increase especially on the side of the RBD (see Fig. 6). Several residues at the binding interfaces

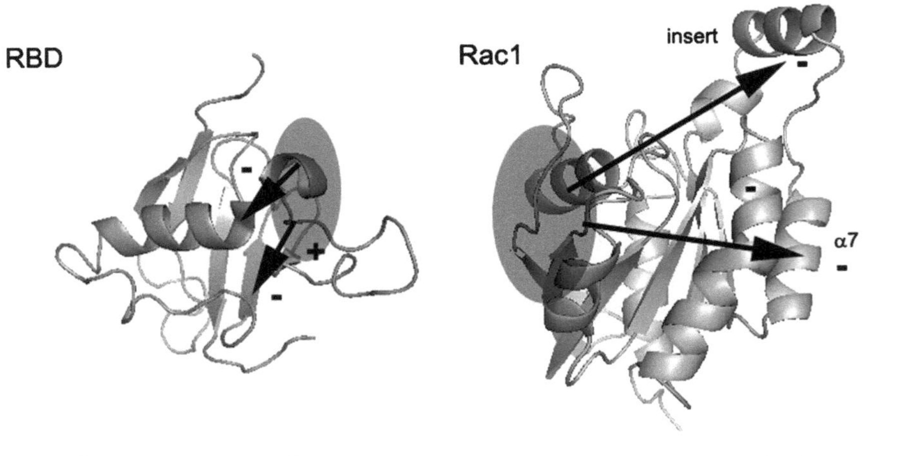

Fig. 6. Changes in ^{15}N order parameter, ΔS^2, in plexin-B1 RBD (*left*) and Rac1 (*right*) upon complex formation mapped onto their respective structures. The plexin domain shows increased as well as decreased main-chain dynamics near the binding interface (*shaded*), while Rac1 becomes more rigid in parts of the protein, some of these regions are opposite to the binding site (*black arrows*). Results adapted from our recent paper (36).

of both proteins experience conformational exchange on the μs–ms timescale, suggesting that these regions fluctuate locally between alternate contacts in the protein complex. Flexible regions of the plexin-B1 RBD, such as residues in the first long loop, residue 16–26, and those at the N- and C-termini of the protein, are not extensively perturbed by binding and have unchanged S^2 order parameters. These regions are distant from the interface that interacts with the GTPase. Upon complex formation two regions in the RBD show reduced fluctuations in terms of S^2. The first (res. 77–84) is at the end of β-strand 4 and at the beginning of the dimerization loop, L4. The second region includes the latter part of another long loop, L2 (residue 50–58), with residues 50, 54, 57, and 62 particularly affected. The first of these two regions is close to one of the interfaces with Rac1, while the second is near the beginning of β3 which is adjacent to β4, a strand that is rigidified on binding. It is likely that the strand is extended by making additional contacts with β3, also making that latter slightly more rigid. These findings show that binding is communicated beyond the region of immediate contact. Specifically, we postulate that the binding interface strand β4 is allosterically coupled, via β3, to sections of loop L2. Support for this notion also comes from cancer mutant, L1815F and L1815P which cause chemical shift perturbations not only locally but also in the start of β3 and in L2. The functional role of these longer-range changes is not yet clear, although we speculate that they may alter the interactions of the RBD with the adjacent domains in the intracellular region of the plexin receptor (41).

Regions of small GTPases that are involved in protein–protein interactions (switch 1 and switch 2) typically experience decreased mobility in the complex, compared to the unbound state, suggesting that protein flexibility may be required for binding (e.g., ref. 42). The ability to form complexes with Rac1 can be highly specific to the state of the nucleotide cofactor which is bound to it. For example, Rac1 loaded with GTP or GTP analogs, binds to the plexin RBD with a K_d of ~6 μM whereas no interaction, even transient, could be detected for the GDP state of Rac1 even at mM concentrations (43). This level of discrimination suggests that several regions which contact the nucleotide – the switch regions 1 and 2 as well as the P-loop (residues 10–17) and parts of two other regions, residues 115–118 and 158–160, are directly or indirectly involved in the interaction with the plexin RBD. Internal dynamics appear overall diminished in Rac1 on binding to plexin (Fig. 6). However, dynamics in the switch 1 and 2 regions of active Rac1 bound to the RBD remain complicated, with chemical exchange precluding an NMR analysis in and around these regions. Nevertheless, a number of observations can be made for other regions. Several residues in the C-terminal half of the protein (e.g., 129) stand out as hinges, with S^2 values lower than the surrounding residues in the free protein (44). In the bound protein these hinges are tightened up. Similarly, a number of residues no longer experience ps–ns dynamics in the bound protein. Moreover, fluctuations for the majority of residues in the region which includes α5 (residues 117–120), the insert helix α6 (res. 123–130) as well as helices α7 and α8 are reduced on binding, possibly because the hinge residues, mentioned above, are no longer flexible. These results are intriguing because the insert helix, α6, is not part of the region that contacts the plexin RBD. However, subtle conformational changes and the dynamical changes at the hinges seem to allow changes in protein dynamics to be propagated over longer distances, specifically to the back of the protein (i.e., α5–α8). It is interesting to compare the NMR-derived results for the RBD–Rac1 interaction with the dynamics changes observed in the Rac1 homologous GTPase, Cdc42 on forming a complex with a 46-residue effector peptide from PAK (45). The switch 1 region is in contact with the peptide and experiences diminished fluctuations upon binding, as does the insert helix and helix α3. The apparent correlation of binding with a change in the flexibility of α3, which is adjacent to the insert helix (α6), was noted and makes these findings in part similar to ours on plexin RBD–Rac1 complex formation. Similar to our work, this study suggests that a "network" of amino-acid contacts exists in the proteins, which serves to couple these fluctuations possibly in all Rho family small GTPases.

2.3. Molecular Dynamics Simulations and Analysis

The structure of the plexin-B1 RBD has been determined by NMR spectroscopy (46), and the structure of the RBD–Rac1 complex has been extensively refined in the Buck laboratory employing HADDOCK and XPLOR-NIH, including the use of residual dipolar coupling and relaxation tensor constraints. The structure of the plexin-B1 RBD in complex with a Rac1 homologous GTPase, Rnd1 was determined by X-ray crystallography (47). Classical all-atom MD simulations were carried out on the two RBD–GTPase complexes and on the free proteins. For simplicity of this book chapter, we report data for the low pH/protonated form of RBD histidines throughout. As a first step, the protein complexes were solvated in a rectangular box of explicitly represented water. Counterions were added to neutralize the system and to represent the ionic charge in the experimental sample. After brief minimization, the initial structures were prepared and the NAMD program was used to run constant pressure (1 atm) and temperature (300 K) dynamics for 55 ns (48). The CHARMM27 all-atom potential function (49) was used with CMAP correction and with the standard Particle-Mesh Ewald method for long-range electrostatic interactions under periodic boundary conditions.

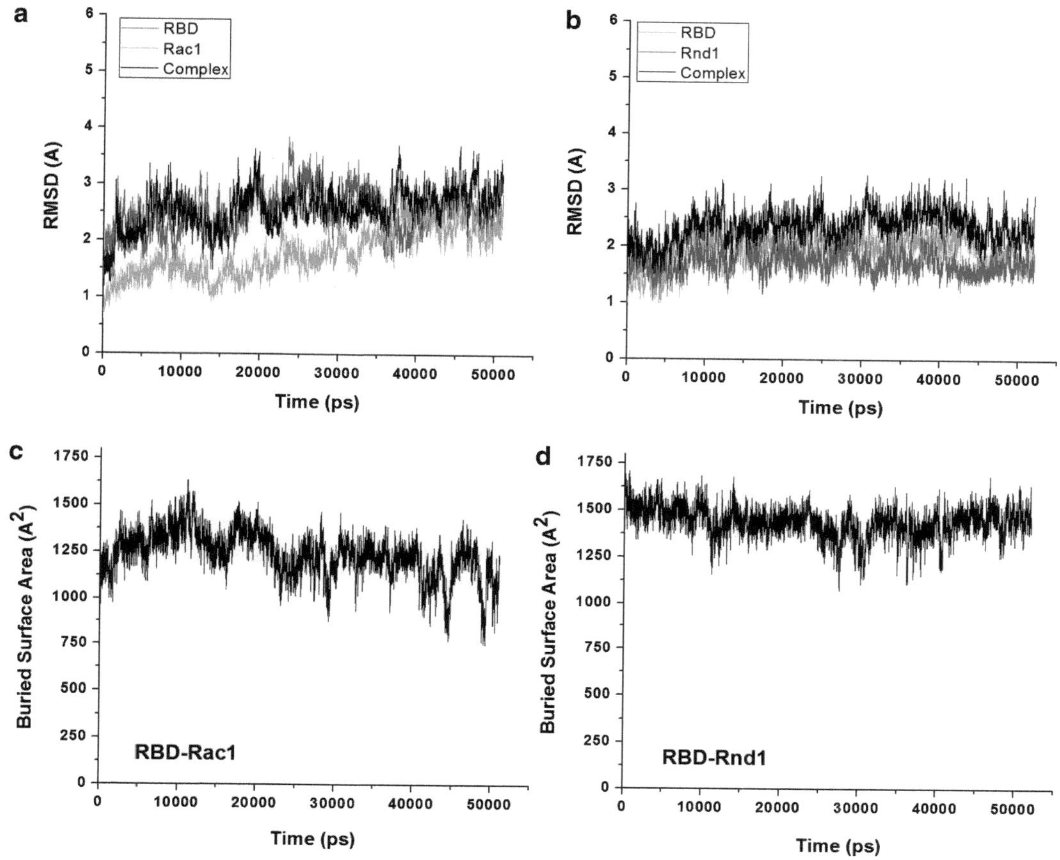

Fig. 7. (**a**, **b**) Plot Cα RMSD of Rac1–RBD and Rnd1–RBD complexes versus their starting structures (superimposed on complex and on single protein in complex). (**c**, **d**) Surface area covered in protein–protein interface (same simulations).

Simple analyses were carried out to indicate the equilibration and stability of the trajectories. Figure 7 shows the RMSD (real mean square deviation) from the starting structure and the accessible surface area (ASA) that is buried in the protein–protein complexes (ΔASA). These are calculated according to the following:

$$\mathrm{Rmsd}(t) = \frac{1}{N}\sum_{i} \mathrm{sqrt}[(r_{\mathrm{i}}(t) - r_{\mathrm{i}}(0))^2]$$

with $r_{\mathrm{i}}(0)$ and $r_{\mathrm{i}}(t)$ as the initial and simulation time evolved coordinates, respectively.

$$\Delta\mathrm{ASA} = \mathrm{ASA\ (free\ proteins)} - \mathrm{ASA\ (complex)},$$

using a probe radius equivalent to that of water of 1.4 Å. The observation that the simulations reach and stay near a low RMSD plateau value shows that the complexes are stable, as a protein that unfolds or proteins that completely dissociate would have an increase to a large RMSD value. The same is true for the buried surface area, which would dramatically decrease if the proteins dissociated.

Order parameters, S^2 and τ_e for main-chain N–H bonds (for each individual residue) are calculated in CHARMM from auto-correlation functions of $C_2(t)$ over trajectory segments.

$$S^2 = \lim_{t \to ns} C_2(t), \quad \text{where} \quad C_2(t) = A\langle P_2[\mu(\tau).\mu(\tau + t)]\rangle \tag{3}$$

A is a such that $C(0) = 1$. The second Legendre polynomial $P_2(x) = 1/2\ (3x^2 - 1)$. The angle brackets (<>) represent the trajectory average over a sliding window of 5 ns (close to the global tumbling in the simulations) (see ref. 50). The unit vectors $\mu(\tau)$ and $\mu(\tau + t)$describe the orientation of the N–H vector at time τ and $(\tau + t)$ in relation to a fixed reference frame. To construct this frame, the global motion of the protein is removed prior to this analysis by Cα superposition on a reference structure (e.g., the starting structure), so the correlation function represents the internal (local) motion only. Again, similarly to the analysis of the experimental data, alternative but more complex approaches are available that do not assume that global and internal motions are separable (and uncorrelated) (e.g., ref. 51). S^2 is the plateau value, calculated for the $C_2(t)$ function in the NMR module of CHARMM. τ is estimated by integration of the correlation function up to time Tconv, the time when first crosses

$$\tau_e = \frac{1}{1 - S^2} \int_0^{T\text{conv}} (C_2(t) - S^2).dt. \tag{4}$$

In some cases, while S_s^2 and S_f^2 are not calculated separately, the τ_e estimated from the trajectories corresponds to τ_s, a slower motion. An issue is the converge of the internal motions of individual bond vectors. This is most easily assessed by examination of the correlation function (see Fig. 8). Running multiple simulations with different assigned starting velocities or even structures allows one to check for convergence and estimate an uncertainty in the calculated S^2 and τ_e values. When this uncertainty is too large, nonconvergence is indicated. In this case, longer simulations and/or enhanced sampling protocols such as replica exchange are required (52). Further alternative strategies exist to sample conformational space (see Note 2).

2.4. Analysis of Motions in Free Proteins and RBD–Rac1/Rnd1 Complexes as Seen in MD

Figure 7 shows two simple analyses carried out on the MD trajectories. Trajectories were calculated for the unbound proteins and complexes. The plexin-B1 RBD binds two homologous GTPases, Rac1 (see experimental results above) and Rnd1 (for which NMR relaxation experiments are in progress). The real mean square deviation (RMSD) shows that the trajectories, with exception of bound Rac1, are equilibrated within 10 ns in terms of the average

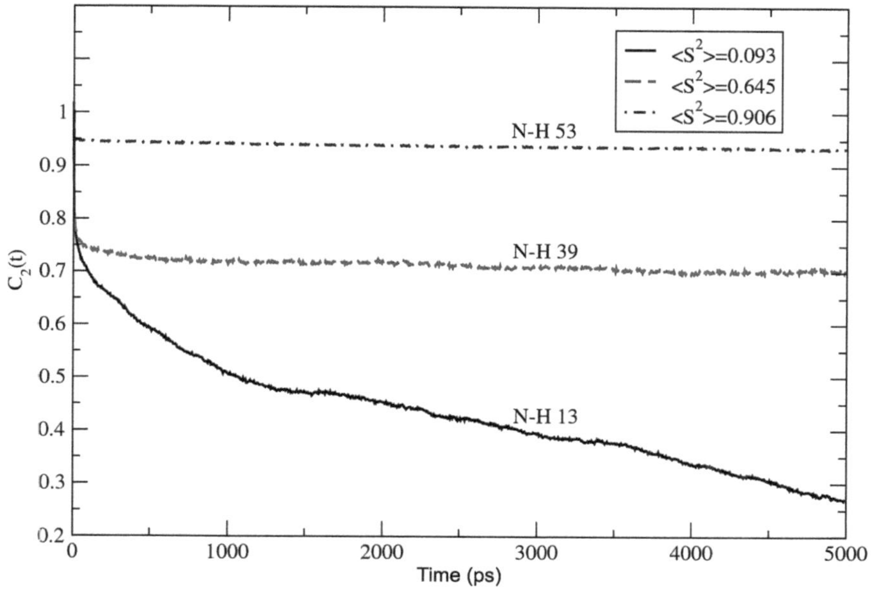

Fig. 8. Selected correlation function decays to illustrate convergence issues. $C_2(t)$ correlation functions are shown for the N–H bond vector of residues 13, 39, and 53 of the RBD. Residue 13, located in loop1, is not fully converged in this simulation.

amplitude of the majority of the internal motions, as the displacement from the starting structure reaches a plateau value. The RMSD for the complexes is larger than for the free proteins, as in the former case superposition is made on both proteins, a process that is very sensitive to global changes of the proteins relative to one another. The results are similar for the RBD–Rnd1 and RBD–Rac1 simulations (Fig. 7a, b respectively). Calculating the surface area that is buried by the interface of the complex also gives an indication of the global stability and dynamics of the protein–protein association (Fig. 7c, d). The RBD–Rac1 complex shows more interface fluctuations than the RBD–Rnd1 complex.

2.5. Comparison of Order Parameters Derived from NMR Experiments and from MD Simulations

Figure 9, as an example for the results and their interpretation of an NMR/MD comparison, illustrates that a good correspondence is possible between experimental and simulated order parameters, suggesting that the simulations are overall accurate. A linear correlation coefficient of 0.74 for S^2 and 0.63 for τ_e, averaging the results from four independent 55 ns simulations of the RBD. To date very few papers have compared experimental and MD derived values for the effective correlation time, τ_e. This parameter is difficult to estimate and is not further discussed here. In the bound state, the RBD N–H bond vector motions remain highly flexible in loop1, 4 and at the N- and C-termini. In the case of the RBD–Rac1 complex, RBD's L3, the interface β4 strand as well as α2 exhibit greater fluctuations than in the RBD–Rnd1 bound

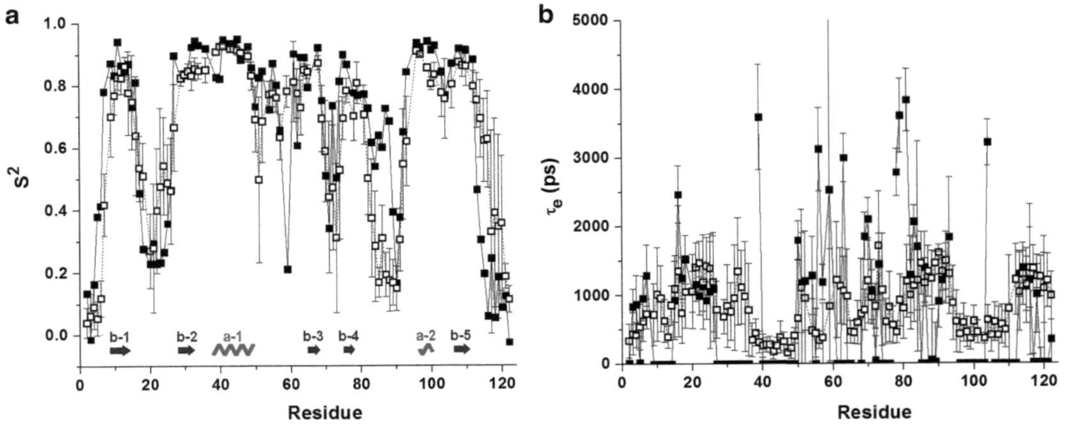

Fig. 9. Order parameters from MD versus experiment (*open* and *closed squares*) showing (**a**) S^2 and (**b**) τ_e, comparison for the unbound RBD.

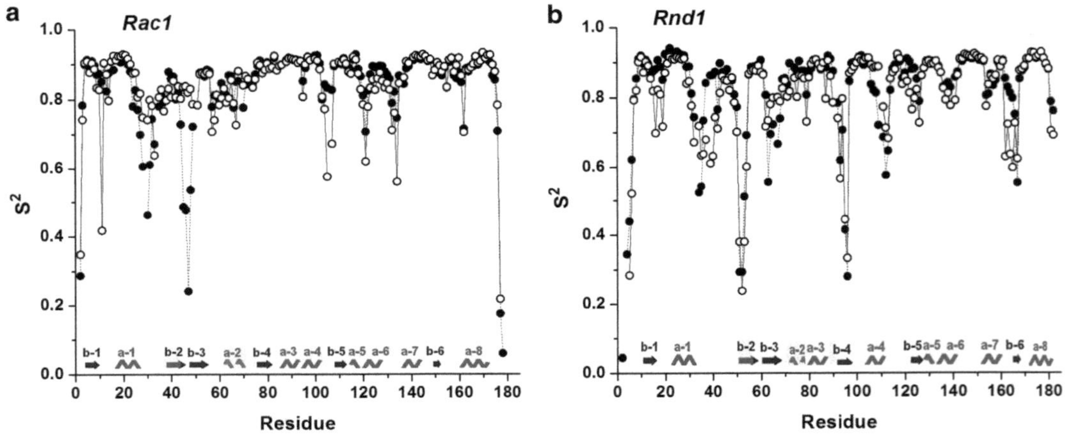

Fig. 10. GTPase order parameters from MD, comparing S^2 unbound (*open circles*) and bound (*closed circles*) for (**a**) Rac1 and RBD–Rac1; (**b**) Rnd1 and RBD–Rnd1.

state (data not shown). The comparison of order parameters between NMR and MD for unbound Rac1 is not as good, mainly because the majority of S^2 values for the GTPase is high, near 0.9–1.0, and small variations are difficult to reproduce in the simulations. Many of the observations made from the experimentally derived order parameters upon Rac1 binding, described above, are, however, confirmed by the simulations (Fig. 10). Specifically, the MD derived S^2 show that upon RBD binding the end of Rac1 helixes α-4 and -6, also switch 2, become more rigid, whereas flexibility in switch 1 increases. In case of binding to Rnd1, an increase in flexibility is observed in switch 2 (switch 1 is already flexible on the ps–ns timescale in the free protein), at the end of the α-4 turn, and the beginning of β-5. Although, no NMR relaxation data is yet available for Rnd1, this observation is intriguing as it confirms a difference in the entropy change seen upon binding of the RBDs to the two GTPases (53). However, this correspondence is necessarily extrapolated as no experimental relaxation data is yet available for side chains. Order parameters are a local probe and are very sensitive to changes in hydrogen bonding, side-chain packing and solvation. NMR measurements of side-chain dynamics and the analysis of side-chain motions from MD trajectories are in progress for both RBD–GTPase complexes. For the detection of a possible coupling between fluctuations, the analyses of cross-correlations are particularly useful, as these represent longer-range dynamics effects.

2.6. Cross-Correlation Analysis to Detect Coincidence of Motions

In order to analyze the atomic fluctuations in proteins, the cross-correlations are obtained from the simulation trajectories for the Cα–Cα pairs in the plexin RBD and in the GTPases. The covariance of the spatial atom displacements for selected atom pairs is calculated based on cross-correlation coefficient calculation using following equation:

$$C_{\mathrm{ij}} = \frac{<\Delta r_{\mathrm{i}} \Delta r_{\mathrm{j}}>}{\sqrt{<\Delta r_{\mathrm{i}}^2><\Delta r_{\mathrm{j}}^2>}} \tag{5}$$

Here, Δr_{i} is the displacement of the mean position of the *i*th atom (here Cα) determined from all configurations in the dynamic trajectory segment (5–55 ns) being analyzed. The numerator in above equation is the covariance for the displacement vectors for atom i and j respectively.

For completely correlated motions, that have the same phase as well as period, $C_{\mathrm{ij}} = 1$; and $C_{\mathrm{ij}} = -1$ for completely anticorrelated motions (have the same period but counter phase). Deviations from 1 or −1 imply the motions of two residues (i and j) are less correlated (or anticorrelated), or mean they deviate from motion along a straight line. When two displacements are in a direction perpendicular to one another, the dot product of the displacement vectors becomes zero, thus the cross-correlation function equals zero, presenting a limitation to this analysis (54). Alternative approaches that do not have these limitations include analysis of correlation between dihedral angle fluctuations or between interresidue energies (e.g., ref. 55) as well as novel computational approaches that probe specific coupling networks (see Note 3).

2.7. Analysis of the Results in the Case of the RBD–Rac1 and RBD–Rnd1 Complexes

The correlation maps were calculated for both the unbound proteins and for the protein complexes (see Fig. 11).

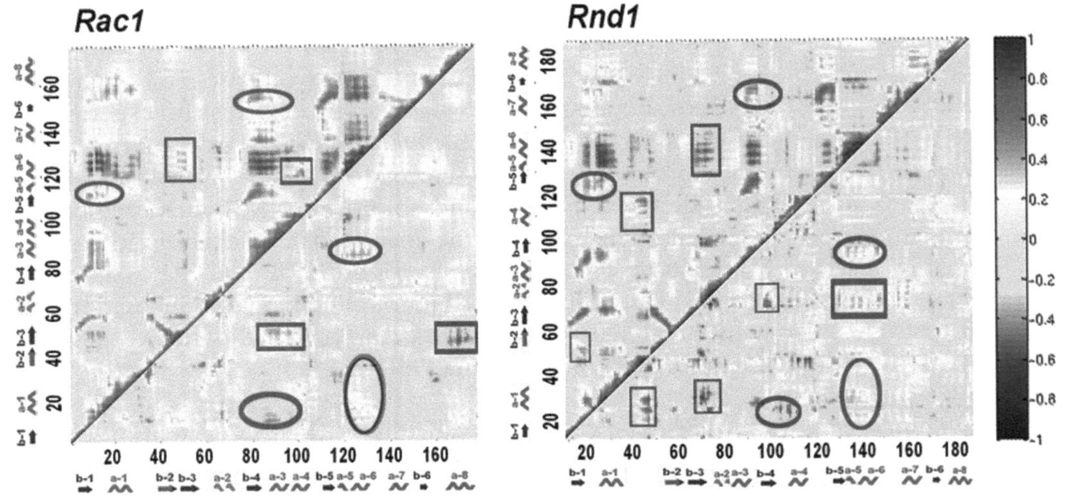

Fig. 11. Correlation analysis of free and RBD bound GTPases. Correlation analysis for free GTPases are shown above the diagonal and the difference maps (bound–free) are shown below the diagonal. *Circled areas* show correlation/changes in correlation common to both structures, in the case the free GTPases these are long range correlations. Areas indicated by *squares* indicate different correlations between the two GTPases.

2.7.1. Unbound Proteins

For the unbound RBD, moderate to strong correlations are seen for the flexible N- and C-termini and all of the loops which are modestly anticorrelated with protein core motions. Correlations are mostly observed between elements of secondary structure that are in contact in the protein (e.g., β1 with β5 and β2). Much more modest correlations are seen between β-strands β1 and -β3 as well as between β4 and β5 which have one strand β5 and β3, respectively to bridge between them. In the case of unbound Rac1 and Rnd1, GTPases structural elements that are spatially close to each other also predominate correlations. There are a few longer-range correlated motions, again through a bridging strand such as β1 with β5 and β4 with β6 in both proteins. In the case of Rnd1, an additional correlation is seen between the β2 region and the end of β1, two regions that are not in direct contact but again bridged. There is also a strong correlation between the switch1 region and β5-strand, two regions that are separated by switch 2, which does not show appreciable correlations with either of them. The region leading to the only remarkable anticorrelation with most of the protein is the insert helix in both GTPases, although here a difference also becomes apparent between the two GTPases. In Rac1 there is considerable correlation between the insert helix motion and the β3, although these elements are not close in the structure. Similarly a correlation between the beginning of the insert helix and α4/a hinge region is only seen in Rac1. In Rnd1 the correlation of the insert helix (α6) region is shifted to α2/the beginning of switch α2. Again, these structures are not close in space. Thus, there are long range correlated motions in the

GTPases, and the several of these as well as some of the next neighbor fluctuations appear to be GTPase specific. These findings are in broad agreement with other MD studies on different members and mutants of the Ras family of small GTPases (53, 56–58).

2.7.2. Intra-protein Motions in the Bound Proteins

In order to visualize the change in correlations for the protein internal motions in the complex, we superimpose the complex either on the bound plexin RBD or GTPase starting structures. In this way the correlations are not affected by fluctuations of the two proteins relative to one another in the complex. For the RBD domain, remarkably the internal motions are changed considerably in certain regions and in a manner that differs between the two complexes. For example, in the RBD bound to Rnd1 next neighbor correlations between β1–β2, β1–α1, β1–β5, are all increased, as are some of the bridged correlations, β2–β5 and β1–β3, whereas surprisingly these correlations are all diminished in the Rac1 bound state. A major difference in the RBD structures is that the RBD–Rnd1 started from a crystal structure that has loop1 bound across the back of the RBD structure to interact with β5, β3 and β4. Even over the course of the 55 ns simulation this loop structure largely persists. Intriguingly, in the RBD–Rac1 structure the loop also moves toward the GTPase, even though our NMR relaxation data suggests it is just as flexible in solution as in the unbound state. Similar to the correlations between β-strand motions, correlations involving the two RBD α-helices also respond differentially to binding of the two GTPases. In the Rnd1 bound RBD new correlations are seen between α1 and strands β3/β5 and the end of GTpase binding strand β4. In the Rac1 bound RBD the latter correlations also exist but β3 and β5, surprisingly move in a manner that is anticorrelated with α1. Instead, the end of β4 and the region that returns from the dimerization loop appears to be correlated with loop2 and the beginning of β3. This is consistent with our experimental finding that mutagenesis of Leu1815 to Pro or Phe in β4 causes chemical shift perturbations at the beginning of β3 and loop2 (46).

A similarly complex picture emerges on the side of the GTPases upon binding. Several changes are common to both GTPases upon RBD binding. Helix turn α3/β4 is also more correlated with the insert helix in both. Motions of a few residues at the center of β2 becomes decorrelated in both GTPases, beginning at switch 2, which becomes less correlated in Rnd1 compared to the unbound state. Correlations between β1–β2 and α1–α3 are extended in bound Rnd1 more than in Rac1. Other differences concern the anticorrelation of residue 95 and insert helix (α6) with switch 2 which changes to a correlated motion in Rnd1, whereas in Rac1 there is a new correlation of α3 with the β2–β3 turn. There is also less anticorrelation and more correlated movement of other regions with the insert helix, suggesting that this latter structure has tightened up, an observation made for both GTPases upon RBD binding and is also evident from NH order parameters. In summary, the pattern of correlated motions and the changes to this pattern in the GTPases upon RBD binding is complex. The analysis by tools such as the cross-correlation matrix reports on changes as observed for the main chain. However, to rationalize and further understand the changes, we need to consider differences in side-chain packing between the two GTPases and also between the bound and free states. These studies are still ongoing and use a different set of analysis tools, as described in a future publication.

2.7.3. Correlated Motions Between Proteins

Motions across the interface are difficult to interpret, since alignment on neither protein in the complex alone presents an unbiased picture. Alignment on the entire complex structure is one option, but the fluctuations of the proteins relative to one another are considerable; this in itself will lead to spurious long-distance correlations. We are currently exploring a number of approaches, including those that are not based on absolute coordinate but instead on relative coordinate or dihedral angle space. These results are reported elsewhere. It is remarkable, however, that the regions of both proteins, RBD and GTPase, that are in contact with one another are also involved in medium range correlated motions. This is consistent with a proposal from statistical coupling analysis (59) that interfaces are not only correlated during protein evolution but that they also connect to protein allosteric networks 1–2 layers beyond the interface into each protein.

3. Notes

1. Classical MD is currently confined to being able to sample motions that occur on the tens, if not hundreds of ns and in some recent cases up to a μs timescale. Thus, to overcome this limitation in sampling, several alternatives, or as above, hybrid computational approaches have been developed. One set of strategies uses biased simulations by applying a force, as seen above. Another derives possible motions from simplified forces between atoms (normal mode, contact or network analysis) (60) and finally coarse-grained models are being used to sample longer timescales (61). Several of these techniques have been used to examine a coupling of residues and motions across protein interfaces, suggesting that coupling networks may connect two proteins across interacting surfaces (61, 62).

2. NMR also has a timescale problem. NMR relaxation measurements can be made on the ps–ns and μs–ms timescale, but this method has until recently been blind to motions that are slower than the overall correlation time of the protein but fast enough not to cause exchange broadening. The analyses of extensive RDC (residual dipolar coupling) datasets have been used to extract μs–ms timescale motions. However, measurements have to date only been done on individual domains (63). These are intensive experiments in terms of time and effort and have not been carried out for μs–ms dynamics in protein complexes. A popular alternative are relaxation dispersion experiments (e.g., see ref. 64 for a paper on Ras) that can monitor a similar timescale of motions and an analysis that can also extract the extent of conformational change (as indicated by chemical shift change) and the population of the different states involved. The interested reader is referred to a number of excellent articles and reviews in this area (65, 66).
3. Once identified, allosteric coupling networks need to be validated by computational modeling and if possible by experiments. A priori, it is not clear what functional consequences the plexin-induced dynamics changes in GTPases have for the signaling mechanism. A surface of the GTPases opposite to the binding site for the RBD is rigidified but a binding partner that would specifically recognize/bind with higher affinity to the reduced entropy of this region remains to be identified. In its absence (and since the binding partner will be present in vivo), we are now disrupting the coupling network by mutagenesis, and anticipate to get a functional readout utilizing cellular assays for plexin function. The challenge is similar to those associated with the interpretation of protein dynamics: Are the particular dynamics simply a consequence of the underlying protein structural features that are wiggling and jiggling in a stochastic solvent environment? Or can the dynamics be defined, say, as part of coupling networks that are important for protein function? Assuming that computer simulations are accurate enough, both in terms of sampling and in the potential functions used, molecular modeling of mutants and recent novel computational methods could provide the details required to answer this question. Simulations have the advantages that conditions can be set up that are difficult, if not impossible to realize in the laboratory. Specifically, Agard and colleagues have recently developed a novel computational method to probe coupling networks by perturbing specific residues within them by individual thermal agitation in conjunction with otherwise classical molecular dynamics (67). By forcing dynamics events to take place in one area of the protein, the stochastic nature of the fluctuations is partly overcome.

Acknowledgments

We thank Shufen Cao, Dr. Prasanta K. Hota, and other members of the Buck laboratory for insightful discussion, as well as Dr. Aron Fenton for editorial help. Some of the molecular dynamics calculations were carried out by Dr. Mehdi Bagheri Hamaneh at the Case Western Reserve High Performance Cluster and at Lonestar (Austin, TX, USA) via a TeraGrid award (to M. B.). The work of M. B. is supported by the NIH grants 1R01GM092851, 1K02HL084384, and 1R01GM73071, which included an ARRA supplement.

References

1. Swain, J. F. and Gierasch, L. M. (2006) The changing landscape of protein allostery. Curr Opin Struct Biol. **16**, 102–8.
2. Gunasekaran, K., Ma, B., and Nussinov, R. (2004) Is allostery an intrinsic property of all dynamic proteins? Proteins **57**, 433–443.
3. Smock, R. G. and Gierasch, L. M. (2009) Sending signals dynamically. Science **324**, 198–203.
4. Cooper, A. and Dryden, D. T. (1984) Allostery without conformational change. A plausible model. Eur Biophys J. **11**, 103–9.
5. Gianni, S., Walma, T., Arcovito, A., Calosci, N., Bellelli, A., et al. (2006) Demonstration of long-range interactions in a PDZ domain by NMR, kinetics, and protein engineering. Structure **14**, 1801–9.
6. Riek, R., Fiaux, J., Bertelsen, E.B., Horwich, A.L. and Wüthrich, K. (2002) Solution NMR techniques for large molecular and supramolecular structures. J. Am. Chem. Soc. **124**, 12144–12153.
7. Kay, L. E. (2005) NMR studies of protein structure and dynamics. J. Magn. Reson. **173**, 193–207.
8. Palmer III, A.G. (2001) NMR probes of molecular dynamics: overview and comparison with other techniques. Annu. Rev. Biophys. Biomol. Struct. **30**, 129–55.
9. Nirmala, N. R. and Wagner, G. (1988) Measurement of 13C relaxation times in proteins by two-dimensional heteronuclear 1H-13C correlation spectroscopy. J. Am. Chem. Soc. **110**, 7557–7558.
10. Kay, L. E., Torchia, D. A. and Bax, A. (1989) Backbone dynamics of proteins as studied by nitrogen-15 inverse detected heteronuclear NMR spectroscopy: application to staphylococcal nuclease. Biochemistry **28**, 8972–8979.
11. Clore, G. M., Driscoll, P. C., Wingfield, P. T. and Gronenborn, A. M. (1990) Analysis of the backbone dynamics of interleukin-1 beta using two-dimensional inverse detected heteronuclear nitrogen-15-proton NMR spectroscopy. Biochemistry **29**, 7387–7401.
12. Palmer III, A. G., Rance, M. and Wright, P. E. (1991) Intramolecular motions of a zinc finger DNA-binding domain from Xfin characterized by proton-detected natural abundance carbon-13 heteronuclear NMR spectroscopy. J. Am. Chem. Soc. **113**, 4371–4380.
13. Peng, J. W. and Wagner, G. (1992) Mapping of the spectral densities of nitrogen-hydrogen bond motions in Eglin c using heteronuclear relaxation experiments. Biochemistry **31**, 8571–8586.
14. Buck, M., Boyd, J., Redfield, C., MacKenzie, D. A., Jeenes, D. J., et al. (1995) Structural determinants of protein dynamics: Analysis of 15N relaxation measurements for mainchain and sidechain nuclei of hen egg-white lysozyme. Biochemistry **34**, 4041–4055.
15. Palmer, A. G., Hochstrasser, R. A., Millar, D. P., Rance, M., Wright, P. E. (1993) Characterization of amino-acid sidechain dynamics in a zinc-finger peptide using C-13 NMR spectroscopy and time-resolved fluorescence spectroscopy. J. Am. Chem. Soc. **115**, 6333–45.
16. Nicholson, L. K., Kay, L. E., Baldisseri, D. M., Arango, J., Young, P. E., et al. (1992) Dynamics of methyl groups in proteins as studied by proton-detected C-13 NMR spectroscopy – application to the leucine residues of staphylococcal nuclease. Biochemistry **31**, 5253–63.
17. Muhandiram, D. R., Yamazaki, T., Sykes, B. D., and Kay, L. E. (1995) Measurement of H-2 T1 and T1p relaxation times in uniformly C-13 labeled and fractionally H-2 labeled

proteins in solution. J. Am. Chem. Soc. **117**, 11536–44.

18. MaCammon, J. A., Gelin, B. R., and Karplus, M. (1977) Dynamics of folded proteins. Nature **267**, 585–590.
19. Klepeis, J. L., Lindorff-Larsen, K., Dror, R. O., and Shaw, D. E. (2009) Long-timescale molecular dynamics simulations of protein structure and function. Curr Opin Struct Biol. **19**, 120–7.
20. Vetter, I. R. and Wittinghofer, A. (2001) The guanine nucleotide-binding switch in three dimensions. Science. **294**, 1299–304.
21. Spoerner, M., Herrmann, C., Vetter, I. R., Kalbitzer, H. R., and Wittinghofer, A. (2001) Dynamic properties of the Ras switch I region and its importance for binding to effectors. Proc Natl Acad Sci **98**, 4944–9.
22. Ford, B., Skowronek, K., Boykevisch, S., Bar-Sagi, D., and Nassar, N. (2005) Structure of the G60A mutant of Ras: implications for the dominant negative effect. J. Biol Chem. **280**, 25697–705.
23. Heo, W. D. and Meyer, T. (2003) Switch-of-function mutants based on morphology classification of Ras superfamily small GTPases. Cell. **113**, 315–28.
24. Hatley, M. E., Lockless, S.W., Gibson, S.K., Gilman, A.G., and Ranganathan, R. (2003) Allosteric determinants in guanine nucleotide-binding proteins. Proc Natl Acad Sci **100**, 14445–50.
25. Abankwa, D., Hanzal-Bayer, M., Ariotti, N., Plowman, S. J., Gorfe, A. A., et al. (2008) A novel switch region regulates H-ras membrane orientation and signal output. EMBO J. **27**, 727–35.
26. Edreira, M. M., Li, S., Hochbaum, D., Wong, S., Gorfe, A. A., et al. (2009) Phosphorylation-induced conformational changes in Rap1b: allosteric effects on switch domains and effector loop. J Biol Chem. **284**, 27480–6.
27. Modha, R., Campbell, L. J., Nietlispach, D., Buhecha, H. R., Owen, D., et al. (2008) The Rac1 polybasic region is required for interaction with its effector PRK1. J Biol Chem. **283**, 1492–500.
28. Pasqualato, S., Renault, L., and Cherfils, J. (2002) Arf, Arl, Arp and Sar proteins: a family of GTP-binding proteins with a structural device for 'front-back' communication. EMBO Rep. **3**, 1035–41.
29. Buhrman, G., Holzapfel, G., Fetics, S., and Mattos, C. (2010) Allosteric modulation of Ras positions Q61 for a direct role in catalysis. Proc Natl Acad Sci USA. **107**, 4931–6.
30. Fenton, A. W. (2008) Allostery: an illustrated definition for the 'second secret of life'. Trends Biochem Sci. **33**, 420–5.
31. Farrow, N. A., Muhandiram, R., Singer, A. U., Pascal, S. M., Kay, C. M., et al. (1994) Backbone dynamics of a free and phosphopeptide-complexed Src homology 2 domain studied by 15N NMR relaxation. Biochemistry **33**, 5984–6003.
32. Lipari, G. and Szabo, A. (1982) Model-free approach to the interpretation of nuclear magnetic resonance relaxation in macromolecules. 1. Theory and range of validity. J. Am. Chem. Soc. **104**, 4546–4559.
33. Clore, G. M., Szabo, A., Bax, A., Kay, L. E., Driscoll, P. C. et al. (1990) Deviations from the simple two parameter model free approach to the interpretation of ^{15}N nuclear magnetic relaxation of proteins. J. Am. Chem. Soc. **112**, 4989–4991.
34. Hall, J. B. and Fushman, D. (2003) Characterization of the overall and local dynamics of a protein with intermediate rotational anisotropy: differentiating between conformational exchange and anisotropic diffusion in the B3 domain of protein G. J. Biomol. NMR **27**, 261–275.
35. Walker, O., Varadan, R. and Fushman, D. (2004) Efficient and accurate determination of the overall rotational diffusion tensor of a molecule from ^{15}N relaxation data using computer program ROTDIF. J. Magn. Reson. **168**, 336–345.
36. Bouguet-Bonnet, S. and Buck, M. (2008) Compensatory and long-range changes in ps-ns mainchain dynamics upon complex formation. 15N relaxation analysis of the free and bound states of the ubiquitin-like domain of human plexin-B1 and the small GTPase Rac1. J. Mol. Biol. **377**, 1474–1487.
37. Akerud, T., Thulin, E., Van Etten, R. L. and Akke, M. (2002) Intramolecular dynamics of low molecular weight protein tyrosine phosphatase in monomer-dimer equilibrium studied by NMR: a model for changes in dynamics upon target binding. J. Mol. Biol. **322**, 137–52.
38. Spyracopoulos, L., Lewis, M. J. and Saltibus, L. F. (2005) Main chain and side chain dynamics of the ubiquitin conjugating enzyme variant human Mms2 in the free and ubiquitin-bound States. Biochemistry **44**, 8770–81.
39. Boehr, D. D., McElheny, D., Dyson, H. J. and Wright, P. E. (2006) The dynamic energy landscape of dihydrofolate reductase catalysis. Science **313**, 1638–42.
40. Finerty, P. J., Mittermaier, A. K., Muhandiram, R., Kay, L. E. and Forman-Kay, J. D.

(2005) NMR dynamics-derived insights into the binding properties of a peptide interacting with an SH2 domain. Biochemistry **44**, 694–703.

41. Tong, Y., Bagheri-Hamaneh, M., Penachioni, J. Y., Hota, P.K., Kim, S., et al. (2009) Structure and Function of the Intracellular Region of the Plexin-B1 Transmembrane Receptor. J. Biol. Chem. **284**, 35962–35972.
42. Ma, B., Kumar, S., Tsai, C. J., and Nussinov, R. (1999) Folding funnels and binding mechanisms. Protein Eng. **12**, 713–20.
43. Tong, Y, Chugha, P., Hota, PK., Li, M., Alviani, RS., et al. (2007) Binding of Rac1, Rnd1 and RhoD to a novel Rho GTPase interaction motif destabilizes dimerization of the plexin-B1 Effector domain. J.Biol.Chem. **282**, 37215–37224.
44. Henzler-Wildman, K. A., Lei, M., Thai, V., Kerns, S. J., Karplus, M., et al. (2007) A hierarchy of timescales in protein dynamics is linked to enzyme catalysis. Nature **450**, 913–6.
45. Gizachew, D. and Oswald, R. E. (2001) Concerted motion of a protein–peptide complex: backbone dynamics studies of a 15N-labeled peptide derived from P21-activated kinase bound to Cdc42Hs, GMPPCP. Biochemistry **40**, 14368–14375.
46. Tong, Y., Hota, P. K., Bagheri Hamaneh, M., and Buck, M. (2008) Insights into Oncogenic Mutations of plexin-B1 based on the Solution Structure of the Rho GTPase Binding Domain. Structure **16**, 246–258.
47. Wang, H., Hota, P. K., Tong, Y., Li, B. Shen, L., et al. (2011) Structural Basis of Rho GTPase Rnd1 Binding to Plexin RBDs J. Biol. Chem. **286**, 26093–2610.
48. James, C. P., Rosemary, B., Wei, W., James, G., Emad, T., et al. (2005) Scalable molecular dynamics with NAMD. Journal of Computational Chemistry **26**, 1781–1802.
49. MacKerell, A. D., Jr, Bashford, D., Bellott, M., Dunbrack, R. I., Jr, Evanseck, J., et al. (1998) All-atom empirical potential for molecular modeling and dynamics studies of proteins. J. Phys. Chem. B **102**, 3586–3616.
50. Buck, M., Bouguet-Bonnet, S., Pastor, R.W., and MacKerell, A. D. (2006) Importance of the CMAP correction to the CHARMM22 protein force field: Dynamics of Hen Lysozyme. Biophys.J. **90**, L36-L39.
51. Zerbetto, M., Polimeno, A., and Meirovitch, E. (2009) General theoretical/computational tool for interpreting NMR spin relaxation in proteins. J Phys Chem B **113**, 13613–25.
52. Best, R. B., Clarke, J., and Karplus, M. (2005) What contributions to protein side-chain dynamics are probed by NMR experiments? A molecular dynamics simulation analysis. J Mol Biol. **349**, 185–203.
53. Hota, P. and Buck, M. (2009) Thermodynamic characterization of two homologous protein complexes: Association of the semaphorin receptor plexin-B1 Rho GTPase binding domain with Rnd1 and active Rac1. Protein Science **18**, 1060–1071.
54. Ichiye, T. and Karplus, M. (1991) Collective motions in proteins: a covariance analysis of atomic fluctuations in molecular dynamics and normal mode simulations. Proteins **11**, 205–217.
55. Kong, Y. and Karplus, M. (2009) Signaling pathways of PDZ2 domain: a molecular dynamics interaction correlation analysis. Proteins **74**, 145–54.
56. Gorfe, A.A., Grant, B.J., and McCammon, J. A. (2008) Mapping the nucleotide and isoform-dependent structural and dynamical features of Ras proteins. Structure **16**, 885–96.
57. Lukman, S., Grant, B. J., Gorfe, A. A., Grant, G. H., and McCammon, J. A. (2010) The Distinct Conformational Dynamics of K-Ras and H-Ras A59G. PLoS Comput Biol. **6**, pii: e1000922.
58. Gohlke, H., Kuhn, L. A., and Case, D. A. (2004) Change in protein flexibility upon complex formation: analysis of Ras-Raf using molecular dynamics and a molecular framework approach. Proteins **56**, 322–37.
59. Lee, J., Natarajan, M., Nashine, V. C., Socolich, M., Vo, T., et al. (2008) Surface sites for engineering allosteric control in proteins. Science **322**, 438–42.
60. Chennubhotla, C. and Bahar, I. (2007) Signal propagation in proteins and relation to equilibrium fluctuations. PLoS Comput Biol. **3**, 1716–26.
61. Moritsugu, K., Kurkal-Siebert, V., and Smith, J. C. (2009) REACH coarse-grained normal mode analysis of protein dimer interaction dynamics. Biophys J. **97**, 1158–67.
62. Dobbins, S. E., Lesk, V. I., and Sternberg, M. J. (2008) Insights into protein flexibility: The relationship between normal modes and conformational change upon protein–protein docking. Proc Natl Acad Sci USA. **105**, 10390–5.
63. Lange, O. F., Lakomek, N. A., Farès, C., Schröder, G. F., Walter, K. F., et al. (2008) Recognition dynamics up to microseconds revealed from an RDC-derived ubiquitin ensemble in solution. Science. **320**, 1471–5.
64. O'Connor, C. and Kovrigin, E. L. Global conformational dynamics in ras. Biochemistry **47**, 10244–6.

65. Korzhnev, D. M. and Kay, L. E. (2008) Probing invisible, low-populated States of protein molecules by relaxation dispersion NMR spectroscopy: an application to protein folding. Acc Chem Res. **41**, 442–51.

66. Palmer, A. G. 3rd and Massi, F. (2006) Characterization of the dynamics of biomacromolecules using rotating-frame spin relaxation NMR spectroscopy. Chem Rev. **106**, 1700–19.

67. Ho, B. K. and Agard, D. A. (2010) Conserved tertiary couplings stabilize elements in the PDZ fold, leading to characteristic patterns of domain conformational flexibility. Protein Sci. **19**, 398–411.

Chapter 14

Hydrogen–Deuterium Exchange Study of an Allosteric Energy Cycle

Dorothy Beckett

Abstract

Elucidation of mechanisms of energy transduction through macromolecules in allosteric systems requires application of a broad range of techniques and approaches. High-resolution structures of the end states in an allosteric system provide invaluable clues about allosteric mechanism. Thermodynamic and kinetic studies reveal the rules that govern the transitions between states in the system. Acquisition of detailed molecular level information about allosteric mechanism requires interrogation of the structural and dynamic properties of both intermediates and end states in the allosteric cycle. Many experimental and computational tools have been developed to probe allostery. Among these are hydrogen–deuterium exchange detected by either NMR spectroscopy or mass spectrometry. This article provides a detailed description of application of hydrogen exchange detected by mass spectrometry (HDX-MS) to investigate an allosteric system.

Key words: Hydrogen–deuterium exchange, Mass spectrometry, Allostery, Dynamics

1. Introduction

In allosteric processes, energy is transmitted from one site to a distal site to change the functional properties of that distal site. Allostery exists in a broad range of biological systems including simple enzymes (1), transcription regulatory proteins (2), transmembrane receptors (3), and complex machines such as GroEL (4). An allosteric effect may result from small ligand binding, posttranslational modification or binding of another macromolecule. Detailed structures of the end states in allosteric systems are available through high-resolution structural studies performed using either NMR spectroscopy or X-ray crystallography. These snapshots provide indispensible information in formulating hypotheses about allosteric mechanism. Tools that allow probing of the dynamics of the end states as well as intermediate species in

Aron W. Fenton (ed.), *Allostery: Methods and Protocols*, Methods in Molecular Biology, vol. 796,
DOI 10.1007/978-1-61779-334-9_14,

an allosteric cycle provide additional information that is necessary for determining allosteric mechanism. Included among these tools is hydrogen–deuterium (H–D) exchange. Development of hydrogen exchange for investigation of allosteric systems was pioneered by Englander and Englander in their extensive studies of hemoglobin (5). In their early work the exchange of backbone hydrogens for tritium in tritiated water was detected by scintillation counting. Currently, hydrogen exchange, which refers the exchange of backbone N–H with solvent deuterium, detected using either NMR spectroscopy or mass spectrometry is commonly used (see Note 1). Results of exchange measurements may be interpreted at structural and/or dynamic levels, depending on other data available for the system of interest.

Hydrogen exchange measurements to investigate protein conformation and function started in the 1950s with the pioneering work of Kai U. Linderstrøm-Lang and his coworkers. Since then, the method has gone through several permutations that are recounted in a review article by Englander (6). Although NMR spectroscopy can be used to monitor exchange, the marriage of mass spectrometry with hydrogen–deuterium exchange (HDX-MS) enables monitoring of high-resolution mapping of exchange data in very large systems that are inaccessible by NMR spectroscopy.

1.1. What Should Be Known Prior to Performing H–D Exchange Measurements on an Allosteric System?

The purpose of performing HDX-MS measurements on an allosteric system is to determine the conformational and/or dynamic changes that accompany the transition of the system from one species in the allosteric cycle to another species in the cycle. Therefore, prior to initiating HDX-MS exchange studies of any allosteric system, it is critical to know the conditions in which the relevant species in an allosteric cycle are populated (see Note 2). Identification of these conditions requires careful thermodynamic and/or kinetic measurements. An example for which this analysis has been performed in our laboratory is the *Escherichia coli* biotin repressor or BirA. This protein is both the enzyme that catalyzes posttranslational biotin addition to biotin-dependent carboxylases and the transcriptional regulator that binds sequence specifically to the biotin operator to prevent transcription initiation at the biotin biosynthetic operon (7–9). For the purposes of this article, the focus is on the transcriptional regulatory function of BirA. The repressor undergoes ligand-linked homodimerization that is a prerequisite to site-specific DNA binding (Fig. 1) and binding of the allosteric ligand, bio-5′-AMP, enhances dimerization by 1,000-fold or −4 kcal/mol in Gibbs free energy (10–12). Extensive ligand binding measurements using ITC and fluorescence spectroscopy have revealed the equilibrium constants governing

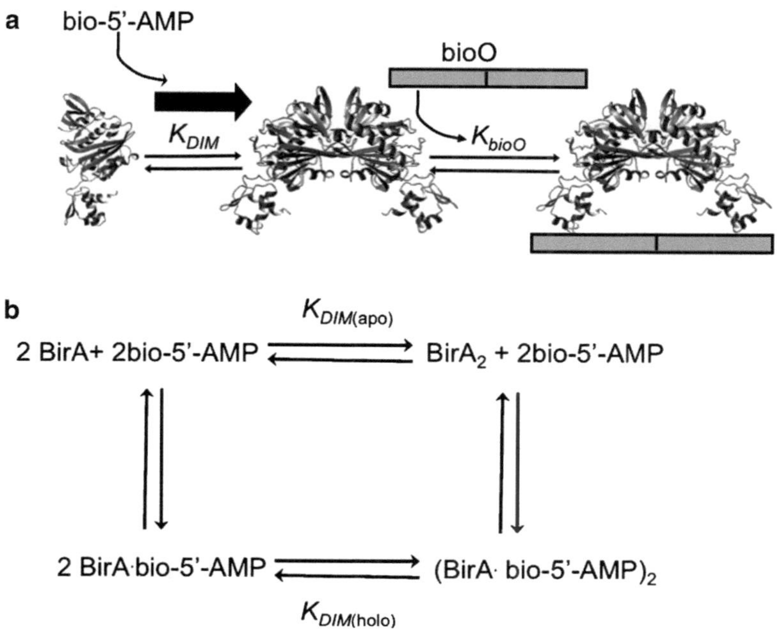

Fig. 1. Linkage of small ligand, bio-5′-AMP, binding to assembly of the BirA·bioO protein–DNA complex. (**a**) Bio-5′-AMP binding drives assembly of the complex by promoting BirA dimerization. (**b**) Thermodynamic cycle illustrating the coupling between bio-5′-AMP binding and dimerization. The equilibrium constant for dimerization of unliganded BirA is much weaker than that for dimerization of BirA·bio-5′-AMP.

allosteric effector binding to the repressor monomer in a broad range of solution conditions (13, 14). Additionally, the dimerization energetics for the unliganded and liganded forms of the protein was measured using equilibrium analytical ultracentrifugation (11). Armed with this information on the energetics governing the coupled ligand binding and dimerization, the system can be manipulated to populate the unliganded monomer, the unliganded dimer, the liganded monomer, or the liganded dimer. The studies described below were designed to elucidate how ligand binding to the BirA monomer activates it for dimerization by monitoring the differences in HDX in the unliganded and liganded monomer. The effects of four ligands, each of which activates BirA dimerization to a different extent, were probed (15). All measurements were performed in conditions in which either the unliganded or liganded monomer is the predominant species in solution.

The HDX-MS methods described in this article were adapted from the publications of several leaders in the field. Included among the methodological references are (16–18).

2. Materials

2.1. All Measurements

1. BirA protein of high purity and specific activity.
2. Ligands: biotin, 5′-*O*-[*N*-(biotinoyl)-sulfamoyl]adenosine (btn-SA) and biotinol-adenylate (btnOH-AMP), biotinyl-5′-AMP (19).
3. Water used in buffer preparation was purified using a mQ (Millipore) purification system and was characterized by a resistivity of 18.2 MΩ.cm.

2.2. Buffers for the Exchange Process

1. H_2O (see Note 3) buffer: the standard buffer for the system prepared using water. In this case it was 10 mM Tris–HCl, pH 7.5 at 20°C, 200 mM KCl, 2.5 mM $MgCl_2$.
2. D_2O buffer: the standard buffer for the system prepared using deuterated water. Note that when preparing the D_2O (99.9% Sigma-Aldrich, St. Louis) the pH adjustment should take into account the relationship between pH and pD.
3. D_2O MOPS buffer: prepared identically to the H_2O buffer with the exceptions that D_2O is the solvent and Tris is replaced by MOPS as the buffering agent (see Note 4).

2.3. Global Exchange Measurements

1. Quench solution: 25 mM Na citrate 25 mM Na succinate pH 2.4.
2. Dry ice–ethanol bath.
3. C18 ZipTip columns (Millipore).
4. 70% Acetonitrile(HPLC grade)/30% 0.1% TFA in H_2O.
5. Mass spectrometer: AccuTOF-CS ESI-TOF mass spectrometer (JEOL, Tokyo, Japan). Please note that other instruments can be used for this experiment.
6. Hamilton syringe.

2.4. Peptide Mapping and Local H–D Exchange Measurements

1. Agarose-immobilized pepsin (Sigma-Aldrich).
2. Ultrafree MC centrifuge filter device (Millipore).
3. 0.1% TFA in water.
4. MALDI matrix α-cyano-4-hydroxycinnamic acid (Sigma-Aldrich).
5. MALDI target plate (from MALDI-ToF MS manufacturer, Shimadzu Biotech, Columbia, MD).
6. Vacuum desiccator with chilled block.
7. Freezer bag.
8. Vacuum pump.

9. Mass spectrometers: Axima-CFR Plus MALDI-TOF mass spectrometer (Shimadzu Biotech, Columbia, MD). PE Sciex API QSTAR Pulsar i Q-TOF mass spectrometer (Concord, ON).
10. Thermal cycler (Thermo).
11. Lyophilization system.

2.5. Data Analysis Software

1. Convolution of ESI-ToF spectra in global exchange measurements: MagTran 1.0 software (20).
2. Nonlinear least squares analysis of time courses of global exchange: Prism 4.0 (Graphpad).
3. Identification of peptides in MS/MS analysis: Mascot software (Matrix Science).
4. Determination of centroids from envelopes of peptides obtained in MALDI-ToF measurements-Kratos Kompact MALDI software for Axima Instruments 2.3.4.

3. Methods

3.1. Global H–D Exchange

Initial measurements of H–D exchange of the different species in an allosteric cycle in the context of the intact protein provide information about the exchange properties of the distinct species. In the experiment, measurements of the protein mass are made as exchange of backbone amide hydrogens with deuterium is allowed to occur over time. The molecular weight data yield the number of deuteriums incorporated versus time and the overall curves provide information about whether protection from or enhancement of exchange occurs as the system shifts from one allosteric state to another.

A flowchart for measurement of global H–D exchange is shown in Fig. 2 (see Note 5).

3.1.1. Protein Preparation and H–D Exchange Reaction

The protein, in conditions corresponding to the different allosteric states is first prepared at a concentration tenfold greater than that at which the exchange will be performed. In the case of BirA the protein was either prepared in its unliganded form or in the presence of saturating amounts of four different allosteric ligands that activate the protein dimerization to different extents. The equilibrated protein (plus or minus saturating amounts of ligand) is diluted tenfold into the D_2O buffer.

3.1.2. Quench and Desalting of Exchange Reaction

Aliquots are removed at specific time intervals and combined with cold succinate/citrate solution to quench the exchange. Both the low pH and the decreased temperature accomplish this goal.

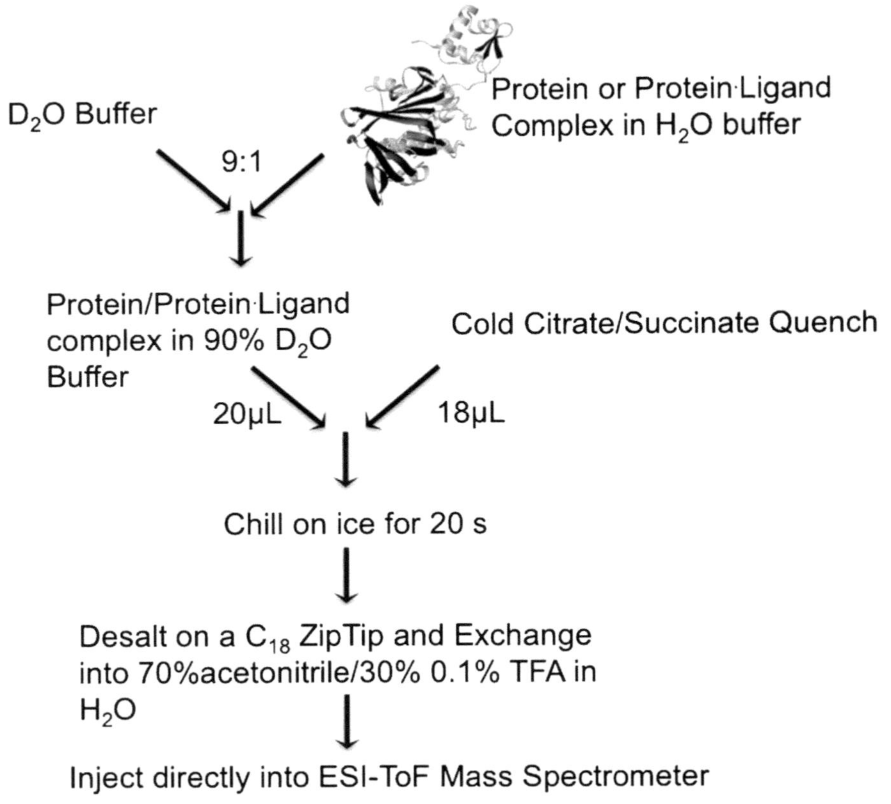

Fig. 2. Flowchart for global hydrogen–deuterium exchange measurements. *Please* see the text for details.

The sample is chilled in a dry ice/ethanol bath and then desalted by rapid chromatography on a cold C_{18} ZipTip column (Millipore) according to the manufacturer's instructions.

3.1.3. Elution and Mass Spectrometry

Elute sample with 70% acetonitrile/30% 0.1% TFA in water and immediately inject, using a Hamilton syringe, into the ESI-ToF mass spectrometer by direct infusion into the electrospray source.

For each spectrum an envelope of peaks corresponding to the different charged states of the protein is observed. The data are convoluted using the MagTran 1.0 (20) software that is provided with the mass spectrometer to obtain the mass at each time point in the exchange reaction. As the protein is incubated in the deuterium buffer, increasing numbers of backbone hydrogens exchange with deuterium and the protein mass increases with the time. These data can be displayed as molecular weight versus time or, alternatively, provided that a zero exchange control measurement is performed, they can be plotted in the form number of deuteriums incorporated versus time.

Time courses obtained with BirA monomer in the absence of ligand and in the presence of saturating concentrations of four

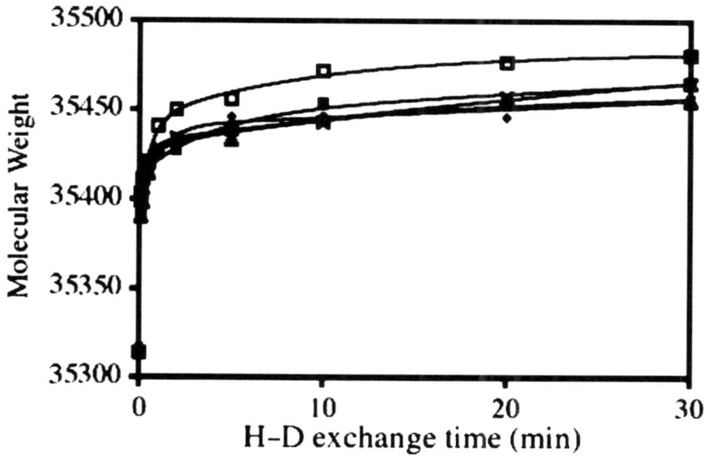

Fig. 3. Time courses of global hydrogen–deuterium exchange determined for ApoBirA (*open squares*), BirA·biotin (*bold x*), BirA·btnSA (*filled triangles*), BirA·btnOH-AMP (*filled squares*), and BirA·bio-5′-AMP (*filled diamonds*). The *solid lines* are the best-fit curves obtained from nonlinear least squares analysis of the data using a triple exponential model with Prism GraphPad. Reprinted from ref. 26 with permission from Elsevier.

different allosteric ligands are shown in Fig. 3. The results indicate, as evidenced by the higher molecular weight of unliganded, exchanged protein relative to any of the liganded species, that binding of each allosteric effector protects the protein from H–D exchange. A second feature of the results is that nonlinear least squares analysis of the data reveals that the simplest model that describes the time courses is a triple exponential model. This result is consistent with previous observations of native state H–D exchange measurements on proteins that group the backbone exchange rates into slow, medium, and fast categories (21). Finally, the results indicate that after approximately 60 min the exchange rate is negligibly small.

3.2. Localizing the H–D Exchange

The global exchange experiments probe H–D exchange in the context of the intact protein, but provide no information about which regions of the protein are altered in exchanged as a consequence of the transition from one allosteric state to another. However, measurement of the masses of proteolytic fragments of the exchanged protein reveals the H–D exchange properties of specific segments of the protein sequence. This method was initially developed for HDX-MS by Smith and colleague (16). Completion of this analysis for different allosteric states yields information about alterations in exchange properties of specific regions of the protein sequence in the distinct states. If a three-dimensional structure of the allosteric protein is known, the peptide exchange data can be used to map regions in the three-dimensional structure that are altered in HDX as the system shifts from one state in the allosteric cycle to another. In the case of the BirA system, the goal was to determine which regions in the three-dimensional structure of the protein monomer are altered in exchange as a consequence of allosteric activation via ligand binding.

3.2.1. Determining the Peptic Digest Map of the Protein

The strategy in localization of H–D exchange is to determine the levels or rates of exchange of local regions of the protein by measuring the masses of individual peptide digestion products of the exchanged protein. The protease used most frequently for these studies is pepsin because it retains activity at the low temperature and pH required to quench the H–D exchange reaction. Another characteristic of pepsin that is advantageous for HDX-MS is its lack of specificity, which results in overlapping cleavage products that help to maximize the fraction of the protein sequence that is accessible to HDX-MS analysis. In order to analyze the amount of exchange in any peptide, it is first necessary to use tandem mass spectrometry to identify peptic digest products of the unexchanged protein with the goal of maximizing coverage of the protein primary sequence. A flowchart of the strategy for identifying peptic digestion products is provided in Fig. 4. There exist several limitations to obtaining full coverage of a sequence in mass spectrometric analysis of pepsin digestion

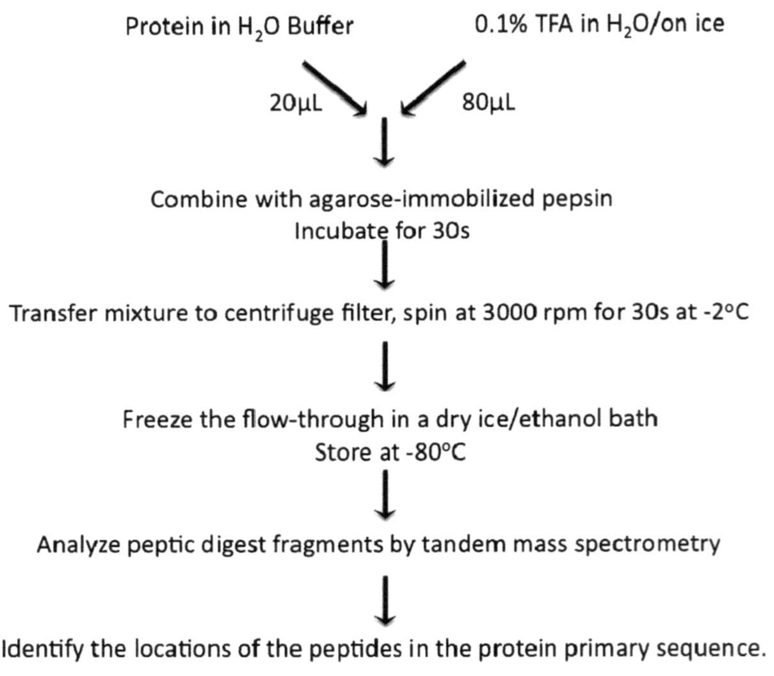

Fig. 4. Flowchart for procedure to identify pepsin digestion products. *Please* see the text for details.

products in HDX measurements. The first is related to the conditions under which the cleavage is carried out. In the HDX measurement, the exchanged sample is quenched into a low pH solution at low temperature. In these conditions, the protein tertiary structure may be characterized as somewhere between native and denatured. Therefore, the proteolytic enzyme may not have access to all possible cleavage sites, thus limiting the sequence coverage. A second limitation is the length of time allowed for the digestion. Once the exchange is quenched, it is critical to bring the sample to mass analysis as rapidly as possible. This is because, once H–D exchange is quenched into an aqueous buffer, despite the low temperature and low pH, the backbone N–H groups can undergo back exchange in which the deuteriums exchange with protons in water. An empirical method to correct for back exchange is performed (see below). Nevertheless, pepsin digestion should be limited in time to minimize the magnitude of the back exchange. Finally, in the mass spectrometric analysis of peptide products some peptides do not ionize as readily as others, thus limiting the ability to detect representative peptides from the entire protein primary sequence. This third limitation depends on the mass spectrometer used for the measurements. The extent of sequence coverage can vary significantly for different proteins. The protease digestion is carried out using agarose-immobilized pepsin, and the digestion conditions to generate peptides for identification by tandem mass spectrometry mimic those that are used in the actual HDX-MS studies (see Note 6).

1. *Preparation of protein sample for pepsin digestion*

   Dilute the protein into the appropriate H_2O buffer at the final desired concentration.

   Incubate for the time required to reach equilibrium.

2. *Preparation of agarose-immobilized pepsin. Prepare just prior to use*

   Weigh out the agarose-immobilized pepsin (10–15 mg).

   Exchange the resin two times with 1.0 mL cold mQ H_2O. Exchange the resin two times with 1 mL cold 0.1% TFA in mQ H_2O.

   Resuspend the resin to a final weight concentration of 25 mg/mL in cold 0.1% TFA in mQ H_2O and keep cold on ice.

3. *Quench and pepsin digestion*

   Combine 80 μL of 0.1% TFA in water with 20 μL of the equilibrated protein on ice. At this point, the pH of the solution should be between 2.0 and 2.5 and can be checked with narrow range pH sticks. Transfer the reaction to 100 μL of the exchanged agarose-immobilized pepsin (95 U) and incubate on ice for 30 s. Transfer the entire reaction to an Ultrafree MC centrifuge filter device (Millipore) and spin at 900*g* for 30 s at −2.0°C. Freeze the flowthrough in a dry ice ethanol bath and store overnight at −80°C for mass spectrometric analysis.

4. *Mass spectrometric analysis*
   The identities of peptides in the peptic digests are determined by liquid chromatography tandem mass spectrometry or LC MS-MS. There are many LC MS-MS systems that can theoretically be used for this purpose. In this work, samples were introduced through an LC Packings Ultimate nanoflow LC system (Abberdaan, the Netherlands) equipped with a C18 PepMap100 column (75 mm, 15 cm, 3 mm, 100 Å) and connected to a nano ionspray source (Protana, Odense, Denmark) or, alternatively, by direct infusion into the ion spray source (PE Sciex). The chromatography step provides limited separation of the peptides. Tandem mass spectrometric (MS-MS) analysis of BirA peptic digest products was performed on a PE Sciex API QSTAR Pulsar i Q-TOF mass spectrometer (Concord, ON), and the data were analyzed using the Mascot software (Matrix Science). In the BirA system, 65% of the total primary sequence was identified in MS-MS analysis of the peptic digestion products (see Note 7).

3.2.2. Determining the Locations of H–D Exchange in the Protein Primary Sequence

In order to determine the extents of H–D exchange in specific regions of the protein sequence the exchanged protein is subjected to proteolytic digestion and the resulting peptides are analyzed by mass spectrometry. A flowchart for the experiment is shown in Fig. 5. The exchange reaction is carried out identically as described for the global exchange measurements and subsequent steps are similar to those described for peptide identification process with the exception that the mass spectrometric analysis is accomplished using MALDI-ToF MS (see Note 8). Exchange reactions are prepared as described for the global exchange measurements. However, in contrast to the global exchange, which was performed as a function of time, the local exchange in studies of BirA was performed at a single time point of 1 h. This incubation time was chosen because at that time in the global exchange time courses the exchange rate was negligible (Fig. 3). Comparison of the results of exchange at the single time point for the different allosteric states allows identification of local differences in extents of deuterium incorporation at the plateau in the exchange process. The linkage of allosteric effector binding to protein dimerization in the BirA system necessitated performing the measurements at a low protein concentration, which precluded performing time-dependent measurements on the individual peptic digest products. However, if protection factors for specific regions of a protein are desired, it is necessary to

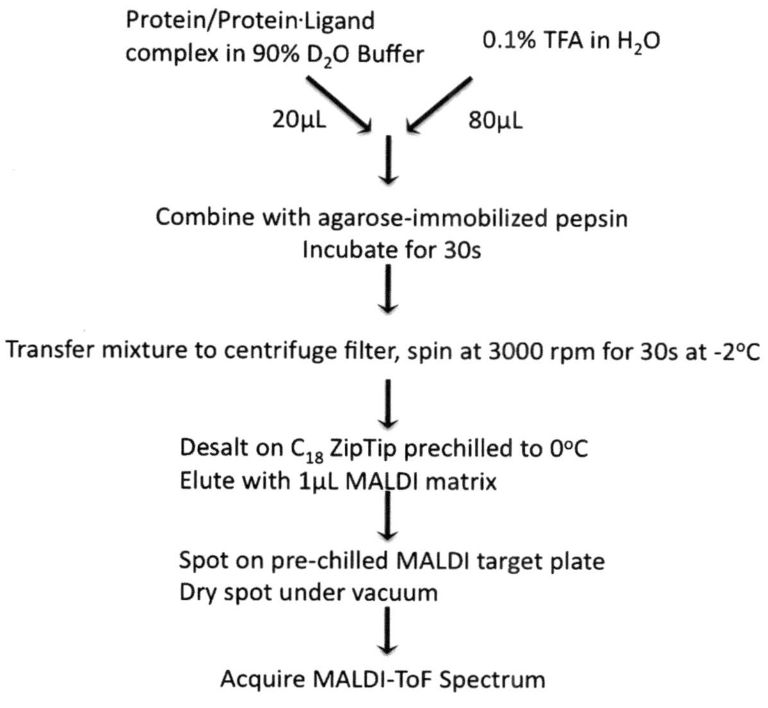

Fig. 5. Flowchart for hydrogen–deuterium exchange, pepsin digestion, and MALDI-ToF MS analysis of the digestion products. *Please* see the text for details.

measure the HDX rate profiles of the individual peptic digestion products (18).

1. *Hydrogen–deuterium exchange reactions*: prepare the protein or protein–ligand complex at 10× the concentration that will be employed in the exchange reaction. Dilute the protein tenfold into the D_2O buffer and incubate for 1 h. Quench the exchange reaction by diluting fivefold into cold 0.1% TFA in H_2O (see Note 9).
2. *Pepsin digestion*: the digestion is performed as described in Subheading 3.2.1. Follow all steps up to and including separation of the cleaved peptides from the agarose-immobilized pepsin by filter centrifugation.
3. *Peptide desalting by chromatography on C_{18} resin and spotting for MALDI-ToF MS*: a C_{18} ZipTip is prepared for chromatography according to the manufacturer's recommendations and chilled on ice. A 20 μL volume of the peptide mixture is loaded onto the column by rapid pipetting 12× up and down. The column is washed by repeated pipetting ten times up and down in 0.1% TFA in mQ H_2O. The sample is eluted in 1 μL of a solution containing 10 mg/mL α-cyano-4-hydroxycinnamic acid matrix in 70% acetonitrile/30% 0.1%

TFA in H_2O at 0°C and immediately spotted onto a cold (0°C) MALDI-target plate.

4. *MALDI-target preparation, sample spotting, and data acquisition*: the target is pre-chilled on a frozen freezer block that is sitting in a vacuum desiccator connected to a vacuum pump. First, break the vacuum and open the desiccator. Spot the eluant from the ZipTip directly onto the target plate as one spot. Immediately close the desiccator and draw a vacuum. Once the spot has dried for approximately 40 s, open the desiccator, remove the target plate, and place immediately into the MALDI-ToF MS sample compartment. Close the door to the sample chamber, apply the vacuum, and acquire the spectrum.

5. *Analysis of Maldi-ToF MS data*: Examples of MALDI-ToF MS data obtained for individual peptic digest products of BirA are shown in Fig. 6. The figure shows the results obtained with two different peptides, including one (001–010) for which the exchange is unaltered by ligand binding and the other (085–101) in which it is altered by the ligand. The data for

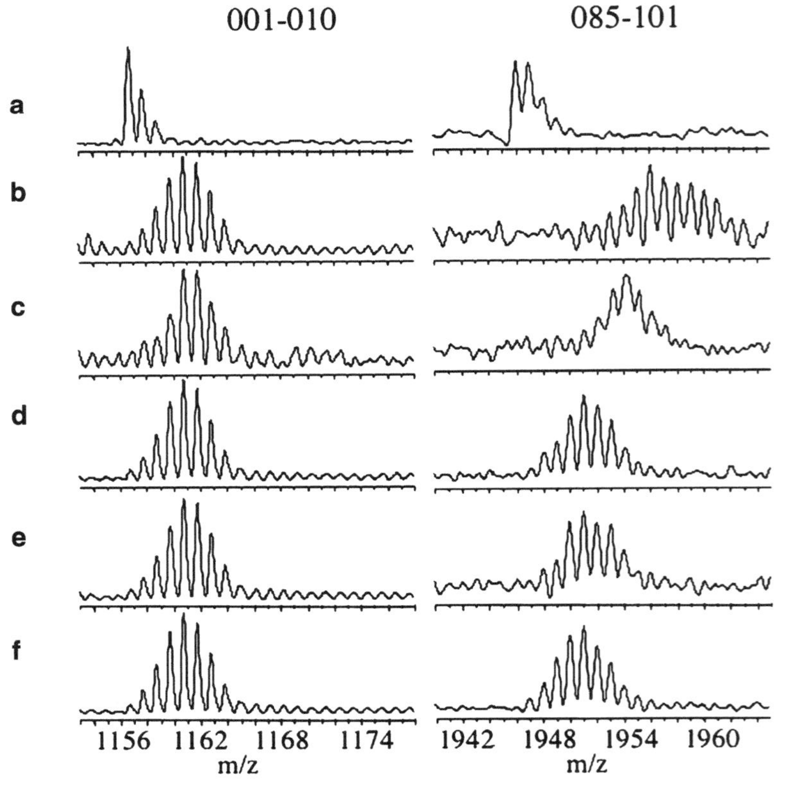

Fig. 6. MALDI-ToF mass envelopes for two pepsin digestion products obtained from protein or protein–ligand complexes subjected to hydrogen–deuterium exchange. (**a**) ApoBirA no H–D exchange and (**b**–**f**) 60 min of deuterium in-exchange in the following conditions: (**b**) no ligand, (**c**) biotin, (**d**) biotinoyl-sulfamoyl adenylate, (**e**) biotinol-5′-AMP, and (**f**) bio-5′-AMP. Reprinted from ref. 26 with permission from Elsevier.

each peptide appears as an envelope of isotopically resolved peaks with the centroid of the peak yielding the average m/z value. In measurements on BirA, the centroids were determined using a tool that is included with the Kratos Kompact MALDI software for Axima Instruments 2.3.4. This average m/z value is compared with that of the unexchanged peptide to estimate the number of deuteriums exchanged into each peptide. However, the number must be corrected to take into account the fact that exchange is carried out in 90% D_2O and for any back exchange that occurs between the time of the quench and the mass spectrometry. The back exchange correction is based on back exchange control measurements that are carried out as described in the next section.

3.2.3. Back Exchange Control

In the time interval between quench of the H–D exchange reaction and the mass spectrometric analysis, some back exchange of the incorporated deuterium inevitably occurs. Furthermore, the extent of back exchange varies among the peptic digest fragments. Therefore, it is important to empirically determine the back exchange for the peptic digestion fragments. There are several methods to measure back exchange. In the method used for the BirA studies, which was adapted from ref. 22, the protein was first subjected to proteolysis, and the resulting peptides were subjected to exchange in conditions in which all backbone hydrogens should be exchanged for deuterium. The resulting fully exchanged peptides were subjected to MALDI-ToF MS as described for the exchange measurements described above. Comparison of the measured mass of each peptide to the mass expected of the fully exchanged peptide provides a measure of the back exchange that occurs in the course of the mass spectrometric analysis.

1. *Preparation of peptide* digest: combine the protein with buffer and equilibrate at the appropriate temperature. Add 80 μL cold 0.1% TFA in H_2O to the reaction on ice.

   Combine this quenched reaction with 100 μL cold 25 mg/mL agarose immobilized pepsin (95 U) in 0.1% TFA in H_2O. Incubate on ice for 30 s. The immobilized pepsin is prepared as described in the section 3.2.1 on peptide identification. Remove the immobilized pepsin by centrifugation in a chilled Ultrafree MC centrifuge filter device at 900*g*, −2°C for 30 s. Desalt the peptides in the effluent by chromatography on a C18ZipTip. Elute with 10 μL of a solution containing acetonitrile/0.1%TFA in H_2O (7:3 v/v) into a 200 μL thin walled thermocycler tube. Dry the peptides by lyophilization.
2. *Deuterium exchange of peptic fragments*: Dissolve the peptides in 20 μL of a buffer containing 10 mM MOPS (see Note 4) pD 8.00 ± 0.02, 200 mM KCl, 2.5 mM $MgCl_2$ prepared with D_2O. The Tris in the original buffer is replaced with MOPS because the temperature coefficient of MOPS is

significantly smaller than that of Tris. Heat the solution at 90°C for 90 min in a PCR Sprint thermal cycler.

3. *Mass spectrometric analysis of peptic fragments*: Desalt the exchanged peptides using the C_{18} ZipTip as described in Subheading 3.2.2 and elute with 1 μL of a solution containing 10 mg/mL a-cyano-4-hydroxycinnamic acid matrix in ACN–0.1% TFA (7:3 v/v). Spot on the cold MALDI target and acquire spectra as described in Subheading 3.2.2.

3.2.4. Data Analysis

The weight average molecular weight obtained from the centroids in Maldi-ToF spectra are corrected for the back exchange and the fact that exchange is carried out in 90% D_2O. The latter correction is made by multiplying the initial estimate of the number of backbone deuterium exchanged in by 100/90. The correction for back exchange is made using the back exchange values measured in the control experiments described above. Each peptide has a characteristic back exchange value associated with it. The equation for correcting for back exchange is:

$$D = \frac{m - m_{0\%}}{m_{100\%} - m_{0\%}} \times N,$$

in which D is the corrected deuterium level in a particular peptide, m is the experimentally determined weight average mass obtained as the centroid of the MALDI-ToF MS peaks for that peptide, $m_{0\%}$ is the mass of the unexchanged peptide, $m_{100\%}$ is the experimentally determined mass for the theoretically fully exchanged peptide, and N is the total number of exchangeable backbone amide hydrogens on the peptide. In the H–D exchange studies of BirA, the back exchange in the peptides varied from 9 to 50%.

3.2.5. Data Presentation

The mass spectrometric data obtained on samples subjected to H–D exchange yield the number of deuteriums that each peptide accumulates in the exchange reaction. In studying allostery, the goal is to understand the dynamic and structural changes that occur as a molecule undergoes an allosteric transition. Thus, it is useful to calculate the difference in the number of deuterons in each peptide for different allosteric states. In carrying out this calculation, one species is designated the reference state. In measurements performed on the BirA system, the unliganded monomer was considered the reference state to which the liganded monomer was compared. In the study described in this article, four ligands were employed, each of which elicits an allosteric response of different magnitude (15). First, peptides of liganded species that showed differences in levels of H–D exchange relative to the unliganded protein were identified from the corrected values of the number of deuteriums per peptide. These peptides were mapped onto the three-dimensional structure of the protein that was determined by X-ray crystallography (23). The number of deuteriums incorporated for each of these peptides in each of the liganded states was then calculated and subtracted from the corresponding values measured for the unliganded protein. These values of delta deuteriums were plotted for each of the liganded states (Fig. 7). In the BirA system, a trend was observed in which

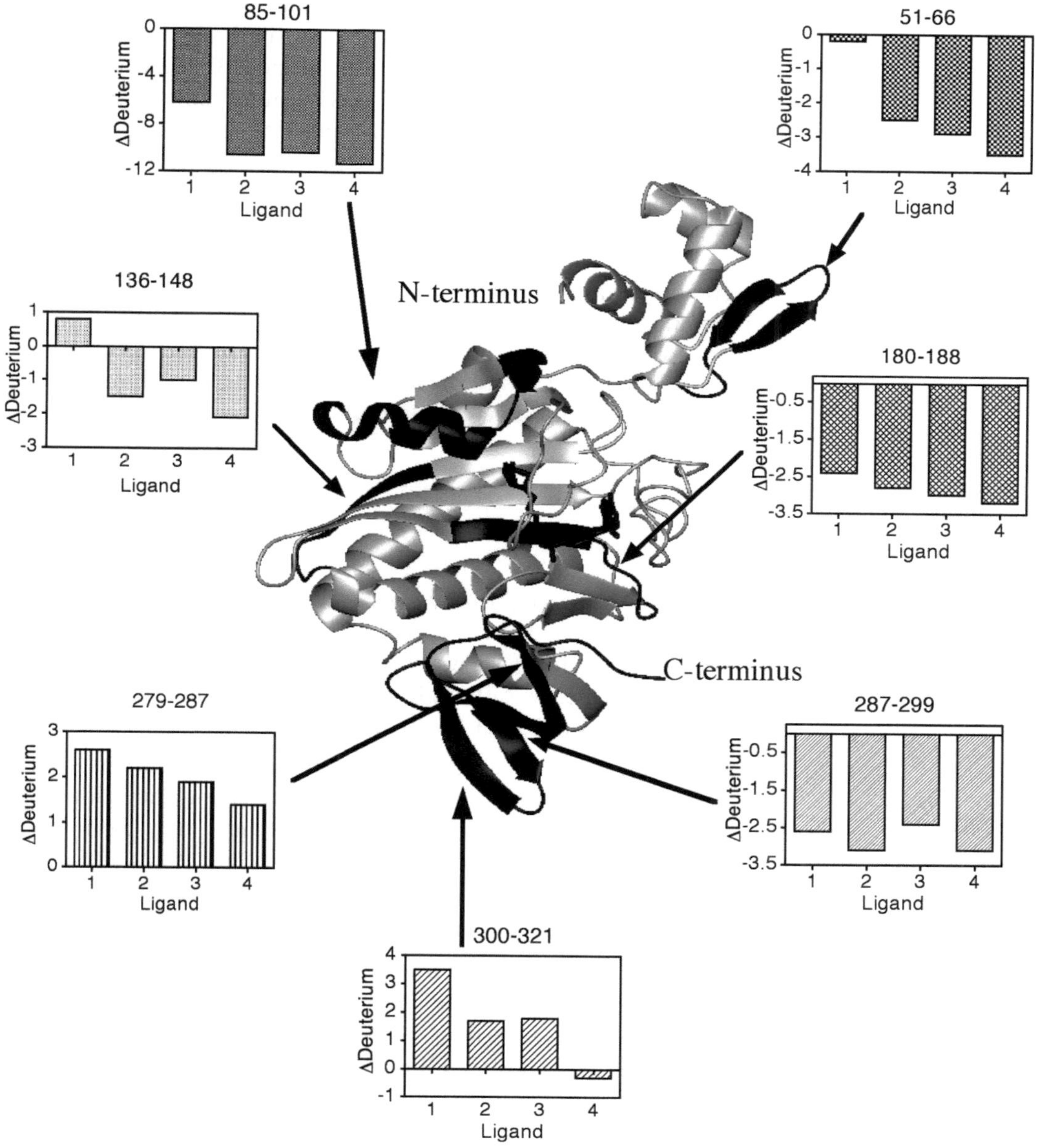

Fig. 7. Results of H–D exchange measurements mapped onto the three-dimensional structure of BirA. *Bar graphs* show the difference in the deuterium content in each peptide in the different liganded states relative to the level found in the same peptide in the unliganded protein. (1) Biotin, (2) biotinoyl-sulfamoyl adenylate, (3) biotinol-5′-AMP, (4) bio-5′-AMP. Adapted from ref. 26 with permission from Elsevier.

the larger the magnitude of the allosteric response, measured in terms of the coupling free energy between ligand binding and homodimerization, the greater the protection from deuterium exchange for the majority of the peptides that were monitored.

3.3. Interpretation of the Exchange Data

Changes in backbone H–D exchange in the distinct states in an allosteric cycle can reflect changes in structure and/or dynamics. However, the availability of high-resolution structural data of the multiple states in the cycle can constrain the interpretation of the exchange data. Structures of the unliganded BirA monomer and the biotin-bound and biotinol-5′-AMP-bound dimers determined by X-ray crystallography reveal that all three of these states of the protein are structurally similar (23–25). Therefore, the decreased H–D exchange that is correlated with allosteric activation of the monomer was interpreted as loss of dynamic motion on the monomer (see Note 10). The magnitude of the functional allosteric response, measured as the increase in the equilibrium dimerization constant triggered by binding of each ligand, correlates with the degree of protection from HDX. Interpretation of this correlation in dynamic terms leads to the hypothesis that decreased dynamics in the entire protein activate the protein for dimerization.

4. Notes

1. Before embarking on hydrogen deuterium exchange measurements, it is advisable to read about the theory of the exchange process and how it may relate to structural and dynamic properties of proteins. Although several suitable references exist, the review by Englander (6) provides an excellent starting point.
2. The importance of these preliminary measurements of the thermodynamics and/or kinetics of allosteric transitions in a system cannot be overemphasized because they alone allow establishment of conditions in which specific states in the allosteric cycle are the predominant species in exchange reactions.
3. The water used in all buffers and solutions for these measurements is purified using a MilliQ system (Millipore) and is characterized by a resistivity of at least 18.2 MΩ.
4. In the back exchange control experiments, the buffering agent, Tris, was replaced by MOPS. This change was necessitated by the large enthalpy of protonation of Tris relative to that of MOPS and the fact that a step in which the sample was heated to 90°C was incorporated in these control experiments.

5. In the Global Exchange measurements, the time between the quench of the H–D exchange reaction and the mass spectrometric analysis must be kept to a minimum to minimize back exchange of the deuteriums.
6. Conditions for pepsin digestion must be optimized empirically for each protein. However, in optimizing the conditions keep in mind that the low temperature and pH cannot be changed and the digestion time should be kept to a minimum. This means that the ratio of pepsin enzyme to protein is, realistically, the only variable that can be changed.
7. It is possible to improve upon this by using mass spectrometers that are capable of greater mass resolution and sensitivity.
8. It is extremely important to limit the time between the quench of the exchange reaction and the mass spectrometric analysis. Optimization of the pepsin digestion step is most helpful in limiting the total time. In the measurements presented in this article, this total time interval was approximately 3.0 min.
9. Once the exchange reactions are quenched, all subsequent manipulations must be carried out at $T < 0°C$.
10. The magnitude of the motions, partial unfolding versus local fluctuations, detected in H–D exchange measurements is open to debate (6).

References

1. Di Cera, E. (2009) Serine proteases, *IUBMB Life 61*, 510–515.
2. Swint-Kruse, L., and Matthews, K. S. (2009) Allostery in the LacI/GalR family: variations on a theme, *Curr Opin Microbiol 12*, 129–137.
3. Kenakin, T. P. (2009) '7TM receptor allostery: putting numbers to shapeshifting proteins, *Trends Pharmacol Sci 30*, 460–469.
4. Yifrach, O., and Horovitz, A. (1995) Nested cooperativity in the ATPase activity of the oligomeric chaperonin GroEL, *Biochemistry 34*, 5303–5308.
5. Englander, S. W., and Englander, J. J. (1994) Structure and energy change in hemoglobin by hydrogen exchange labeling, *Methods Enzymol 232*, 26–42.
6. Englander, S. W. (2006) Hydrogen exchange and mass spectrometry: A historical perspective, *J Am Soc Mass Spectrom 17*, 1481–1489.
7. Barker, D. F., and Campbell, A. M. (1981) The birA gene of *Escherichia coli* encodes a biotin holoenzyme synthetase, *J. Mol. Biol 146*, 451–467.
8. Barker, D. F., and Campbell, A. M. (1981) Genetic and biochemical characterization of the birA gene and its product: evidence for a direct role of biotin holoenzyme synthetase in repression of the biotin operon in *Escherichia coli*, *J Mol Biol 146*, 469–492.
9. Cronan, J. E., Jr. (1989) The *E. coli* bio operon: transcriptional repression by an essential protein modification enzyme., *Cell 58*, 427–429.
10. Eisenstein, E., and Beckett, D. (1999) Dimerization of the *Escherichia coli* biotin repressor: corepressor function in protein assembly., *Biochemistry 38*, 13077–13084.
11. Streaker, E. D., Gupta, A., and Beckett, D. (2002) Th biotin repressor: thermodynamic coupling of corepressor binding, protein assembly, and sequence-specific DNA binding., *Biochemistry 41*, 14263–14271.
12. Streaker, E. D., and Beckett, D. (2003) Coupling of protein assembly and DNA binding:biotin repressor dimerization precedes biotin operator binding., *J. Mol. Biol 325*, 937–948.
13. Xu, Y., Johnson, C. R., and Beckett, D. (1996) Thermodynamic analysis of small ligand binding to the *Escherichia coli* repressor of biotin biosynthesis, *Biochemistry 35*, 5509–5517.

14. Brown, P. H., and Beckett, D. (2005) Use of binding enthalpy to drive an allosteric transition, *Biochemistry 44*, 3112–3121.
15. Brown, P., Cronan, J.E., Grotli, M., and Beckett, D. (2004) The biotin repressor: Modulation of allostery by corepressor analogs., *J. Mol. Biol 337*, 857–869.
16. Zhang, Z., and Smith, D. L. (1993) Determination of amide hydrogen exchange by mass spectrometry: a new tool for protein structure elucidation, *Protein Sci 2*, 522–531.
17. Mandell, J. G., Falick, A. M., and Komives, E. A. (1998) Measurement of amide hydrogen exchange by MALDI-TOF mass spectrometry, *Anal Chem 70*, 3987–3995.
18. Hoofnagle, A. N., Resing, K. A., and Ahn, N. G. (2004) Practical methods for deuterium exchange/mass spectrometry, *Methods Mol Biol 250*, 283–298.
19. Brown, P. H., Cronan, J. E., Grotli, M., and Beckett, D. (2004) The biotin repressor: modulation of allostery by corepressor analogs, *J Mol Biol 337*, 857–869.
20. Zhang, Z. MAG-TRAN, Amgen, Inc., Thousand Oaks, CA.
21. Hoofnagle, A. N., Resing, K. A., and Ahn, N. G. (2003) Protein analysis by hydrogen exchange mass spectrometry, *Annu Rev Biophys Biomol Struct 32*, 1–25.
22. Resing, K. A., and Ahn, N. G. (1998) Deuterium exchange mass spectrometry as a probe of protein kinase activation. Analysis of wild-type and constitutively active mutants of MAP kinase kinase-1, *Biochemistry 37*, 463–475.
23. Wilson, K. P., Shewchuk, L. M., Brennan, R. G., Otsuka, A. J., and Matthews, B. W. (1992) *Escherichia coli* biotin holoenzyme synthetase/bio repressor crystal structure delineates the biotin- and DNA-binding domains, *Proc Natl Acad Sci USA 89*, 9257–9261.
24. Weaver, L. H., Kwon, K., Beckett, D., and Matthews, B. W. (2001) Corepressor-induced organization and assembly of the biotin repressor: a model for allosteric activation of a transcriptional regulator, *Proc Natl Acad Sci USA 98*, 6045–6050.
25. Wood, Z. A., Weaver, L. H., Brown, P. H., Beckett, D., and Matthews, B. W. (2006) Co-repressor induced order and biotin repressor dimerization: a case for divergent followed by convergent evolution, *J Mol Biol 357*, 509–523.
26. Laine, O., Streaker, E. D., Nabavi, M., Fenselau, C. C., and Beckett, D. (2008) Allosteric signaling in the biotin repressor occurs via local folding coupled to global dampening of protein dynamics, *J Mol Biol 381*, 89–101.

Chapter 15

Ensemble Properties of Network Rigidity Reveal Allosteric Mechanisms

Donald J. Jacobs, Dennis R. Livesay, James M. Mottonen, Oleg K. Vorov, Andrei Y. Istomin, and Deeptak Verma

Abstract

The distance constraint model (DCM) is a unique computational modeling paradigm that integrates mechanical and thermodynamic descriptions of macromolecular structure. That is, network rigidity calculations are used to account for nonadditivity within entropy components, thus restoring the utility of free-energy decomposition. The DCM outputs a large number of structural characterizations that collectively allow for quantified stability–flexibility relationships (QSFR) to be identified. In this review, we describe the theoretical underpinnings of the DCM and introduce several common QSFR metrics. Application of the DCM across protein families highlights the sensitivity within the set of protein structure residue-to-residue couplings. Further, we have developed a perturbation method to identify putative allosteric sites, where large changes in QSFR upon rigidification (mimicking ligand-binding) detect sites likely to invoke allosteric changes.

Key words: Network rigidity, Distance constraint model, Quantitative stability–flexibility relationships, Allostery

1. Introduction

Static glimpses of proteins provided by X-ray crystal structures routinely shown in biochemistry textbooks do not convey how proteins flex and wiggle over a spectrum of timescales that can span more than 12 orders of magnitude (1). In fact, dynamical motions associated with conformational changes on long timescales are generally paramount to protein function. These motions are modulated by thermodynamic and environmental conditions that a protein is subjected to, where pH and other environmental conditions are determined by the cellular environment. A given protein will slosh around in a mixture of other proteins, complex polymers, and myriad small molecules in the midst of tens of

Aron W. Fenton (ed.), *Allostery: Methods and Protocols*, Methods in Molecular Biology, vol. 796,
DOI 10.1007/978-1-61779-334-9_15,

thousands of chemical reactions taking place. Being in thermal contact with a heat reservoir (the cell), proteins are subject to thermal fluctuations that give rise to statistical interactions, and yet, they perform a wide range of functions with exquisite precision. For example, enzymes catalyze chemical reactions with a remarkable degree of specificity, control, and efficiency (2). Decades of painstaking structure–function biochemical and structural biology studies have uncovered many physiochemical principles that consist of the interplay of a large number of different types of interactions. Typically, proteins have thousands of atoms that interact through covalent bonding, hydrogen bonding, and salt bridges, and weak nonbonding forces (3).

An open challenge is to develop a computational method that accurately predicts protein thermodynamics and flexibility for specified solvent and thermodynamic conditions in computing times fast enough for high-throughput applications. We review here crucial theoretical aspects that enable new types of algorithms to be successfully developed to accurately predict protein thermodynamics. Despite considerable complexity, we describe how progress has been made in predicting protein properties and behavior by focusing on the essential physics, which subsequently greatly simplifies model details and calculations. Central to this task is to reconcile conformational flexibility as a critical link relating structure to stability. After discussing key properties of mechanical networks, upon which our approach is based, a Distance Constraint Model (DCM) (4, 5) is defined based on a paradigm that combines constraint theory with free-energy decomposition. With the theoretical framework laid out, the DCM is solved using an efficient graph-rigidity algorithm in conjunction with a hybrid Monte Carlo and mean field approximation (6). The consequence of this approach is the ability to calculate a large number of mechanical network properties in a thermodynamically meaningful way, including pairwise residue-to-residue couplings intrinsic to the native structure ensemble (7–11) and putative allosteric sites via a perturbation method (12).

2. Generating Protein Ensembles

Generating conformational ensembles is a necessary first step to describe protein stability, dynamics and function, including allosteric mechanisms. Molecular dynamics (MD) simulation is the most common approach used to explore protein dynamics and investigate detailed atomic mechanisms (13). Unfortunately, MD is severely limited when trying to describe thermodynamic properties because robust and statistically significant sampling of atomic configurations is required to accurately estimate conformational entropy. Even in the next decades to come, there is little hope that an all-atom brute force MD simulation will be able to robustly explore conformational space to make free-energy calculations meaningful. However, this is not to say that MD cannot be used, quite the contrary. One realizes that much of the motion of a protein is organized though cooperative behavior and local constraints that restrict much of the motion. When this information is incorporated into a model, then the simulation time can be greatly reduced by effectively working with less number of degrees of freedom (DOF). Multiscale modeling is such an approach, allowing much more efficient sampling through a hierarchical characterization of protein structure and dynamics (14). Combined with advance sampling algorithms (15), free-energy calculations using MD simulations are tractable (16), although they will probably remain prohibitively expensive for a high-throughput workflow. Moreover, the free-energy estimates will always have statistical errors associated with the limited and/or biased sampling, and systematic errors are introduced through specific coarse-graining approximations made within the multiscale model (17).

Proteins exhibit a high degree of fidelity in function, and their ensemble of conformations cluster into well-defined thermodynamic states (18). Consequently, conformational sampling can be localized into key regions in configuration space, characterized by subensembles. A multiscale approach deals with the process of coarse graining to reduce the number of dynamical variables to accurately describe a system within the most relevant subensembles. That being said, a coarse-grained description should be able to reduce sampling errors to a point where they pose no concern, and the model approximation becomes the only relevant factor. With this appreciation, an alternate computational strategy emerges when considering tradeoffs between statistical sampling error (controlled by CPU time) and systematic error (controlled by model approximations).

Two alternative methods that explore the dynamics of proteins within their native basin are the elastic network model (ENM) (19) and the Framework Rigidity Optimized Dynamics Algorithm (FRODA) (20). Both methods rely on a known three-dimensional structure as the primary determinate of characterizing the essential dynamics of a protein. The ENM approach (on its own) is not suitable for generating conformations that deviate far from the starting point, and this poses too great of a limitation for further discussions here. By contrast, FRODA is in principle able to generate conformations that deviate far from the native basin. Recently, pathways have been generated between conformational states as demonstrated with a new improved version called FRODAN (21), which is no longer an acronym. Both FRODA and FRODAN are based on a specific type of coarse-grained molecular mechanics potential, and they use Monte Carlo sampling rather than propagating dynamical equations of motion to more efficiently generate conformational ensembles. The most relevant features pertinent to our discussions here is that they are an all-atom-based model that runs much faster than MD (perhaps 10^4 times faster) by employing concepts of rigidity to naturally coarse-grain the protein structure into rigid subunits based on chemical bonding and atomic packing. The tremendous gain in speed derives from a reduction in DOF and because of the method of geometrical simulation, where the potentials are greatly simplified. It is prudent at this point to briefly highlight some properties about network rigidity, which is the central concept that enables us to estimate conformational entropy and restore the utility of free-energy decomposition, as discussed below.

2.1. Network Rigidity

Classical mechanics textbooks define a "rigid body" as a set of material points with positions, whose mutual separations are fixed. There is no room for an ensemble of conformations in a rigid system; rather, it can only execute trivial motions: translations and rotations. In other words, there can be no internal motions within a rigid object. A very simple but powerful approach is to realize that constant separations between atoms result from chemical bonds. In mechanical models, bonds are represented by distance constraints, connected at atoms, which are treated as universal pivot joints. The bars and atoms define a graph consisting of vertices (atoms) connected by edges (bars). Rigidity theory determines how the number of internal independent DOF depends on the number of edges (bars) and their distribution within the network.

An analysis of network rigidity allows the determination of all continuous deformations of the network that are possible by checking if relative atomic motions are allowed while all bar lengths are fixed. A brute force mathematical procedure of counting the DOF is similar to normal mode analysis in molecular systems (22). If a system is simple, intuition can be used to determine whether it is rigid or not. For purpose of discussion and defining terms, we consider a quadrangle in two dimensions. It can be seen by inspection that a quadrangle is flexible having one internal DOF in two dimensions. The allowed displacements of the particles are shown in Fig. 1 by arrows. Using this example, we work through the important exercise of constraint counting. If there are no constraints, the total number of DOF is the number of particles, N, times the number of independent displacements of each particle, which is equal to dimensionality, d. Each distance constraint eliminates a single DOF. This implies that the constraints are all independent, which is the case for the quadrangle shown in Fig. 1 (redundant constraints are discussed later). Having K constraints, the total number of DOF is given by: $N_t = d \times N - K$, which is equal to $N_t = 2 \times 4 - 4 = 4$ for a quadrangle in a two-dimensional plane.

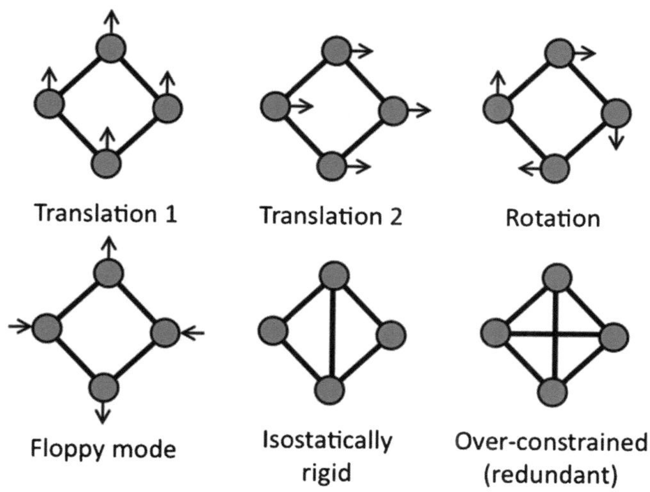

Fig. 1. Examples of floppy modes in a quadrangle network. In two dimensions, all networks posses the three trivial modes (two translations + rotation), which are indicated by the *arrows*. A quadrangle also has one deformable floppy mode. Addition of a cross-linking constraint fixes the distance between the connected vertices, thus freezing out the floppy mode. This marginally rigid network is referred to as isostatic, indicting $N_f = N_t - G = 0$. Adding a second cross-linking constraint, which overconstrains the network, does not affect the number of floppy modes because the structure was already rigid, meaning it is redundant. Removal of a redundant constraint has no effect on the number of floppy modes.

Not all of the four DOF describe continuous deformation of the network because any body in a plane can be displaced in two orthogonal directions, say x and y. It can also be rotated around the z-axis that is perpendicular to the plane. Under these global translations and rotations, all the particle separations remain the same. These *global motions* always contribute to the total number of DOF of a constrained network, N_t. In particular, the global motions are the only available DOF for a rigid body. Thus, to get the number of internal DOF that describes the intrinsic flexibility of a network, we need to subtract the number of global motions from the total number of DOF. In a d dimensional space, a body can rotate in $d(d-1)/2$ independent planes. It can also be translated in d independent directions. The number of global motions is, therefore, given by $G = d + d(d-1)/2 = d(d+1)/2$. For $d = 2$, $G = 3$. The number of internal DOF governing the conformation of the quadrangle in a plane is, therefore, $N_f = N_t - G = 4 - 3 = 1$, corresponding to only one possible mode for continuous (no energy cost) deformation shown in Fig. 1. For a triangle in the plane, the same counting gives $N_t = 2 \times 3 - 3 = 3$, and the number of conformational DOF is $N_f = 3 - 3 = 0$, meaning a triangle is rigid.

Adding an additional constraint along one of the diagonals removes the final DOF (locking the angle), and results in a rigid quadrangle, $N_f = N_t - G = 4 - 4 = 0$. Whenever there are just enough constraints in a network to make it rigid, then the network is said to be marginally rigid or isostatic. Adding a second distance constraint along the other diagonal leads to an overconstrained network with one ***redundant*** constraint. That is, there are more constraints present than possible internal DOF, resulting in some constraints being redundant. Since the distance constraints are modeling atomic interactions in the physical system, strain energy will reside in any region within a network identified as overconstrained by constraint counting. In other words, some distances will have to stretch or compress to accommodate adding a distance constraint between a pair of atoms in which the distances between the atoms are already predefined based on network rigidity. A redundant constraint can be removed from the network without affecting the number of DOF, cf. Fig. 1. When all constraints are uniformly distributed, the number of DOF is given by $N_f = \max[N_t - G, 0]$, which is called Maxwell counting based on his profound insight (23). While Maxwell constraint counting can be quite powerful as a mean field approximation, the method fails when constraint density is not uniform, which can be seen clearly by inspection in Fig. 2. In general, a network will consist of regions that are rigid that interconnect through flexible mechanisms. It is possible to decompose a network into a set of rigid substructures (or clusters), and identify rigid clusters as isostatic or overconstrained. In large networks, counting the available conformational DOF is possible using graph-rigidity algorithms to identify the independent and redundant constraints in practical computing times.

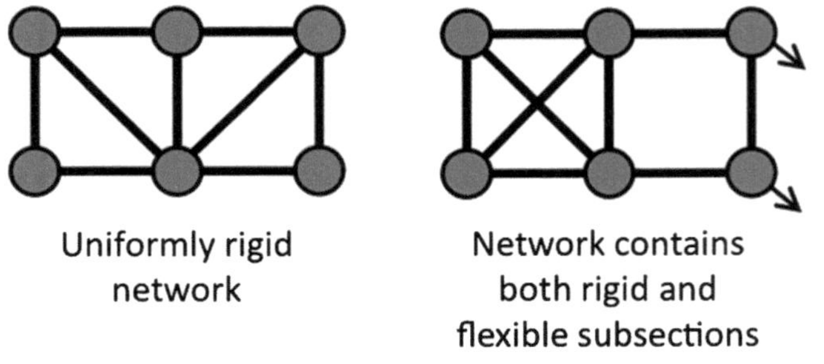

Fig. 2. Two example face-sharing quadrangle networks. The example on the *left* is isostatically rigid with no redundant constraints, whereas the example on the *right* is more complicated. That is, based on a heterogeneous constraint distribution, it possesses both a redundant constraint and an internal degree of freedom. While it is possible to identify the redundant constraints and internal degrees of freedom by inspection in this case, sophisticated graph-rigidity algorithms are needed as networks become more complicated.

For a protein consisting of thousands of atoms, an algorithm to count the internal DOF is required, for which a number of methods can be employed (24). The pebble game is one such algorithm (25), which quickly and accurately calculates network rigidity properties by implementing combinatorial constraint counting by visualizing the DOF as pebbles. Each pebble corresponds to a single DOF of each atom. By tracking pebble movements according to simple rules, the pebble game determines how DOF are lost to constraints, and it exactly identifies all rigid clusters, redundant constraints and overconstrained regions. The program FIRST (floppy inclusion and rigid substructure topography) (26) analyzes a protein structure, and maps out all rigid and flexible regions. In performing this analysis, it should be noted that the calculation is valid for a given set of constraints, and furthermore, the way in which distance constraints model interactions is not unique. This FIRST approach has proven to be a powerful tool to describe protein rigidity and flexibility.

When thermal fluctuations are taken into account, some interactions will break while others will form. In the implementation of FIRST, native contacts were diluted (removed) in the order from weakest to strongest, which simulated the process of protein unfolding (27). Moreover, FRODA (20) also only considers native contacts. In both cases, the primary criticism of using network rigidity is that thermal fluctuations are not modeled. The important point that is critical for obtaining a robust description of protein stability and calculating free energy is to find a way to model protein ensembles that span the range from folded to unfolded states, as well as any intermediate states (or partially unfolded states) and the transition state. In terms of rigidity, this means the number of constraints must be allowed to fluctuate. When many constraints are present, the protein motions are greatly restricted, and as distance constraints are removed, motion is increased. FRODAN (21) improves upon this problem, but in doing so, it has dramatically rendered its dependence on FIRST as a preprocessing step. Our approach to modeling the process of constraints breaking and forming is through an Ising-like model. Ising-like models have been employed with great success to capture the gross features of protein thermodynamics and kinetics (28).

2.2. Ising-Like Models and Native Contacts

The Ising model was originally created to study ferromagnetism based on discrete spin variables that can be in one of two states (spin up or down) representing magnetic moments. The spins interact with one another, and with an external magnetic field. The Ising model has become a hallmark paradigm to describe phase transitions for all kinds of phenomena, including protein folding. The first example of this goes back to the classic Zimm–Bragg (29) and Lifson–Roig models (30) for the helix–coil transition. In the Zimm–Bragg model, backbone hydrogen bonds (H-bonds) are considered as formed (spin up) or broken (spin down). In the helix state, almost all H-bonds are formed, while in the coil state almost all of the H-bonds are missing. In the Lifson–Roig model, the residues are considered as being in a helical conformation (spin up) or coil conformation (spin down). The nature of the polypeptide is completely described by the spin configuration at a coarse-grained level. These two models differ from one another in terms of details, but they both apply free-energy decomposition (FED). That is, as more native interactions form (either H-bonds along the backbone, or a consecutive sequence of residues in an alpha-helix state), both enthalpy and entropy is lowered. The entropy reduction is a critical part of the contribution.

With respect to the original Ising-model that only represented the Hamiltonian (energy) of the system, there is a fundamental departure using Ising models for the helix–coil transition or protein folding. Now, the local spin variables represent local conformational states that add certain amounts of free energy that consist of both enthalpy and entropy contributions. For a given spin configuration of the system, the total enthalpy and entropy contributions are simply added together. Despite this difference, a partition function is calculated as a sum over all Boltzmann factors as done in the original Ising-model, where one should in principle include all spin configurations (microstates). However, these spin configurations actually represent macrostates of a protein because they describe atomic conformations at a coarse-grained level. Together, all spin configurations define the complete ensemble from which all thermodynamic properties can be calculated, including metastable states, if present.

As applied to proteins, the three-dimensional structure of the protein is assumed as known, and residues are assigned as being folded or unfolded. When neighboring residues are in a folded-state, the native contacts that connect these two residues form, otherwise they are broken. Two popular Ising-like models are COREX (31, 32) and the Wako–Saitô–Muñoz–Eaton (WSME) model (33), which have been reviewed recently (34) in the context of methods that generate protein ensembles. The important aspect that we mention here is that both models assume additivity within the FED, just as do the Zimm–Bragg and Lifson–Roig models. The meaning of additivity in the context used here is mathematically precise. It means for a given spin configuration, the total free energy is the sum over all parts, where both the enthalpy and entropy contributions are individually additive. It is worth noting that once the partition function is calculated, the thermodynamic entropy will always be a nonadditive function that reflects all the spin configurations within the ensemble, or sub-ensemble of interest.

What is often overlooked is that a given spin configuration represents a subensemble of a protein, and the process of adding free-energy contributions from each spin variable implies each local state acts independently from each other. If each component operates independently from one another, then cooperativity between the local units is lost by definition! Cooperativity can be built into an Ising-like model through spin–spin coupling terms, as done in the Zimm–Bragg, Lifson–Roig and WSME models, or through an external field using 1-body interactions by relating the solvation of residues to solvent accessible surface area (SASA) as done in COREX. In general, long-range cooperativity within a given spin-configuration is lost whenever an additive model is employed. As recently discussed in the context of the helix–coil transition (35, 36), cooperativity can also be explicitly calculated using the properties of network rigidity. Network rigidity is essential for restoring the utility of a FED because of the fundamental problems that occur with free-energy reconstitution, next reviewed.

2.3. Free-Energy Decomposition and Reconstitution

The free energy of a protein determines its thermodynamic state. Accurate calculation of the free energy for a protein, consisting of hundreds of amino acids connected by covalent bonds and other types of chemical bonds and weak interactions is formidably difficult. One reason for persistent inaccuracies in these calculations is the assumption of additivity. Therein, it is assumed that the free energy of a protein can be obtained as a sum of free energies of each individual amino acid. Unfortunately, while additivity models have been applied to proteins for decades, they are fundamentally flawed (37). In their seminal 1989 paper (38), Karplus and coworkers coined the tantalizing term "hidden thermodynamics," reflecting the observation that some aspect or key element of the FED scheme was missing. Nevertheless, it was, at this point, absolutely clear that some essential physics was not being accounting for, leading to the search for better FED schemes that somehow included the missing physics.

Attempts to analyze protein thermodynamics have revealed that additivity principles applied to a FED scheme in terms of specific interactions breakdown when cooperativity is present in the system. In view of the fact that most proteins exhibit some level of cooperativity, Mark and van Gunsteren (39) write: "In regard to the detailed separation of free-energy components, we must acknowledge that the hidden thermodynamics of a protein will, unfortunately, remain hidden." Further, Hallerbach and Hinz (40) analyzed the cold denaturation and found inconsistencies in heat capacity predictions. They conclude that models that assume additivity in the conformational entropy due to hydration effects violate the second law of thermodynamics. These and many other studies point to inconsistencies that arise when additivity principles are applied to the FED approach. Dill (37) notes: "Perhaps some of our models in computational biology are based on flawed assumptions. Thermodynamic additivity principles are the foundations of chemistry, but few additivity principles have yet been found successful in biochemistry." The best way to understand why additive models are unlikely to succeed is to write the consequence of such a condition whenever it is found to be true. Given that the k-th component of free energy is given by G_k, and assuming the total free energy is indeed additive over all components, such that $G_{tot} = \sum_k G_k$ then it follows that the total partition function is a product of individual partition functions, where $Z_{tot} = \prod_k Z_k$ and the standard relations follow: $G_k = -RT \ln(Z_k)$ and $G_{tot} = -RT \ln(Z_{tot})$, where R is the ideal gas constant and T is absolute temperature. This complete factorization of the Z_k partition functions can only be true when there is strictly no coupling between any of the subsystems. Said another way, a set of generalized coordinates must be determined that are separable across all subsystems. As it routinely happens, because of the well-separated types of forces involved in chemical bonding contributing to vibrational motions in bond-lengths, bond-angles, and dihedral angle rotations, this factorization approximately holds in small molecules, where additive models are found to work markedly well (but never exactly).

In general, the enthalpy and entropy contributions are never exactly additive due to coupling terms that arise in part due to local variation in microenvironments. In macromolecules such as proteins, the nonadditivity in the entropy is much more significant than that found in enthalpy. The main reason for this is the formation of loop connections in the otherwise chain-like structure of these molecules. Once a protein starts to collapse into its native structure from a (hypothetical) completely extended state, even a single H-bond can dramatically reduce the entropy of the chain. The situation is not too bad with only a few cross-linking H-bonds, but additivity becomes problematic when there is a high density of cross-linking interactions. As the cross-links are added, more constraints are imposed on the motions of the atoms within the various parts of the protein that have been identified as subunits or components. The source of nonadditivity that Mark and van Gunsteren (39) have shown definitively has to do with the geometrical, and as it turns out, topological, properties of the protein structure. As such, once coupling occurs, the problem related to reconstituting the total free energy of a protein, based on knowing the free-energy component parts presents itself. In other words, the key ingredient missing is structural information, relating to the available DOF in the system, which directly determines cooperativity found in the protein. The root of nonadditivity lies in mechanics. One way to avoid the nonadditivity problem is to only work with energies directly, and to calculate conformational entropy by sampling over conformations using molecular simulation techniques, such as molecular dynamics or Monte Carlo. A much bolder alternative is to apply FED, but modify the reconstitution process by developing a quantitative way to account for nonadditivity of entropy, as discussed below using a Distance Constraint Model.

3. Distance Constraint Model

Many models employ a FED, where it is assumed that the free-energy components can be added to calculate the total free energy of a protein structure. As explained above, unless the generalized coordinates that are used to describe the protein are 100% separable, and independent over all accessible conformations that can take place, this type of approach will fail. An alternative is to use an overspecified coordinate system. In this case, the model applies distinct coordinates for all intramolecular interactions, thus obtaining an overspecification of the atomic positions within a protein. In this situation, the pertinent question becomes which of these interactions (modeled as a set of distance constraints) are independent or redundant. Using the properties of network rigidity, a new paradigm has been formulated that adds entropy contributions only from independent distance constraints.

3.1. Naive Estimate of Conformational Entropy from DOF

Proteins interact with their environment through energy exchange, and therefore, statistical mechanics must be employed to predict any kind of measurable property that a protein may exhibit. In equilibrium with a heat bath at temperature T, the probability to find the protein in a given macrostate v is given by:

$$\Pr(v) = \Omega(v) \exp\left[\frac{-E(v)}{RT}\right], \qquad (1)$$

where $E(v)$ is the energy of a macrostate and $\Omega(v)$ is the number of microstates which correspond to the macrostate v (41). The relevant part of a protein's energy in a given macrostate, $E(v)$, is mainly determined by the set of H-bonds present in the structure. The number of microstates, $\Omega(v)$, is proportional to the number of all possible atomic configurations that the protein can explore at the same constant energy, $E(v)$, which depends on the details of the topological arrangement of all the chemical bonds that defines a macrostate. To first order, a macrostate conserves the same bonding topology. These bonds are modeled as distance constraints that define a distance constraint network. As H-bonds fluctuate, different constraint topologies will be realized and will lead to different set of rigid substructures. In other words, depending on the topology of chemical bonding (covalent bonding plus salt bridges and H-bonding), the protein will be flexible in different regions, and different types of correlated motions will be accessible.

Proteins respond like typical condensed matter systems. Namely, as $E(v)$ increases, $\Omega(v)$ rapidly increases, and the Boltzmann factor rapidly decreases. Note that as energy increases, this means H-bonds break, and more DOF appear, resulting in greater conformational entropy. As a result, the probability to find the system with a given energy, $\Pr(E)$, is a sharply peaked function. The state of a system is decided by the competition between entropy and energy. At room temperature, the folded state of a protein is favorable. As temperature is elevated, the H-bonds break so both energy and entropy increases. At high enough temperatures, the additional DOF compensate the energy increase, and the protein is found predominantly in an unfolded state. It is, therefore, necessary to know how many microstates contribute to a given macrostate of a protein.

The macrostate, v, effectively specifies a conformation. The concept of conformation requires some type of coarse-graining procedure, which can vary based on the model used. In the approach considered here, the macrostate, v, is well defined by specifying a particular set of constraints in the network. Then, the macrostate, v, consists of all atomic configurations that are consistent with the continuous motions of the mechanical network defined by the set of constraints. As a rough estimate, the degeneracy $\Omega(v)$ scales as $\Omega(v) \sim \omega^{N_f}$, where ω is a typical phase space volume of a single DOF, and N_f is the number of internal DOF. The conformational entropy for macrostate, v, is given as: $S(v) = R\ln[\Omega(v)] = RN_f\ln(\omega)$. Each DOF participates in the continuous deformation of a network. Formally, these DOF are related to the low frequency normal modes of vibration. We also assume that all atomic configurations associated with the same macrostate have approximately the same energy. Thus, macrostate v has constant energy, $E(v)$, and conformational entropy that is proportional to N_f. Better estimates will be made below, because one should question why all DOF should be assigned the same phase space factor. Also, as discussed below, there will be a question as to when an interaction should be viewed as a constraint or not.

3.2. Network Rigidity Hierarchically Applied to Protein Structure

To identify the number of DOF within a protein, the constraint network must first be determined based on the H-bond network. To see how this might be done, it is useful to consider the following potential energy (enthalpy) decomposition into terms corresponding to several kinds of interactions:

$$U_{\text{tot}} = U_{\text{CF}} + U_{\text{BB}} + U_{\text{SB}} + U_{\text{HB}} + U_{\text{DA}} + U_{\text{weak}}, \qquad (2)$$

where CF, BB, SB, HB, and DA correspond, respectively, to central force, bond bending, salt bridge, H-bond, and dihedral angle forces. The U_{CF}, U_{BB}, and U_{DA} terms describe covalent bonding. Note that covalent bonds freeze out many DOF at the temperatures of interest. The first two terms are particularly strong, meaning that they are modeled well as distance constraints. The dihedral angle term characterizes torsion forces governing the twist motion through a dihedral angle. Torsion forces are weak in most cases, allowing peptide chains to be flexible. However, torsion forces can be very strong when associated with the partial double bond character of the peptide bond, and in this case, this type of torsion force is also modeled by a distance constraint. Weak torsion force cases are considered to be DOF. In addition to the covalent bond interaction terms, U_{SB} is the potential energy corresponding to salt-bridges, and U_{HB} is the potential energy for H-bonds. Finally, there are many other weak interactions present that are related to nonbonding forces (i.e., van der Waals interactions). These weak forces are not modeled as distance constraints. The terms in Eq. 2 are shown in the order of their decreasing magnitude, left to right. The strongest interactions are the covalent bonds making up the primary structure of the protein. The covalent bond interactions are never broken under physiological conditions. Therefore, the role of covalent bonding is mechanical in nature, and do not contribute to energy fluctuations. In particular, the forming and breaking of salt bridges and H-bonds are among the main processes found in protein folding and unfolding.

In order to map chemical interactions described above to a network of distance constraints, one must decide how strong the bond has to be to get modeled by a constraint. Since covalent bonds are never broken, the central-force and bond-bending interactions, in addition to strong (double-bond) dihedral angle interactions, are always modeled in terms of quenched constraints. On the contrary, all rotatable dihedral angles are left as free angles of rotation, and thus are defined as DOF. Interestingly, some H-bonds are very weak, while others are very strong. (Note: salt bridges are considered as special types of H-bonds going forward.) For this reason, a threshold energy used to determine when an H-bond should be modeled by a set of distance constraints appears necessary to define. If the threshold binding energy is too high, there will be few H-bonds represented by a constraint and the protein structure will be very flexible. On the contrary, if the threshold binding energy is too small, too many weak H-bonds will also be represented by constraints and the protein structure will be almost completely rigid.

In the attempt to model H-bonds as distance constraints, based on an energy criterion (26), many conceptual problems were created. Although having a sliding energy cut-off allows one to adjust the number of H-bonds, each constraint is treated as being infinitely strong (a true constraint), whereas not having the H-bond present is equivalent to ignoring it completely. However, constraints are used to model interactions that are not actual constraints in the first place. Therefore, it is better to think about a constraint as having a characteristic strength in terms of the curvature of the bottom of a potential well. The flatter the well is (low curvature), the weaker the constraint. Thus, it is not the value of the energy (depth of the well) that determines how effective the constraint is in maintaining a certain fixed distance between two atoms, rather, it is the amount of phase space (wiggle room) associated with that constraint. The energy criteria works reasonably well, but this is due to the high correlation between bonding potentials that are deep (very low energy) have high curvature. In addition, as $\varepsilon \rightarrow 0$, why should a H-bond with a binding energy of $E_c + \varepsilon$ be considered as an infinitely strong constraint, whereas another H-bond with binding energy of $E_c - \varepsilon$ is not considered as a constraint? It is clear that a proper measure for the amount of phase space (wiggle room) must be assigned to all constraints, associated with some sort of interaction. A weak torsion force now becomes one type of constraint with more phase space associated with it, rather than considered a DOF. With these basic problems in mind, a statistical mechanical treatment involving network rigidity has been developed.

3.3. Integrating Rigidity and Statistical Mechanics

Lord Kelvin wrote: "I never satisfy myself until I can make a mechanical model of a thing. If I can make a mechanical model I can understand it!" The DCM is based on formulating a mechanical model to capture the essential elements that govern the properties of protein thermodynamics. That is, we implement a FED scheme by employing network rigidity as an underlying mechanical interaction, which is determined by the topology of distance constraints (4). We assume that there are a finite number of constraint types, t, that can be encountered within the structure. These constraint types will be assigned an energy and entropy that will in general depend on the local geometrical details of the atomic configuration in the vicinity of where the constraint is found in the network. The main elements of the DCM are given below:

- For each constraint type, t, assign a molecular free-energy contribution given by: $G_t = E_t - TS_t$.
- $E_{\text{tot}} = \sum_t N_t E_t$, where the total energy is obtained by summing over the energy contributions over all N_t constraints of each type t.
- $S_{\text{tot}} = \sum_t I_t S_t$, where the total entropy is obtained by summing only over the I_t *independent* constraints of type t, as

determined by the network rigidity calculation, thus accounting for entropy nonadditivity.

- $G_{tot} = E_{tot} - TS_{tot}$, which is the free energy of a specified constraint topology.

As described above, rigidity theory allows us to calculate the total number of independent constraints within a given network. Although this number is unique, the identification of which constraint is independent and which is redundant is not unique. As such, different values I_t will be determined based on the ordering of which constraints are placed first in the network during the graph-rigidity algorithm (pebble game). Nevertheless, the above expression for S_{tot} is very useful because it provides a rigorous upper bound estimate for the total conformational entropy. The reason for the upper bound is because the set of independent constraints are not necessarily orthogonal. Only summing over orthogonal components will lead to additivity in entropy. Because of the lack of orthogonality between the constraints that form a linear vector space, there remains some "double counting" in phase space. Nevertheless, any upper bound is better than a straight additivity model (i.e., using N_t instead of I_t). So, simply placing constraints down at random, and using the pebble game algorithm, one can be assured to obtain a better estimate for conformational entropy, albeit this is only an upper bound estimate. The calculated S_{tot} is dependent on the ordering of constraint placement in the network, because ordering effects how the set of I_t is identified. The best we can do then is to find the lowest possible upper bound. A greedy algorithm achieves this. Namely, recursively place the constraints in the network in order from those having the lowest entropy parameters, $\{S_t\}$, to the highest. Then, by implementing an initial sorting procedure, a rigorous lowest upper bound is obtained with virtually no additional computational cost in the calculations. This procedure is called preferentially placing constraint types with the lowest S_t values first.

Building the free-energy function in this way effectively accounts for the nonadditivity of entropy through the mechanical interaction. Because in practice the pebble game performs linearly in the number of atoms, in a tiny fraction of a second for a moderately large protein, the *free energy* for a fixed constraint topology can be readily calculated. This calculation must be repeated again and again for a large ensemble of diverse constraint topologies (folded to unfolded) to construct the complete thermodynamic response of the protein. In a protein, the free energy, $G(F)$, energy, $E(F)$, and entropy, $S(F)$, of a mechanical framework F (i.e., a constraint network) provide a thermodynamic description of a macrostate of conformations sharing the same free energy. With these definitions, we can formally proceed to construct a full partition function of the protein. We define a dimensionless entropy, $\gamma_t = S_t/R$, whose labels are in order from smallest to largest, such that $\gamma_1 < \gamma_2 < \gamma_3 < \ldots$; note that the ordering of energies is irrelevant. Then,

$$E = N_1\varepsilon_1 + N_2\varepsilon_2 + N_3\varepsilon_3 + \cdots$$
$$S = R\tau, \text{ where } \tau = I_1\gamma_1 + I_2\gamma_2 + I_3\gamma_3 + \cdots, \quad (3)$$

The full partition function of the protein is then written as: $Q = \sum_F \exp[\tau(F)] \exp[-E(F)/RT]$ and the summation is evaluated over all possible topologies, F. The term $E(F)$ gives the total energy at fixed constraint topology, whereas $\tau(F)$ is a total pure entropy. Note that $\exp[\tau(F)]$ is the conformational degeneracy associated with all conformations with a fixed constraint topology.

Once constraints are assigned entropy values that reflect their strength, there is a gray scale – no longer is there a binary split between infinitely strong or nothing. The entropy can be assigned because we are dealing with a coarse-grained model. This scale allows us to provide a measure for the conformational entropy in a molecular network. In particular, it is important to note that very weak interactions, having potential energy functions with characteristically shallow curvature, are counted as constraints, albeit very feeble. The set of constraint types used to model interactions in the protein need to be complete, such that after all constraints are placed in the network, it is completely rigid. The concept of rigidity used in this context is with respect to the number of independent constraints that will always equal $3N - 6$, with precisely no DOF remaining in the network. Paradoxically, once we assign entropy values to constraints, what can be said about the difference between a constraint or a DOF?

The paradox is resolved by noting that flexibility is defined in terms of the set of constraints having very large values of entropy associated with them. Below, we will introduce the notion of native and disordered torsion constraints. A native torsion constraint confines the structure to be near that of the template structure. A disordered torsion means that the entropy value is very high, reflecting a relatively much larger degree of flexibility that allows the structure to deviate far from the template structure. Despite a large entropic contribution, a disordered torsion is regarded as a constraint to adhere to the formulation used to estimate conformational entropy. Concurrently, disordered torsion interactions are interpreted as effectively defining internal DOF when they are independent. This is possible because they facilitate large conformational variations within the coarse-grained description that maintains a constant constraint topology. In other words, DOF have finite measure of entropy, and so do constraints. A native torsion constraint has much lower entropy than a disordered constraint. Thus, the gray scale of characterizing entropy contributions creates this interesting lack of distinction, but well-defined quantification of entropy.

The advantage of employing network rigidity within a DCM is that a good estimate for the conformational entropy can be obtained without moving any atoms. By contrast, most of the computational expense in standard methods is used to simulate motions of protein atoms to geometrically sample configurational space. In the DCM, for each particular constraint topology, *F*, the protein will remain at approximately constant energy, while it wiggles geometrically to different degrees depending on constraint topology. Thus, the DCM shifts the focus from considering all possible geometries that a protein (or macromolecule) can explore to all possible constraint topologies. Since the constraint topologies are generated in terms of graphs, computational times are better than 10^{10} times faster than methods that rely on simulation. The constraint types can be of diverse variety, and as a result the effects of hydration can be modeled in a natural way, so as to accurately describe cold denaturation (42). Cold denaturation falls out of the calculation as the most probable constraint topology changes, as the temperature changes.

3.4. Mean-Field Theory and Calculating the Free-Energy Landscape

Of course, nothing in life is really for free. In order to evaluate the free energy for a given thermodynamic condition, the sum over an astronomical number of constraint topologies is required. Rather than doing this, we collect all possible microstates corresponding to different constraint topologies into well-defined macrostates. Each macrostate consist of an ensemble of all possible conformations that are consistent with the same number of H-bonds, N_{hb}, and same number of native torsion constraints, N_{nt}. Then the partition function can be calculated by performing a double summation over N_{hb} and N_{nt} in the two-dimensional constraint space shown in Fig. 3a. A grid of nodes is formed within this defined constraint space, where each node specifies a macrostate given by (N_{hb}, N_{nt}). We then apply a mean field approximation by assuming the probability for each H-bond and each native torsion constraint is independently distributed. These probabilities are thought of as occupation probabilities, where the constraint is present or not. The occupation probabilities are worked out analytically. In the case of native torsion constraints, a simple probability is assigned to all native torsions, given by $\Pr(\text{nt}) = N_{nt}/N_{max}$, where N_{max} is the maximum number of native torsions possible within the protein. All torsion interactions are treated the same, with the probability of being in a native or disordered state is independent, because the mean-field approximation is employed. In reality, we know the probabilities should depend on residue type and its local environment. This drastic approximation is possible because we will introduce effective parameters that are not transferable. Without difficulty, this extreme

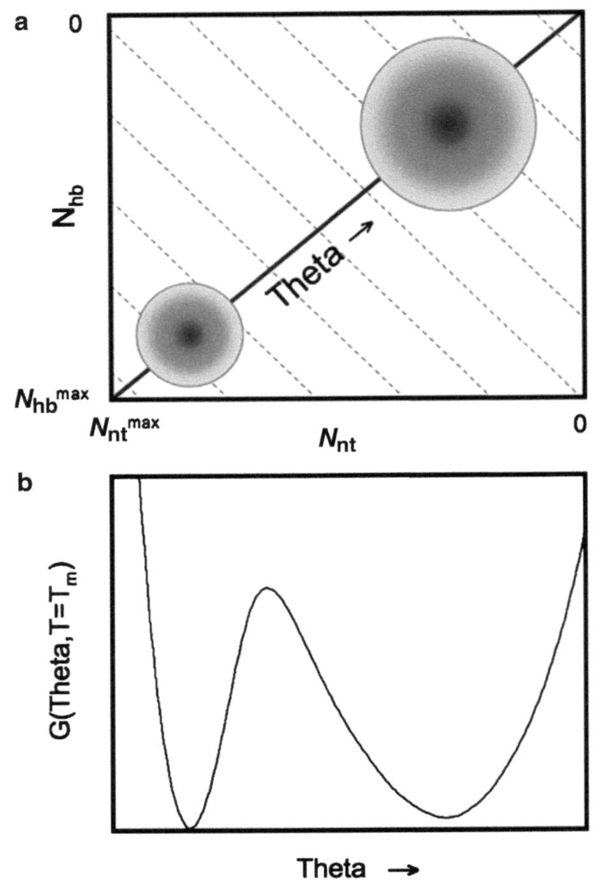

Fig. 3. (**a**) Cartoon of the free-energy landscape using the number of hydrogen bonds and native torsions as order parameters. Each point grid defines a macrostate (N_{nt}, N_{hb}), where the free energy $G(N_{nt}, N_{hb})$, is calculated. The circles identify the native (*lower-right*) and unfolded (*upper-left*) basins. (Notice that the axes are decreasing from bottom to top and left to right.) At times, it is convenient to express the free energy as a function of a single global flexibility order parameter, $\theta(N_{nt}, N_{hb})$. *Gray dashed* lines represent (approximate) fronts of constant global flexibility due to tradeoff between two constraints types. The *thick diagonal line* denotes the shortest path crossing a single saddle from the unfolded to folded basins, which explains why θ is a useful order parameter to consider. (**b**) An example one-dimensional free-energy landscape highlights the straddling barrier that must be crossed as the protein transitions between folded and unfolded. This figure is reproduced from Livesay et al., 2008, *Chem. Central J.* 2:17.

approximation can be lifted, and furthermore, the DCM is amendable to many generalizations, with only minimal additional cost in computations.

In what we refer to as the minimal DCM (mDCM) (6, 7), torsion interactions are treated the same throughout the protein. However, we do treat H-bonds with more care, where local environmental differences are considered. In other words, the

H-bond occupation probabilities will depend on local geometry, which determines the energy of the H-bond. We want to do this in the context of mean field theory to make the problem tractable. Consideration of local variance in a system creates a problem for normal mean field theories because they were developed for homogeneous systems. Even a system with disorder that breaks the homogeneity assumption is transformed into an effective medium that represents the average effect of the disorder. However, in the protein, it is not homogeneous. We focus on the H-bond pattern because it is well known that H-bonds govern specificity in the system, and are critical to protein stability and function.

To proceed, we have developed a heterogeneous mean field theory. Since the H-bond occupation probability depends on local environment, we use a Fermi-Dirac distribution function that models a two level system (H-bond present or not) consistent with a global constraint, which is satisfied through a Lagrange multiplier, which is essentially the same thing as a chemical potential. That is, the Lagrange multiplier is adjusted to put precisely N_{hb} into the network, and thus satisfies the desired global constraint. At this point, we have used mean field theory to efficiently define all occupation probabilities of the constraints while controlling the macroscopic global properties of the protein (system). The ensemble of constraint topologies will depend on the set of these occupation probabilities. Because we require the pebble game algorithm to calculate the long-range interactions that couple entropic contributions of the constraints, we use the a priori known probabilities determined from the heterogeneous mean field theory to generate typical constraint topologies associated with a specified macrostate.

From the precalculated occupation probabilities, we sample over an ensemble of microstates consistent with the specified macrostate. Each macrostate corresponds to a node within a grid that is introduced to define a free-energy landscape (in constraint space). Within each node, an astronomical number of distinct constraint topologies remain. To get an estimate of the free energy of a particular node within this grid, we use Monte Carlo sampling to find an average energy and entropy, consistent with the global constraints as dictated by the node. This Monte Carlo sampling is not a simulation; rather, it is a means to estimate averages, given the a priori known probability distribution functions. This hybrid approach is much more efficient than what a normal Monte Carlo simulation would entail, including bias sampling techniques. It turns out that because network rigidity is a long-range interaction, only a small number of Monte Carlo samples are necessary to find reasonably accurate ensemble averages due to self-averaging. We find that one can average over N samples. If $N = 200$, one obtains averages that are about the same as for $N = 10{,}000$. It is quite amazing that a sample of only 200 networks provides good estimates for astronomically large number of constraint topologies, which reveals that the total number of constraints in the network is the most important aspect, rather than the details of what type of constraints there are.

The atomic coordinates for a protein and experimental data for heat capacity, C_p, serve as an input to the mDCM. Three parameters are optimized to fit the predicted C_p curves to the experimental data. Two of the parameters, v_{nat} and δ_{nat}, respectively correspond to the energy and entropy associated with a native constraint. Note that the energy and entropy values for a disordered constraint are treated as transferable parameters. The energy of a H-bond is calculated using an empirical potential (43), which is linearly related to the H-bond entropy. The third fitting parameter, u_{sol}, describes the energy of forming a H-bond to solvent when intramolecular H-bonds break. Once these three parameters have been determined, one can easily produce a free-energy landscape in the constraint space for a given temperature. For proteins that exhibit two-state thermodynamic behavior, two basins will form corresponding to folded and unfolded states (cf. Fig. 3a).

Being that the total number of constraints appears to be of essential importance, rather than the specific types of constraints, it is of interest, therefore, to map the two-dimensional constraint space into a one-dimensional graph showing the free energy as a function of a single global order parameter. The global flexibility, θ, is defined as the average number of independent DOF per residue. Figure 3b provides an example of the one-dimensional free-energy landscape for $T = T_m$, where the unfolded and folding basin are equally probable. The deepest minimum corresponds to the native state basin characterized by a small value of the global flexibility at $T < T_m$, whereas the unfolded basin is more stable at $T > T_m$. Many of the details of the calculations and simulation are found in references (6, 9).

4. DCM Descriptions of Allostery

As discussed above, the DCM provides a computationally tractable approach to quantify the give-and-take between thermodynamic and mechanical response. As a consequence, quantified stability-flexibility relationships (QSFR) can be predicted (8–11), which provides a high dimensional characterization of protein stability, flexibility and their interrelationships. Each QSFR metric integrates mechanical and thermodynamic descriptions of structure. For example, the concept of QSFR is best exemplified

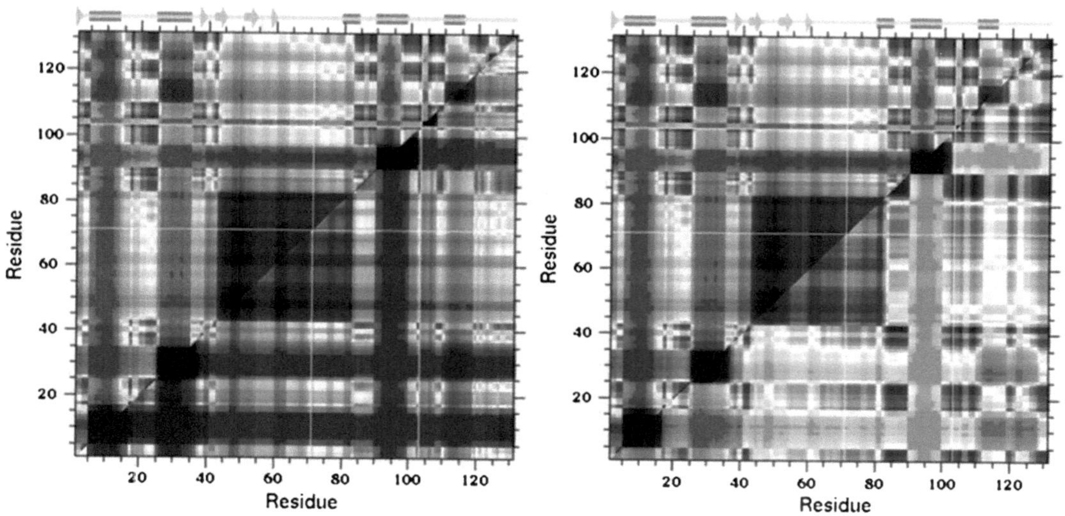

Fig. 4. Cooperativity correlation plots identify all pairwise residue-to-residue mechanical couplings within a given protein structure as a specified thermodynamic condition (*dark* indicates stronger coupling). In each example, the *upper triangle* corresponds to the wild-type human c-type lysozyme, whereas the *lower triangles* correspond to two different mutants. In the V100A mutant (*left*), the structure is rigidified with respect to the wild type, whereas in the Y45F mutant (*right*) results in increased flexibility. In each case, stark differences are observed upon mutation.

by the one-dimensional free-energy landscapes in Fig. 3b that directly relate mechanical and thermodynamic quantities. Other useful QSFR metrics include local descriptions of flexibility, which are based on a thermodynamic average across the ensemble. As such, these flexibility descriptions appropriately adjust with temperature – a protein is more rigid at low T and more flexible at high T. Two common descriptions of local flexibility are the probability to rotate and the flexibility index, each describing the flexibility within rotatable backbone (PHI and PSI) torsions. The probability to rotate is simply the probability for a given torsion to be disordered (vs. native), whereas the flexibility index quantifies how far from isostatic (marginally rigid) a given torsion is. While providing some orthogonal information, the metrics are highly correlated. Of particular interest here, cooperativity correlation (CC) plots identify mechanical couplings within protein structure, again thermodynamically averaged across the ensemble. That is, CC identifies all pairwise residue-to-residue mechanical couplings within the structure (cf. Fig. 4). While not shown here, CC actually identifies both rigid couplings (residue pairs that are likely to be included within the same rigid cluster) and correlated motions (8–11).

Our previous work has used the described QSFR metrics to compare stability and flexibility mechanisms across evolutionarily related proteins, and to assess the consequences of ligand-binding. For example, our results (8) on a mesophilic/thermophilic RNase H pair reproduce experimental conclusion (44) that a balance between stability and flexibility is required for function. Moreover, our collective results over several different protein families demonstrate that backbone flexibility is mostly conserved across families, which is not surprising due to fold conservation. Nevertheless, despite qualitative agreement, many quantitative differences in flexibility are observed. Conversely, protein families show surprising diversity and richness. Specifically, our results identify drastic CC differences across datasets of the RNase H pair (8), four bacterial periplasmic binding proteins (10) and nine oxidized thioredoxin mutants (11). In addition, even greater CC changes are observed across pairs of holo/apo binding proteins (10) and oxidized/reduced thioredoxins (11). Together, these results highlight the sensitivity within the set of pairwise allosteric couplings present in protein structures.

Ongoing work is attempting to quantify the extent and frequency of changes in QSFR upon minimal perturbation. That is, we are considering a dataset of 14 human c-type lysozyme point mutants, which is the same dataset considered in Verma et al. (45). QSFR characterizations of each mutant are compared to the wild-type structure. Across this dataset, our results further underscore the highly sensitive nature of QSFR. For example, while small in scale compared to the changes in CC, changes in backbone flexibility are frequent and occur over long ranges, highlighting the frequency of multiple allosteric pathways within protein structure (46). In addition, increases in flexibility are mostly balanced by increases in rigidity across the dataset. Similar results are observed when considering CC. Because CC plots can be thought of as a snapshot of all pairwise allosteric couplings present in structure, changes in CC upon mutation identify how the *set* of allosteric mechanisms change. As a typical example, Fig. 4 compares the wild-type (upper triangles) lysozyme CC to two mutants (lower triangles). Myriad changes occur, many at structural locations distal to the mutation site. Interestingly, the V100A mutant tends to rigidify the structure, whereas the Y45F mutant tends to increase its flexibility.

With these results in mind, we have recently developed a mechanical perturbation method (MCM) to identify putative allosteric sites using the mDCM (12). Therein, we introduce a small number of quenched native torsion constraints to mimic the locally rigidifying effect of ligand-binding, and then recalculate all QSFR properties. Thereafter, large changes in the calculated QSFR properties identify sites likely to initiate allostery. While any QSFR metric can be considered, our work thus far has focused on changes to ΔG_{fold}, flexibility index and CC (cf. Fig. 5). Across three CheY orthologs, our results demonstrate an intriguing mix of conservation and variability. For example, the $\beta 4/\alpha 4$ loop, which has been demonstrated experimentally for relaying the required allosteric signal upon phosphorylation of Asp57

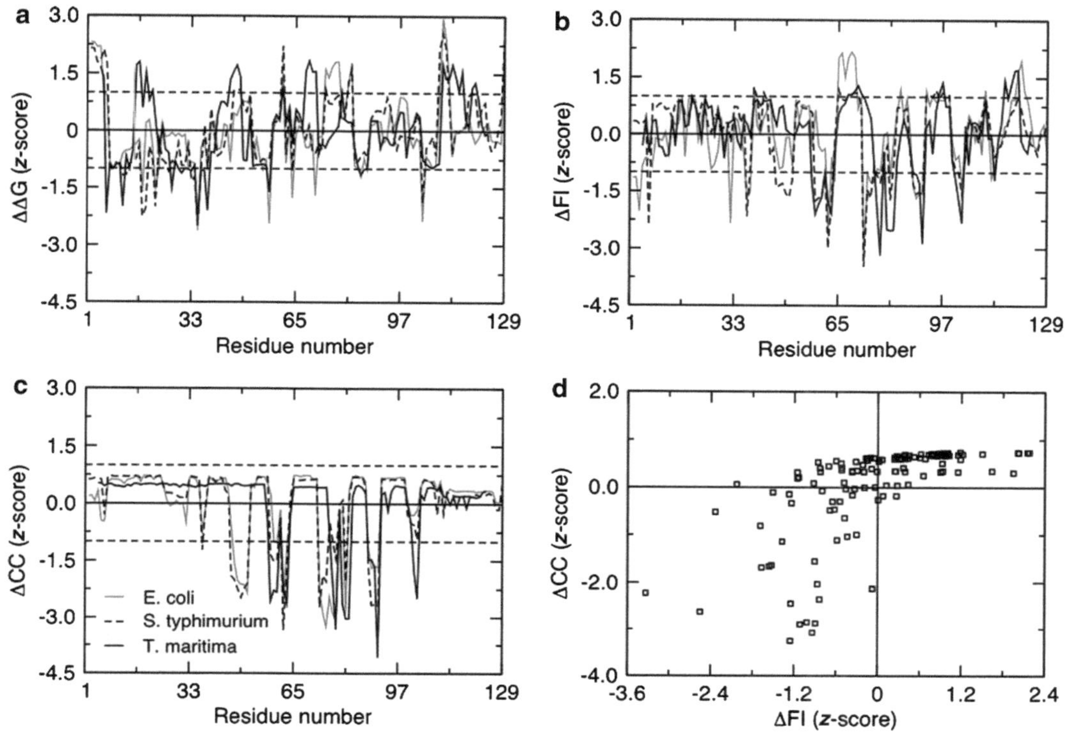

Fig. 5. Summary of the three MPM results applied to three CheY orthologs. The first *three panels* plot the results for each QSFR response function ($\Delta\Delta G_{\text{fold}}$, ΔFI, and ΔCC). To facilitate analysis, the raw scores have been converted to statistical *z*-scores, where minima always indicate a larger allosteric response. The horizontal dashed lines ($z = \pm 1$) are used to highlight strong signals. While there are substantial differences between any pair, it is immediately apparent that the results on the *Escherichia coli* and the *Salmonella typhimurium* orthologs are much more similar than either is to the *T. maritime* ortholog, which parallels the evolutionary relationships. In addition, the ΔFI and ΔCC results are similar, yet quite distinct from $\Delta\Delta G_{\text{fold}}$. The scatter plot in the *lower right-hand panel* compares the results of the two mechanical QSFR response functions (correlation coefficient = 0.63). A similar figure in color can be found in Mottonen et al., 2010, *Biophys J*, 99:2245–2254.

(47, 48), is the only allosteric site conserved in all three orthologs. As expected, similarity within the allosteric responses strongly parallels evolutionary relationships; however, more than 50% of the best scoring putative sites are only identified in a single ortholog. These results suggest that detailed descriptions of intraprotein communication are substantially more variable than structure and function, yet do maintain some evolutionary relationships.

5. Conclusions

The motivation behind the DCM is based on the need for computational methods to quickly and accurately characterize protein stability and flexibility, which is a complex problem due to many-body effects where interactions compete with one another simultaneously. As a result, both theoretical and computational methods must balance inclusion of all relevant interactions with computational feasibility. Our ability to maintain this balance has been demonstrated in our early works. Moreover, the DCM is ideally suited to characterize allostery because it robustly characterizes how mechanical and thermodynamic quantities change upon perturbation. Across the various examples presented herein, our collective results advocate that allosteric response is highly sensitive to differences in structure. Finally, it is worth nothing that, while not discussed here, there is also strong experimental evidence for sensitivity within allosteric mechanisms across protein families (the interested reader is referred to the Livesay et al. chapter on correlated mutation algorithms, which is also included in this book).

Acknowledgement

The work described here has been primarily supported by the NIH since 2003. We acknowledge continuing support from NIH NIGMS grant R01 GM073082, which is allowing us to extend these ideas more fully.

References

1. Mittermaier, A., and Kay, L. E. (2006). New tools provide new insights in NMR studies of protein dynamics. *Science* **312**, 224–228.
2. Fersht, A. R. (1974). Catalysis, binding and enzyme-substrate complementarity. *Proc R Soc Lond B* **187**, 397–407.
3. Robertson, A. D., and Murphy, K. P. (1997). Protein Structure and the Energetics of Protein Stability. *Chem Rev* **97**, 1251–1268.
4. Jacobs, D. J., Dallakyan, S., Wood, G. G., and Heckathorne, A. (2003). Network rigidity at finite temperature: relationships between thermodynamic stability, the nonadditivity of entropy, and cooperativity in molecular systems. *Phys Rev E Stat Nonlin Soft Matter Phys* **68**, 061109.
5. Jacobs, D. J. 2006. *Predicting protein flexibility and stability using network rigidity: a new modeling paradigm.* In Recent Research Developments in Biophysics. Transworld Research Network, Trivandrum, India. 71–131.
6. Jacobs, D. J., and Dallakyan, S. (2005). Elucidating protein thermodynamics from the three-dimensional structure of the native state using network rigidity. *Biophys J* **88**, 903–915.
7. Livesay, D. R., Dallakyan, S., Wood, G. G., and Jacobs, D. J. (2004). A flexible approach for understanding protein stability. *FEBS Lett* **576**, 468–476.
8. Livesay, D. R., and Jacobs, D. J. (2006). Conserved quantitative stability/flexibility relationships (QSFR) in an orthologous RNase H pair. *Proteins* **62**, 130–143.
9. Jacobs, D. J., Livesay, D. R., Hules, J., and Tasayco, M. L. (2006). Elucidating quantitative stability/flexibility relationships within thioredoxin and its fragments using a distance constraint model. *J Mol Biol* **358**, 882–904.
10. Livesay, D. R., Huynh, D. H., Dallakyan, S., and Jacobs, D. J. (2008). Hydrogen bond networks determine emergent mechanical and thermodynamic properties across a protein family. *Chem Cent J* **2**, 17.
11. Mottonen, J. M., Xu, M., Jacobs, D. J., and Livesay, D. R. (2009). Unifying mechanical

and thermodynamic descriptions across the thioredoxin protein family. *Proteins* **75**, 610–627.

12. Mottonen, J. M., Jacobs, D. J., and Livesay, D. R. (2010). Allosteric Response is Both Conserved and Variable Across Three CheY Orthologs. *Biophys J* **99**, 2245–2254.
13. Salsbury, F. R., Jr. (2010). Molecular dynamics simulations of protein dynamics and their relevance to drug discovery. *Curr Opin Pharmacol* **10**, 738–744.
14. Tozzini, V. (2010). Multiscale modeling of proteins. *Acc Chem Res* **43**, 220–230.
15. Zwier, M. C., and Chong, L. T. (2010). Reaching biological timescales with all-atom molecular dynamics simulations. *Curr Opin Pharmacol* **10**, 745–752.
16. Meirovitch, H. (2007). Recent developments in methodologies for calculating the entropy and free energy of biological systems by computer simulation. *Curr Opin Struct Biol* **17**, 181–186.
17. Rodinger, T., and Pomes, R. (2005). Enhancing the accuracy, the efficiency and the scope of free energy simulations. *Curr Opin Struct Biol* **15**, 164–170.
18. Kumar, S., Ma, B., Tsai, C. J., Sinha, N., and Nussinov, R. (2000). Folding and binding cascades: dynamic landscapes and population shifts. *Protein Sci* **9**, 10–19.
19. Chennubhotla, C., Rader, A. J., Yang, L. W., and Bahar, I. (2005). Elastic network models for understanding biomolecular machinery: from enzymes to supramolecular assemblies. *Phys Biol* **2**, S173–180.
20. Wells, S., Menor, S., Hespenheide, B., and Thorpe, M. F. (2005). Constrained geometric simulation of diffusive motion in proteins. *Phys Biol* **2**, S127–136.
21. Farrell, D. W., Speranskiy, K., and Thorpe, M. F. (2010). Generating stereochemically acceptable protein pathways. *Proteins* **78**, 2908–2921.
22. Herzberg, G. 1945. *Infrared and Raman spectra of polyatomic molecules*. D. Van Nostrand Company, New York.
23. Maxwell, J. C. (1864). On the calculation of the equilibrium and stiffness of frames. *Phil Mag* **27**, 294–299.
24. Thorpe, M. F., and Duxbury, P. M. 1999. *Rigidity Theory and Applications*. Kluwer Academic/Plenum Publishers, New York.
25. Jacobs, D. J., and Hendrickson, B. (1997). An algorithm for two-dimensional rigidity percolation: The pebble game. *J. Comp. Phys.* **137**, 346–365.
26. Jacobs, D. J., Rader, A. J., Kuhn, L. A., and Thorpe, M. F. (2001). Protein flexibility predictions using graph theory. *Proteins* **44**, 150–165.
27. Rader, A. J., Hespenheide, B. M., Kuhn, L. A., and Thorpe, M. F. (2002). Protein unfolding: rigidity lost. *Proc Natl Acad Sci USA* **99**, 3540–3545.
28. Munoz, V. (2001). What can we learn about protein folding from Ising-like models? *Curr Opin Struct Biol* **11**, 212–216.
29. Zimm, B. H., and Bragg, J. K. (1959). Theory of the phase transition between helix and random coil in polypeptide chains. *J. Chem. Phys.* **31**, 526–535.
30. Lifson, S., and Roig, A. (1961). On the theory of helix-coil transitions in polypeptides. *J. Chem. Phys.* **34**, 1963–1974.
31. Hilser, V. J., and Freire, E. (1996). Structure-based calculation of the equilibrium folding pathway of proteins. Correlation with hydrogen exchange protection factors. *J Mol Biol* **262**, 756–772.
32. Hilser, V. J., Garcia-Moreno, E. B., Oas, T. G., Kapp, G., and Whitten, S. T. (2006). A statistical thermodynamic model of the protein ensemble. *Chem Rev* **106**, 1545–1558.
33. Zamparo, M., and Pelizzola, A. (2006). Kinetics of the Wako-Saito-Munoz-Eaton model of protein folding. *Phys Rev Lett* **97**, 068106.
34. Jacobs, D. J. (2010). Ensemble-based methods for describing protein dynamics. *Curr Opin Pharmacol* **10**, 760–769.
35. Vorov, O. K., Livesay, D. R., and Jacobs, D. J. (2009). Helix/coil nucleation: a local response to global demands. *Biophys J* **97**, 3000–3009.
36. Wood, G. G., Clinkenbeard, D. A., and Jacobs, D. J. (2011). Nonadditivity in the alpha-helix to coil transition. *Biopolymers* **95**, 240–253.
37. Dill, K. A. (1997). Additivity principles in biochemistry. *J Biol Chem* **272**, 701–704.
38. Gao, J., Kuczera, K., Tidor, B., and Karplus, M. (1989). Hidden thermodynamics of mutant proteins: a molecular dynamics analysis. *Science* **244**, 1069–1072.
39. Mark, A. E., and van Gunsteren, W. F. (1994). Decomposition of the free energy of a system in terms of specific interactions. Implications for theoretical and experimental studies. *J Mol Biol* **240**, 167–176.
40. Hallerbach, B., and Hinz, H. J. (1999). Protein heat capacity: inconsistencies in the current view of cold denaturation. *Biophys Chem* **76**, 219–227.

41. Huang, K. 1987. *Statistical mechanics*. Wiley, New York.
42. Jacobs D. J., and Wood G. G. (2004). Understanding the α-Helix to Coil Transition in Polypeptides Using Network Rigidity: Predicting Heat and Cold Denaturation in Mixed Solvent Conditions. Biopolymers 75, 1–31.
43. Dahiyat, B. I., Gordon, D. B., and Mayo, S. L. (1997). Automated design of the surface positions of protein helices. *Protein Sci* **6**, 1333–1337.
44. Hollien, J., and Marqusee, S. (1999). A thermodynamic comparison of mesophilic and thermophilic ribonucleases H. *Biochemistry* **38**, 3831–3836.
45. Verma, D., Jacobs, D. J., and Livesay, D. R. (2010). Predicting the melting point of human c-type lysozyme mutants. *Curr Protein Pept Sci* **11**, 562–572.
46. del Sol, A., Tsai, C. J., Ma, B., and Nussinov, R. (2009). The origin of allosteric functional modulation: multiple pre-existing pathways. *Structure* **17**, 1042–1050.
47. Zhu, X., Amsler, C. D., Volz, K., and Matsumura, P. (1996). Tyrosine 106 of CheY plays an important role in chemotaxis signal transduction in Escherichia coli. *J Bacteriol* **178**, 4208–4215.
48. Cho, H. S., Lee, S. Y., Yan, D., Pan, X., Parkinson, J. S., Kustu, S., Wemmer, D. E., and Pelton, J. G. (2000). NMR structure of activated CheY. *J Mol Biol* **297**, 543–551.

Part IV

Macromolecular and Ligand Engineering Allosteric Functions

Chapter 16

An In Vivo Approach to Isolating Allosteric Pathways Using Hybrid Multimeric Proteins

Cuijuan Tie and Gregory D. Reinhart

Abstract

Hybrid tetramers of *Escherichia coli* phosphofructokinase (EC 2.7.1.11; EcPFK) have been used to dissect the complicated allosteric interactions within the native tetramer. The method used previously to generate hybrids in vitro involves dissociation of the parent proteins with KSCN followed by re-association as KSCN is removed via dialysis. However, this procedure is time consuming and is plagued with low hybrid yields. Consequently, we have attempted to produce hybrids more quickly and with potentially higher yields in vivo by co-expressing the parental EcPFK protein in *E. coli*. Wild-type EcPFK gene was cloned into pALTER-Ex2 and pALTER-1, respectively. Site-directed mutagenesis was performed to make mutant EcPFK gene in pALTER-1. Since each vector has a different origin of replication and antibiotic selection marker, we were able to co-transform both plasmids to competent *E. coli* cells. Following an affinity purification column, anion-exchange chromatography was used to separate the five hybrid species (4:0, 3:1, 2:2, 1:3, 0:4). While all five hybrid species were obtained, the amount 1:3 and 0:4 hybrids were very small. By changing the expression vector for the mutant EcPFK protein from pALTER-1 to pALTER-Ex1 and the charge-tag mutations from K2E/K3E to K90E/K91E, the yield of 1:3 hybrid was substantially increased. The in vivo method does increase the yield of the hybrids produced while decreasing the time required for their isolation.

Key words: Hybrid enzymes, Co-expression, Phosphofructokinase, Allosteric interactions

1. Introduction

Although some monomeric enzymes and proteins exhibit allosteric properties, it is by far more common that allosteric enzymes are multimeric in composition, even when all subunits are identical. This composition invariably leads to multiple ligand binding sites, usually including multiple copies of the same binding sites, in the whole functional protein. If one is to probe the mechanism of allosteric communication in such a system beyond a merely phenomenological description of the overall behavior of the whole protein, it is necessary to consider how the individual components

Aron W. Fenton (ed.), *Allostery: Methods and Protocols*, Methods in Molecular Biology, vol. 796,
DOI 10.1007/978-1-61779-334-9_16, © Springer Science+Business Media, LLC 2012

and/or binding at each individual site communicates with each of the other components and/or binding events of the protein.

It can often be useful in the latter circumstance to construct hybrid enzymes to isolate specific allosteric interactions (1–3). For example, in a homotetramer such as bacterial phosphofructokinase (PFK – see Note 1 for other abbreviations), on average each subunit contains a single active site and a single allosteric site per subunit. Hybrids have been constructed in which each unique heterotropic interaction that exists in the native enzyme between a single substrate site and a single allosteric ligand binding site has been isolated (4, 5). Since the native tetramer contains four substrate sites and four allosteric sites, the isolation of a single pair-wise interaction requires the formation and purification of a 1:3 hybrid in which one native subunit is combined with three subunits which contain modifications that prevent binding to their respective sites. By modifying the mutations utilized, each of the four unique heterotropic interactions in both BsPFK and EcPFK have been isolated (6–8). Allosteric behavior overall can then be explained by a combination of various pair-wise interactions (9).

In order to construct hybrids of this type, two procedures are required. First, one must be able to produce the hybrid species, and second, one must be able to purify the single species one wants to study further. We and others have typically utilized an approach for producing hybrids that employs a mild denaturant that causes the native oligomeric enzyme to dissociate to monomers (1–5). It is important that the denaturant not irreversibly unfold the proteins, or at least not do so with an appreciable rate. In our case, we have dissociated the tetramers of PFK from *Escherichia coli* and *Bacillus stearothermophilus* with the addition of KSCN, first employed for this purpose by Garel and co-workers (10). This dissociation is then followed by re-association of the tetramers by removing KSCN through dialysis. When the original solution contains a mixture of the purified parent proteins dissociated with KSCN, re-association results in a mixture of hybrid tetramers. If the parents have the same subunit–subunit association constants, and equal concentrations of each parent are utilized, a statistical mixture of parents and hybrids results.

It is often then necessary to isolate the desired hybrid for further studies. We have found this to be accomplished straightforwardly by altering one or more surface charge residues of one parent, followed by ion exchange chromatography. The systematic alteration in charge with subunit composition within the hybrid mixture then affords ready separation. As an added benefit, this change also allows for the analytical separation and identification of species in native gel electrophoresis. Since the size is common to all, separation is achieved on the basis of the differing charge on each hybrid species.

Production of hybrids in this manner, however, can be time consuming and can be plagued with low yield given the requirement to purify each of the parent proteins and then to purify the desired hybrid. An alternative approach is to produce the hybrid proteins in vivo by co-expressing both parent proteins within the cell. In this way, hybrids are formed at the time of protein production, in principle requiring only a single purification procedure to be performed upon releasing the cell contents. We describe below how we have utilized this approach for producing hybrids between wild-type EcPFK and a mutant EcPFK and then isolating the hybrid with one wild-type subunit and three mutant subunits for further study.

2. Materials

Ampicillin- and chloramphenicol-resistant transformants were selected on LB agar medium containing the appropriate antibiotic at 100 μg/mL. SOC medium was used during transformation. All chemical reagents were analytical grade, purchased from Fisher Scientific (Hampton, NH) or Sigma-Aldrich (St. Louis, MO). Mimetic Blue 1 agarose resin from Prometic Biosciences (Montreal, Canada) was used in protein purification. Oligonucleotides were synthesized by IDT (Coralville, IA). Qiagen (Hilden, Germany) products were used for plasmid purifications. pALTER-1, pALTER-Ex1, and pALTER-Ex2 were purchased from Promega. pGDR16 contains wild-type *E. coli* PFK gene in pALTER-1, which is expressed under the lac promoter. pGDR147 and pGDR148 contain wild-type *E. coli* PFK-1 gene in pALTER-Ex1 and pALTER-Ex2, respectively. Plasmids were expressed in *E. coli* strain RL257 (MQ Δ*pfkB::FRT* Δ*pfkA::FRT*). MQ is a *lac*$^+$ *lacI*^{q} derivation of MC4100 (11).

3. Methods

3.1. Mutagenesis

pGDR147 and pGDR16, containing the wild-type EcPFK gene in pALTER-EX1 or pALTER-1, respectively, were used with the Altered Sites II in vitro Mutagenesis System (Promega) according to the manufacturer's instructions to construct mutations. Mutagenesis primers were as follows:

K2E/K3E, 5'-GCT TGT CAA CAC ACC GAT TTC CTC AAT CAT GAC TAC CTC TGA AGC-3'

K90E/K91E, 5'-CAG CGC ATC GAT CCC ACG TTC TTC CAG GTT TTC GAT AGC CAC-3'

The mutated DNA was sequenced across the modified site to confirm the desired mutation. Plasmids containing wild-type and mutant EcPFK genes were co-transformed into RL257 cells for protein expression. Other mutations were introduced similarly as required for functional analysis of allosteric communication as described previously using the traditional hybridization procedure (5–8).

3.2. Protein Purification and Hybrid Isolation

All proteins were purified from RL257 cells as previously described (8). The presence of all five hybrid tetramers was verified on native polyacylamide gel. A Mono-Q 10/10 (Pharmacia) anionic exchange column was used to separate the five hybrid tetramers based on the charge differences. The hybrids were eluted with a linear NaCl gradient. The amount of each hybrid was quantified using BCA protein assay (Pierce).

3.3. Co-expression of Wild-Type and Charge-Tag PFK

The wild-type EcPFK gene was cloned into pALTER-Ex2 and pALTER-1 to give pGDR148 and pGDR16, respectively. The EcPFK gene in pALTER-Ex2 was kept as wild-type form. Using site-directed mutagenesis, the second and the third codons of EcPFK in pGDR16 were mutated from lysine to glutamate to change the surface charge of the protein. pALTER-Ex2 and pALTER-1 have different origins of replication, p15A and ColE1, respectively. In addition, the antibiotic selection markers are different and are chloramphenicol and ampicillin, respectively. They are able to be co-transformed into competent RL257 cells (Fig. 1).

Following an affinity purification column, anion-exchange chromatography was used to separate the five hybrid species (4:0, 3:1, 2:2, 1:3, 0:4) based on the charge differences among the hybrid tetramers. While all five hybrid species were obtained, they were not well separated. In addition, the amounts of the 1:3 and 0:4 hybrids were relatively low based on the elution profile and native gel electrophoresis (Fig. 2). The yield of these species was later enhanced (see Note 2). For our purposes, we were most interested in isolating the 1:3 hybrid which contains only a single native heterotropic interaction between the substrate, Fru-6-P, and the allosteric ligand, either MgADP or PEP.

Previously in our laboratory, hybrid tetramers of PFK were successfully created in vitro (4–8). Among the five different hybrid species possible, the 1:3 hybrid, which has one native active site and one native allosteric site, presents a specific heterotropic

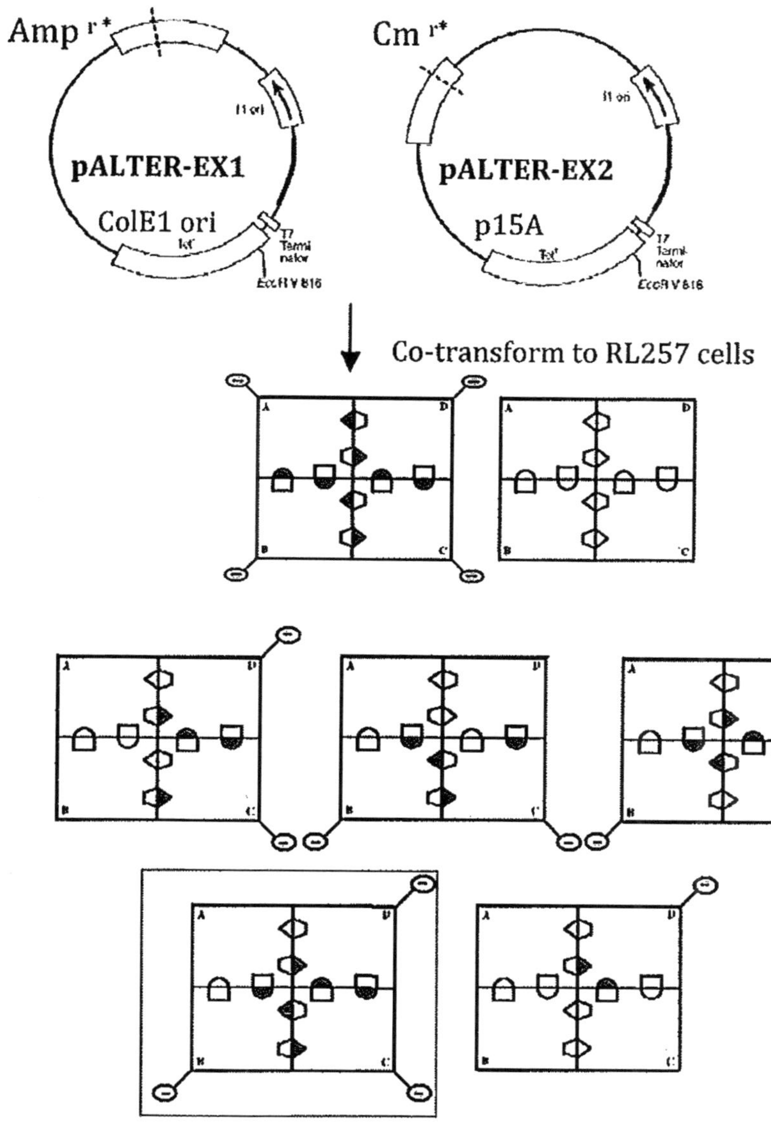

Fig. 1. Schematic graph of in vivo hybrid formation. Wild-type and mutant EcPFK genes were in two plasmids with different selection markers and origins of replication. After co-expression in RL257 cells, the different hybrid tetramers were formed in vivo. The 1:3 hybrid that represents a specific heterotropic interaction is shown in the *box*.

interaction. However, the in vitro method has been plagued with low yield and can be time consuming. In addition, KSCN, a mild chaotropic reagent, is used in the process to dissociate PFK tetramers. Thus, the hybrid formation is not a natural process.

In the current study, we co-expressed the two parent proteins of the hybrid in vivo so that the chemical treatment is unnecessary. This process cuts the hybrid production time in half. The efficiency of making hybrid proteins is also improved using the in vivo

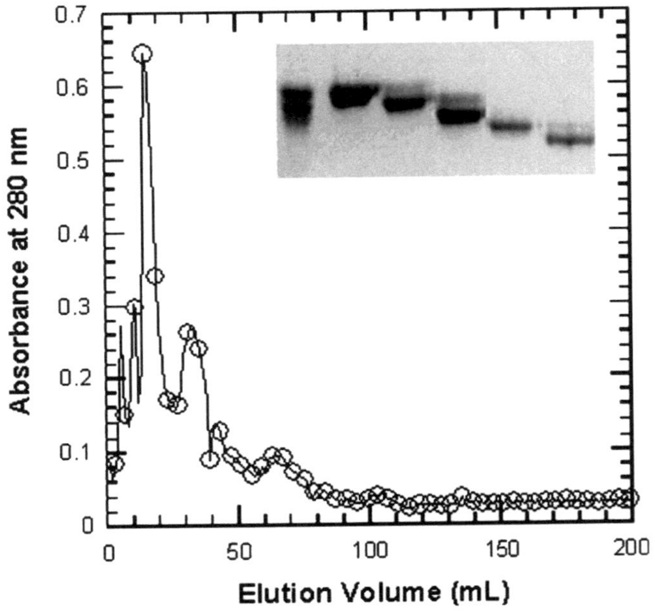

Fig. 2. Elution profile for the hybrid species between wild-type and charge-tag (K2E/K3E) EcPFK by anion-exchange column using the in vivo hybrid formation method. *Inset*: 7.5% native PAGE of the different hybrid species obtained from this separation. *Lane 1*, hybrid mixture; *lane 2*, 4:0 hybrid; *lane 3*, 3:1 hybrid; *lane 4*, 2:2 hybrid; *lane 5*, 1:3 hybrid; *lane 6*, 0:4 hybrid. The 1:3 hybrid eluted at 102 mL from this anion-exchange column.

method as can be seen by the increased yield for the 1:3 hybrid (see Note 3). Eventually, the five hybrid species were successfully made by this in vivo method. In addition, the yield of the 1:3 hybrid is increased threefold by changing the vector to pALTER-Ex1 and charge-tag mutations to K90E/K91E. The in vivo method also avoids using any denaturing reagents that might alter the tertiary structure of the protein while at the same time saving time relative to the in vitro method. The characteristics of the 1:3 hybrids containing each heterotropic interaction produced in vivo were similar to those produced in vitro allowing them to be used for future studies of allosteric regulation in EcPFK (see Note 4).

4. Notes

1. Abbreviations: *PFK* phosphofructokinase, *EcPFK E. coli* phosphofructokinase, *BsPFK B. stearothermophilus* phosphofructokinase, *LB* Luria–Bertani media, *MOPS* 3-(*N*-morpholino) propanesulfonic acid, *DTT* dithiothreotol, *ATP* adenosine

5′-triphosphate, *ADP* adenosine 5′-diphosphate, *Fru-6-P* fructose 6-phosphate, *Fru-1,6-BP* fructose 1,6-bisphosphate.

2. The initial low yield of the 1:3 hybrid was due to the low level of expression of the K2E/K3E charge-tag mutations. There are two likely reasons. First, pALTER-1 is not an expression vector strictly speaking. Although other proteins express well in pALTER-1, the expression level of two proteins at the same time cannot be controlled in vivo. Consequently, the pALTER-1 vector was changed to pALTER-Ex1, which has the strong tac promoter. Second, the K2E/K3E charge-tag mutations are at the beginning of EcPFK gene and thus may influence the expression of EcPFK protein (12). To address this possibility, alternative charge-tag mutations, K90E/K91E, were substituted for K2E/K3E. These charge-tag mutations have been used successfully in *B. stearothermophilus* (BsPFK) (4, 7). New constructs were created using the pALTER-Ex1 and the K90E/K91E charge-tag mutations. After the purification and separation, the yield of the 1:3 hybrid was about three times higher than that of the in vitro method (Table 1). In addition, the separation resolution of the five hybrid species is significantly improved compared with the K2E/K3E charge-tag (Fig. 3).
3. The comparison between the in vivo and in vitro method is shown in Fig. 4. The in vivo method improves yield and saves time. In addition, the in vivo method avoids exposing the enzyme to denaturing conditions, which might produce subtle changes to the protein structure.
4. Based on the results we obtained from the above experiments, the in vivo method was applied to make the 1:3 hybrid tetramers that each contains one of the four unique heterotropic interactions. Using K90E/K91E as the charge-tag mutations, site-directed mutagenesis was performed on EcPFK gene in pALTER-Ex1 (pGDR147). The four mutant EcPFKs that

Table 1
Yield comparison between in vitro and in vivo making hybrids methods

	In vitro	In vivo
Load on Mono-Q (mg)	10	10
Yield of all the hybrids (mg)	4.0	6.8
Yield of 1:3 hybrid (mg)	0.5	1.6
Yield of 1:3 hybrid (%)	5.0	16.0

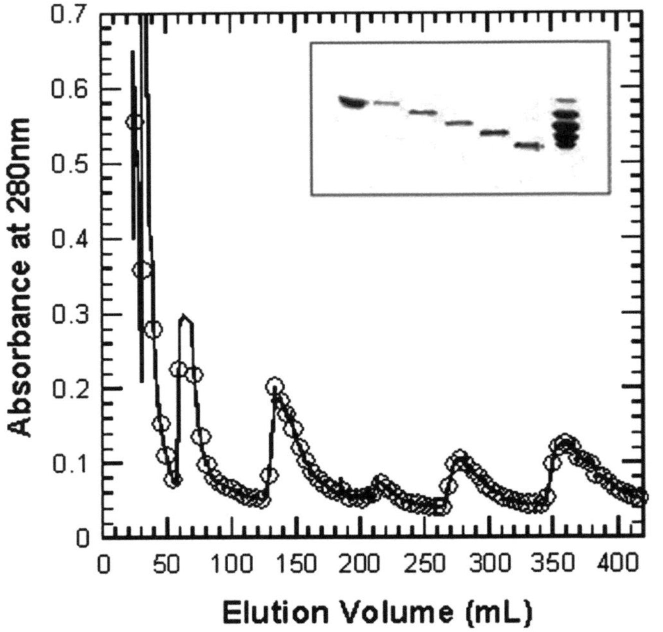

Fig. 3. Elution profile for the hybrid species between wild-type and charge-tag (K90E/K91E) EcPFK by anion-exchange column using the in vivo hybrid formation method. *Inset*: 7.5% native PAGE of the different hybrid species obtained from this separation. *Lane 1*, hybrid mixture; *lane 2*, 4:0 hybrid; *lane 3*, 3:1 hybrid; *lane 4*, 2:2 hybrid; *lane 5*, 1:3 hybrid; *lane 6*, 0:4 hybrid. The 1:3 hybrid eluted at 280 mL from this anion-exchange column.

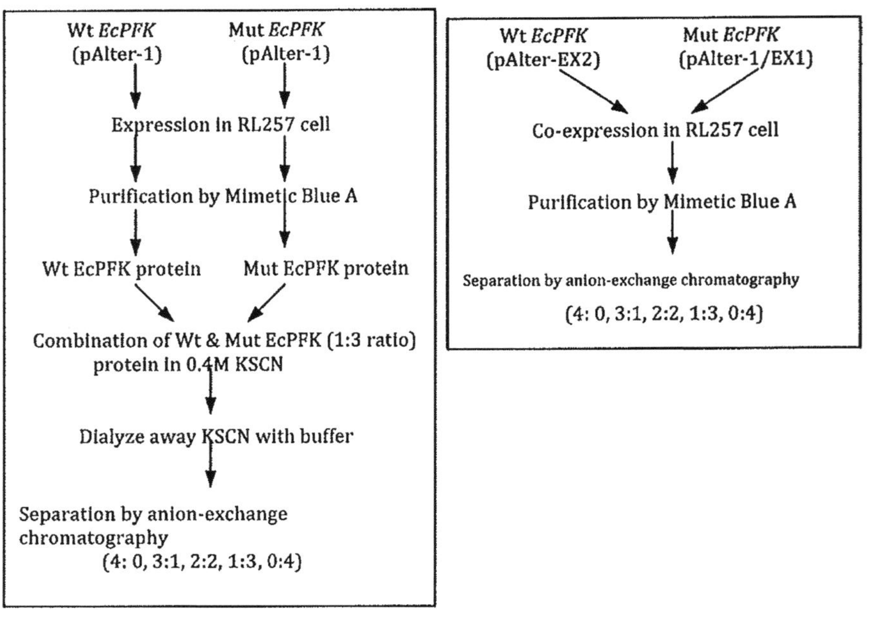

Fig. 4. Comparison of the in vivo and in vitro methods for forming hybrid tetramers of EcPFK.

will be used to make the 1:3 hybrid tetramer containing each of the four heterotropic interactions were obtained. Each mutated plasmid and pGDR148, which has the wild-type EcPFK gene, was co-transformed in RL257 cells. All four of the mutated proteins are able to form the five hybrid tetramers in vivo. The amount of each hybrid species is different. The yield of 1:3 hybrid is relatively higher than the other hybrids except 0:4 (Fig. 3).

Acknowledgments

We thank Dr. Michelle Lovingshimer for the construction of pGDR147 and pGDR148. This work was supported by grant GM033216 from the NIH and by grant A1543 from the Robert A. Welch Foundation.

References

1. Huang, Y., Doyle, M. L., and Ackers, G. K. (1996) The Oxygen Binding Intermediates of Human Hemoglobin: Evaluation of their contributions to Cooperativity Using Zinc-Containing Hybrids, *Biophys J.* **71**, 2094–2105.
2. Nelson, S. W., Honzatko, R. B., and Fromm, H. J. (2002) Hybrid Tetramers of Porcine Liver Fructose-1,6-bisphosphatase Reveal Multiple Pathways of Allosteric Inhibition, *J. Biol. Chem.* **277**, 15539–15545.
3. Evans, P. R., Farrants, G. W., and Hudson, P. J. (1981) Phosphofructokinase: Structure and control, *Phil. Trans. R. Soc. Lond. B.* **293**, 53–62.
4. Grant, G. A., Xu, X. L., and Hu, Z. (2004) Quantitative Relationships of Site to Site Interactions in *Escherichia coli* D-3-Phosphoglycerate Dehydrogenase Revealed by Asymmetric Hybrid Tetramers, *J. Biol. Chem.* **279**, 13452–13460.
5. Fenton, A. W., and Reinhart, G. D. (2002) Isolation of a Single Activating Interaction in *E. coli* Phosphofructokinase, *Biochemistry* **41**, 13410–13416.
6. Fenton, A. W., Paricharttankul, N. M., and Reinhart, G. D. (2004) Disentangling the Web of Allosteric Communication: Quantification of the Four Unique Heterotropic Activating Interactions in *E. coli* Phosphofructokinase, *Biochemistry* **43** 14104–14110.
7. Ortigosa, A. D., Kimmel, J. L., and Reinhart, G. D. (2004) Disentangling the web of allosteric communication in a homotetramer: heterotropic inhibition of phosphofructokinase from *Bacillus stearothermophilus*, *Biochemistry* **43**, 577–586.
8. Fenton, A. W., and Reinhart, G. D. (2009) Disentangling the Web of Allosteric Communication in a Homotetramer: Heterotropic Inhibition in Phosphofructokinase from *Escherichia coli*, *Biochemistry* **48**, 12323–12328.
9. Reinhart, G. D. (1988) Linked Function Origins of Cooperativity in a Symmetrical Dimer, *Biophys. Chem.* **30**, 159–172.
10. Le Bras, G., Auzat, I., and Garel, J.-R. (1995) Tetramer-Dimer Equilibrium of Phosphofructokinase and Formation of Hybrid Tetramers, *Biochemistry* **34**, 13203–13210.
11. Lovingshimer, M. R., Siegele, D., and Reinhart, G. D. (2006) Construction of an inducible, *pfk*A and *pfk*B deficient strain of *Escherichia coli* for the expression and purification of phosphofructokinase from bacterial sources, *Protein Expression and Purification* **46**, 475–482.
12. Gold, L. and Stormo, G. (1987) Translational initiation. *Escherichia coli* and *Salmonella typhimurium* cellular and molecular biology, Vol 2. F. C. Neidhardt, editor. American society for microbiology. Washington, D.C. (1987) 1302–1307.

Chapter 17

Mutations in the $GABA_A$ Receptor that Mimic the Allosteric Ligand Etomidate

Stuart A. Forman and Deirdre Stewart

Abstract

Etomidate is a hydrophobic molecule, a potent general anesthetic, and the best understood drug in this group. Etomidate's target molecules are $GABA_A$ receptors, its site of action has been identified with photolabeling, and a quantitative allosteric coagonist model has emerged for etomidate effects on $GABA_A$ receptors. We have shown that when methionine residues that are thought to be adjacent to the etomidate site are mutated to tryptophan, that the bulky hydrophobic side-chains alter mutant $GABA_A$ receptor function in ways that mimic the effects of etomidate binding to wild-type receptors. Furthermore, these mutations reduce receptor modulation by etomidate. Both of these observations support the hypothesis that these methionine residues form part of the etomidate binding pocket.

Key words: GABA-A receptor, Nicotinic acetylcholine receptor, Ion channel, Allosteric model, Etomidate, General anesthesia, Site-directed mutagenesis, Electrophysiology

1. Introduction

The potent general anesthetic etomidate, acts via ionotropic γ-aminobutyric acid type A ($GABA_A$) receptors containing $\beta 2$ and $\beta 3$ subunits to produce sedation, loss-of-righting reflexes, and immobilization in the presence of noxious stimuli (1, 2). $GABA_A$ receptors are ligand-gated chloride channels formed from five subunits arranged around a central transmembrane pore. Etomidate acts as a positive allosteric modulator of $GABA_A$ receptors, enhancing chloride channel activation in the presence of low GABA concentrations, lowering GABA EC_{50}, and prolonging inhibitory post-synaptic currents in the brain (3–5). High concentrations of etomidate also directly activate (i.e., agonize) $GABA_A$ receptors in the absence of GABA (3, 4). Using quantitative electrophysiology in *Xenopus* oocytes expressing

Aron W. Fenton (ed.), *Allostery: Methods and Protocols*, Methods in Molecular Biology, vol. 796,
DOI 10.1007/978-1-61779-334-9_17, © Springer Science+Business Media, LLC 2012

α1β2γ2L $GABA_A$ receptors with stoichiometry 2α:2β:1γ (6), we found that both GABA potentiation and direct activation by etomidate are consistent with a single class of allosteric agonist sites (7). Our functional analysis suggested that two equivalent etomidate sites best explained the results. Subsequently, a radiolabeled photoreactive etomidate analog, azi-etomidate, was synthesized (8) and used to identify etomidate binding residues in purified bovine $GABA_A$ receptors (9). The two photomodified residues that accounted for nearly all the incorporated radioactivity were conserved methionines in transmembrane domains, one on α subunits (αM236), and one on β subunits (βM286). Azi-etomidate photoincorporation at both αM236 and βM286 is inhibited in parallel by etomidate, suggesting that these residues both contribute to a single class of binding sites. Structural studies of $GABA_A$ receptors confirm that these methionines are located at the transmembrane interface where α and β subunits abut (10) and that two such interfacial sites are present on each receptor-channel complex.

For the experiments described here, we hypothesized that because etomidate is a bulky hydrophobic molecule, mutating the photolabeled methionines to bulky hydrophobic tryptophans would mimic the presence of etomidate in its binding pocket. We created expression plasmids encoding α1M236W and β2M286W mutant subunits, coexpressed these in *Xenopus* oocytes with other wild-type subunits to produce α1M236Wβ2γ2L and α1β2M286Wγ2L receptors, and characterized their gating behavior and sensitivity to both GABA and etomidate (11). Gating behavior have been quantified under multiple conditions: (1) basal gating in the absence of ligands, (2) GABA-dependent gating, (3) maximal GABA efficacy, (4) etomidate direct activation and efficacy, and (5) etomidate modulation of GABA sensitivity (left-shift or EC_{50} ratio). Our results demonstrated that tryptophan mutations at these two residues produce channels with enhanced sensitivity to GABA and spontaneous gating activity, changes that mimic the GABA-potentiating and direct activating effects of etomidate. In addition, left-shift analysis showed that, compared to wild-type, the mutant receptors are far less sensitive to etomidate, possibly because they partially occupy the space where etomidate normally binds.

Here, we describe the methods for *Xenopus* oocyte channel expression and electrophysiology, detail the procedures used to quantify each of the functional characteristics of $GABA_A$ receptors, and the quantitative analysis and model-fitting procedures.

2. Materials

2.1. Site-Directed Mutagenesis and Messenger RNA Synthesis

1. Plasmids encoding human α1, β2, and γ2L $GABA_A$ receptor subunits. Cloning vectors must have properly oriented transcription initiation sites for in vitro synthesis of mRNA. We subcloned cDNAs into pCDNA3.1 vectors (Invitrogen), which include a T7 transcription site and a BGH polyadenylation sequence that is important for stability of mRNA transcripts.
2. Quick-Change mutagenesis kit (Stratagene).
3. Oligonucleotides for mutagenesis (see Note 1). The oligonucleotide sequence used for the Quick-change mutagenesis were:

   α1M236W:TATTCAAACATACCTGCCGTGCATA**TG-G**AC AGTTATTCTCTCCCAAGTCTCAT

   β2M286W: TGAAGGCCATTGACATGTACCTG**TGG**GG GTGCTTTGTCTTCGTTTTCATG.
4. DNA Miniprep kits (Qiagen).
5. DNA purification kits: Wizard SV Gel and PCR Cleanup (Promega Corporation).
6. Endonuclease enzymes and buffers for linearization (New England Biolabs).
7. Agarose Gel Electrophoresis (minigel) equipment.
8. DNA Sequencing facility.
9. mMessage Machine and mRNA Purification kits (Ambion/Applied Biosystems).

2.2. Xenopus Oocyte Harvest, Injection, and Culture

1. Frogs. Adult female *Xenopus laevis*, injected with hcg, were purchased from Xenopus One. They are housed in a veterinary-supervised temperature-controlled environment with 12-h light–dark cycles.
2. Calcium-free OR2 (82 mM NaCl, 2 mM KCl, 1 mM $MgCl_2$, 5 mM HEPES, pH 7.6) and ND96 (96 mM NaCl, 2 mM KCl, 1.8 mM $CaCl_2$, 1 mM $MgCl_2$, 5 mM HEPES, pH 7.6) supplemented with Gentamicin 50 μg/ml (see Note 2).
3. Tricaine (Ethyl 3-aminobenzoate methanesulfonate salt) is purchased from Sigma Aldrich.
4. Fine forceps (Dumont #4 or #5 from Fine Science Tools).
5. Standard forceps (Dumont #2 or #3 from Fine Science Tools).
6. Collagenase Type 2 (Worthington Biochemical) solution (1 mg/ml in OR2).

7. Plastic-ware: Cell-culture dishes (10 and 15 cm diameter); Centrifuge tubes (15 and 50 ml); Disposable plastic transfer pipettes; Disposable small plastic culture loops.
8. Dissecting microscope and illuminator.
9. Manual 3D micromanipulator with pipette holder (World Precision Instruments; see Note 3).
10. Micropipettes for injection (10 μl).
11. Microinjector (Drummond Digital Microdispenser).
12. Cell-culture dish (5 cm) with plastic mesh (or Velcro) insert. The plastic mesh insert, cut to fit snuggly in the bottom of the cell-culture dish, provides a grid of appropriately sized holes (about 0.5 mm square) to keep the oocytes from rolling during injection.
13. Mineral oil.
14. Messenger RNA mixtures in RNAase-free water.

2.3. Oocyte Electrophysiology (see Note 4)

1. Vibration damping and faraday cage. While commercial tabletop vibration-damping devices are available, we achieve adequate damping using a heavy iron plate (12″ × 16″ × 0.75″) mounted on four large rubber stoppers. A faraday cage adequate for these experiments can be built from conductive metal screening on a frame (copper is ideal, but aluminum is adequate).
2. Dissecting microscope and fiberoptic illuminator (both from World Precisions Instruments).
3. Multireservoir remote-controllable superfusion system (ALA Scientific Instruments).
4. Low-volume flow chamber. Figure 1 depicts a simple linear flow cell with a low-volume chamber for an oocyte, a guard chamber for use of a salt-bridge, and a depressed “drain” area where outflow solution is removed into a suction canister using a house vacuum or peristaltic pump.
5. Two microelectrode voltage clamp amplifier (Warner Instruments).
6. Micromanipulators for electrodes.
7. Digitizer and data collection software: Digidata 1440 digitizer interface and pClamp software, both from Molecular Devices.
8. Microelectrodes (borosilicate pipettes and pipette puller; see Note 5).
9. 3 M KCl for microelectrodes.
10. Bleach for treating Ag–$AgCl_2$ electrodes.
11. GABA (Sigma-Aldrich) stock solution: 1 mM in ND96.
12. Picrotoxin (Sigma-Aldrich): 2 mM in ND96.

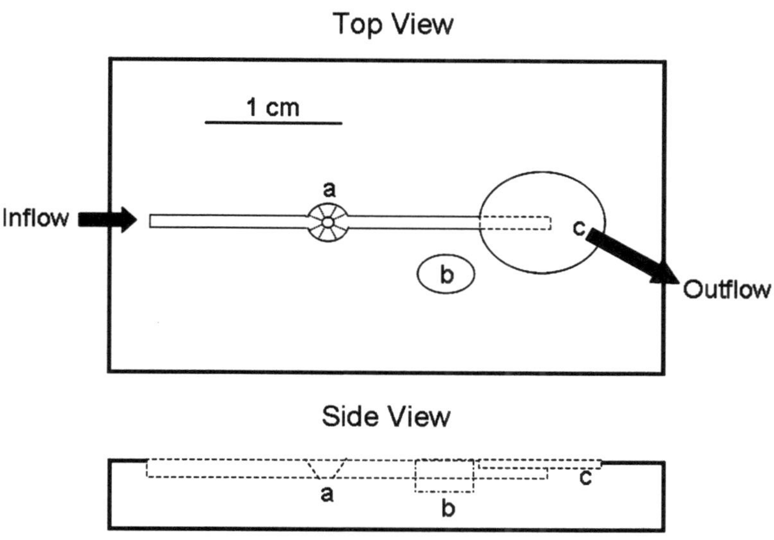

Fig. 1. Diagram of a simple flow-cell design for oocyte electrophysiology. The inflow channel is sized to accommodate small PTFE tubing (OD 0.12″). The oocyte chamber (**a**) is beveled to allow microelectrodes to enter at an angle from either side of the flow channel. The guard chamber (**b**) enables the use of a salt bridge to reduce grounding variability. The outflow incorporates a stepped chamber design that allows suction to remove excess superfusate without removing all the fluid in the flow channel if inflow stops.

13. Etomidate (Bedford Laboratories): clinical preparation is 2 mg/ml (8.2 mM) in 35% (v/v) propylene glycol.
14. Alphaxalone (MP Biomedical): 2 mM in DMSO.

2.4. Data Analyses

1. Clampfit 9.0 software (Molecular Devices).
2. Prism 5.0 software (GraphPad Software Inc.).
3. Origin 6.0 software (OriginLab Corp.).

3. Methods

3.1. Site-Directed Mutagenesis and Messenger RNA Synthesis

1. Perform site-directed mutagenesis with appropriate template and mutagenic oligonucleotides.
2. Confirm sequence. We sequence the entire cDNA region with a set of sequencing oligonucleotides.
3. Linearize DNAs: Plasmids containing alpha and gamma GABA receptor wild type and mutant cDNAs were linearized with Xma I, and beta cDNAs were cut with Stu I. Linearized DNA is visualized on 0.8% agarose gel and extracted from excised gel regions with Wizard cleanup kits.

4. mRNAs were synthesized using the mMessage mMachine T7 kit and were purified using NucAway Spin Columns (both from Ambion). mRNA yields were quantified using A260 measurements.
5. mRNAs were mixed for injection at a ratio of 1 alpha:1 beta:2–5 gamma subunit at a total RNA concentration of 0.2 μg/μl using Nuclease-free water (Ambion).
6. DNAs and mRNAs were visualized on 1% agarose gels using ethidium bromide and documented using a digital camera.

3.2. *Xenopus* Oocyte Harvest, Injection, and Culture

1. Prior to mini-laparotomy (0.5 cm incision) to extract oocytes, frogs are anesthetized by immersion in ice cold 1.5% tricaine, pH 7.0. We prepare this solution using 1 L of deionized water in a styrofoam ice bucket fitted with a cover.
2. Adequate anesthesia is assessed by both handling the frogs and pinching their abdomens and limbs with forceps prior to surgery. Modified aseptic technique is used in accordance with recent ARAC guidelines (http://oacu.od.nih.gov/ARAC/XenopusOocyte_101007_Fnl.pdf). Instruments are sterilized with heat or alcohol and flame, but no antiseptic prep is used on frog skin, which secretes peptides with broad antibiotic activity.
3. The abdominal skin is retracted using fine forceps or the tip of a sterile 19 gauge needle and a 0.5 cm incision is made in skin only using small sharp scissors (heat sterilized). The abdominal wall is then held with heavy forceps and punctured using the 19 gauge needle. The abdominal opening through muscle and fascia is enlarged by inserting the tips of the small scissors and opening them in a roughly cephalad-to-caudad direction (parallel to the muscle fibers). The heavy forceps are inserted into the wound and used to gently retract a lobe of ovary (see Note 6). The lobe (volume approximately 1–2 ml) is excised with sharp scissors and placed in OR2 solution in a 10 cm diameter plastic cell-culture dish. The remaining ovary is pushed back into the abdominal cavity. The wound is closed in two layers (muscle/fascia first, then skin) with resorbable suture (3.0 or 4.0 Vicryl), using simple or figure-of-eight stitches.
4. Following surgery, frogs are recovered in room temperature shallow deionized water, oriented to keep their snouts above water. It is important to keep a heavy lid on the recovery chamber, to keep the frogs from jumping out and running for cover in the lab (they prefer the dark areas under equipment). Frogs are returned to the animal care facility following 1–2 h of recovery, when they display vigorous activity.

5. Using one pair of heavy forceps and one pair of fine forceps, the excised portion of ovary is gently torn into small pieces, each containing four to ten oocytes (see Note 7).
6. Transfer the small portions of ovary from the culture dish to 15-ml centrifuge tubes, approximately 1 ml of oocytes per tube, each containing 5 ml OR2 (see Note 8). The ovaries will sink to the bottom of the tube. Gently aspirate the OR2 above the oocytes and wash the oocytes by gently agitating with additional 5 ml aliquots of OR2 until clear.
7. Add 5 ml of Collagenase solution to each centrifuge tube. Place on a rocking platform or gentle orbital shaker and incubate for 1 h at room temperature (22°C).
8. After 1 h, aspirate collagenase solution from above the oocytes and replace with fresh collagenase solution. Incubate for an additional 30–60 min (see Note 9)
9. When oocytes are mostly free of connective tissue, aspirate collagenase solution and wash six times with 5 ml aliquots of OR2, gently agitating, then aspirating after each rinse. Resuspend oocytes in ND96 supplemented with antibiotics, and transfer to a 15 cm diameter cell-culture dish filled to a 3–4 mm depth with ND96 supplemented with antibiotics. While transferring, arrange the oocytes in a line across the diameter of the dish.
10. Examine the oocytes under a dissecting microscope. Identify healthy-appearing stage V and VI oocytes (these have a distinct band between the dark and light poles), and using a small plastic culture loop, separate these oocytes to another part of the culture dish, then use a disposable pasteur pipette to transfer them to a small culture dish containing ND96 supplemented with antibiotics. These oocytes will be used for injection. The remaining oocytes can be discarded or maintained in culture, in ND96 plus antibiotics at 17°C for several days.
11. Prepare mRNA mixtures for injection in nuclease-free water. We usually use mRNA concentrations of 0.2–0.5 ng/nl and w/w ratios of 1α:1β:$2–5\gamma$ to improve incorporation of γ subunits into receptors (12). These mixtures are stored at −80°C and kept on ice during use.
12. Injection pipettes are pulled to a sharp tapered needle and broken to a sharp bevel on a glass bead on the microforge (see Note 10). A small amount of mineral oil is inserted into the back end of the injection pipette, leaving no air space, and the pipette is carefully slid onto the plunger needle of the microinjector. After sealing the pipette onto the microinjector with the locking nut/flange, the plunger is advanced until a small droplet of oil is expressed at the tip. The microinjector is

then mounted on the micromanipulator. A small depression is created in a piece of parafilm using a disposable pipette tip and a 2 μl droplet of mRNA mixture is deposited in the depression – it should appear clear with no particulate matter in it. Under the microscope, the injection pipette tip is positioned so that the tip is aiming at the center of this droplet, and then advanced until the tip is in the droplet. Slowly draw the mRNA into the injection pipette by withdrawing the plunger, making sure that the solution is flowing, and that no air bubbles enter the injection pipette. When all of the solution is loaded into the injection pipette, retract the entire injector to keep the tip from being damaged while setting up oocytes for injection.

13. Transfer about 10–20 oocytes in ND96 plus antibiotics to the cell-culture dish fitted with a circle of plastic screen on its base. Arrange the oocytes in a line with their dark sides or the equatorial bands oriented toward the injection pipette. Advance the pipette until the tip enters the ND96 and is near the oocytes. Under microscope guidance, advance the pipette tip until it touches the first oocyte. Continue advancing until the pipette tip enters the oocyte. Slowly inject 25–50 nl (this is only a fraction of a turn on the Drummond injector) while observing the oocyte. It should visibly "plump up" as the injection occurs. Wait 5–10 s and withdraw the pipette tip to a position above the oocytes. Manually move the cell-culture dish until the next oocyte is positioned below the injector and continue injecting until all the oocytes are injected.
14. Transfer injected oocytes into labeled cell-culture trays in ND96 supplemented with gentamicin. Incubate at 17°C for 24–96 h. Examine daily for unhealthy oocytes (bloated with pale opalescent membranes) and remove these. Change ND96 solution daily.

3.3. Oocyte Electrophysiology

1. *GABA concentration response.* Starting with a fresh aliquot of 1 M GABA stock solution (see Note 11), prepare a range of ND96 (antibiotic-free) superfusion solutions with GABA concentrations ranging from 0.1 μM to 1 mM. We typically use half-log (3.2-fold) concentration steps. Concentration responses should include at least 6 concentrations: one or two producing less than 10% activation, two or three defining the steep rising portion of the response curve, and at least two defining the maximal plateau. Recordings are performed at room temperature with oocytes clamped at −50 mV. Software-controlled experiments typically record 5–10 s of baseline data (ND96 solution) before switching superfusate to a GABA solution for 20–60 s, and then

returning to ND96 for 20–30 s. The time required to reach peak current depends on the GABA concentration and flow characteristics in the chamber holding the oocyte. After each sweep, oocytes are "washed" in ND96 for at least 3 min before the subsequent experiment. This wash time is needed to allow desensitized receptors to return to activatable states (the current should return to a value similar to baseline current prior to GABA application), and is longer after long exposures to high GABA concentrations (13). To facilitate normalization of responses at different GABA concentrations, we measure a standard response every second or third sweep (see Note 12).

2. *Spontaneous channel activation*. To assess whether spontaneously active $GABA_A$ receptors are present in the absence of agonists, we apply picrotoxin (see Note 13), a potent inhibitor, and determine whether the "leak" current observed during superfusion with ND96 alone can be reduced. Data from Chang and Weiss (14) suggest that, in comparison to wild-type, spontaneously active mutant $GABA_A$ receptors may be less sensitive to picrotoxin. We therefore use a high (2 mM) concentration of picrotoxin. The picrotoxin current (I_{PTX}) must be normalized to maximal GABA responses (I_{max}^{GABA}) measured before and after the picrotoxin experiment. It is important to note that spontaneously active channels are often extremely sensitive to even nanomolar concentrations of GABA, so we make sure that the tubing and valves used for picrotoxin experiments are entirely GABA-free. In some cases, we have used "virgin" tubing and valves and manually apply picrotoxin, in order to avoid GABA contamination in these experiments.

   Wild-type $GABA_A$ receptors have no detectable spontaneous activation (see Note 14). In mutants where no spontaneous activity is detectable using picrotoxin, we have used known gating mutations and mutant-cycle analysis to estimate the mutant effect on spontaneous gating (15).

3. *Maximal GABA efficacy.* We use positive allosteric modulators to detect whether the maximal current response to GABA alone can be enhanced. In wild-type receptors, we have used 3 μM etomidate or 2 μM alphaxalone as allosteric enhancers. In channels with mutations that create etomidate insensitivity (such as α1M236W and β2M286W), we have found that alphaxalone retains strong enhancing activity (see Note 15). The experiment is performed as a single sweep with three superfusate solutions: a baseline period in ND96, a period in maximally activating GABA until desensitization begins, and then a period in maximal GABA plus enhancing compound. If the open probability at maximal GABA (I_{max}^{GABA}) is

significantly less than 1.0, addition of enhancer (i.e., $I^{GABA}{}_{+Alphax}$) will open channels that are closed.

4. *Etomidate direct activation.* Etomidate concentration-response data is obtained similarly to GABA concentration responses. Our etomidate (FW = 244 g/mol) stock is the clinical preparation at 2 mg/ml (8.2 mM) in 35 vol% propylene glycol. We use concentrations up to 1 mM in experiments (see Note 16) and have shown that the resulting propylene glycol concentration does not affect $GABA_A$ receptors. Etomidate is extremely hydrophobic and binds nonspecifically to lipids and proteins in the oocyte cytoplasm (see Note 17). As a result, washout of etomidate can be extremely slow, requiring wash periods of up to 15 min between exposures (longer with higher etomidate concentrations). As with other experiments, normalization sweeps establishing maximal GABA responses are intermittently performed along with the etomidate responses.
5. *Etomidate modulation of GABA responses (left-shift).* To measure the GABA-enhancing activity of etomidate, we use left-shift analysis. This simply compares the ratios of GABA EC_{50} in the absence of etomidate to GABA EC_{50}s when etomidate is present. At least one etomidate concentration should be assessed this way. We chose a low concentration (3.2 μM) that is clinically relevant and produces large shifts in wild-type GABA sensitivity. Because control GABA EC_{50} can vary significantly from oocyte to oocyte, we prefer to measure left-shift by performing both control and etomidate-shifted GABA concentration responses in individual oocytes, generating an EC_{50} ratio for each cell studied. Control GABA concentration responses are measured as described above (Subheading 3.3, step 1). We then repeat the GABA concentration response, using three-solution sweeps where oocytes are preexposed to 3.2 μM etomidate in ND96 for 30 s, then to solutions that contain both variable GABA and 3.2 μM etomidate (see Note 18). Normalization sweeps establishing maximal GABA responses are intermittently performed along with the GABA concentration responses.

3.4. Data Analyses

We have evaluated our data using a two-state equilibrium Monod-Wyman-Changeux (MWC) coagonist model. Although this mechanism oversimplifies the number of states accessible to $GABA_A$ receptors (for instance, we ignore desensitized states), it provides a remarkably good quantitative fit as well as a useful framework for interpretation of results (7, 11, 15). This model assumes that an equilibrium between inactive and active protein conformations is present in the absence of agonists or modulating ligands. This lower boundary of receptor activity is estimated using experiments described in Subheading 3.3, step 2. The theoretical upper bound of activation (100%) is assumed to be approximated when maximal GABA is enhanced with the positive allosteric modulator, alphaxalone. The efficacies of GABA (Subheading 3.3, step 3), etomidate (Subheading 3.3, step 4) and the combination of GABA plus etomidate can be estimated by comparison with the maximal GABA + alphaxalone response.

1. *Leak correction.* We use Clampfit 9.0 for leak correction (see Note 19). The software has a function that allows identification of a baseline region between two cursors, and then corrects the entire sweep by subtracting the average current between the two cursors. A third cursor is positioned at the peak response, which is recorded along with the activating ligand concentrations.
2. *Agonist concentration-response analysis.* For either GABA or etomidate activation, descriptive analysis is performed using nonlinear least squares regression fits to logistic functions (Eq. 1):

$$\frac{I}{I_{\max}^{\mathrm{GABA}}} = \mathrm{Max} \times \frac{[\mathrm{Agonist}]^{nH}}{[\mathrm{Agonist}]^{nH} + \mathrm{EC}_{50}{}^{nH}} \quad (1)$$

Normalized concentration-response data is pooled into a Prism worksheet (log agonist concentration vs. normalized responses, one column per oocyte). The data are plotted in a semi-log plot with mean ± sem points. Nonlinear least squares regression fitting to a four parameter logistic function [log(agonist) vs. response with variable slope] will result in the following fitted parameters and errors (SEM): Minimum (bottom), Maximum (top), Hill slope (nH), and log(EC_{50}). Because the data are leak-corrected, it is appropriate to constrain the minimum to be zero in the fit, which will reduce the errors in other fitted parameters. For left-shift analyses, data for both GABA concentration responses can be combined in the Prism worksheet (as A and B data sets) and simultaneously fitted. This results in fitted parameters for both data sets as well as a statistical analysis determining the probability that the results are identical (the null hypothesis). The fits can also be constrained so that only the EC_{50} values are compared (see Note 20).
3. *Spontaneous activation and maximum GABA efficacy.* Normalized values for $I_{\mathrm{PTX}}/I_{\max}^{\mathrm{GABA}}$ and $I^{\mathrm{GABA+alphax}}/I_{\max}^{\mathrm{GABA}}$ are simply pooled and the mean ± sem calculated. $I_{\mathrm{PTX}}/I_{\max}^{\mathrm{GABA}}$ is an initial estimate of spontaneous open probability, and assuming that maximal GABA plus alphaxalone open 100% of receptors, $I^{\mathrm{GABA+alphax}}/I_{\max}^{GABA}$ is the inverse of GABA efficacy.

4. *Data renormalization for MWC coagonist model fitting.* Whereas descriptive fitting is performed on leak-corrected data that is normalized to maximal GABA responses, the MWC coagonist model is fit to P_{open} values (Fig. 2). Thus, if receptors are spontaneously active ($P_0 > 0$), this basal activity must be added as a nonzero baseline. Furthermore, since GABA may activate less than 100% of receptors, responses must be renormalized so that $I^{\text{GABA+alphax}}$ represents the maximum open probability. To do this, normalized data are transformed into estimated P_{open} values using Eq. 2:

$$P_{\text{open}}^{\text{est}} = \frac{I + I_{\text{PTX}}}{I^{\text{GABA+}Alphax} + I_{\text{PTX}}} = \frac{\frac{I}{I_{\text{max}}^{\text{GABA}}} + \frac{I_{\text{PTX}}}{I_{\text{max}}^{\text{GABA}}}}{\frac{I^{\text{GABA+}Alphax}}{I_{\text{max}}^{\text{GABA}}} + \frac{I_{\text{PTX}}}{I_{\text{max}}^{\text{GABA}}}} \tag{2}$$

We transfer average normalized response data from GABA concentration responses (with and without etomidate) and etomidate concentration response to an Excel spreadsheet, organizing the data into three columns: GABA concentration;

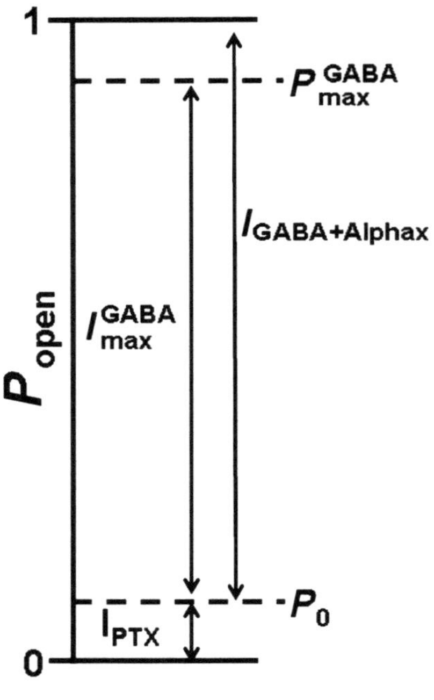

Fig. 2. Diagram illustrating corrections for spontaneous gating and maximal GABA efficacy. The diagram depicts an absolute scale of channel open probability from 0 to 1. In mutant GABA_A receptors with spontaneous activation, P_0 is greater than 1 and the current carried by open channels can be blocked with picrotoxin (I_{PTX}). In mutant GABA_A receptors where maximal GABA efficacy is significantly less than 1, closed GABA-bound channels can be opened using allosteric enhancers such as alphaxalone. Rescaling leak-corrected, normalized data to the full P_{open} range is necessary before fitting with the allosteric mechanism.

Etomidate concentration; Normalized Response. We input the mean values for I_{PTX}/I_{max}^{GABA} and $I^{GABA+alphax}/I_{max}^{GABA}$ into two cells and then calculate P_{open}^{est} in a fourth column from these values and the normalized responses.

5. *Fitting MWC coagonist parameters.* The following procedure describes using Origin 6.1 for this nonlinear least-squares fit. Prism 5.0, which supports global fitting procedures to multiple data sets, can also be used for this fit once the new equation is added to its nonlinear least-squares repertoire.

Data for GABA concentrations, etomidate concentrations, and P_{open}^{est} are transferred or imported from an Excel spreadsheet into a three-column Origin Worksheet with the following headers: GABA, ETO, and Popen. GABA concentration-response data can be plotted on a semi-log plot of GABA vs. Popen. The nonlinear least squares fitting module is opened and a new user-defined equation is created to enable fits to the MWC coagonist model (Eq. 3):

$$P_{open} = \frac{1}{1 + L_0\left(\frac{1+[GABA]/K_G}{1+[GABA]/cK_G}\right)^2\left(\frac{1+[ETO]/K_E}{1+[ETO]/dK_E}\right)^2} \quad (3)$$

In the function editor, we named our function "MWC_coagonist" and added it to the "Growth/Sigmoidal" group of functions that also contains Logistic equations. The Parameter names are L0, Kg, c, Ke, d, and n. Independent variables are GABA and ETO. Dependent variable is Popen. The equation in Y-script is defined as the following:

$$\text{Popen} = 1/(1 + \text{L0} * (((1 + \text{gaba/Kg})/(1 + \text{gaba/Kg/c})) \wedge 2) * (((1 + \text{eto/Ke})/(1 + \text{eto/Ke/d})) \wedge \text{n})); \quad (4)$$

Note that Eq. 4 has one more free parameter than Eq. 3: the parameter n represents the number of equivalent etomidate sites in the model. We constrain this value to 2 when fitting data for etomidate, but we can also use this equation for other allosteric coagonists with different numbers of sites (16). L_0 in Eq. 3 is a dimensionless basal equilibrium gating variable, approximately ${P_0}^{-1}$. K_G and K_E are equilibrium dissociation constants for GABA and etomidate binding to inactive states, and c and d are dimensionless allosteric efficacy parameters representing the respective ratios of binding constants in active vs. inactive states.

To perform the nonlinear least-squares fit in Origin, open the nonlinear least squares fitting module. Select the user-defined "MWC_coagonist" function. Click on the "Select Dataset" button and assign Popen, gaba, and eto data columns and ranges. Under "More/Options/Constraints" provide lower bounds of 0 for KG, c, KE, and d. Under "Options/Control" set all the significant digits to 3, tolerance to 0.01, and the maximum number of iterations to 50. Click on "Basic Mode" and insert initial parameters, including the constraint $n = 2$. Then hit "Start Fitting." When the fit is satisfactory and you hit "Done," the fitted equation will be plotted overlaying the GABA concentration-response plot.

4. Notes

1. Quick-change uses only a single oligonucleotide to replicate circular plasmid DNA. An alternative method we frequently use for site-directed mutagenesis is overlap extension, using two mutagenic oligonucleotides, one for each strand, each paired with a flanking oligonucleotide that encompasses an endonuclease site that can be used to transfer the double-stranded mutated product into a double-cut (gapped) wild-type plasmid.
2. All solutions are prepared using sterile deionized water with a resistivity of at least 10 MΩ cm. Oocyte preparation and electrophysiology buffers are prepared as 20× stocks, stored at 4°C.
3. Our oocyte injection station is mounted on a steel plate, providing a solid base for the magnetic stand holding the micromanipulator. The Drummond injector was modified with a custom-built brass locking nut that is much longer than the plastic piece delivered with the injector. Without this modification, we found that injection pipette tips moved as the needle plunger was advanced, perhaps because the rotational movement that drives the plunger was translated to the tip. The extra-long locking nut prevents this movement.
4. A number of commercial superfusion and digitizer/data acquisition systems are available to build a two-microelectrode voltage clamp setup for oocyte electrophysiology. Here, we describe a setup that incorporates mostly commercially available equipment and a flow cell that is easily built with access to a small milling machine. With a tight budget and access to a workshop, one can build an inexpensive multireservoir superfusion system with manual selection valves and one or two electrically actuated "T" valves downstream. Low-cost digitizer equipment and software to both capture data and control experiments can also be obtained from various sources, (e.g., National Instruments Corp.).

5. While we fabricate our microelectrodes, prefabricated microelectrodes may be purchased from World Precision Instruments.
6. Retracting ovary through a small abdominal incision requires continuous gentle traction back and forth so that small portions of the ovary emerge at a time. If the ovary is very scarred, there may be little option but to enlarge the abdominal opening further.
7. Try to avoid rupturing oocytes by using forceps to retract connective tissue only about 5 mm apart and pulling the two forceps in opposite directions.
8. For transfer of clumped oocytes, we use a disposable plastic transfer pipette and cut the end at a bevel to create a larger opening that can accommodate the clumps of oocytes.
9. Finding the right type (and batch) of collagenase can be a tedious business. Different batches of collagenase have different activities and adjustment in the total incubation time will likely be needed. Worthington Biochemical Corp has a sample program specifically designed for testing different lots of collagenase. When most, but not all of the oocytes appear to be free of connective tissue (i.e., they are no longer in clumps, but move individually), the collagenase treatment is complete. Overtreatment with collagenase can damage membrane proteins and reduce oocyte viability in culture. Undertreatment with collagenase may leave connective tissue that may require other methods of removal, such as rolling on plastic (while immersed in ND96) or manual dissection under a dissecting scope.
10. This is achieved by first positioning the tip of the pipette next to the bead and then translating the pipette sideways so that a small portion of the tip is snapped off. Alternatively, a pipette tip can be used to break the tip of the glass injector pipette while observing under high magnification on the dissecting microscope.
11. We prepare a 100 ml batch of 1 M GABA in water and then store 1 ml aliquots in labeled microcentrifuge tubes at −80°C. A fresh aliquot is removed each day and kept on ice to minimize the breakdown.
12. The number of active $GABA_A$ receptors on the oocyte surface can change over time. We usually use a maximally activating GABA concentration for our normalization. Another option is to use a lower nondesensitizing agonist concentration for normalization to shorten the wash-time required after each normalization sweep. A second normalization step is then required to convert responses to fractions of maximal.

13. Picrotoxin (FW ≈ 600 g/M) is light sensitive and must be stored at room temperature away from windows. It is slow to go into solution – we add 60 mg picrotoxin to 50 ml ND96 electrophysiology solution in a capped centrifuge tube, cover with aluminum foil, and place on a gently rocking surface for 30 min, or until fully dissolved.
14. We and others have estimated that the open probability of wild-type α1β2γ2L $GABA_A$ receptors is 0.0001 or less. The typical dynamic range for oocyte electrophysiology equipment is between 5 nA and 10 μA. Thus, in oocytes where maximal GABA currents are high, a quiet baseline enables detection of spontaneous activity levels of approximately 0.0005 × maximal GABA response.
15. We test the strength of enhancing activity by adding the enhancer to a low concentration of GABA, such as EC_{10} from GABA concentration responses. Addition of enhancer should produce at least a threefold (hopefully higher) increase in the GABA EC_{10} current response. Note also that alphaxalone is very hydrophobic and should be stored in glass.
16. Etomidate concentrations over 1 mM inhibit $GABA_A$ receptors.
17. For this reason, we also use glass containers for etomidate and alphaxalone solutions. These hydrophobic drugs will adsorb into plastic.
18. A simple switch to etomidate plus GABA will likely underestimate the etomidate enhancement, as etomidate at low concentrations equilibrates slowly with the oocyte membrane.
19. An attractive feature of the pClamp/Clampfit suite (Molecular Devices) is that both programs can be open during data acquisition. As soon as a sweep is completed and saved, the resulting data file is automatically opened in Clampfit for analysis while the oocyte is undergoing its ND96 wash for several minutes.
20. Prism also has other useful fitting equations for these analyses including an "Allosteric EC50 shift" equation.

References

1. Jurd R, Arras M, Lambert S, Drexler B, Siegwart R, Crestani F, Zaugg M, Vogt KE, Ledermann B, Antkowiak B and Rudolph U (2003) General anesthetic actions in vivo strongly attenuated by a point mutation in the GABA(A) receptor beta3 subunit. *FASEB J* 17:250–252.
2. Reynolds DS, Rosahl TW, Cirone J, O'Meara GF, Haythornthwaite A, Newman RJ, Myers J, Sur C, Howell O, Rutter AR, Atack J, Macaulay AJ, Hadingham KL, Hutson PH, Belelli D, Lambert JJ, Dawson GR, McKernan R, Whiting PJ and Wafford KA (2003) Sedation and anesthesia mediated by distinct GABA(A) receptor isoforms. *J Neurosci* 23:8608–8617.
3. Hill-Venning C, Belelli D, Peters JA and Lambert JJ (1997) Subunit-dependent interaction of the general anaesthetic etomidate with the gamma-aminobutyric acid type A receptor. *Br J Pharmacol* 120:749–756.

4. Yang J and Uchida I (1996) Mechanisms of etomidate potentiation of GABAA receptor-gated currents in cultured postnatal hippocampal neurons. *Neuroscience* 73:69–78.
5. Zhong H, Rusch D and Forman SA (2008) Photo-activated azi-etomidate, a general anesthetic photolabel, irreversibly enhances gating and desensitization of gamma-aminobutyric acid type A receptors. *Anesthesiology* 108:103–112.
6. Chang Y, Wang R, Barot S and Weiss DS (1996) Stoichiometry of a recombinant GABAA receptor. *J Neurosci* 16:5415–5424.
7. Rüsch D, Zhong H and Forman SA (2004) Gating allosterism at a single class of etomidate sites on alpha1beta2gamma2L GABA-A receptors accounts for both direct activation and agonist modulation. *J Biol Chem* 279:20982–20992.
8. Husain SS, Ziebell MR, Ruesch D, Hong F, Arevalo E, Kosterlitz JA, Olsen RW, Forman SA, Cohen JB and Miller KW (2003) 2-(3-Methyl-3H-diaziren-3-yl)ethyl 1-(1-phenylethyl)-1H-imidazole-5-carboxylate: A derivative of the stereoselective general anesthetic etomidate for photolabeling ligand-gated ion channels. *Journal of medicinal chemistry* 46:1257–1265.
9. Li GD, Chiara DC, Sawyer GW, Husain SS, Olsen RW and Cohen JB (2006) Identification of a $GABA_A$ receptor anesthetic binding site at subunit interfaces by photolabeling with an etomidate analog. *J Neurosci* 26:11599–11605.
10. Bali M, Jansen M and Akabas MH (2009) GABA-induced intersubunit conformational movement in the GABAA receptor alpha1M1-beta2M3 transmembrane subunit interface: Experimental basis for homology modeling of an intravenous anesthetic binding site. *J Neurosci* 29:3083–3092.
11. Stewart DS, Desai R, Cheng Q, Liu A and Forman SA (2008) Tryptophan mutations at azi-etomidate photo-incorporation sites on α1 or β2 subunits enhance GABAA receptor gating and reduce etomidate modulation *Mol Pharmacol* 74:1687–1695.
12. Boileau AJ, Baur R, Sharkey LM, Sigel E and Czajkowski C (2002) The relative amount of cRNA coding for gamma2 subunits affects stimulation by benzodiazepines in GABA(A) receptors expressed in Xenopus oocytes. *Neuropharmacology* 43:695–700.
13. Chang Y, Ghansah E, Chen Y, Ye J and Weiss DS (2002) Desensitization mechanism of GABA receptors revealed by single oocyte binding and receptor function. 22:7982–7990.
14. Chang Y and Weiss DS (1999) Allosteric activation mechanism of the alpha1beta2gamma2 gamma-aminobutyric acid type A receptor revealed by mutation of the conserved M2 leucine. *Biophys J* 77:2542–2551.
15. Desai R, Ruesch D and Forman SA (2009) Gamma-Amino Butyric Acid Type A Receptor Mutations at beta2N265 Alter Etomidate Efficacy While Preserving Basal and Agonist-dependent Activity. *Anesthesiology* 111:774–784.
16. Rüsch D and Forman SA (2005) Classic benzodiazepines modulate the open-close equilibrium in alpha1beta2gamma2L gamma-aminobutyric acid type A receptors. *Anesthesiology* 102:783–792.

Chapter 18

Allosteric Regulation of Human Liver Pyruvate Kinase by Peptides that Mimic the Phosphorylated/Dephosphorylated N-Terminus

Charulata B. Prasannan, Qingling Tang, and Aron W. Fenton

Abstract

An advantage of studying allosteric regulation over covalent modification is that allostery allows the experimentalist to vary the concentration of effector, thereby allowing independent quantification of effector binding and allosteric coupling. In turn, this capacity allows the use of effector analogues to determine which regions of the effector contribute to effector binding and which contribute to allosteric regulation. Like many other proteins, human liver pyruvate kinase (hL-PYK) is regulated by phosphorylation. The phosphorylation of hL-PYK occurs on Ser12 of the N-terminus. Phosphorylation appears to interrupt an interaction (distant from the active site) between the N-terminus and the main body of the protein. Since this interaction increases the affinity of hL-PYK for the substrate (phosphoenolpyruvate, PEP), phosphorylation-dependent interruption of the N-terminus/main-body interaction results in an antagonism of PEP binding. Due to the advantages of studying an allosteric system, we detail a protocol to express and purify N-terminal peptides of hL-PYK using a SUMO-fusion system. We further demonstrate that these peptides act as allosteric regulators that modulate the affinity of hL-PYK for PEP.

Key words: Allostery, Pyruvate kinase, hL-PYK, Linked equilibrium, N-terminus phosphorylation

1. Introduction

Although covalent modification and allosteric regulation of a protein often result in similar regulatory consequences (e.g. impact on protein affinity for substrate), there is an experimental advantage to the study of allosteric effectors. Our working definition of allosteric regulation is how one ligand (A) binds to a protein (E) in the presence versus absence of a second ligand (X) (1). This definition defined a thermodynamic energy cycle (Fig. 1). It also defines the allosteric coupling constant (Q_{ax}) as a ratio of binding constants:

$$Q_{ax} = \left(\frac{K_{ia}}{K_{ia/x}}\right) = \left(\frac{K_{ix}}{K_{ix/a}}\right), \tag{1}$$

Aron W. Fenton (ed.), *Allostery: Methods and Protocols*, Methods in Molecular Biology, vol. 796,
DOI 10.1007/978-1-61779-334-9_18,

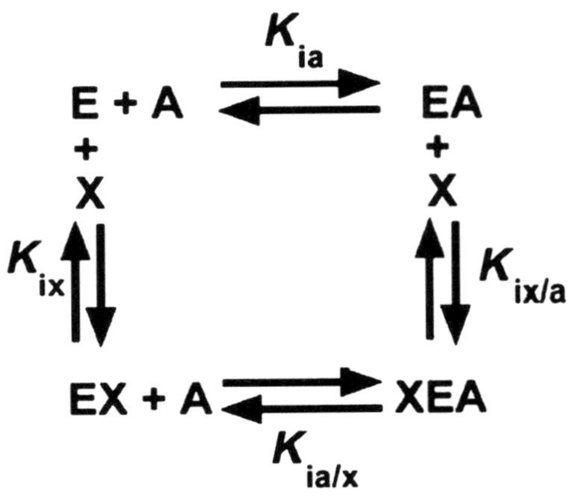

Fig. 1. Thermodynamic cycle defines the binding enzyme (E) to substrate (A) and allosteric effector (X).

where K_{ia} = the dissociation constant for the first ligand, A, binding to the protein in the absence of the second ligand, X; $K_{ia/x}$ = the dissociation constant for A binding to the protein with X pre-bound, K_{ix} = the dissociation constant for X binding to the protein in the absence of A; and $K_{ix/a}$ = the dissociation constant for molecule X binding to the protein with A pre-bound. Since Q_{ax} is a ratio, the magnitude of this allosteric coupling is independent of the magnitude of any one ligand dissociation constant. It follows that individual atom–atom interactions between atoms from protein residues and atoms from chemical moieties of the ligand may contribute uniquely to ligand binding versus allostery. Several studies have now utilized this knowledge in combination with an analogue series to detail which region(s) of the effector contributes to ligand binding and which region(s) contribute to determining the magnitude of the allosteric coupling, Q_{ax} (2–5). Although mutations are commonly used to mimic covalent modifications (e.g. phosphorylation (6–8)), such studies do not distinguish if (1) the modification alters the ability of the modified protein region to interact with other regions of the protein (i.e. an internal binding event between different regions of the protein), versus (2) the mutation has little effect on the internal binding event, but modifies the ability of this binding event to modify protein function (i.e. allosteric effect).

Although used under many different names, chemical fragmentation and fragment complementation have often been employed to identify functional roles for increasingly small chemical moieties. Such fragmentation studies include: (1) the use of substrate analogues to study enzymatic mechanisms; (2) the use of effector analogues (referenced above) to study allosteric effectors (2–5); (3) the use of "fragment-based" screening in conjunction with structural determination to identify ligand-binding sites on protein surfaces (9); (4) the addition of small molecules to rescue mutant side chain function (chemical rescue) (10); and (5) the complementation of protein fragments to recover protein activity (11–14). However, Carty et al. was the first to propose that a protein derived phosphopeptide could regulate glycogen phosphorylase in the absence of being covalently linked with the main body of the protein (15) (Note: The results supporting this early work have now been questioned (16).). Following this suggestion, the use of peptides (with and without mutations that mimic covalent modification) as allosteric effectors is the focus of the current work.

Like glycogen phosphorylase, human liver pyruvate kinase (hL-PYK) is regulated by phosphorylation. Pyruvate kinase isozymes, including hL-PYK, catalyze the final step in glycolysis:

$$\text{Phosphoenolpyruvate (PEP)} + \text{ADP} \rightleftarrows \text{Pyruvate} + \text{ATP}.$$

Regulation of hL-PYK maintains glucose homeostasis, preventing hyperglycemia and hypoglycemia. This regulation is accomplished both by the phosphorylation of the N-terminus at Ser 12 (17–19) and by allosteric regulation by fructose-1,6-bisphosphate (Fru-1,6-BP), Ala, and ATP (20). Introducing the S12D mutation mimics the impact of phosphorylation, decreasing PEP affinity to an extent comparable to the effect caused by phosphorylation (18). We have previously used a truncation series to outline which regions of the N-terminus are energetically coupled with PEP binding in the active site (18). Truncating residues 1–6 has little impact on PEP affinity. In contrast sequentially removing residues 7–10 decreases PEP affinity to an extent equivalent to that caused by phosphorylation. However, this effect is not apparent when S12D has already been introduced. Taken together, these observations are consistent with an activating interaction between the main-body (all parts of the protein except the N-terminus) and the unphosphorylated N-terminus. Phosphorylation may simply interrupt this activating interaction. Unfortunately, the first 26 residues of hL-PYK are not defined in the currently available crystal structure (21). Therefore, it remains unclear where or how the N-terminus of hL-PYK interacts with the main-body of the protein.

To further elucidate which regions of the N-terminus contribute to "binding" to the main body of the protein and which contribute to energetic coupling with the active site, we would like to use the advantages available via an allosteric system. This approach requires that the N-terminus be added as a peptide that is not covalently linked to the main body of the protein.

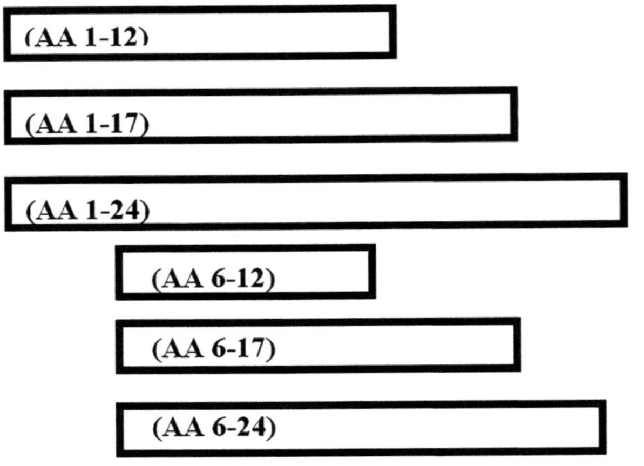

Fig. 2. A representation of the peptides tested in Figs. 5 and 6 as allosteric effectors of hL-PYK. Peptides are labeled by the residue positions of hL-PYK represented in the peptide.

Therefore, the ability of N-terminal peptides (Fig. 2) to allosterically modify the affinity of S12D-hL-PYK (hL-PYK with the S12D mutation) for PEP was tested. The magnitude of allosteric coupling, Q_{ax}, was measured by determining ligand affinity of the protein for PEP over a concentration range of the peptide effector (1, 22, 23). From the onset of our experimental design, we expected that high concentrations of peptides would be required for binding; binding between peptide and main-body was expected to be much weaker due to the removal of an entropic contribution (i.e. removal of covalent attachment). Therefore, this study required large quantities of peptides. This need for high peptide concentration (at high purity) challenged the cost-effectiveness of commercial peptide synthesis. Furthermore, our focus on regulation required a convenient purification tag that does not modify the regulatory properties of the peptide. Therefore, we report the use of an *Escherichia coli* SUMO-fusion system to express and purify peptides of interest (see Note 1). This expression system has two advantages; high yields and specific tag cleavage using SUMO-protease (24); the latter property allows the removal of the purification tag. A detailed method for the purification of the peptides and the use of these peptides as allosteric effectors is described in this chapter. Based on this proof-in-principle study demonstrating allosteric regulation by the N-terminal peptides, future work can use modified peptides in an effort to map which regions of the N-terminus of hL-PYK contribute to binding and which regions participate in allosteric coupling.

2. Materials

6×His-SUMO-tagged peptides were constructed for the ease of purification and tag removal. Other systems such as 6×His-tagged systems are also commonly used for the purification of proteins and peptides. However, in hL-PYK, we observed that the introduction of the 6×His-tag to the N-terminus of the protein changes the $K_{app\text{-}PEP}$ (see Note 2). Since this might indicate that the 6×His-tag modifies the interaction of the N-terminus with the main body of the protein (i.e. the property we wish to study), this tag was not used in further studies. Based on the potential for specific site cleavage for tag removal, we choose to use the SUMO expression system (Life Sensors) to express peptides.

2.1. Peptide Purification

1. Resuspension buffer: 25 mM HEPES/KOH pH 7.5, 100 mM NaCl, 10 mM imidazole, 10% glycerol, 5 mM β-mercaptoethanol, and 0.1% Triton-X.
2. Wash buffer: 25 mM HEPES/KOH pH 7.5, 100 mM NaCl, 10 mM imidazole, and 5 mM β-mercaptoethanol.
3. Elution buffer: 25 mM HEPES/KOH pH 7.5, 100 mM NaCl, 300 mM imidazole, and 5 mM β-mercaptoethanol.
4. Dialysis buffer: 25 mM HEPES/KOH pH 7.5, 100 mM NaCl, and 5 mM β-mercaptoethanol.

2.2. Enzymatic Assay

1. Assay buffer: A 10× assay buffer was prepared with 500 mM bicine, 100 mM $MgCl_2$, 1 mM EDTA. The pH of the buffer was adjusted to pH 7.5 using KOH and stored at 4°C.
2. 100 mM NADH (Sigma) stock solution was made and stored at −20°C.
3. L-Lactate dehydrogenase (LDH) (Type III bovine heart) from Calzyme laboratories, Inc., was dialyzed in 50 mM Bicine pH 7.5, 10 mM $MgCl_2$, 0.1 mM EDTA and 100 mM KCl buffer. The dialyzed LDH was stored at 4°C until use.
4. 400 mM stock of ADP(Na) was prepared and the pH of the solution was adjusted to pH 7.5 and stored at −20°C.
5. Potassium-PEP (Chem-IMPEX International) was used to prepare 480 mM stock solution, pH of the solution was adjusted to 7.5 using KOH and the stock solution of PEP was stored at −20°C. The PEP solution was diluted serially (1:2) in 147 mM KCl to obtain the working concentrations.
6. The effector (peptides prepared as described later) was serially (1:2) diluted in a buffer to maintain constant K^+ of 150 mM in the final reaction.

3. Methods

3.1. Construction and Expression of SUMO Peptides

1. The N-terminal peptides were created by first cloning the human liver pyruvate kinase (*l-pyk*) from pLC11 (20, 21) into the pE-SUMOpro Kan vector (Life sensors), a vector that contains the 6×His-SUMO gene immediately in front of the multiple cloning site. BbsI and SacI restriction sites were added at the 5′- and 3′-ends of the *l-pyk* gene, respectively, using PCR. PCR product cleaved with both BbsI and SacI was cloned into BsaI-SacI cleaved pE-SUMOpro Kan vector. The constructed plasmid containing the *l-pyk* gene was named pSUMO-LPYK.
2. Once cloned, a second construct (pSUMO-LPYK/Δ1–5) was created by deleting codons for the initial five residues of hL-PYK. This deletion was based on our observation that the initial five N-terminal amino acids do not impact PEP affinity (18). Both pSUMO-LPYK and pSUMO-LPYK/Δ1–5 were used in further steps.
3. Stop codons were introduced at different locations in the *l-pyk* coding sequence of pSUMO-LPYK and pSUMO-LPYK/Δ1–5 in order to obtain desired peptides. Termination sites introduced at position 13 were chosen based on the observation that residues 7–10 in the N-terminus are energetically coupled with PEP binding (18). Termination at position 18 was based on the hL-PYK residues removed from the protein used in crystallization studies (21). Termination at position 25 was chosen because residue position 26 was the first observable residue in the crystallized protein (21). Termination at position 36 removes the initial helix in the N-terminal helix-turn-helix observed in the crystal structure.
4. The SUMO-peptide gene constructs were transformed into BL21(DE3) *E. coli* cells (Stratagene) for expression.
5. Cultures were grown to an OD_{600} of 0.6 before inducing protein expression by the addition of 1 mM IPTG. After induction, cultures were allowed to grow for 4 h (see Note 3).

3.2. Purification of N-Terminal Peptides

1. Resuspend pellet in resuspension buffer.
2. Lyse the cells using sonication.
3. Centrifuge the protein sample at 9,681 × *g* (9,000 rpm in Sorvall SS-34 rotor) for 15 min.
4. Collect cell lysate and add 0.1% protamine sulfate. Mix for approximately 1 h at 4°C (see Note 4).
5. Centrifuge at 23,426 × *g* (14,000 rpm in Sorvall SS-34 rotor) for 10 min. Collect the supernatant. A clear solution

should be obtained. Repeat the procedure if required. Continue with the supernatant.

6. Nickel resin (Biorad) should be washed and pre-equilibrated with wash buffer. To promote protein binding to the Ni-resin, mix the supernatant from step 5 in Subheading 3.2 with 500 μL of Ni-resin for approximately 2 h at 4°C.
7. Pour the resin into a column using gravity flow. Collect the flow-through and store until the purification is complete. Wash the resin with wash buffer with a minimum of 20 column volumes or until the absorbance is less than 0.1.
8. Elute the protein with elution buffer (i.e. 300 mM imidazole). Add 1 mL of the elution buffer on the resin. Incubate for 5 min. Collect elutions (see Note 5).
9. Check the fractions on an SDS gel for purity (Fig. 3). Pool fractions with highest purity and dialyze in dialysis buffer to remove traces of imidazole (see Note 6).
10. Add 0.5 mL of 5 mg/mL 6×His-SUMO protease (24) and mix at 4°C (see Note 7). The SUMO protease used in this step has been modified to contain an N-terminal 6×His-tag. This modification aids in removing the protease from purified peptides.

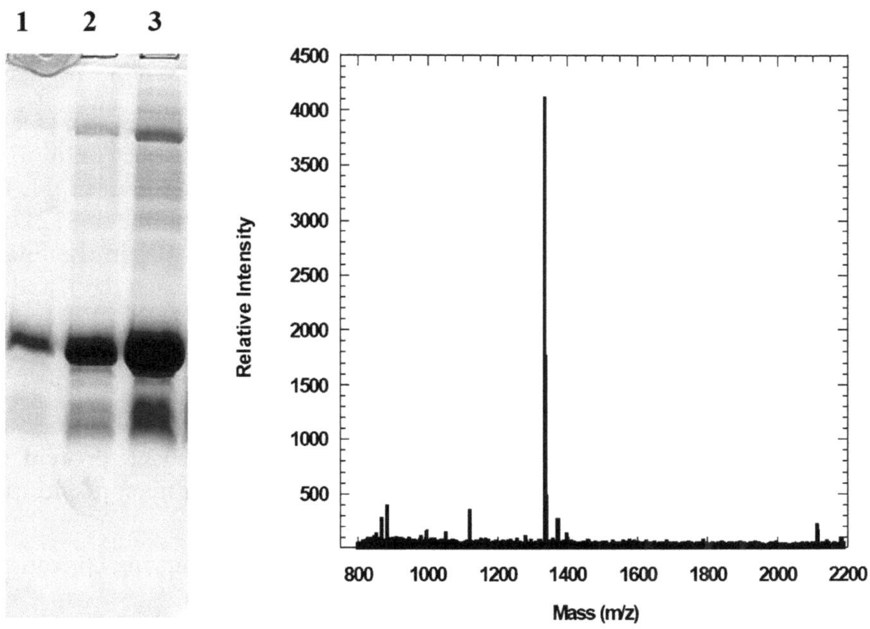

Fig. 3. The purity of the purified peptides analyzed using SDS-PAGE gel and MALDI. Panel (**a**): shows a gel with purified protein from cells grown with a 4-h induction time; *lanes 1–3* represent fractions 1–3 for the elution of His-Sumo-AA 6–17 from the initial Ni-column step. Panel (**b**): MALDI data for purified peptide AA 6–17.

11. To allow the cleaved 6×His SUMO-tag and the 6×His-SUMO protease to bind the Ni-resin, allow the protein sample to mix with the 500 μL Ni-resin (50% slurry provided by the manufacturer) and pour the resin on the column using gravity flow, collect the flow-through.
12. Cleaved peptides run through the column and the 6×His-SUMO-tag and 6×His-SUMO protease bind to the column. Wash the resin with a minimum volume of dialysis buffer.
13. Check protein fraction by SDS gel and MALDI analysis (Fig. 3).
14. Measure the absorbance at 214 nm for the amide bonds and a peptide-specific calculated extinction coefficient (25) to measure the concentration of the purified peptide.

A schematic of the peptide purification steps described in this section is shown in Fig. 4.

3.3. Assay to Test the Peptides as Effectors

1. Expression of hL-PYK in FF50 *E. coli* and the purification of hL-PYK from this expression system have previously been described (20). Our hypothesis of hL-PYK regulation is that phosphorylation interrupts an activating interaction between the non-phosphorylated N-terminus and the main body of the protein (18). Therefore, all tests for interaction of isolated peptides reported in the current study were with the S12D-hL-PYK mutant protein. This mutation mimics phosphorylation and is expected to prevent competition between binding of the isolated peptide and binding of the covalently attached N-terminus.
2. For the enzymatic assays, a cocktail was prepared (see Note 8) from the stock solutions as described above (3.96 mL of 10× assay buffer, 87 μl of 100 mM NADH, 8 μl of LDH, 198 μl of 400 mM ADP, 6 mg of fresh DTT, a volume of S12D hL-PYK that results in a ΔA_{340}/min equal to ~0.1 in the final in-well assay (determined at the beginning of each day), and distilled water to a final volume of 22 mL). Both PEP and effector solutions were serially diluted to obtain the required concentrations.
3. Figure 5 shows that all designed N-terminal peptides increase the apparent affinity of the S12D-hL-PYK protein for PEP. Initial velocities as a function of substrate concentration were fit as described earlier (23).
4. The $K_{\text{app-PEP}}$ values are plotted with varying concentration of the effector and fitted to the equation,

$$K_{\text{app}-\text{PEP}} = K_{\text{a}}\left(\frac{K_{\text{ix}} + [\text{Effector}]}{K_{\text{ix}} + Q_{\text{ax}}[\text{Effector}]}\right)$$

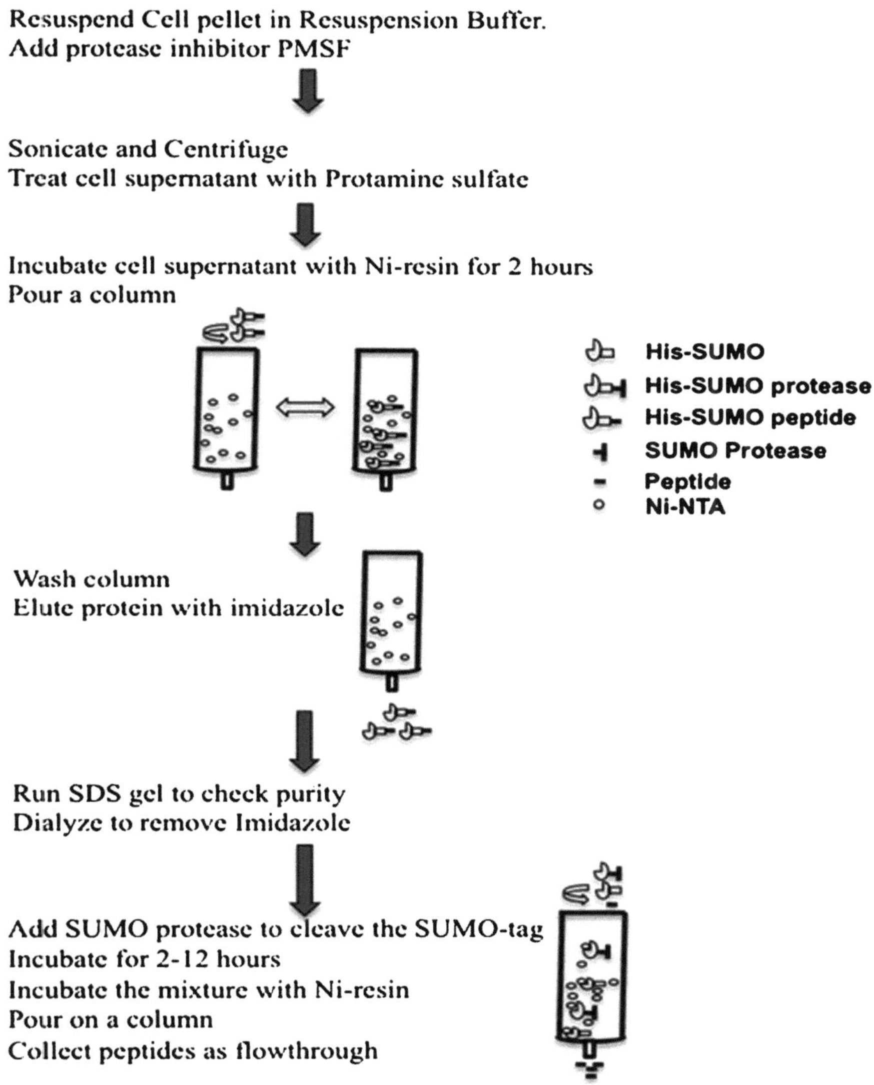

Fig. 4. Flow chart representing the peptide purification steps.

where $K_a = K_{app\text{-}PEP}$ in the absence of effector, K_{ix} is the dissociation constant for effector (X) in the absence of substrate, and Q_{ax} is the coupling constant as introduced in the earlier section (1, 22, 26, 27).

All the N-terminal peptides show an allosteric regulation when added as effectors to the S12D-hL-PYK protein (Fig. 5). S12D mutation mimics the phosphorylated protein, both the mutation and the covalent modification reduce affinity for the substrate PEP. The increased PEP apparent affinity of S12D-hL-PYK upon the addition of the N-terminal peptides emphasizes the energetic coupling between the N-terminus of the protein and

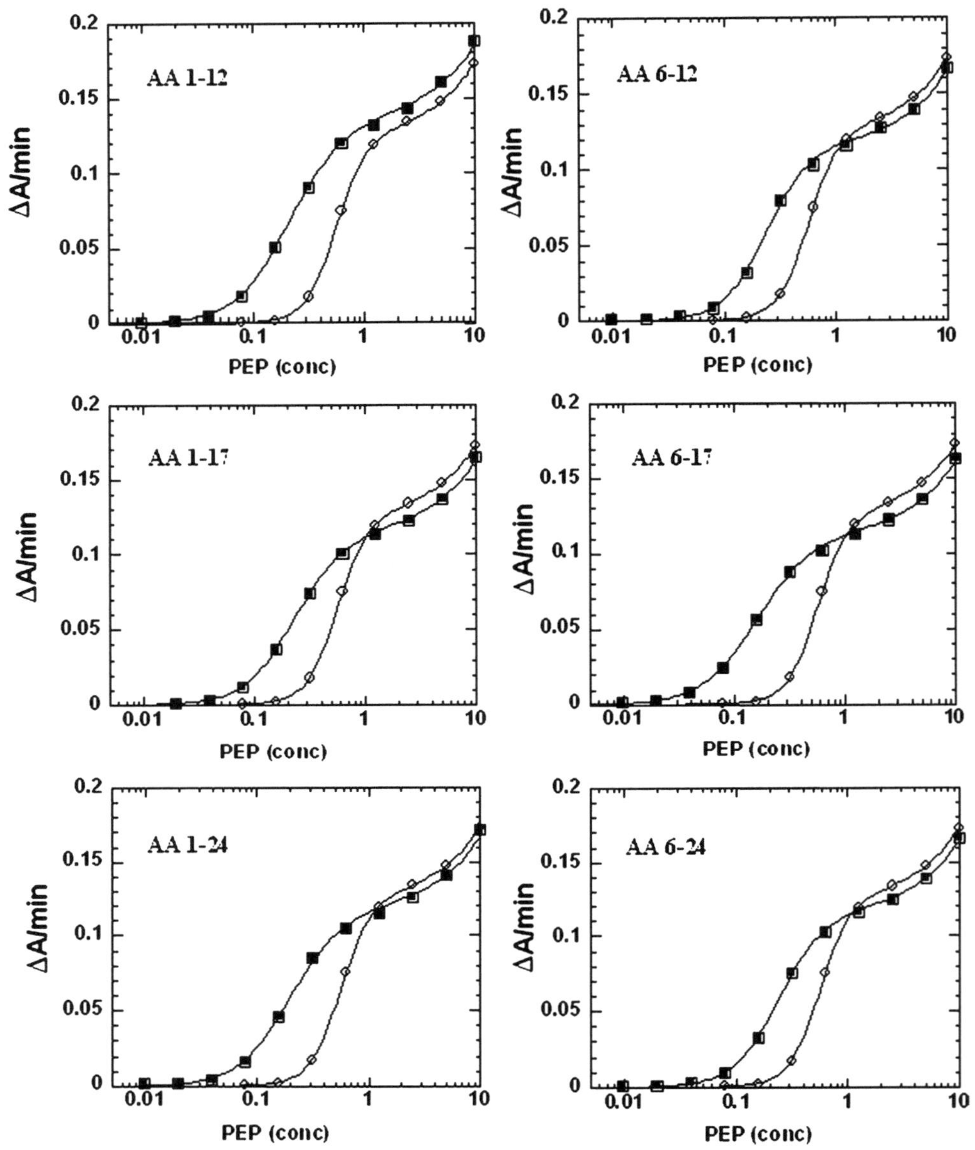

Fig. 5. An initial screen for the effect of N-terminal peptides on S12D-hL-PYK. In each panel, *open circles* represent the S12D protein with no effector. *Filled squares* represent the data for PEP binding for S12D protein in the presence of peptide as labeled, AA 1–12, AA 6–12, AA 1–17, AA 1–24, AA 6–24, and AA 6–17. All the peptide used in this screen increases the PEP affinity of the S12D-hL-PYK protein.

PEP affinity. In addition to the initial screening at a single concentration of peptide (Fig. 5), a quantitative analysis of peptide affinity and allosteric coupling between peptide and substrate binding can be considered. Figure 6 exemplifies this quantitative approach for the AA 6–17 peptide. With increasing concentration of peptide, the PEP affinity of the S12D protein approaches that for the wild type protein (23). Restated, the loss

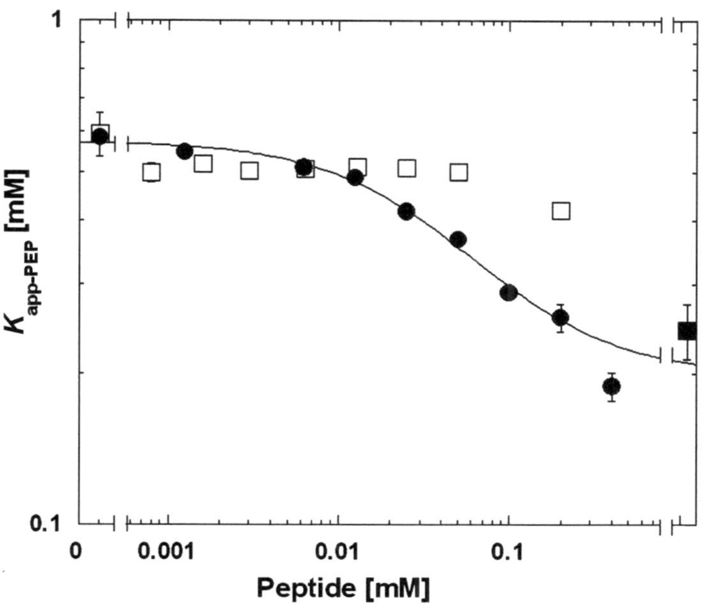

Fig. 6. $K_{app\text{-}PEP}$ versus peptide concentration to monitor the effect of N-terminal peptide on the affinity of PEP. *Filled circles* are for the AA6–17 peptide. *Open squares* are for the AA6–17-S12D control peptide. The wild type $K_{app\text{-}PEP}$ (*solid square*) data point (in the absence of effector) was included in the fit (representing the "infinite concentration") for the response to the AA 6–17 peptide.

of affinity due to the phosphorylation mimic is rescued by the addition of a non-covalent N-terminal peptides. Furthermore, this is not a non-specific effect due to high concentrations of peptide; although only initial data have been collected for the impact of the AA 6–17 peptide containing the S12D mutation (Fig. 6), it is clear that if this peptide elicits any impact on the affinity of the protein for PEP, it does so only at much higher peptide concentrations.

Using the quantitative approach exemplified in Fig. 6, the utility of creating peptide that mimic covalent modification can be appreciated. Consider that we now have a working hypothesis that phosphorylation interrupts an activating interaction between the N-terminus and the main body of the hL-PYK protein (18). Furthermore, residues 7–10 are required for regulation by the N-terminus. In the full length protein it might be possible to mutate residues 7–10 to determine which replacement residues and at which residue positions modify PEP affinity. However, this approach would not distinguish whether a mutation alters PEP affinity by altering the binding between the N-terminus and the main body of the protein, or by altering the energetic coupling (Q_{ax}). As presented in Fig. 6, Q_{ax} is the distance between the horizontal plateau at low peptide concentration and the horizontal plateau approached at high peptide concentration (1, 22, 26, 27). In contrast to making mutations in the full length protein, mutating residues 7–10 in the peptide AA 6–17 will allow a systematic partitioning of how each of these residues contribute to N-terminal/main body binding and to allosteric coupling between N-terminal and PEP binding. This general strategy should find broad application in the study of mechanisms by which proteins are regulated by covalent modification.

4. Notes

1. There are other systems for the expression and purification of peptides in bacterial system (28, 29).
2. The addition of 6×His-tag modified the affinity for PEP when tested using a single pass assay. Table 1 shows various designs

Table 1
Different construct for hL-PYK-N-terminal peptides with the 6×His tag, compared to the wild type N-terminal sequence[a]

	Sequence
1	MDHHHHHHGGGG M E G P A G Y L R R A S V A Q L T Q...
2	MDHHHHHHGG M E G P A G Y L R R A S V A Q L T Q...
3	MDHHHHHH M E G P A G Y L R R A S V A Q L T Q...
4	MHHHHHH M E G P A G Y L R R A S V A Q L T Q...
Wild type	**M E G P A G Y L R R A S V A Q L T Q**...
5	MDHHHHHH G G P A G Y L R R A S V A Q L T Q...
6	MDHHHHHH G P A G Y L R R A S V A Q L T Q...
7	MDHHHHHH P A G Y L R R A S V A Q L T Q...
8	MDHHHHHH A G Y L R R A S V A Q L T Q...
9	MDHHHHHH G G Y L R R A S V A Q L T Q...
10	MDHHHHHH G Y L R R A S V A Q L T Q...
11	MDHHHHHH Y L R R A S V A Q L T Q...

[a]Each of these modifications slightly perturbed the $K_{app\text{-}PEP}$ as compared to that of the wild type protein. Based on a single assay, the smallest perturbation was the #3 construct that shifted the $K_{app\text{-}PEP}$ to 0.34 mM. However, regulation by phosphorylation only shifts $K_{app\text{-}PEP}$ from 0.2 to 0.6 mM. Therefore, the impact of adding the 6xHis-tag caused a change that was on the order of half that caused by phosphorylation. Proteins with these labels were not considered further

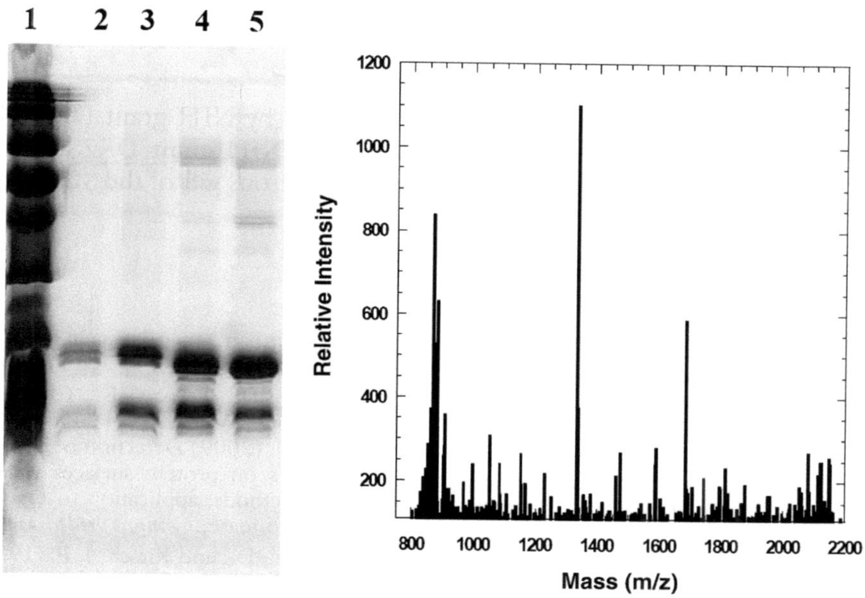

Fig. 7. A longer time of induction shows degradation of peptides. Panel (**a**) shows a gel with purified protein from cells grown with 24-h induction. *Lane 1* is the high molecular weight protein marker. *Lanes 2–5* represent elutions of His-Sumo-AA 1–12, His-Sumo-AA 1–17, His-Sumo-AA 1–24, and His-Sumo-AA 1–35, respectively. Panel (**b**): MALDI data for the AA1–25 with 24 h induction show degradation of peptide.

which were attempted. The sequence in bold represents the WT sequence for the hL-PYK gene.

3. Longer growth times after induction caused degradation of the peptides (Fig. 7).
4. Protamine sulfate is used to precipitate the nucleic acids. It is important to obtain a clear solution after the centrifugation step.
5. Additional purity of the sample/elutions can be obtained by using an imidazole gradient.
6. Sample dialysis removes traces of imidazole so that when subjected to the Ni-column for the second time, the cleaved 6×His-SUMO protein and the 6×His-SUMO protease will bind to the column. In the second Ni-column step, all proteins except the cleaved peptides should bind to the Ni-resin. The desired peptide will be in the flow-through after the second Ni-column (Fig. 4).
7. At this step, we have left the samples stored at 4°C overnight with no adverse effects.
8. Preparation of a reaction cocktail reduces the number of additions to a well in the 96-well plate in an effort to minimize pipetting error. In the current study, only three additions (cocktail including hL-PYK, peptide effector, and PEP) were added to each well.

Acknowledgments

This work was supported in part by NIH grant DK78076. This work was supported in part by NIH grant DK78076. We also thank Dr. Chris Lima for his generous gift of the SUMO protease clone.

References

1. Fenton, A. W. (2008) Allostery: an illustrated definition for the 'second secret of life', *Trends Biochem Sci 33*, 420–425.
2. Williams, R., Holyoak, T., McDonald, G., Gui, C., and Fenton, A. W. (2006) Differentiating a Ligand's Chemical Requirements for Allosteric Interactions from Those for Protein Binding. Phenylalanine Inhibition of Pyruvate Kinase, *Biochemistry 45*, 5421–5429.
3. Frieden, C. (1959) Glutamic dehydrogenase. II. The effect of various nucleotides on the association-dissociation and kinetic properties, *J Biol Chem 234*, 815–820.
4. Cheng, A., Fitzgerald, T. J., Bhatnagar, D., Roskoski, R., Jr., and Carlson, G. M. (1988) Allosteric nucleotide specificity of phosphorylase kinase: correlation of binding, conformational transitions, and activation. Utilization of lin-benzo-ADP to measure the binding of other nucleoside diphosphates, including the phosphorothioates of ADP, *J Biol Chem 263*, 5534–5542.
5. Brown, P. H., and Beckett, D. (2005) Use of binding enthalpy to drive an allosteric transition, *Biochemistry 44*, 3112–3121.
6. Reddy, Y. V., Ding, Q., Lees-Miller, S. P., Meek, K., and Ramsden, D. A. (2004) Non-homologous end joining requires that the DNA-PK complex undergo an autophosphorylation-dependent rearrangement at DNA ends, *J Biol Chem 279*, 39408–39413.
7. Mayo, L. D., Seo, Y. R., Jackson, M. W., Smith, M. L., Rivera Guzman, J., Korgaonkar, C. K., and Donner, D. B. (2005) Phosphorylation of human p53 at serine 46 determines promoter selection and whether apoptosis is attenuated or amplified, *J Biol Chem 280*, 25953–25959.
8. Yoo, D., Fang, L., Mason, A., Kim, B. Y., and Welling, P. A. (2005) A phosphorylation-dependent export structure in ROMK (Kir 1.1) channel overrides an endoplasmic reticulum localization signal, *J Biol Chem 280*, 35281–35289.
9. Landon, M. R., Lieberman, R. L., Hoang, Q. Q., Ju, S., Caaveiro, J. M., Orwig, S. D., Kozakov, D., Brenke, R., Chuang, G. Y., Beglov, D., Vajda, S., Petsko, G. A., and Ringe, D. (2009) Detection of ligand binding hot spots on protein surfaces via fragment-based methods: application to DJ-1 and glucocerebrosidase, *Comput Aided Mol Des.*
10. Toney, M. D., and Kirsch, J. F. (1989) Direct Bronsted analysis of the restoration of activity to a mutant enzyme by exogenous amines, *Science 243*, 1485–1488.
11. Gutte, B., Lin, M. C., Caldi, D. G., and Merrifield, R. B. (1972) Reactivation of des(119-, 120-, or 121–124) ribonuclease A by mixture with synthetic COOH-terminal peptides of varying lengths, *J Biol Chem 247*, 4763–4767.
12. Hofmann, K., Smithers, M. J., and Finn, F. M. (1966) Studies on polypeptides. XXXV. Synthesis of S-peptide 1–20 and its ability to activate S-protein, *J Am Chem Soc 88*, 4017–4019.
13. van Gent, D. C., Vink, C., Groeneger, A. A., and Plasterk, R. H. (1993) Complementation between HIV integrase proteins mutated in different domains, *Embo J 12*, 3261–3267.
14. Park, K., Yi, S. Y., Lee, C. S., Kim, K. E., Pai, H. S., Seol, D. W., Chung, B. H., and Kim, M. (2007) A split enhanced green fluorescent protein-based reporter in yeast two-hybrid system, *Protein J 26*, 107–116.
15. Carty, T. J., Tu, J. I., and Graves, D. J. (1975) Regulation of glycogen phosphorylase. Role of the peptide region surrounding the phosphoserine residue in determining enzyme properties, *J Biol Chem 250*, 4980–4985.
16. Bigley, A. N., and Reinhart, G. D. (2010) The N-terminus of glycogen phosphorylase b is not required for activation by adenosine 5′-monophosphate, *Biochemistry 49*, 4760–4765.
17. El-Maghrabi, M. R., Haston, W. S., Flockhart, D. A., Claus, T. H., and Pilkis, S. J. (1980) Studies on the phosphorylation and dephosphorylation of L-type pyruvate kinase by the catalytic subunit of cyclic AMP-

dependent protein kinase, *J Biol Chem 255*, 668–675.

18. Fenton, A. W., and Tang, Q. (2009) An activating interaction between the unphosphorylated n-terminus of human liver pyruvate kinase and the main body of the protein is interrupted by phosphorylation, *Biochemistry 48*, 3816–3818.
19. Engstrom, L. (1978) The regulation of liver pyruvate kinase by phosphorylation – dephosphorylation, *Curr Top Cell Regul 13*, 28–51.
20. Fenton, A. W., and Hutchinson, M. (2009) The pH dependence of the allosteric response of human liver pyruvate kinase to fructose-1,6-bisphosphate, ATP, and alanine, *Arch Biochem Biophys 484*, 16–23.
21. Valentini, G., Chiarelli, L. R., Fortin, R., Dolzan, M., Galizzi, A., Abraham, D. J., Wang, C., Bianchi, P., Zanella, A., and Mattevi, A. (2002) Structure and function of human erythrocyte pyruvate kinase. Molecular basis of nonspherocytic hemolytic anemia, *J Biol Chem 277*, 23807–23814.
22. Reinhart, G. D. (2004) Quantitative analysis and interpretation of allosteric behavior, *Methods Enzymol 380*, 187–203.
23. Fenton, A. W., and Alontaga, A. Y. (2009) Chapter 5: The Impact of Ions on Allosteric Functions in Human Liver Pyruvate Kinase, *Methods Enzymol 466*, 83–107.
24. Reverter, D., and Lima, C. D. (2009) Preparation of SUMO proteases and kinetic analysis using endogenous substrates, *Methods Mol Biol 497*, 225–239.
25. Kuipers, B. J., and Gruppen, H. (2007) Prediction of molar extinction coefficients of proteins and peptides using UV absorption of the constituent amino acids at 214 nm to enable quantitative reverse phase high-performance liquid chromatography-mass spectrometry analysis, *J Agric Food Chem 55*, 5445–5451.
26. Reinhart, G. D. (1983) The determination of thermodynamic allosteric parameters of an enzyme undergoing steady-state turnover, *Arch Biochem Biophys 224*, 389–401.
27. Reinhart, G. D. (1988) Linked-function origins of cooperativity in a symmetrical dimer, *Biophys Chem 30*, 159–172.
28. Skosyrev, V. S., Kulesskiy, E. A., Yakhnin, A. V., Temirov, Y. V., and Vinokurov, L. M. (2003) Expression of the recombinant antibacterial peptide sarcotoxin IA in Escherichia coli cells, *Protein Expr Purif 28*, 350–356.
29. Zorko, M., and Jerala, R. (2010) Production of Recombinant Antimicrobial Peptides in Bacteria, in *Methods in Molecular Biology* (Giuliani, A., and Rinaldi, A. C., Eds.).

Chapter 19

In Silico-Screening Approaches for Lead Generation: Identification of Novel Allosteric Modulators of Human-Erythrocyte Pyruvate Kinase

Ashutosh Tripathi and Martin K. Safo

Abstract

Identification of allosteric binding site modulators have gained increased attention lately for their potential to be developed as selective agents with a novel chemotype and targeting perhaps a new and unique binding site with probable fewer side effects. Erythrocyte pyruvate kinase (R-PK) is an important glycolytic enzyme that can be pharmacologically modulated through its allosteric effectors for the treatment of hemolytic anemia, sickle-cell anemia, hypoxia-related diseases, and other disorders arising from erythrocyte PK malfunction. An in-silico screening approach was applied to identify novel allosteric modulators of pyruvate kinase. A small-molecules database of the National Cancer Institute (NCI), was virtually screened based on structure/ligand-based pharmacophore. The virtual screening campaign led to the identification of several compounds with similar pharmacophoric features as fructose-1,6-bisphosphate (FBP), the natural allosteric activator of the kinase. The compounds were subsequently docked into the FBP-binding site using the programs FlexX and GOLD, and their interactions with the protein were analyzed with the energy-scoring function of HINT. Seven promising candidates were obtained from the NCI and subjected to kinetics analysis, which revealed both activators and inhibitors of the R-isozyme of PK (R-PK).

Key words: Allostery, Allosteric inhibitor, Active site, In silico screening, Docking, Pyruvate kinase

1. Introduction

Allosteric proteins regulate some of the most important biochemical pathways in organisms. The allosteric process involves the binding of an allosteric effector to a site on the protein other than the catalytic or active site with a concomitant change in active site function. There is a growing interest in understanding the structures, functions, and regulatory mechanisms of the allosteric binding sites and their potential as drug targets for a range of diseases. In more recent years, there have been increased efforts to identify

Aron W. Fenton (ed.), *Allostery: Methods and Protocols*, Methods in Molecular Biology, vol. 796,
DOI 10.1007/978-1-61779-334-9_19,

novel allosteric modulators that can manipulate the regulatory behavior of a target protein (1–5). This re-evaluation of the target protein in terms of its "druggability" will result in a new class of divergent chemotypes and will afford opportunities for design of more selective drugs with relatively less side effects. However, identification of allosteric binding site and regulatory mechanism of its allosteric modulation is still not well understood which severely undermines the efforts in designing novel drugs. A major bottleneck is the complexity of relationship between the structural (conformational) and functional change induced by allosteric effectors to the topography of active site and binding of substrate (6, 7).

A critical step before starting any campaign to screen for allosteric modulators is to identify and characterize the allosteric binding site on a target protein and their corresponding natural ligand. Another crucial step is to establish a relationship or correlation between the active site–substrate complex and effect of effector binding (i.e. inhibition or activation). To date, new allosteric sites and their corresponding effectors have been identified by combination of traditional high-throughput screening followed by NMR or X-ray crystallography. Nonetheless, several other chemical tools and concepts have also played key roles in the identification and analysis of allosteric sites and their modulators (7).

1.1. Experimental Screening to Identify Allosteric Effectors/ Binding Sites

Efforts in the discovery of allosteric effectors have largely relied on enzymatic high-throughput screens of chemical libraries. This method typically warrants rigorous kinetic experiments to establish the mechanism of allosteric regulation before the site of interaction is determined by structural studies, involving X-ray crystallography or NMR (8–10). High-throughput screens of chemical libraries are not the only method for identifying new allosteric drugs. Hemoglobin (Hb) represents a well-defined allosteric protein and several allosteric states, as well as ligand- or effector-induced conformational states have now been identified using crystallography and NMR (11, 12). Therefore, a structure-based drug design approach has played a significant role in the design of several allosteric effectors of Hb that are potentially useful for treating ischemic-related diseases or sickle cell disease (13–18). Latter two techniques have now become vital in unraveling the structural and functional relationship between allosteric and active sites. They have also become very important in providing us with a detailed view of the different allosteric states of target protein and conformational flexibility induced by its modulators in the presence and absence of natural substrate.

In recent years, NMR relaxation method describing the fractional population of each allosteric state and the rate of interconversion between them has been applied to study the allosteric system in proteins (19–21). Novel experimental methods such as a peptide phage display approach in conjunction with X-ray crystallography have been applied for identification of allosteric inhibitors for factor VIIa, an enzyme involved in the blood-clotting cascade (22–24). Tethering, a fragment-based drug design method, is another pioneering approach that has been applied to identify new allosteric site at the dimer interface of caspase-3 and caspase-7. The method involves a site-directed approach to probe new allosteric site by trapping thiol-containing small molecule ligands on native or cystine-mutated protein surfaces. The advantage of this approach is that it can be combined with peptide mapping to verify the position of the novel allosteric site without having to solve an X-ray crystal structure (25).

The aforementioned approaches have been supported by binding and thermodynamic studies via isothermal titration calorimetry (ITC), electron paramagnetic resonance, and fluorescence quenching (steady state and time resolved), as well as other biochemical and biophysical methods (26, 27). Although these methods are rigorous and demand substantial resources and time, they have been widely applied, resulting in the discovery of several newly identified allosteric sites and their modulators (7, 28).

A lingering and important question is whether an allosteric binding site on a protein and its effectors can be predicted computationally from the structure of a protein (7). While this aspiration is far from being achieved, various advances are being made that are contributing to the foundational principles upon which an in silico screening campaign would rely.

1.2. In Silico Approaches to Identify Allosteric Effectors/Binding Sites

The solution to the problem of cross-talk between two functionally associated site/pockets of a protein depends on how accurately and efficiently the local or global conformational changes are correlated between the two ligand-binding events. One of the ways to address this issue was proposed by Freire et al. where cross-talk between two distal sites can be estimated based on calculation of probability distribution functions of different states of a protein in the presence and absence of ligand. This approach has been applied in a systematic structure-based algorithm, COREX that generates distribution of states existing under equilibrium and when implemented with a ligand linkage algorithm, CORE_BIND, can predict the propagation of binding affect of a ligand on stability of distal binding site (29–31). The ensemble generated and the changes related to their distribution provide an opportunity to examine the long range effects of ligand binding and allosteric regulation. Another promising approach based on statistical coupling analysis between the sequences of a family of proteins can also reveal residues involved in allostery. Ranganathan and co-workers applied this technique to reveal the network of interactions between residues at two distal sites by mapping the evolutionarily conserved network of residues (32–35). Although these approaches facilitate identification of allosteric site and an understanding of the signal propagation to some extent, these methods are still unable to directly aid in screening large databases to identify allosteric effectors. To overcome this, approaches applying a combination of computational docking, biochemical, biophysical and kinetic analysis have been implemented to identify allosteric binding sites and screen for their modulators simultaneously (1,2). Encouraging results from these studies demonstrate that novel allosteric effectors can be identified using computational screening and underscore the potential of this strategy for drug discovery.

The above approaches involving computational screening employ a rational approach to identify a drug-like molecule based on relevant, indicative computational data that direct the design either based on similarity to a known ligand or placement of a complementary ligand within the allosteric site. Several approaches which can be broadly described as ligand-based and/or structure-based drug design have been developed using different computational methodologies (36–40). Herein, these methodologies are briefly discussed to screen large chemical space for a drug-like lead molecule, and new emerging principles in the field are explained.

1.2.1. Ligand-Based Approaches

Ligand-based approaches typically rely on the description of compounds in the so-called multidimensional chemical space, i.e., compound representation by multiple chemical descriptors. These descriptors describe global properties of molecules and the atoms/fragments constituting them. New molecules can be conceived on the basis of the similarities they share with the known lead compounds, and can be described using 1D-, 2D, and/or 3D descriptors (41–43). Molecules can also be represented as keyed 2D fingerprints where each bit in a bit string (a sequence of "0" and "1" or any integer) is associated with a specific chemical feature or fragment (44). Based on their level of enumeration (count/size), databases can be screened using 2D substructure searches or similarity searches (45, 46). Similar compounds based on their properties can be searched and retrieved by using hashed 2D fingerprints, where properties are mapped to overlapping bit segments (42, 47). Multiple-point 3D pharmacophore fingerprints have also been developed based on 3D descriptors (48, 49). In a slightly different approach, 3D steric and electronic features can be calculated from the 3D structure of a set of molecules and a pharmacophore hypothesis can be generated (50). Large database of drug-like or lead-like molecules can be searched using these ligand-based pharmacophore as query. In recent years, many machine-learning methods have evolved and statistically robust, predictive QSAR models can be generated by mapping the n-dimensional descriptor space. These models can be used to virtually screen large databases for novel chemotypes with desired pharmacological activity (47, 51–53). Similarity searching and predictive QSAR modeling have found wide applications and have become the most popular ligand-based techniques (44, 47, 51–53).

1.2.2. Structure-Based Approaches

Structure-based approaches use the 3D structural information from a macromolecular target and/or ligand–protein complex to design novel drugs that may modulate the target protein for desired activity (54–56). The ability to accurately predict the binding mode of a ligand to a target protein from computer simulations is an invaluable resource in understanding biochemical process and drug action. Several aspects of molecular recognition can be discerned from the computational modeling of macromolecule- or protein–ligand complex, and predicting the affinity of a new putative inhibitor for a protein has become essential for designing of more potent analogs.

The first step in structure-based drug design is the elucidation of 3D structure of the target macromolecule (protein or nucleic acid) by X-ray crystallography or NMR. In some cases, where direct structural information is unavailable, a homology model can be used as the starting point. Given the 3D structure of a receptor, known ligands can be docked to examine how they fit so that a protein–ligand complex can be characterized in detail and modifications that improve binding can be suggested. The docking methodology has found widespread applications in drug discovery since a large number of small molecules can be virtually screened by docking compounds from a database into a receptor structure (57).

The success of whole molecule docking into a target site or screening large virtual combinatorial libraries is ultimately dependent on the accuracy of the scoring function that ranks the compounds. Several scoring functions have been implemented that can score and rank a pose/orientation of ligand within the binding site accurately (58, 59). The scoring methods that are used in high-throughput settings dealing with thousands of diverse compounds can reasonably prioritize compounds for further testing.

1.2.3. Virtual Screening: UNITY Searches/Docking/ Scoring

Millions of compounds in available chemical databases from chemical vendors and billions of compounds in synthetically feasible chemical libraries are available for virtual screening. Traditionally virtual screening methods involved docking and scoring tools to screen database of compounds (57–59). However, orienting and scoring all the hundreds and thousands of poses for millions of compounds from several and large databases has become a formidable challenge. In practice, structure-based and small molecule-based approaches are used in combination.

New approaches involving ligand-based and structure-based techniques, followed by virtual screening, are successfully being applied to screen large databases of compounds (60–62). The databases are first screened based on similarity constraints or based on Quantitative Structure Activity Relationship (QSAR) models to narrow down the search space from millions of compounds to few thousand compounds. Further, ligand-based or receptor-based pharmacophore constraints can be generated and the "sorted" database is screened to narrow down the list of compounds from few thousands to few hundreds. These compounds can then be easily virtually screened using docking and scoring tools to prioritize them for experimental testing (63).

Herein, we illustrate through an example from our work on human erythrocyte pyruvate kinase (R-PK), the application of in silico-screening approach to identify novel allosteric regulators for this enzyme (1).

1.3. Pyruvate Kinase

Glycolysis is a sequence of reactions that converts glucose into pyruvate with a concomitant production of ATP. Pyruvate kinase (PK) is a key enzyme of the glycolytic pathway and catalyzes the phosphorylation of ADP with phosphoenolpyruvate (PEP) to ATP and pyruvate respectively. Several PK isozymes have been identified from different tissues, including R-PK (expressed in erythrocyte), M1-PK (expressed in skeletal muscle, brain, and heart), M2-PK (expressed in kidney, lungs, fetal tissues, adipose tissues, platelets, and spleen), L-PK (expressed in liver, renal cortex, and small intestines). With the exception of M1-PK, the other three isozymes exhibit physiological relevance of allosteric properties (64, 65). Activation of PK increases the ATP concentration in cell whereas its down regulation allows for accumulation of phosphometabolite pools, including 2,3-diphosphoglycerate (DPG) preceding the PK reaction (66). The PK catalyzed reaction is also involved in the regulation of glucose consumption of the cell (64).

In most cases, PK exists and functions as a tetramer (66). While the amino acid residues at the active site of PK are conserved for almost all species, the allosteric site displays a much lower degree of sequence conservation (67–69). The 3D structures from different tissues/species show similar fold, with each monomer arranged in three main domains; A, B, and C. The catalytic site is at the interface of the A- and B-domains, while the allosteric-binding site is entirely located within the regulatory C-domain, approximately 40 Å from the catalytic binding site (Fig. 1a). The multidomain architecture is an essential structural

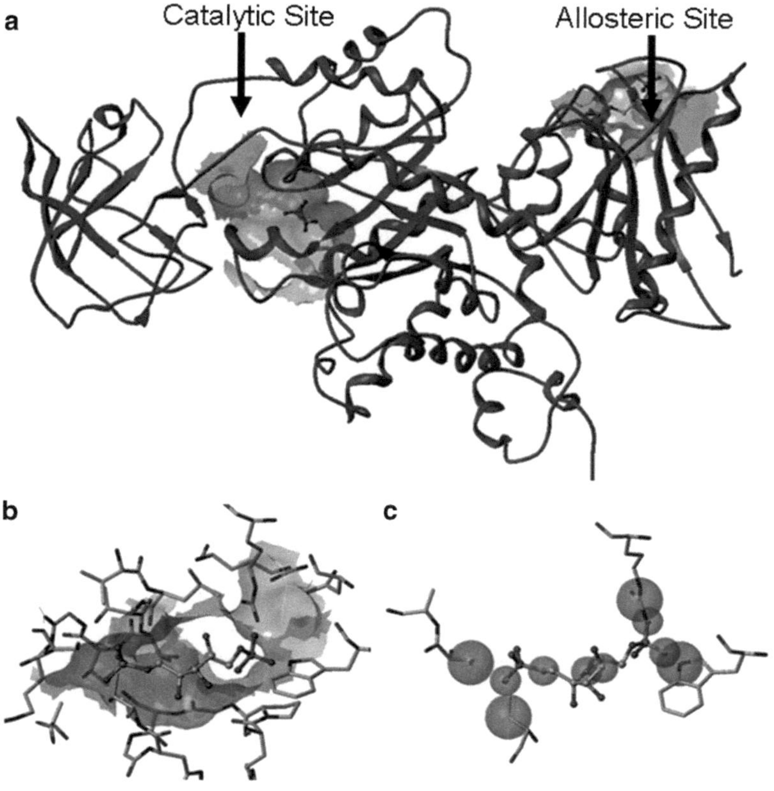

Fig. 1. (**a**) The catalytic site is at the interface of the A- and B-domains, while the allosteric-binding site is entirely located within the regulatory C-domain, approximately 40 Å from the catalytic binding site. (**b**) Allosteric site of R-PK with bound FBP. (**c**) A UNITY pharmacophore query shows hydrogen bond acceptor points and hydrophobic center. *Grey spheres* show spatial point constraints (hydrogen bond acceptor region and hydrophobic centers) and excluded volume constraint.

property for the function and regulation of the activity of the enzyme (66). Enzyme activity involves both subunit and domain movements that allows communication between the active and allosteric site.

Erythrocyte pyruvate kinase, R-PK is an allosteric protein, and its property can be modulated by several indigenous ligands, such as fructose-1,6-bisphosphate (FBP), L-alanine (Ala), and L-phenylalanine (Phe) to either activate or inhibit its activity (66, 70–74). The former acts as an activator to increase ATP production, while Ala and Phe act as inhibitors that result in metabolic block in the glycolytic pathway and disruption of ATP production (75–80). Figure 1b shows the allosteric site of R-PK with bound FBP. In addition to the indigenous effectors, the allosteric equilibrium of R-PK can also be modulated by synthetic effectors. This regulatory behavior offers immense therapeutic potential for treatment of hemolytic anemia, sickle-cell anemia, hypoxia-related diseases, and other disorders arising from erythrocyte PK malfunction (75–77).

1.3.1. Implications and Significance

One of the most-common enzymatic defects of red blood cells (RBCs) is caused by pathogenic mutations in the *PK-LR* gene coding for R-PK, which leads to inability of R-PK to supply ATP in red blood cells (RBC) (81). Lack of ATP leads to mature RBCs unable to function properly and/or maintain their cellular integrity; concomitantly undergoing lyses and causing hemolytic anemia (82). R-PK deficiency and/or inhibition of its activity also causes accumulation of DPG in RBC (82). DPG is an allosteric effector of Hb that binds to Hb, decreases its affinity for oxygen, and resulting in an enhancement of oxygen delivery to tissues.

From the foregoing, it seems that R-PK activators have potential application for the treatment of PK-deficiency-causative hemolytic anemia. Additionally, R-PK activators that cause a decrease in the levels of DPG, leading to an increase in hemoglobin affinity for O_2 could potentially be useful for the treatment of sickle-cell anemia, since unlike deoxygenated sickle hemoglobin, oxygenated sickle hemoglobin does not polymerize (19, 20, 83). In contrast, R-PK inhibitors that cause accumulation of DGP levels has potential application for the development of new treatments for hypoxic conditions, particularly stroke, angina, surgical bypass operations (13). These diverse mechanisms of allosteric regulation indicate that regulatory sites might be attractive drug targets for organism- and tissue-specific antagonists or deregulators.

This chapter describes in detail the in silico approach for the identification of R-PK allosteric regulators, based on the FBP-binding site.

2. Materials

The atomic coordinates of the of R-PK crystal structure in complex with FBP (*PDB* code 1LIU) were retrieved from the RCSB (Brookhaven) Protein Data Bank (84) in PDB format. The NCI database, library of small molecules (comprising ca. 250,000 compounds) was used to search for potential R-PK allosteric lead molecules. Molecular modeling was performed using the Sybyl 7.0 program suite (http://www.tripos.com) on Linux workstations. The program UNITY, present in the SYBYL 7.0 molecular-modeling software (see http://www.tripos.com) was used to perform 3D searches. UNITY is a pharmacophore-based chemical-structure database-screening program that searches for hit compounds against specific queries. Automatic docking programs FlexX (85) and GOLD (86) were used to position the molecule within the binding site and to ascertain the binding mode. FlexX is a flexible docking program (85) that uses an incremental construction algorithm to dock different conformations of a ligand into a receptor active site. GOLD uses a genetic optimization algorithm to predict different binding conformations and orientations of a ligand. The program then scores the different poses and generates a ranking list. The energy-scoring function of HINT (Hydropathic INTeractions) (87–89) was used to analyze the interactions between the docked molecules and R-PK. The HINT module is available as a Sybyl add-on or as a programmable toolkit from eduSoft LC (http://www.edusoft-lc.com). The program calculates a score for each interaction between atoms of protein and ligand, and the cumulative HINT score represents the total interaction score for the biomolecular complex.

3. Methods

3.1. Preparation of Protein Models

To simplify the computation, only the coordinates of subunit A of the tetrameric structure of R-PK (*PDB* code 1LIU) was used. There are a number of experimental factors and uncertainties, as well as flaws in the commonly used structure file formats, which require some "preparation" before they can be effectively used in computational studies. The following steps illustrate a minimal set of steps that may be applied to "clean up" a structure. More extensive preparation such as unrestrained energy minimization may be necessary if there are errors in the experimental structure data, e.g., non-hydrogen atoms that are too close to one another, etc.

Prepare the protein and protein–ligand structures by removing all water molecules and ions associated with the structure. Extract the ligands, and check the bond orders and atom types (hybridization and environment). Add hydrogen atoms to complete valences if necessary. Re-merge the ligand structures with the protein models. Add hydrogen atoms to the protein structures using the tool within the Sybyl Biopolymer module. Optimize the positions of all hydrogens in the system to an energy gradient of 0.005 kcal Å/mol with the Tripos force field (in Sybyl 7.0) while keeping heavy atom positions fixed (i.e., heavy atoms treated as an immobile aggregate). Extract the co-crystallized ligand from the crystal structure. Save the energy minimized protein model (without the ligand) and the ligand separately for docking at later stage.

3.2. Building a UNITY Query: Searching for a HIT/LEAD

The UNITY program is used to screen the NCI small-molecule library for compounds with similar chemical features as FBP, the natural allosteric effector of R-PK. UNITY is a computational search tool integrated within Sybyl ((http://www.tripos.com)) to explore chemical and biological databases. UNITY uses 2D substructure and similarity search to retrieve compounds from the database that have exact or similar structural features. 3D structure searches can also be undertaken based on pharmacologically active ligand molecules, molecular fragments from active molecules, ligand-based pharmacophore or pharmacophore based on the receptor to screen databases against specific queries.

1. Identify key molecular features responsible for favorable interaction between ligand (FBP) and its R-PK binding site, including hydrophobic and hydrophilic interactions from the above prepared R-PK/FBP complex structure and incorporate into a UNITY query. These features would include H-bond acceptors, where the two phosphate groups of FBP are attached, as well as a hydrophobic center at the central hydrophobic ring.
2. Use a defined query (Fig. 1c) to perform various searches on the public release of the NCI compound library, containing ca. 250,000 compounds to identify about 4,500 hits.
3. Reduce the number of hits to about 2,000 by incorporating into the query-specific functional moieties, such as carboxylate, phosphate, hydroxy, and sulfate groups.

   Further reduce the number of compounds by excluding those representatives that show incompatible size and/or shape with the FBP binding site. Do this by imposing a volume constraint, taking into account the size and shape of the FBP binding cavity.
4. Add volume constraint to the query by defining coordinates of protein atoms lining the binding site. These will include Arg 532 (guanidinium C-atom), Ser 480 (OH group), Trp 625 (ring N-atom), and Thr 476 and Arg 559 (O-atoms of C=O groups).

   Perform another UNITY search to further reduce the number of hits to about 327 compounds.
5. Exclude compounds that are without doubt too big to fit the FBP binding pocket, duplicate hits, and linear-chain molecules. This should reduce the list to about 30 potential allosteric effectors of R-PK.

3.3. Virtual Screening: Docking and Scoring

In order to prioritize compounds for experimental testing, and analyze their possible interactions with R-PK, the 30 identified compounds should be docked into the FBP binding site using three different procedures, followed by analyzing their interaction with the protein.

1. Use the automatic docking program FlexX to dock and maximize the interactions between the small molecules and the FBP binding site of the protein. Before any docking run, it is important to analyze the binding site and identify and delineate the site for docking. This is a critical step as it defines and sets the constraints for positioning the ligand in the defined binding region. Use a pocket radius and a sub-pocket radius of 10.5 and 2.5 Å, respectively around the reference ligand (FBP) to define the protein allosteric site during the computation.
2. Use FBP as a reference for docking. Check the accuracy of the docking, by computing the root mean square deviation between the reference structure and the final docked conformation.
3. Prepare and check ligands for their correct atom and bond types.
4. Run FlexX docking to generate 30 poses for each ligand. Save the output as mol2 (default file format for FlexX) file in Sybyl's mdb database. The results are also outputted to MSS (Molecular SpreadSheet) and can then be graphically and visually analyzed.
5. Use a second automatic docking program, GOLD (version 3.0) to dock the ligands at the FBP site. Specify the active site for the docking experiment using the coordinate of the FBP ring O-atom.
6. Select an approximate radius of 10 Å around the reference atom coordinates using the GOLD cavity detection algorithm.
7. Perform the GOLD docking without constraints to get an unbiased result and to explore all possible binding modes of the ligands.
8. Docking can also be carried out with template similarity constraints. Predefined constraint biases the conformation of docked ligands toward a given solution.
9. In this study, perform 100 GOLD genetic algorithm runs, as opposed to the default of 10. Switch off early termination of ligand docking. Use default settings for all other parameters.
10. Evaluate and validate the GOLD performance, using the co-crystallized ligand FBP for docking. GOLD should accurately reproduce the experimentally observed binding mode of FBP. Dock the remaining 29 molecules and optimize their binding.
11. Score the docked ligands using the HINT force field scoring function (vide infra).

12. For the third docking procedure, maintain the position and orientation of the ligands from the UNITY search. Place the hits in the FBP binding pocket, and then energy-minimize with a subset of the binding-site residues. Use a "hot radius" of 6 Å for the protein–ligand subset, and a surrounding radius of 12 Å for computation. Specify other conditions for the minimization process include *Gasteiger–Hückel* charges, a dielectric constant *e* of 1.0, a gradient of 0.005 kcal/mol/Å, and a total of 10,000 iterations. The coordinates of all the atoms *outside* the subset should be kept fixed.

3.4. HINT Interaction Analysis

Following the docking of compounds (using the three different procedures) at the FBP binding site, the program HINT (version 3.10S; eduSoft LC), is used to analyze the interactions between the compounds and the protein. Octanol/water partition coefficient (LogP), also referred to as the *hydrophobicity*, is a key factor in drug design, and is the first step in all HINT calculations as it provides the primary atom-based metric for analyses. The HINT force field is based on empirically-derived Log $P_{o/w}$ values that intuitively estimate free energy of binding, ΔG (88, 89).

1. Use the "dictionary" partition method to calculate log P of the protein, and the "calculate" method, based on the CLOG-P method of *Leo*, to partition the small molecules. The latter option uses the molecules' Sybyl atom types and bond connections as input data to the HINT partition algorithm.
2. To simplify the calculations, explicitly exclude some atoms. For the protein and ligand, only "essential" hydrogens are treated explicitly. Essential hydrogens are bonded to polar atoms and may potentially become involved in hydrogen bond interactions. All other hydrogen atoms are to be treated implicitly, e.g., a methyl is treated as a methyl "atom" rather than an sp^3 carbon plus three distinct hydrogen atoms. Set the "solvent condition" to "inferred," which sets the protonation state for each protein amino acid monomer based on the number and connectivity of the hydrogens present.
3. Correct the SASA (solvent accessible surface area) for protein-backbone N-atoms with the "+30" option. Optimize OH and NH_2 groups to maximize H-bond formation.
4. The Intermolecular Table module tabulates all the interaction between a protein and a ligand and data is written to Hint Table files (.tab). All interactions, within a cut-off radius of typically 8 Å are recorded and summed. The .tab file reports, however, a subset of these interactions defined by the table "extents", which excludes interactions outside a range of

(usually) 6 Å, and table "value", which excludes interactions with absolute value of HINT score below a threshold of (usually) 10.

5. There are four interaction classes represented in an interaction .tab file: favorable hydrophobic–hydrophobic interactions, favorable polar interactions, unfavorable polar interactions and unfavorable hydrophobic-polar interactions. Polar interactions are all those that do not involve hydrophobic atoms. These include hydrogen bond, acid–base, acid–acid, and base–base. Hydrogen bonding interactions are special cases of (favorable) acid–base interactions where the interacting atoms and interatomic distances are consistent with hydrogen bonds being formed. Both acid–acid and base–base interactions are unfavorable interactions and contribute negative-valued grid points and are thus unfavorable.
6. The total HINT interaction score (H_{TOT}) is an indicator of the binding efficiency between the inhibitor and enzyme.

The above exercise would show the selected NCI molecules, as well as FBP, fructose-2,6-bisphosphate and ribulose-1, 5-bisphosphate (RBP) to give variable positive HINT scores, suggesting favorable, but different, modes of interaction with the protein. It will also show that the comparative HINT scores for the three docking procedures are, in general, consistent. It is of note that these compounds were found to possess a wide range of R-PK activities, ranging from inhibition to activation at a final concentration of 1.0 mM. The identified compounds can further be optimized for higher potency and better pharmacokinetic properties. Nevertheless, the identification of these lead compounds is a step forward for future development of potential therapeutic agents.

4. Notes

1. First step in any structure-based virtual screening campaign is to locate and identify the binding pocket on a protein. Many web-based cavity detection software are available where pockets can be graphically visualized. Many of them are freely available for academia.

   Pocket Finder: http://www.modelling.leeds.ac.uk/pocketfinder/

   Q-Site finder: http://www.modelling.leeds.ac.uk/qsitefinder/

   Ligsite^{csc}: http://projects.biotec.tu-dresden.de/cgi-bin/index.php

2. Several automated docking and scoring software are available. Dock and prioritize hundreds and thousands of compounds for experimental testing.

   DOCK: http://dock.compbio.ucsf.edu/

   Auto Dock: http://autodock.scripps.edu/

   VINA: http://vina.scripps.edu/

3. Small molecule database. ZINC database is a freely available database of (ca. 13 million) commercially available compounds for virtual screening in ready-to-dock 3D format.

   http://zinc.docking.org/

4. Supersite: a chemobioinformatics database of protein and ligands. The web-server integrates several database and programs to mine information based on structural data, evolutionary conservation, and similarity.

   http://bioinf-tomcat.charite.de/supersite/index.jsp

5. The aforementioned list of web-servers, docking software, and database is not a comprehensive list. Many other commercial and freely available software can be searched from the following site.

   http://cheminformatics.org/

References

1. Kharalkar, S. S., Joshi, G. S., Musayev, F. N., Fornabaio, M., Abraham, D. J., Safo, M. K. (2007) Identification of novel allosteric regulators of human-erythrocyte pyruvate kinase. *Chem Biodivers.* **4**, 2603–2617.
2. Bond, C. J., Jurica, M. S., Mesecar, A., Stoddard, B. L. (2000) Determinants of Allosteric Activation of Yeast Pyruvate Kinase and Identification of Novel Effectors Using Computational Screening. *Biochemistry.* **39**, 15333–15343.
3. Conn, P. J., Arthur Christopoulos, A., Lindsley, C. W. (2009) Allosteric modulators of GPCRs: a novel approach for the treatment of CNS disorders. *Nature Reviews Drug Discovery.* **8**, 41–54.
4. Yanamala, N., Tirupula, K. C., Seetharaman, J. K. (2008) Preferential binding of allosteric modulators to active and inactive conformational states of metabotropic glutamate receptors. *BMC Bioinformatics.* **9**(Suppl 1):S16.
5. Espinoza-Fonseca, L. M., Trujillo-Ferrara, J. G. (2005) Identification of multiple allosteric sites on the M1 muscarinic acetylcholine receptor. *FEBS Letters.* **79**, 6726–6732.
6. Kern, D., Zuiderweg, E. R. (2003) The role of dynamics in allosteric regulation. *Curr. Opin. Struct. Biol.*, **13**, 748–757.
7. Hardy, J. A., Wells, J. A. (2004) Searching for new allosteric sites in enzymes. *Curr. Opin. Struct. Biol.* **14**, 706–715.
8. Laskowski, R. A., Gerick, F., Thornton, J. M. (2009) The structural basis of allosteric regulation in proteins. *FEBS Lett.* **583**, 1692–8.
9. Kay, L. E. (2005) NMR studies of protein structure and dynamics. *J. Magn. Reson.* **173**, 193–207.
10. Hummer, G., Schotte, F., Anfinrud, P. A. (2004) Unveiling functional protein motions with picosecond x-ray crystallography and molecular dynamics simulations. *Proc. Natl. Acad. Sci. USA.* **101**, 15330–15334.
11. Safo, M. K., Abraham, D. J. (2005) The enigma of the liganded hemoglobin end-state: A novel Quaternary structure of human carbonmonoxy hemoglobin, *Biochemistry*, **44**, 8347–8359.
12. Jenkins, J. D., Musayev, F. N., Danso-Danquah, R., Abraham, D. J., Safo, M. K. (2009) Structure of relaxed-state human hemoglobin: insight into ligand uptake, transport and release. *Acta Crystallogr D Biol Crystallogr.*, **65**, 41–48.
13. Safo, M. K., Moure, C. M., Burnett, J. C., Joshi, G. S., Abraham, D. J. (2001) High-

resolution crystal structure of deoxy hemoglobin complexed with a potent allosteric effector. *Protein Sci.* **10**, 951–957.

14. Safo, M. K., Boyiri, T., Burnett, J. C., Danso-Danquah, R., Moure, C. M., Joshi, G. S., Abraham, D. J. (2002) X-ray crystallographic analyses of symmetrical allosteric effectors of hemoglobin: compounds designed to link primary and secondary binding sites. *Acta Crystallogr D Biol Crystallogr.* **58**, 634–644.
15. Marden, M. C., Cabanes-Macheteau, M., Babes, A., Kiger, L., Griffon, N., Poyart, C., Boyiri, T., Safo, M. K., Abraham, D. J. (2002) Control of the allosteric equilibrium of hemoglobin by cross-linking agents. *Protein Sci.* **11**, 1376–1383.
16. Safo, M. K., Abdulmalik, O., Danso-Danquah, R., Burnett, J. C., Nokuri, S., Joshi, G. S., Musayev, F. N., Asakura, T., Abraham, D. J. (2004) Structural basis for the potent antisickling effect of a novel class of five-membered heterocyclic aldehydic compounds. *J. Med. Chem.* **9**, 4665–4676.
17. Abdulmalik, O., Safo, M. K., Chen, Q., Yang, J., Brugnara, C., Ohene-Frempong, K., Abraham, D. J., Asakura, T. (2005) 5-Hydroxymethyl-2-furfural modifies intracellular sickle haemoglobin and inhibits sickling of red blood cells. *Br. J. Haematol.* **128**, 552–561.
18. Nnamani, I. N., Joshi, G. S., Danso-Danquah, R., Abdulmalik, O., Asakura, T., Abraham, D. J., Safo, M. K. (2008) Pyridyl derivatives of benzaldehyde as potential antisickling agents. *Chem. Biodivers.* **9**, 1762–1769.
19. Palmer, A. G. III, Grey, M. J., Wang, C. (2005) Solution NMR spin relaxation methods for characterizing chemical exchange in high molecular-weight systems. *Methods Enzymol.* **394**, 430–465.
20. Korzhnev, D. M., Kloiber, K., Kay, L. E. (2004) Multiple-quantum relaxation dispersion NMR spectroscopy probing millisecond time-scale dynamics in proteins: theory and application. *J. Am. Chem. Soc.* **126**, 7320–7329.
21. Korzhnev, D. M., Salvatella, X., Vendruscolo, M., Di Nardo, A. A., Davidson, A. R., Dobson, C. M., Kay, L. E. (2004) Low-populated folding intermediates of Fyn SH3 characterized by relaxation dispersion NMR. *Nature.* **430**, 586–590.
22. Dennis, M. S., Eigenbrot, C., Skelton, N. J., Ultsch, M. H., Santell, L., Dwyer, M. A., O'Connell, M. P., Lazarus, R. A. (2000) Peptide exosite inhibitors of factor VIIa as anticoagulants. *Nature.* **404**, 465–470.
23. Dennis, M. S., Roberge, M., Quan, C., Lazarus, R. A. (2001) Selection and characterization of a new class of peptide exosite inhibitors of coagulation factor VIIa. *Biochemistry.* **40**, 9513–9521.
24. Roberge, M., Santell, L., Dennis, M. S., Eigenbrot, C., Dwyer, M. A., Lazarus, R. A. (2001) A novel exosite on coagulation factor VIIa and its molecular interactions with a new class of peptide inhibitors. *Biochemistry.* **40**, 9522–9531.
25. Hardy, J. A., Lam, J., Nguyen, J. T., O'Brien, T., Wells, J. A. (2004) Discovery of an allosteric site in the caspases. *Proc. Natl. Acad. Sci. USA.* **101**, 12461–12466.
26. Heddle, J. G., Okajima, T., Scott, D. J., Akashi, S., Park, S. Y., Tame, J. R. (2007) Dynamic allostery in the ring protein TRAP. *J. Mol. Biol.* **371**, 154–167.
27. Yu, P., Lasagna, M., Pawlyk, A.C., Reinhart, G.D., Pettigrew, D.W. (2007) IIAGlc inhibition of glycerol kinase: a communications network tunes protein motions at the allosteric site. *Biochemistry.* **46**, 12355–12365.
28. Swain, J. F., Gierasch L. M. (2006) The changing landscape of protein allostery. *Current Opinion in Structural Biology.* **16**,102–108.
29. Hilser, V. J., Dowdy, D., Oas, T. G., Freire, E. (1998) The structural distribution of cooperative interactions in proteins: analysis of the native state ensemble. *Proc Natl Acad Sci USA.* **95**, 9903–9908.
30. Freire, E. (2000) Can allosteric regulation be predicted from structure? *Proc Natl Acad Sci USA,* **97**, 11680–11682.
31. Freire, E. (1999) The propagation of binding interactions to remote sites in proteins: analysis of the binding of the monoclonal antibody D1.3 to lysozyme. *Proc Natl Acad Sci USA.* **96**, 10118–22.
32. Lockless, S. W., Ranganathan, R. (1999) Evolutionarily conserved pathways of energetic connectivity in protein families. *Science.* **286**, 295–299.
33. Suel, G. M., Lockless, S. W., Wall, M. A., Ranganathan, R. (2003) Evolutionarily conserved networks of residues mediate allosteric communication in proteins. *Nat. Struct. Biol.* **10**, 59–69.
34. Hatley, M. E., Lockless, S. W., Gibson, S. K., Gilman, A. G., Ranganathan, R. (2003) Allosteric determinants in guanine nucleotide-binding proteins. *Proc. Natl. Acad. Sci. USA.* **100**, 14445–14450.
35. Shulman, A. I., Larson, C., Mangelsdorf, D. J., Ranganathan, R. (2004) Structural determinants of allosteric ligand activation in RXR heterodimers. *Cell.* **116**, 417–429.

36. Bleicher, K. H., Böhm, H., Müller, K., Alanine, A. I. (2003) Hit and lead generation: beyond high-throughput screening. *Nature Reviews Drug Discovery.* **2**, 369–378.
37. Walters, W. P., Murcko, A., Murcko, M. A. (1999) Recognizing molecules with drug-like properties. *Curr. Opin. Chem. Bio.* **3**, 384–387.
38. Walters, W. P., Stahl, M. T., Murcko, M. A. (1998) Virtual screening – an overview. *Drug Discovery Today.* **3**, 160–178.
39. Smith, A. (2002) Screening for drug discovery: the leading question. *Nature.* **418**, 483–459.
40. Schneider, G., Böhm, H. (2002) Virtual screening and fast automated docking methods. *Drug Discovery Today,* **7**, 64–70.
41. Oprea, T. I. (2002) On the information content of 2D and 3D descriptors for QSAR. *J. Braz. Chem. Soc.* **13**, 811–815.
42. Todeschini, R., Consonni, V. (2000) Handbook of molecular descriptors. Wiley-VCH: Weinheim.
43. Brown, R.D., Martin, Y.C. (1997) The information content of 2D and 3D structural descriptors relevant to ligand–receptor binding. *J. Chem. Inf. Comput. Sci.* **37**, 1–9.
44. Willet, P. (2006) Similarity-based virtual screening using 2D fingerprints. *Drug Discovery Today.* **11**, 1046–1053.
45. Willet, P. (1998) Chemical similarity searching. *J. Chem. Inf. Comput. Sci.* **38**, 983–996.
46. Sheridan, R. P., Kearsley, S. K. (2002) Why do we need so many chemical similarity search methods? *Drug Discovery Today.* **7**, 903–911.
47. Eckert, H, Bajorath, J. (2007) Molecular similarity analysis in virtual screening: foundations, limitations and novel approaches. *Drug Discovery Today.* **12**, 225–233.
48. Mason, J.S., Morize, I., Menard, P.R., Cheney, D.L., Hulme, C., and Labaudiniere, R.F. (1999) New 4-point pharmacophore method for molecular similarity and diversity applications: Overview of the method and applications, including a novel approach to the design of combinatorial libraries containing privileged substructures. *J. Med. Chem.* **42**, 3251–3264.
49. Matter, H. & Pötter, T. (1999) Comparing 3D pharmacophore triplets and 2D fingerprints for selecting diverse compound subsets. *J. Chem. Inf. Comput. Sci.* **39**, 1211–1225.
50. Patel, Y., Gillet, V. J., Bravi, G., Leach, A. R. (2002) A comparison of the pharmacophore identification programs: Catalyst, DISCO and GASP. *J. Comput. Aided Mol. Des.* **16**, 653–81.
51. McGregor, M. J., Muskal, S. M. (1999) Pharmacophore Fingerprinting. 1. Application to QSAR and Focused Library Design. *J. Chem. Inf. Comput. Sci.* **39**, 569–574.
52. McGregor, M. J., Muskal, S. M. (2000) Pharmacophore Fingerprinting. 2. Application to Primary Library Design. *J. Chem. Inf. Comput. Sci.* **40**, 117–125.
53. Tropsha, A, Golbraikh, A. (2007) Predictive QSAR Modeling Workflow, Model Applicability Domains, and Virtual Screening. *Curr. Pharm. Des.* **13**, 3494–3504.
54. Bissantz, C., Folkers, G., Rognan, D. (2000) Protein-based virtual screening of chemical databases. 1. Evaluation of different docking/scoring combinations. *J. Med. Chem.* **43**, 4759–4767.
55. Joseph-McCarthy, D. (1999) Computational approaches to structure-based ligand design. *Pharmacology & Therapeutics.* **84**, 179–191.
56. Anderson, A.C. (2003) The Process of structure-based drug design. *Chemistry & Biology.* **10**, 787–797.
57. Kitchen, D. B.; Decornez, H.; Furr, J. R.; and Bajorath, J. (2004). Docking and scoring in virtual screening for drug discovery: methods and applications. *Nat. Rev. Drug Discov.* **3**, 935–949.
58. Ferrara, P., Gohlke, H., Price, D. J., Klebe, G., Brooks III, C. L. (2004) Assessing scoring functions for protein–ligand interactions. *J. Med. Chem.* **47**, 3032–3047.
59. Gohlke, H., Klebe, G. (2002) Approaches to the description and prediction of the binding affinity of small-molecule ligands to macromolecular receptors. *Angew. Chem. Int. Ed.* **41**, 2644–2676.
60. Oloff, S., Zhang, S., Sukumar, N., Breneman, C., Tropsha, A. (2006). Chemometric analysis of ligand receptor complementarity: identifying Complementary Ligands Based on Receptor Information (CoLiBRI). *J. Chem. Inf. Model.* **46**, 844–851.
61. Reid, D., Sadjad, B. S., Zsoldos, Z., Simon, A. (2008) LASSO – ligand activity by surface similarity order: a new tool for ligand based virtual screening. *Journal of Computer-Aided Molecular Design.* **22**, 479–487.
62. Zsoldos, Z., Reid, D., Simon, A., Sadjad, S. B., Johnson, A. P. (2006) eHiTS: a new fast, exhaustive flexible ligand docking system. *J. Mol. Graph. Model.* **7**, 421–435.
63. Bajorath, J. (2002) Integration of virtual and high-throughput screening. *Nat. Rev. Drug Discov.* **1**, 882–894.
64. Larsen, T. M., Laughlin, L. T., Holden, H. M., Rayment, I., Reed, G. H. (1994) Structure of rabbit muscle pyruvate kinase

complexed with Mn2+, K+, and pyruvate. *Biochemistry.* **24**, 6301–6309.

65. Muirhead, H., Clayden, D. A., Barford, D., Lorimer, C. G., Fothergill-Gilmore, L. A., Schiltz, E., Schmitt, W. (1986) The structure of cat muscle pyruvate kinase. *EMBO J.* **5**, 475–481.
66. Valentini, G., Chiarelli, L., Fortin, R., Speranza, M. L., Galizzi, A., Mattevi, A. (2000) The allosteric regulation of pyruvate kinase. *J. Biol. Chem.* **275**, 18145–18152.
67. Laughlin, L. T., Reed, G. H. (1997) The monovalent cation requirement of rabbit muscle pyruvate kinase is eliminated by substitution of lysine for glutamate 117. *Arch. Biochem. Biophys.* **1997**, 348, 262–267.
68. Early, C. N., Britt, B. M. (1998) Sequence similarities of glyceraldehyde-3-phosphate dehydrogenases, phosphoglycerate kinases, and pyruvate kinases are species optimal temperature-dependent. *Eur. Biophys. J.* **27**, 409–410.
69. Ernest, I., Opperdoes, F. R., Michels, P. A. (1994) Cloning and sequence analysis of the gene encoding pyruvate kinase in Trypanoplasma borelli. *Biochem. Biophys. Res. Commun.* **201**, 727–732.
70. Ikeda, Y., Noguchi, T. (1998) Allosteric regulation of pyruvate kinase M2 isozyme involves a cysteine residue in the intersubunit contact. *J. Biol. Chem.* **273**, 12227 12233.
71. Fothergill-Gillmore, L. A., Rigden, D. J., Michels, P. A., Phillips, S. E. (2000) Leishmania pyruvate kinase: the crystal structure reveals the structural basis of its unique regulatory properties. *Biochem. Soc. Trans.* **28**, 186–190.
72. Mattevi, A., Bolognesi, M., Valentini, G. (1996) The allosteric regulation of pyruvate kinase. *FEBS Lett.* **389**, 15–19.
73. Jurica, M. S., Mesecar, A., Heath, P. J., Shi, W., Nowak, T., Stoddard, B. L. (1998) The allosteric regulation of pyruvate kinase by fructose-1,6-bisphosphate. *Structure.* **6**, 195–210.
74. K. J. Johannes, K. J., Hess, B. (1973) Allosteric kinetics of pyruvate kinase of *Saccharomyces carlsbergensis. J. Mol Biol.* **76**, 181–205.
75. Palmer, T. N., Odedra, B. R. (1982) L-Phenylalanine inhibition of muscle pyruvate kinase. *Biosci. Rep.* **2**, 825–833.
76. Mapungwana, S. M., Davies, D. R. (1982) The effect of fructose on pyruvate kinase activity in isolated hepatocytes. Inhibition by allantoin and alanine. *Biochem. J.* **208**, 171–178.
77. Del Valle, P., Busto, F., De Arriaga, D., Soler, J. (1990) ATP inhibition of Phycomyces pyruvate kinase: a kinetic study of the inhibitory effects on the allosteric kinetics shown by the enzyme. *J. Enzyme Inhib.* **3**, 219–228.
78. Lee, J. C. (2008) Modulation of allostery of pyruvate kinase by shifting of an ensemble of microstates. *Acta Biochim Biophys Sin.* **40**, 663–669.
79. Black, J. A., Henderson, M. H. (1972) Activation and inhibition of human erythrocyte pyruvate kinase by organic phosphates, amino acids, dipeptides and anions. *Biochimica et Biophysica Acta (BBA) – Enzymology.* **284**, 115–127.
80. Fenton, A. W., Blair, J. B. (2002) Kinetic and Allosteric Consequences of Mutations in the Subunit and Domain Interfaces and the Allosteric Site of Yeast Pyruvate Kinase. *Archives of Biochemistry and Biophysics.* **397**, 28–39.
81. Valentine, W. N., Tanaka, K. R., Miwa, S. (1961) A specific erythrocyte glycolytic enzyme defect (pyruvate kinase) in three subjects with congenital non-spherocytic hemolytic anemia. *Trans. Assoc. Am. Physicians.* **74**, 100–110.
82. Zanella, A., Bianchi, P. (2000) Red cell pyruvate kinase deficiency: from genetics to clinical manifestations. *Baillieres Best Pract. Res. Clin. Haematol.* **13**, 57–81.
83. Abraham, D. J., Mehanna, A. S., Wireko, F. C., Whitney, J., Thomas, R. P., Orringer, E. P. (1991) Vanillin, a potential agent for the treatment of sickle cell anemia. *Blood* **77**, 1334–1341.
84. Berman, H. M., Westbrook, J., Feng, Z., Gilliland, G., Bhat, T. N., Weissig, H., Shindyalov, I. N., Bourne, P. E. (2000) The protein data bank. *Nucl. Acids. Res.* **28**, 235–242.
85. Rarey, M., Kramer, B., Lengauer, T., Klebe, G. (1996) A fast flexible docking method using an incremental construction algorithm. *J. Mol. Biol.* **23**, 470–489.
86. Jones, G., Willett, P., Glen, R. C., Leach, A. R., Taylor, R. J. (1997) Development and validation of a genetic algorithm for flexible docking. *J. Mol. Biol.*, **267**, 727–748.
87. Kellogg, G. E., Abraham, D. J. (2000) Hydrophobicity: is LogP(o/w) more than the sum of its parts? *Eur. J. Med. Chem.* **35**, 651–661.
88. Cozzini, P., Fornabaio, M., Marabotti, A., Abraham, D. J., Kellogg, G. E., Mozzarelli, A. (2002) Simple, intuitive calculations of free energy of binding for protein–ligand complexes. 1. Models without explicit constrained water. *J. Med. Chem.* **45**, 2469–2483.
89. Spyrakis, F., Amadasi, A., Fornabaio, M., Abraham, D. J., Mozzarelli, A., Kellogg, G. E., Cozzini, P. (2007) The consequences of scoring docked ligand conformations using free energy correlations. *Eur. J. Med. Chem.* **42**, 921–933.

Chapter 20

Identification of Allosteric-Activating Drug Leads for Human Liver Pyruvate Kinase

Aron W. Fenton

Abstract

There is a growing appreciation of the beneficial attributes of allosteric drugs. However, the development of this special class of drugs has in large part been via serendipitous findings from high-throughput screens of drug libraries. Limited success at deliberately identifying allosteric drugs may be due to a focus on enzyme inhibitors, a parallel to the historic focus on competitive inhibitors. In contrast to inhibition, activation of an enzyme by a small molecule can only occur through a limited number of mechanisms, mainly allosteric regulation. Activation of human liver pyruvate kinase (hL-PYK) in an effort to create a glycolytic/gluconeogenic futile cycle is one potential mechanism to counteract hyperglycemia. Using hL-PYK, we demonstrate the potential of drug library screens to identify allosteric-activator drug leads.

Key words: Allosteric, Allostery, Pyruvate kinase, High-throughput, Drug screen

1. Introduction

Literature from the past 5 years has shown a growing interest in developing allosteric drugs (1–3). The defining features we use to describe allosteric drugs are: (1) the effector is not chemically identical to the substrate, (2) the effector elicits a change in a functional property of the protein (i.e. binding of a second ligand or altered catalytic properties), and (3) the effector binds at a site that is topographically distinct from the functional site of the protein (4). Allosteric drugs can mimic the natural regulatory mechanisms responsible for maintaining homeostasis in non-disease states. These drugs can provide specificity and selectivity in the regulation of related isozymes (2, 3). In addition, allosteric drugs (and allosteric effectors in general) have the unique property that upon reaching a concentration that allows the allosteric site to be saturated, no additional effects are observed as the concentration of the drug is raised even higher. This concentration-independent limit

Aron W. Fenton (ed.), *Allostery: Methods and Protocols*, Methods in Molecular Biology, vol. 796,
DOI 10.1007/978-1-61779-334-9_20, © Springer Science+Business Media, LLC 2012

to molecular function adds to the safety of allosteric drugs (5). Furthermore, the magnitude of the allosteric function can be modulated by effector chemistry (6). Allosteric drugs also have the potential to provide the required efficacy only in the presence of appropriate substrate (2, 7). It has been postulated that all proteins have the potential to be allosterically regulated (8, 9). Unfortunately, much of the current allosteric drug advancements are a result of serendipitous findings from high-throughput screens (1–3).

The goal of this chapter is to demonstrate a design to identify allosteric drugs from a high-throughput screen of a 100,000 compound small molecule library. The primary emphasis is placed on allosteric activation rather than inhibition (see Note 1). The pharmacological usefulness of allosteric activators can easily be appreciated when considering that there is an opposing pathway to most metabolic and signaling pathways (e.g. activation of a phosphatase is likely to accomplish the same goal as inhibiting the opposing kinase). However, the greater benefit to our goal of specifically seeking allosteric drug leads is a result of the limited number of mechanisms that can give rise to activation. To further demonstrate this point, we will consider our target enzyme, human liver pyruvate kinase (hL-PYK).

Pyruvate kinase catalyzes the final reaction of glycolysis, the transfer of phosphate from phosphoenolpyruvate (PEP) to ADP to produce ATP and pyruvate. Activation of hL-PYK may facilitate a futile cycle between glycolysis and gluconeogenesis in the liver as a means of treating hyperglycemia. Like other isozymes, hL-PYK requires divalent cation for activity. In addition, the presence of a monovalent ion in the active site greatly facilitates PEP affinity (10). Pyruvate kinase activity is typically measured by coupling pyruvate production with NADH oxidation using lactate dehydrogenase. Therefore, any compound that inhibits lactate dehydrogenase, chelates Mg^{2+} or K^{+}, or that binds to hL-PYK competitively with ADP, Mg^{2+}, K^{+}, or PEP would result in inhibition of activity. None of these mechanisms would be considered allosteric in nature. Therefore, library compounds might inhibit observed activity through a number of mechanisms. In contrast, activation of enzymatic activity can only occur through a limited number of mechanisms, mainly allostery.

Since our goal is activation of hL-PYK as a potential strategy to counteract hyperglycemia, we have designed our screen to explore effectors that increase activity through a number of allosteric mechanisms. Therefore, allosteric inhibitors (alanine and ATP) were included in the assay cocktail, such that activation would be observed if a library compound binds competitively with one of the allosteric inhibitors. Furthermore, the initial screen of library compounds was with the mutant protein containing the S12D mutation to mimic inhibition by phosphorylation. Therefore, increased activity in the initial screen used in this study might indicate either allosteric activation or the prevention of allosteric inhibition. The use of the natural allosteric activator, fructose-1,6-bisphosphate (Fru-1,6-BP) was limited to additions as a positive control.

2. Materials

2.1. Identification and Confirmation of Allosteric Activators

1. S12D mutant protein: This protein was expressed in FF50 *E.coli* and purified using ammonium sulfate fractionation and DEAE column chromatography as previously described (11). Once purified, the protein was stored at 4°C in 10 mM MES pH 6.8, 5 mM $MgCl_2$, 10 mM KCl and 2 mM DTT.
2. L-lactate dehydrogenase (LDH): (Type III bovine heart) from Calzyme Laboratories[2], Inc. dialyzed into 50 mM HEPES pH 7.5, 10 mM $MgCl_2$, 100 mM KCl, and 0.1 mM EDTA. The final 6 KU/ml LDH stock was stored at 4°C.
3. NADH stock: 100 mM, stored at −20°C.
4. Alanine stock: 1.5 M stock, stored at −20°C.
5. 10× screening-assay buffer: 500 mM HEPES, 100 mM $MgCl_2$, 1 M KCl, and 1 mM EDTA, pH adjusted to pH 7.5 using KOH, stored at 4°C.
6. Sodium-ATP stock: 471 mM, pH adjusted to pH 7.5 with KOH, stored at −20°C.
7. ADP stock: Due to the large quantity of ADP required, and a temporary shortage of ADP from most commercial sources, a mixture of sodium and potassium salts of ADP was used. After adjusting pH of the final stock solution with KOH, concentrations were: 400 mM ADP, 681 mM Na^+, and 562 mM K^+. This stock solution was stored at −20°C.
8. Fru-1,6-BP stock: 450 mM in 2.5% DMSO, used as a positive control to show activation.
9. DMSO: 2.5% added to control wells that lack an activator (i.e. negative control).
10. Small Molecule Library: The small molecule library maintained by the University of Kansas High-Throughput Screening Laboratory consists of 45,000 compounds from the ChemBridge Library, 56,000 from the ChemDiv Library, 880 the Prestwick Library, and 2,000 the MicroSource Library. Library compounds, 25 μg/ml in 2.5% DMSO (see Note 2), were pre-dispensed into columns 3–12 of Perkin-Elmer 384-well View plates (see Note 3) and stored at −20°C. No compound additions were made to columns 1 and 2 of each plate so that these wells could be used for negative and positive controls, respectively.
11. Potassium-PEP stock: 270 mM PEP, stored at −20°C.

2.2. Distinguishing K-Type vs. V-Type Effectors

Unless noted, materials and preparation are the same as those listed in Subheading 1. For all solutions in this section, the volume and concentration of KOH added to adjust pH was recorded; K^+ from both KOH additions and the counterion added with ligands were summed and KCl was supplemented to maintain a constant 150 mM K^+ for all assays. As indicated, the most straightforward method for maintaining this constant K^+ is to dilute variable assay components (i.e. substrate and effector) in the respective concentration of KCl.

1. Wild type hL-PYK: expressed, purified, and stored as listed for S12D in Subheading 1 and ref. 11.
2. 10× assay buffer: 500 mM Bicine, 100 mM $MgCl_2$, and 1 mM EDTA, pH was adjusted to 7.5 using KOH, stored at 4°C. Note that data collected in Bicine and HEPES (used in the initial library screen) are interchangeable (10).
3. Potassium-ADP stock: 400 mM, pH was adjusted with KOH, stored at −20°C.
4. Potassium-PEP stock: 27 mM. After the addition of KOH to adjust pH to 7.5, the total K^+ in this stock was 147 mM. Therefore, this PEP solution was serially diluted 1:2 in 147 mM KCl.
5. Fru-1,6-BP stock: 92 μM, 177.5 mM KCl and either 1% DMSO or 1% ethanol, used as a positive control to show activation. This stock was diluted 1:5 in 177.5 mM KCl and the respective 1% organic additive.
6. Allsoteric compounds: Four compounds identified from the high-throughput screen of the small molecule library were contract synthesized by Obiter Research, LLC. Stock solutions of these compounds were made in 177.5 mM KCl with either 1% DMSO or 1% ethanol, depending on the solubility of the compound. This stock was diluted 1:5 in 177.5 mM KCl and the respective 1% organic additive.
7. Initial velocity data were collected in Greiner Bio-one UV-Star flat bottomed 96-well plates.

3. Methods

3.1. Identification of Allosteric Activators

The substrate concentration was of primary importance in this screen (see Note 4). To fully appreciate this, consider a plot of initial velocity as a function of substrate concentration (Fig. 1a). Since we are attempting to identify compounds that either allosterically activate, or that prevent the pre-existing inhibition, let us consider an enzyme that is fully inhibited. In the case of hL-PYK,

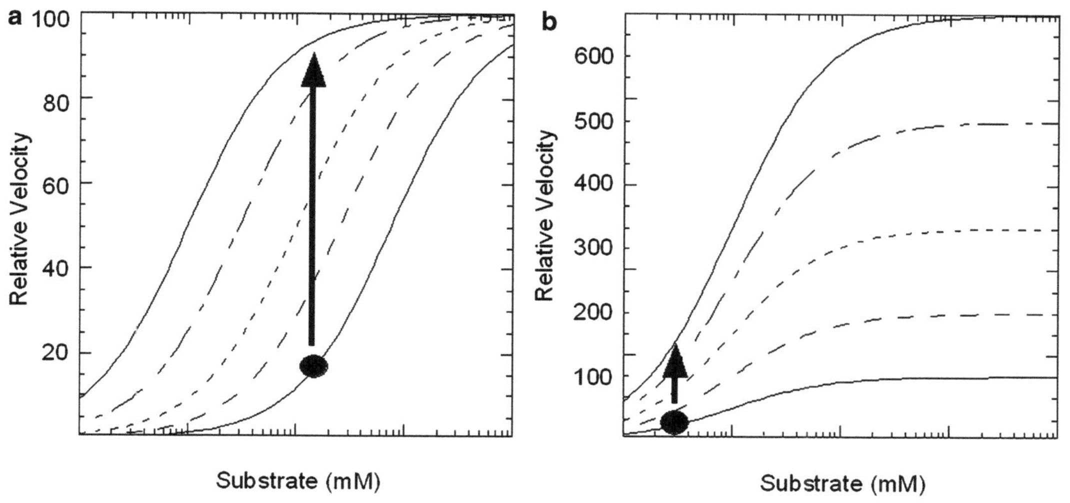

Fig. 1. Modeled titrations of activity as a function of substrate, illustrated at five concentrations of activator or of ligands that bind competitively with inhibitors (increasing effector concentration from *right* to *left* in (**a**) and *bottom* to *top* in (**b**)). (**a**) For maximum sensitive within the screen, substrate concentration should be chosen to give the maximum change in initial velocity upon addition of the activator control, Fru-1,6-BP in our example. (**b**) Since our design is primarily focused on K-type activation, the assay conditions are less sensitive to effectors that increase V_{max}.

inhibition is achieved by the addition of S12D (a mimic of phosphorylation) and concentrations of Ala and ATP that saturate the effect of these ligands on the enzyme's affinity for PEP (for this determination, see previous work (10–12)). In Fig. 1a, the fully inhibited form of the enzyme is represented by the bottom/right most response curve. However, bear in mind that our screen uses only a single concentration of PEP, as represented by the black dot in Fig. 1a. By choosing a PEP concentration that gives the maximum increase in initial velocity upon addition of the positive control (i.e. Fru-1,6-BP), the screen will be maximally sensitive to any form of activation or removal of inhibition. This point is illustrated by the black arrow that passes through representative response curves reflecting different PEP affinities. However, the example in Fig. 1a is representative of an effector that modifies substrate affinity (K-type activator). Such a response might be expected for hL-PYK given that this K-type control is elicited by currently known forms of regulation (phosphorylation, allosteric regulation by ATP, Ala, and Fru-1,6-BP). However, a second potential outcome of our screen is the identification of effectors that modify V_{max} activity (i.e. a V-type activator). Our assay design is likely less sensitive to a V-type activator (Fig. 1b). The final assay concentration of 1.4 mM PEP was empirically determined by monitoring the level of activation upon the addition of 0.1 mM Fru-1,6-BP (the in-well concentration used as a positive control) in the presence of the S12D mimic of protein phosphorylation and in the presence of allosteric inhibitors (Fig. 2).

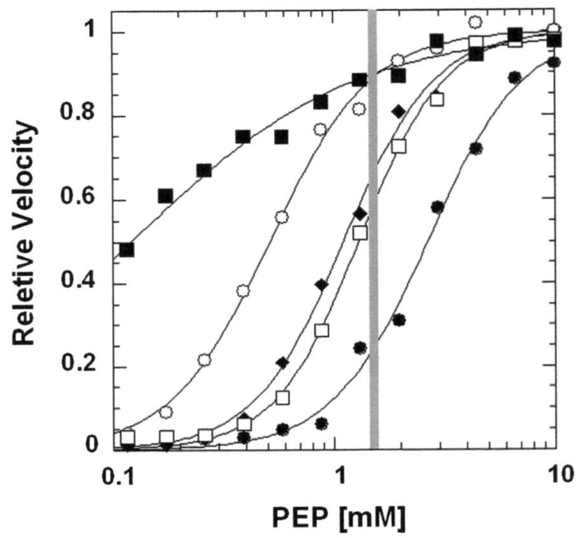

Fig. 2. Expected increases in initial velocity due to an allosteric activation or removal of pre-existing inhibition. A *vertical gray bar* is included at the single PEP concentration used in the high-throughput screen. The response representative of the negative control (the S12D mutant mimic of phosphorylation, in the presence of Ala and ATP) included in the screen is in *solid circles*. Prevention of inhibition by phosphorylation is mimicked by the response of the wild type protein to varying PEP concentration, but in the presence of Ala and ATP (*open square*). The response curve of S12D in the presence of Ala, but in the absence of ATP (*filled diamond*), illustrates an expected response, should ATP binding/inhibition be interrupted by a library compound. The removal of Ala (S12D assayed in the presence of ATP) demonstrates the expected outcome should a compound interrupt binding of Ala (*open circle*). Finally, the addition of Fru-1,6-BP when assaying S12D in the presence of both ATP and Ala serves both as the positive control and an example of allosteric activation (*filled square*). By comparing the %-change as each response curve cross the *gray bar*, an appreciation can be gained for the potential POC caused by library compounds.

1. Immediately before initiation of the high-throughput screen, an enzyme cocktail was mixed. The total volume of this cocktail to be mixed depended on the number of plates to be read. Volumes mixed per plate were 3.96 ml of 10× screening-assay buffer, 87 μl of 100 mM NADH, 8 μl of LDH, 198 μl of 400 mM ADP, 188 μl of 471 mM ATP, 64 μl of 1.5 M Ala, 6 mg of fresh DTT, a volume of S12D hL-PYK that results in a ΔA_{340}/min equal to ~0.015 in the final in-well assay (determined at the beginning of each day), and water to a final volume of 22 ml.
2. Also immediately before use, the 270 mM stock of PEP was diluted to 6.3 mM in 0.045% Triton-X-100 (see Note 5).
3. Cocktail was pre-incubated before use. As noted below, a zero time point and a 1-h time point were taken for each well. However, we noted a slight variation in linearity of the activity over the first 45 min. Although we could speculate a number

of potential mechanisms for this slight variation (e.g. protein conformation shift in the cocktail with detergent and ligands, loss of ATP due to ATPase activity of hL-PYK, etc.), we found that pre-incubating the cocktail for 2 h at room temperature before initiating the reaction reduced this non-linearity.

4. For each plate containing aliquots of library compounds, 20 μl of 2.5% DMSO was added to column 1 using a 16-channel pipettor. These wells act as a negative control with no allosteric activation.
5. Likewise, 20 μl of 450 mM Fru-1,6-BP in 2.5% DMSO was added to column 2 of each plate using a 16-channel pipettor. These wells act as a positive control with known allosteric activation.
6. 50 μl of pre-mixed, pre-incubated cocktail was added to each well using a Thermo Scientific Multidrop 384.
7. After a 4 min delay, the reaction was started via the addition of 20 μl of 6.3 mM PEP in 0.045% Triton-X-100 using a second Multidrop 384.
8. Immediately after the addition of PEP, the A_{340} (see Notes 6 and 7) of each well of each plate was read (i.e. zero time point; A_{340}^{0}) using a Perkin Elmer Envision plate reader and using an internal P340 filter (340 nm ± 14 nm, $T_{\max} = 20\%$).
9. After completion of the zero time point read for all 50 plates (see Note 8), A_{340} of each well of each plate was determined a second time (i.e. the 1-h time point; $A_{340}^{1\text{-h}}$).
10. Data from an individual plate was only used if the Z' score for that plate was >0.3, where

$$Z' = 1 - \frac{3((\text{STDEV of negative control}) + (\text{STDEV of positive control}))}{(\text{Ave of negative control}) - (\text{Ave of positive control})}. \quad (1)$$

Data to calculate the Z' scores is contained in the first and second column of the plate. For each well of each plate (i.e. for each compound in the library)

$$\Delta A_{340} = A_{340}^{1\text{-h}} - A_{340}^{0}. \quad (2)$$

11. Within each plate, the average ΔA_{340} for the positive control wells (i.e. in the presence of Fru-1,6-BP) was set equal to 100% and the average ΔA_{340} for the negative control wells was set equal to 0%. Data for each compound was expressed as a percentage of control (POC; Fig. 3) using

$$\text{POC} = \frac{\Delta A_{340}(\text{sample}) - \text{ave } \Delta A_{340}(\text{negative control})}{\text{ave } \Delta A_{340}(\text{positive control}) - \text{ave } \Delta A_{340}(\text{negative control})}. \quad (3)$$

12. Compounds were sorted based on POC.

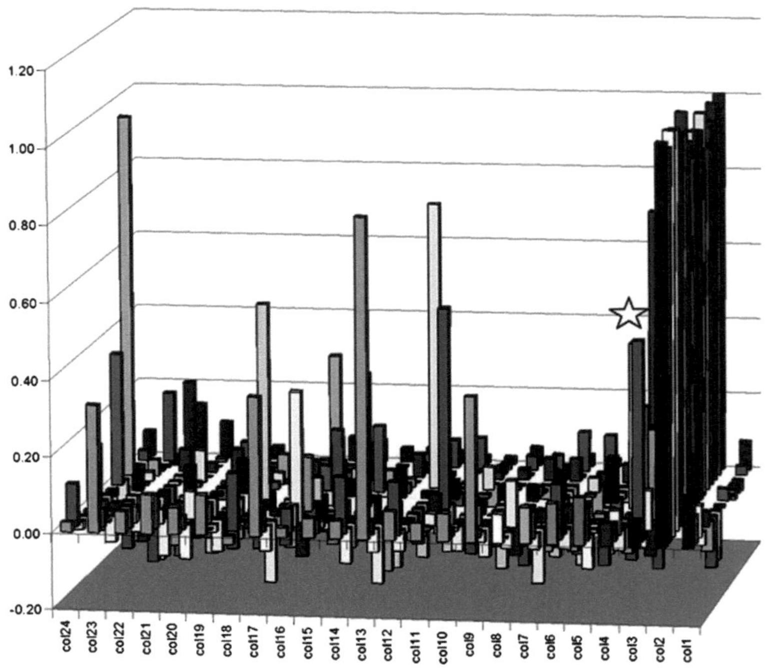

Fig. 3. A bar graph representation of a single 384-well plate. All data is presented as POC, as defined in the text. Plate columns (col#) are marked on the front *x*-axis and POC is marked on the *y*-axis. Plate rows on the *z*-axis are not labeled. "Col1" on the *far right* contains the negative control series. "Col2" contains the positive control series. This is an unusual plate that contained several hit compounds; most plates contained no hit compounds. One of the compound discussed as an example below is labeled with a *star* above the representative bar.

3.2. Confirmation Assay

The goal of this study was to identify potential drug leads from the compound library, not to identify all possible interacting compounds represented in the library. Therefore, false negatives are not a concern. In contrast, false positive are a major concern. Potential sources for false positives could arise if a library compound (included, but may not be limited to): (1) changes color over the time course of the assay, (2) slowly reacts with NADH, thereby increasing the rate of A_{340} reduction, (3) causes LDH to use NADH in a pyruvate-independent manner, (4) binds to one active site of hL-PYK and mimics the homotropic cooperativity to increase the PEP affinity in a second active site of the protein, and (5) interacts synergistically with DMSO or Triton-X-100 to modify enzymatic activity, but which would not have an impact in the absence of DMSO or Triton-X-100.

Although it is common to run confirmation assays for the top 0.3% of hit compounds, this threshold contained only compounds that showed POC scores greater than 70. As demonstrated in Fig. 2, the removal of inhibition due to phosphorylation or ATP will, at best, result in a POC score of 50. Therefore, the top 0.3%

is not likely to contain compounds that bind competitively with ATP or that prevent inhibition by phosphorylation. In an effort to include these other types of drug leads, the top 0.6% compounds were selected. The same assay used for initial compound identification was used for the confirmation assay, with the one exception that hit compounds were transferred by hand from stock solutions. Four hundred and seventy two compounds were confirmed to cause activation, as defined by a POC greater than 40 (This value allows for at least 20% error of the POC score of 50 used in selecting the top 0.6% of hit compounds) in the confirmation assay.

At the same time that confirmation assays were read, additional control plates were analyzed. (1) To test for PEP-independent oxidation, the screen assay was read with the modification that 0.45% Triton-X-100 was added to columns 1 and 3–24, in place of the 6.3 mM PEP in 0.45% Trition-X100 previously used (see Note 9). PEP was added to column 2 to maintain the positive control. (2) hL-PYK independent oxidation of NADH was evaluated by repeating the screen assay for hit compounds; with the exception that hL-PYK was not included for columns 1 and 3–24. hL-PYK was added to column 2 to maintain the positive control. Only nine compounds were removed due to PEP/hL-PYK-independent oxidation of NADH.

Once hit compounds have been identified, there are a number of software options to aid in grouping compounds with similar chemical features. A common next step in drug screen strategies is to screen customized libraries containing compounds with the summarized chemical features obtained from the initial library screen. However, the multiple potential mechanisms of action anticipated by our design (see Fig. 2) does not lend itself to this cyclic screen approach.

3.3. Distinguishing K-Type vs. V-Type Effectors

To exemplify how one might further characterize the allosteric impact of a hit compound (i.e. a drug lead) after initial identification, the experimental approach to distinguish if a hit compound functions through an impact on the affinity for PEP (K-type response) vs. on V_{max}, i.e. k_{cat}, (V-type response) is detailed here.

Activity assays were used to determine initial velocity over a concentration range of PEP. From this, data maximum velocity (V_{max}) and substrate affinity/apparent affinity ($K_{app\text{-}PEP}$) are determined. In turn, by determining V_{max} and $K_{app\text{-}PEP}$ over a concentration range of the effector the influence of the effector on these parameters can be quantitatively assessed. Activity was measured by coupling the pyruvate production by hL-PYK with NADH oxidation catalyzed by lactate dehydrogenase.

1. Immediately before use, an enzyme cocktail was mixed. This cocktail included 36.96 ml of 10× assay buffer, 960 μl of 100 mM NADH, 85 U/ml of lactate dehydrogenase,

1.848 ml of 400 mM ADP, and solid DTT to result in a final 2 mM concentration. Although the volume of wild type hL-PYK to be added to the enzyme cocktail was dependent on the activity of the individual protein preparation; the goal was to add sufficient activity to result in 0.08 ΔA_{340}/min in the cuvette once effector and PEP were added. As a final addition to the enzyme cocktail, water was added to a total volume of 79 ml.

2. Three additions were made to each well in a 96-well plate: (1) 195.5 μl of the appropriate effector concentration; (2) 74.5 μl of the enzyme cocktail; (3) the reaction was initiated by the addition of 80 μl of the appropriate PEP concentration. Since PEP and effectors were made as serial dilutions, 12 PEP concentrations were varied across columns 1–12 of the plate and 8 concentrations of effector were varied across rows A–H of the plate.

3. Changes in A_{340} of NADH over time were monitored using a Molecular Device Spectramax Plus384 spectrophotometer. Assays were at 30°C. This temperature satisfies the ambient + 5°C temperature requirement of the Molecular Devices Spectramax Plus384.

4. Once the initial velocity dependence on PEP concentration has been fit to a Michaelis–Menten or Hill equation, the fit parameters (V_{max} and $K_{app\text{-}PEP}$) can be plotted as a function of effector concentration (Fig. 4). This data presentation alone can allow for qualitative data interpretation. However, equations for fitting this data to allow for quantitative interpretation are also available (13–15). To exemplify such fits, the control Fru-1,6-BP data included in Fig. 4a is fit to:

$$K_{app-PEP} = K_a\left(\frac{K_{ix} + [\text{Effector}]}{K_{ix} + Q_{ax}[\text{Effector}]}\right) \quad (4)$$

where $K_a = K_{app\text{-}PEP}$ when [Effector] = 0, K_{ix} = the dissociation constant for effector when [PEP] = 0 and Q_{ax} = the coupling constant between PEP and the effector.

We can draw a number of conclusions about the four compounds represented in Fig. 4. Each "activates" by influencing PEP apparent affinity, without modifying V_{max} activity. Each binds to hL-PYK with much lower affinity than Fru-1,6-BP. The full activation (quantitative analysis of Q_{ax} from Eq. 3) for each of the four compounds cannot be assessed with this data because the lower plateau is not defined by the effector concentration range included. Since the data in Fig. 4 were collected with the wild type enzyme and in the absence of alanine and ATP, it seems likely that the representative four compounds increased the activity in

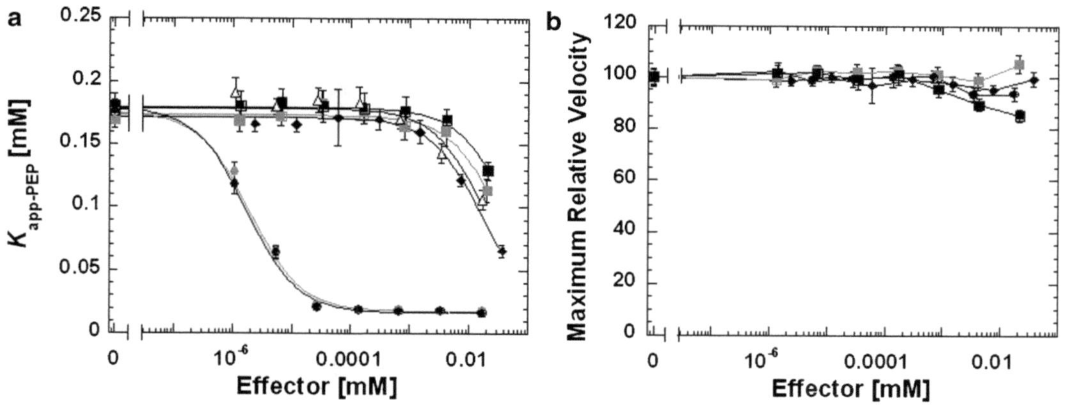

Fig. 4. (**a**) $K_{app\text{-}PEP}$ and (**b**) maximal relative velocity as a function of effector concentration. To improve solubility of compounds, three of the compounds were used in the presence of 1% ethanol (data in *black*). A fourth compound was used in the presence of 1% DMSO (data in *gray*). $K_{app\text{-}PEP}$ as a function of Fru-1,6-BP (*solid circle*), either with 1% ethanol or with 1% DMSO is included in (**a**) as a control and lines running through data for this control represent the best fit to Eq. 4. Without differentiating the four compounds included here, it is clear that each modifies $K_{app\text{-}PEP}$ without modifying maximal velocity.

the initial library screen by an allosteric activating mechanism, rather than by removing inhibition by alanine, ATP or the phosphorylation mimic.

4. Notes

1. The use of the words "activation" and "inhibition" should be further explored in the context of allosteric regulators. Traditionally, activation and inhibition refer to increased or decreased rates of catalysis. In the context of a simple Michaelis–Menten model of substrate binding, two parameters are necessary to describe initial velocities of enzymatic reactions plotted as a function of substrate concentration; V_{max} and $K_{app\text{-}PEP}$. It follows that allosteric effectors that increases or decreases V_{max} are considered activators or inhibitors, respectively. It may be less intuitive why effectors that modulate $K_{app\text{-}PEP}$ are also commonly referred to by the terms activator and inhibitor. This more liberal use of these terms can best be appreciated by considering a simplistic experimental design that focus on assays at only a single substrate concentration (Fig. 5). When assaying at a single substrate concentration, an increase or a decrease in activity is observed for effectors that either facilitate or antagonize substrate affinity, respectively.
2. Library compounds are often in a low concentration of DMSO. The stability of enzyme activity in the equivalent concentration of DMSO should be confirmed before initiating a screen.

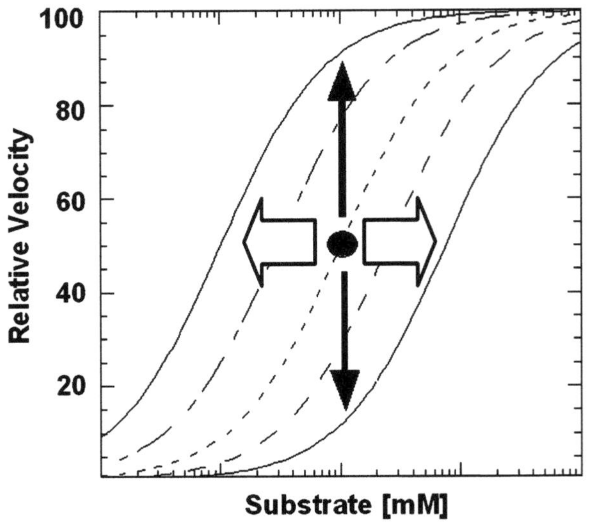

Fig. 5. "Activation" and "inhibition" in an enzyme with allosteric regulation of substrate affinity. The *open block arrows* indicate effector influences on substrate affinity (i.e. left or right displacement of the curve.). However, if assays include only a single substrate concentration (i.e. the concentration at the *black dot*), the only observation is increase or decrease in initial velocity, as represented by the *solid vertical arrows*. As a result, allosteric effectors that facilitate substrate affinity are commonly referred to as activators and those that antagonize substrate affinity are commonly referred to as inhibitors.

3. Compounds in libraries are often pre-dispensed into the plates to be used. When using a spectroscopic technique, such as change in absorbance, it is necessary to confirm that the plates in use allow transmission of the proper wavelength.
4. Considering assays at one concentration of effector is a primary concern for high-throughput screening. Due to the limited number of measurements that can be collected for each compound when screening in a high-throughput setting, these screens are typically performed at one, or a limited number of (16), substrate concentrations. The detection used in our study was an increase in initial velocity when assayed at the same concentration of PEP as control experiments.
5. Evidence supports that colloidal aggregates of library compounds may be a cause of nonspecific effects (17, 18). Colloid formation and the resulting nonspecific effects of these colloids can be prevented by the addition of Triton-X-100 (19, 20). However, before screening the library, it should be confirmed that the enzyme being screened retains activity in the presence of the detergent. Follow-up assays are also necessary to ensure that identified compound leads impact enzyme function in the absence of Triton-X-100 and

observed effects are not due to a synergistic effect between Triton-X-100 and the library compound. In the presence of the 0.01% Triton-X-100 (final in well concentration) used in our assays, the use of a Multidrop Micro to make additions to plates resulted in the formation of foam. This problem was avoided by using the larger diameter/lower flow rate of the Multidrop 384.

6. Precautions should be taken when using detections other than enzymatic activity. Detection such as protein conformational shifts detected by changes in tryptophan fluorescence that result from the binding of a library compound can indicate binding with no influence on active site function (4).
7. Although single point assays are common in high-throughput screens, the use of such an approach with an optical detection fails to consider that the library compound may also contribute to spectroscopic properties. Therefore if using an optical property of the protein or assay mix, it is best to use a detection based on a change in the optical property as a function of time.
8. Timing between the first and a second read may be (as was true in the experiments described here) determined by the rate of data collection. In our study, 50 plates were read in 1 h by the Envision.
9. Although not included in this study, it is conceivable to assay in the absence of Triton-X-100 to confirm that hit compounds do not act synergistically with this detergent to result in activation, as well as assaying in the absence of phosphorylation or one of the allosteric inhibitors (Ala and ATP) as a means of distinguishing the mechanism by which each hit compound increase activity.

Acknowledgments

This work was supported in part by NIH grant DK78076. Service for screening of a chemical library will be purchased from the High-Throughput Screening Laboratory at The University of Kansas. I would like to thank Byron Taylor for his aid during screening of the 100,000 compound library. I also appreciate the contributions from Dr. Rathnam Chagaturu and Ashleigh Price.

References

1. Hardy JA, Wells JA (2004) Searching for new allosteric sites in enzymes. *Curr Opin Struct Biol* 14:706–715
2. Groebe DR (2006) Screening for positive allosteric modulators of biological targets. *Drug discovery today* 11:632–639
3. Groebe DR (2009) In search of negative allosteric modulators of biological targets. *Drug discovery today* 14:41–49
4. Fenton AW (2008) Allostery: An illustrated definition for the 'second secret of life'. *Trends Biochem Sci* 33:420–425
5. Soudijn W, Van Wijngaarden I, AP IJ (2004) Allosteric modulation of g protein-coupled receptors: Perspectives and recent developments. *Drug discovery today* 9:752–758
6. Williams R, Holyoak T, McDonald G, Gui C, Fenton AW (2006) Differentiating a ligand's chemical requirements for allosteric interactions from those for protein binding. Phenylalanine inhibition of pyruvate kinase. *Biochemistry* 45:5421–5429
7. Christopoulos A (2002) Allosteric binding sites on cell-surface receptors: Novel targets for drug discovery. *Nat Rev Drug Discov* 1:198–210
8. Gunasekaran K, Ma B, Nussinov R (2004) Is allostery an intrinsic property of all dynamic proteins? *Proteins* 57:433–443
9. Swain JF, Gierasch LM (2006) The changing landscape of protein allostery. *Curr Opin Struct Biol* 16:102–108
10. Fenton AW, Alontaga AY (2009) Chapter 5: The impact of ions on allosteric functions in human liver pyruvate kinase. *Methods Enzymol* 466:83–107
11. Fenton AW, Hutchinson M (2009) The ph dependence of the allosteric response of human liver pyruvate kinase to fructose-1,6-bisphosphate, atp, and alanine. *Arch Biochem Biophys* 484:16–23
12. Fenton AW, Tang Q (2009) An activating interaction between the unphosphorylated n-terminus of human liver pyruvate kinase and the main body of the protein is interrupted by phosphorylation. *Biochemistry* 48:3816–3818
13. Reinhart GD (2004) Quantitative analysis and interpretation of allosteric behavior. *Methods Enzymol* 380:187–203
14. Reinhart GD (1983) The determination of thermodynamic allosteric parameters of an enzyme undergoing steady-state turnover. *Arch Biochem Biophys* 224:389–401
15. Reinhart GD (1988) Linked-function origins of cooperativity in a symmetrical dimer. *Biophys Chem* 30:159–172
16. Inglese J, Auld DS, Jadhav A, Johnson RL, Simeonov A, Yasgar A, Zheng W, Austin CP (2006) Quantitative high-throughput screening: A titration-based approach that efficiently identifies biological activities in large chemical libraries. *Proc Natl Acad Sci U S A* 103:11473–11478
17. Feng BY, Simeonov A, Jadhav A, Babaoglu K, Inglese J, Shoichet BK, Austin CP (2007) A high-throughput screen for aggregation-based inhibition in a large compound library. *J Med Chem* 50:2385–2390
18. Feng BY, Shelat A, Doman TN, Guy RK, Shoichet BK (2005) High-throughput assays for promiscuous inhibitors. *Nature chemical biology* 1:146–148
19. McGovern SL, Helfand BT, Feng B, Shoichet BK (2003) A specific mechanism of nonspecific inhibition. *J Med Chem* 46:4265–4272
20. Feng BY, Shoichet BK (2006) A detergent-based assay for the detection of promiscuous inhibitors. *Nature protocols* 1:550–553

Part V

Computational Methods/Aids in the Study of Allostery

Chapter 21

A Critical Evaluation of Correlated Mutation Algorithms and Coevolution Within Allosteric Mechanisms

Dennis R. Livesay, Kyle E. Kreth, and Anthony A. Fodor

Abstract

The notion of using the evolutionary history encoded within multiple sequence alignments to predict allosteric mechanisms is appealing. In this approach, correlated mutations are expected to reflect coordinated changes that maintain intramolecular coupling between residue pairs. Despite much early fanfare, the general suitability of correlated mutations to predict allosteric couplings has not yet been established. Lack of progress along these lines has been hindered by several algorithmic limitations including phylogenetic artifacts within alignments masking true covariance and the computational intractability of consideration of more than two correlated residues at a time. Recent progress in algorithm development, however, has been substantial with a new generation of correlated mutation algorithms that have made fundamental progress toward solving these difficult problems. Despite these encouraging results, there remains little evidence to suggest that the evolutionary constraints acting on allosteric couplings are sufficient to be recovered from multiple sequence alignments. In this review, we argue that due to the exquisite sensitivity of protein dynamics, and hence that of allosteric mechanisms, the latter vary widely within protein families. If it turns out to be generally true that even very similar homologs display a wide divergence of allosteric mechanisms, then even a perfect correlated mutation algorithm could not be reliably used as a general mechanism for discovery of allosteric pathways.

Key words: Allostery, Thermodynamic coupling, Correlated mutation, Covariance, Coevolution, Mutual information

1. Introduction

Starting with the proposal by Horovitz et al. in 1994 (1), there has been a deep interest in predicting allosteric couplings within proteins based on coevolutionary processes. The intuitive approach is based on multiple sequence alignment column pairs displaying correlated mutations, which have been interpreted to reflect coordinated changes that maintain pairwise intramolecular couplings. The premise is that when a mutation occurs within a protein, a compensating mutation can occur elsewhere, and conserved patterns of these pairs across a multiple sequence alignment are

Aron W. Fenton (ed.), *Allostery: Methods and Protocols*, Methods in Molecular Biology, vol. 796,
DOI 10.1007/978-1-61779-334-9_21, © Springer Science+Business Media, LLC 2012

interpreted as a signal of coevolutionary processes. The seminal paper by Lockless and Ranganathan (2) argued that such an approach predicts thermodynamic coupling in proteins. Specifically, they demonstrated that the extent of non-additive binding energy from double mutant cycles within PDZ domains is linearly related to the strength of the correlated mutation signal. Since then, the explosion of publically available protein sequences has ensured that coevolutionary analysis has continued to develop as a staple within the field of protein bioinformatics. However, despite more than 10 years of subsequent vigorous research, there remains no statistically significant demonstration of the ability for correlated mutation algorithms to predict intramolecular couplings over long distances (3–5).

In their study, Lockless and Ranganathan utilized a novel algorithm (named SCA) for detecting correlated mutations. In the following decade since the publication of their paper, extraordinary progress has been made in the study of correlated mutation algorithms, which is highlighted by dramatic improvements in the ability of correlated mutations to predict structure contacts. Conversely, little progress has been made related to prediction of allosteric mechanisms. In the first part of this review, we summarize this progress and the remaining algorithmic problems that need to be addressed. In the second part of our review, we present our arguments for why we believe that the enterprise of predicting allosteric couplings from correlated mutations may be based on flawed assumptions of allostery. We suggest that there is no underlying reason to believe that coevolution of allosteric mechanisms actually occurs routinely. While allosteric mechanisms can be conserved within close taxonomic groups (6), anecdotal reports indicating that allostery is not a strong evolutionary driving force are increasingly commonplace (7–9). Surprisingly, diversity of response is even evident in hemoglobin (10, 11), an archetype of long-range intramolecular communication. Clearly, conservation of allostery and intramolecular couplings is nowhere near that of structure and/or catalytic function (12). Moreover, allosteric pathways are both frequent (13) and mechanistically plastic (14, 15). As a consequence, even if the algorithms were perfect, we remain to be convinced of the underlying notion that allosteric couplings can be recovered from correlated mutations.

2. Improved Correlated Mutation Algorithms

The limiting factor of correlated mutation analysis is a low signal-to-noise ratio (16). (See Note 1 for a survey of common algorithms used to identify correlated mutations.) In a 2004 comparison of four different covariance algorithms, Fodor and Aldrich found that Mutual Information (MI) was the worst performing algorithm in predicting residue contacts (17). Based on their most widespread application, the predictive power of correlated mutation algorithms was assessed by the ability to predict structural contacts. The poor performance of MI was ascribed in part to its tendency to give high scores to random (or poorly conserved) columns. Nevertheless, MI remains an appealing approach because it is stated in the simple, formal language of the application of information theory to entropy and is, therefore, easy to understand and calculate. As such, a good deal of recent research has found ways to dramatically improve the performance of MI. As a result, the past 5 years has produced a number of updated MI methods that are now the most powerful correlated mutation algorithms. However, it is interesting to note that some of the algorithmic improvements applied to MI are also applicable to other approaches (16). Below, we summarize this recent exciting progress.

2.1. Reducing Phylogenetic Noise Improves Predictive Power

A critical factor limiting the predictive power of covariance methods originates from tangling of phylogenetic and correlated mutation relationships (18). Consider two subfamilies in which the sequences within each are closely related, but the intersubfamily relationships are more distant. An alignment that consists of sequences from both subfamilies will have a great deal of apparent covariation due to the many changes in each column that are correlated with changes in other columns, where each alignment column identifies orthologous positions across the dataset. However, these changes reflect a phylogenetic artifact of the way the alignment was constructed and do not reflect the underlying structure or functional constraints on the protein. This problem has been widely recognized but is difficult to correct because the true phylogenetic history of a family is unknown. Several algorithms have attempted to reduce bias by using rigorous phylogenetic approaches that take into account evolutionary distances within the family (19–21). These methods have been demonstrated to improve structural contact identification, but they are computationally intensive and are, therefore, not appropriate for the many cases where alignments are made up of thousands or even tens of thousands of protein sequences. Simple ad hoc methods, such as removing overly similar sequences from the alignment, can be easily employed, but this is a blunt approach with arbitrary parameters. Similarly, owing to disparate evolutionary forces, others have attempted to remove paralog sequences; however, the presence of paralogs has been shown to actually improve correlated mutation identification in some cases (22). Alternatively, others have developed methods based on physiochemical properties of the identified residue pairs (23) or a complicated number of algorithmic filters (24). In both cases, improvements in contact prediction accuracy have been reported. However, the ad hoc nature of these methods suggests that they are particularly tuned to prediction of structural contacts, and it is unclear the degree to which they actually filter phylogenetic biases.

In 2008 (25), a simple and computationally efficient method to suppress phylogenetic bias that dramatically improves contact prediction performance was introduced. Rather than attempting a phylogenetic reconstruction, this method normalized the observed covariance of a pair of columns by the background covariance of the columns, where background covariance is measured as the average covariance score of a column with all other columns (see Note 2). With this correction in place, MI went from the worst performing algorithm (17) to the best, easily outperforming previously described methods (25). For the rest of this chapter, we refer to MI with this correction as MI_p. Application of the MI_p correction to other correlated mutation algorithms also improves predictive power (16), highlighting that phylogenetic bias is a general problem and is not limited to just MI.

2.2. There Is No Generally Accepted Method to Produce "Correct" Alignment Inputs

Correlated mutation algorithms obviously require a multiple sequence alignment as input. As such, collation and alignment of the sequence dataset are critical first steps in this process. Unfortunately, there is little consensus on dataset and alignment protocols. Alignment algorithms are especially prone to errors at alignment ends and in gapped positions. Recently, it has been demonstrated that the introduction of even modest alignment errors could produce a substantial number of false positives when using MI_p (26). The poor performance of MI is caused in part by sensitivity to the background conservation of each column in the alignment (17), and the MI_p correction only partially removes this sensitivity (27). To correct for this, Little and Chen (27) introduced a further correction to MI_p that regressed background and observed MI scores against one another and used residuals from the regression model normalized as *Z*-scores as the covariance score. Using a separate mathematical formulation of the algorithm that produces essentially identical results, Dickson et al. (26) demonstrated that this further improvement to MI_p reduces sensitivity to alignment artifacts. Even for this algorithm, however, alignment errors can still substantially mask true covariance hampering the sensitivity of the correlated mutation approach (26). A fully automated multiple sequence alignment procedure that produces alignments without errors is still beyond the reach of current bioinformatics (28), and as such this issue remains a potential problem that can hinder successful application of covariance techniques. Interestingly, covariance algorithms may, themselves, be an important tool for determining when an alignment has an error (26). Clearly, there is much work left to be done in this area.

Another largely unsolved problem is how many sequences are required for successful application of covariance techniques. Gloor et al. have demonstrated that at least 125 sequences must be considered before the random MI signal is surpassed by the true correlated mutation signal (29). Others go further; for example, it has been suggested by Hamacher et al. that 200–300 sequences are required (30), whereas Nielsen et al. suggest that at least 400 sequence "clusters" are needed (31) where a cluster indicates groups of sequences that are related to one another by some similarity threshold, typically ~60% identity. Interestingly, application of the joint alignment background correction used by MI_p to McBASC improves predictive power for alignments up to ~100 sequences (16), although MI_p outperforms the corrected McBASC in larger sequence alignments. Guidelines for the minimal number of required sequences remain empirically derived and development of a more rigorous theory to guide algorithm choice for a given alignment may be helpful. Recent work looking at the impact of how different assumptions used in calculating a probability of a residue in a column from the number of times that residue is observed in a column may be a first step in this direction (32). Also showing great promise is a bootstrap approach that randomly divides an alignment into subsets many times over, and asks how often the same set of covarying pairs is observed in each permutated subalignment (33). Intriguingly, this approach found that some covariance algorithms were more accurate while others were more reproducible, although neither the accuracy nor reproducibility of any of the algorithms was perfect. Going forward, explicit consideration of the trade-offs in power vs. sensitivity to alignment artifacts should help guide future algorithm design.

2.3. Moving Beyond Pairwise Covariance

Even with the above improvements made to covariance algorithms, nearly all protein families contain a large number of covarying pairs that are not in close physical contact (34). These "covarying but distal" pairs may reflect algorithmic errors or phylogenetic biases in the alignment that still remain uncorrected for by the covariance algorithms. Alternatively, a more intriguing hypothesis is that the covarying residues are functionally linked through space by energetic couplings, which includes allosteric mechanisms. A recent paper that rigorously demonstrates that most distal coevolving pairs are simply explained by coevolving contact chains provides considerable support for this view (34). Therein, using a Bayesian network model, Burger and van Nimwegen demonstrate that most covarying distal pairs are in fact connected by chains of residues that are also covarying. Because of the Bayesian formalism, this approach is generally computationally efficient, especially as compared to methods with similar intentions such as Weigt et al. (35). Instead of using the MI_p score to rank residues, Burger and van Nimwegen propose using a posterior probability reflecting the strength of the MI_p score between the two residues relative to the covariance scores of all possible residues that link the two residues in a chain. Remarkably, this procedure dramatically improves the performance of MI_p, reflecting the second major improvement to the performance of correlated mutation algorithms that has been described in the last 3 years (25, 27, 34). Considering the fact that these algorithms have been actively studied since the early 1990s (36, 37), this progress is both unexpected and exciting.

3. A Critical View of the Underlying Concept of Conserved Allosteric Pathways

Functional allosteric mechanisms involve a subset of the energetic coupling present within a protein structure. Nevertheless, we now expand our discussion to consider couplings between all residue pairs, whether they are related to functional efficiency or not, which allows us to consider double mutant cycles. The stability of a protein, ΔG, compares the free energy of the folded vs. unfolded state, and the stability of a protein double mutant is given by $\Delta G_{ij} = \Delta G_i + \Delta G_j + \delta_{ij}$. The δ_{ij} term quantifies the amount of nonadditivity within the cycle relative to the sum of the constituent single mutants, which identifies thermodynamic coupling between the pair of single mutants (see Note 3). It has been appreciated for over 25 years that nonadditivity within double mutant cycles is trivially expected within structural neighbors (38). Conversely, when nonadditivity occurs within distal pairs, the physical origins of the coupling is less obvious. While correlated mutations can be used to identify functionally important residues (39), there is little evidence to suggest that thermodynamically coupled pairs are limited to correlated mutations (3). Rather, no correlation is observed between δ_{ij} and correlated mutation scores in three example protein families with good double mutant cycle coverage (3). In fact, long-range thermodynamic coupling is, in itself, quite rare across the three datasets, which is consistent with the much larger double mutant dataset considered by Istomin et al. (40).

Nevertheless, the findings of Burger and van Nimwegen that consideration of covarying chains improves predictive power over considerations of residue pairs alone seems, at first glance, consistent with ideas put forward by the Ranganathan lab (2, 41, 42). Perhaps these chains of correlated mutations that we can now rigorously identify with Bayesian statistics reflect long-range couplings, including the allosteric couplings proposed by Lockless and Ranganathan (2). Suel et al. (41) assert a "sparse network" of allosteric interactions, which may be related to the chains of covarying residues identified by Burger and van Nimwegen (34). We hope that investigators will use the new and improved tools for detecting covarying chains to test this relationship. While the veracity of this link will be ultimately decided by how well predictions derived from these new algorithms match mutagenesis experiments, our suspicion is that the information that can be gleaned from multiple sequence alignments is unlikely to reflect the plasticity of allosteric mechanisms (43), even with the improved covariance detection algorithms.

Our skepticism arises from our sense that the physical basis of the long-range intramolecular couplings that underlie allosteric response remains ambiguous (44). Consistent with the idea of coevolving chains, *molecular wires* describe allostery as a cascade of local induced fit events that sequentially propagate over long distances (45–47), like a series of dominos falling in a line. Conversely, concerted *population shift* models describe pairwise couplings based on global changes in the free energy landscape (48, 49), which is akin to the conformer selection model of ligand binding (50). Regardless of which model is "correct," both stress the importance of protein dynamics in allostery (51, 52). That is, upon perturbation of an allosteric site, a signal is propagated to the effected site via a complex and dynamic change in structure. It is exactly this point that calls into question the notion that allosteric pathways are precisely conserved across a family. The literature includes nearly countless examples demonstrating that protein structures and their dynamics are highly sensitive to small perturbations, regardless of whether the perturbation is mutation (53), ligand binding (54), or simply changing the type of metal ion bound to the protein (55). Related, several reports stress the diversity within dynamic signatures across protein families (56, 57). If point mutations can perturb allostery in multiple ways, then it stands to reason that even closely related orthologs that have as much as 90–95% sequence identity may also have drastically different allosteric pathways. If this is true, it seems unlikely to us that the information required to predict these allosteric pathways could be contained in a multiple sequence alignment.

An appreciable number of sites are frequently identified as being critical to intramolecular communication within a given structure (58). The large amount of variation within intramolecular couplings makes it difficult to uncover general "traffic rules" regarding allosteric mechanisms across a protein family. For example, as discussed by del Sol et al. (13), the plasticity within an allosteric response is such that nearby residues can easily functionally substitute for one another. Related, it has been shown in several systems (e.g., pyruvate kinase (59)) that mutations at highly variable positions distal to the active site can have dramatic impacts on catalytic activity. As such, based on the ubiquitous diversity of allosteric mechanisms, Kuriyan and Eisenberg (43) have intriguingly suggested that allosteric diversity is responsible for the complexity of life. They argue for a "*rule of varied allosteric control*" where sensitivity in allosteric response is a fundamental evolutionary mechanism used to discover new pathways and functions. We are not arguing that such allosteric control does not ultimately arise from sequence or that it is not subject to selection. Rather, we are asserting that allosteric control is so sensitive to context that it is unlikely to be recovered from sequence alignment information. The underlying assumption of a sequence alignment is that the sequences within that alignment share something in common evolutionarily. If sequences within an alignment are unlikely to share a common allosteric pathway, then that pathway cannot be reconstructed from the alignment.

As an instructive example, consider the case of cyclic-nucleotide gated ion channels. The opening of these channels can be modeled as two independent steps: a ligand binding step, followed by a fully-liganded allosteric transition from the closed to the open state (60). Consistent with our arguments of the general complexity of allostery in proteins, mutations throughout the channel sequence can alter the free energy of the allosteric transition without affecting ligand binding kinetics (60). In the PFAM database, there are 11,189 sequences in the cNMP-binding family (PF00027) that contain a cyclic nucleotide binding domain. We can be reasonably confident that nearly all of these protein domains share the same fold and that most will bind cyclic nucleotides. One might expect a functional covariance analysis to find residues that are involved in, for example, the discrimination of cAMP vs. cGMP binding, whereas the allosteric mechanisms found in these proteins will vary widely and depend on many factors not described by the alignment such as interactions with other protein domains and the microenvironment in which the protein is expressed. We assert that there is little reason to suppose a common allosteric mechanism shared by all, or even a significant fraction, of these 11,000 sequences.

On the contrary, allostery is likely conserved across very short evolutionary timescales. For example, one of us has recently demonstrated using computational modeling that, while residue-specific differences within CheY allostery are large and frequent, there is a general tendency for residues that initiate allostery to be structurally clustered (61). Furthermore, we also observe stronger correlation within allostery across two closely related *Escherichia coli* and *Salmonella typhimurium* CheY orthologs, whereas there is greater diversity when compared to the more divergent *Thermotoga maritima* ortholog. In addition, quantitative differences in allosteric regulation have been used as a molecular basis of taxonomic assignments within the 3-deoxy-D-arabino-heptulosonate-7-phosphate synthetase family (6, 62, 63).

That is, while the extent of allosteric response is overall variable across the family, response is conserved within taxonomic groups, highlighting that conservation of allostery can occur over short evolutionary distances.

As such, a challenge to our assertion that allosteric information cannot be gleaned from sequence alignments will, therefore, require a departure from the current practice of building alignments based on every sequence that can be found with sufficient conservation to the experiential protein of interest. If there is any hope of detecting allostery in proteins using multiple sequence alignments, we assert that it will require restricting alignments to a subset of proteins for which there is reason to believe there is a common allosteric mechanism. However, as we have discussed above covariance algorithms tend to perform better on alignments containing large numbers of sequences. Unfortunately, there is no rigorous theory to guide the minimum number of sequences required for covariance analysis. It will be interesting to see in the next few years how the tension between the requirement for limiting alignment to proteins with well-described functions balances against a requirement for sequence depth in covariance analysis.

4. Conclusions

In conclusion, the last few years have seen rapid development in the sophistication and power of algorithms for discovery of correlated mutations. Application of this new generation of algorithms should improve the utility of covariance methods in structure determination and the discovery of functionally and structurally constrained residues. The future of these techniques for discovery of prediction of allosteric pathways remains less clear, although we expect continued experimental effort focused on this question.

5. Notes

1. A large number of algorithms predict correlated mutations, see ref. 64 for a good review. However, in this chapter we focus on the four commonly applied approaches: (1) mutual information, (2) linear correlation, (3) phylogenetic perturbation, and (4) chi-square analysis. Below, we briefly summarize each approach.

   Linear correlation. One of the most popular correlated mutation identification algorithms is based on the Pearson correlation coefficient (36). Juxtaposed to the other methods described, this approach is unique in that it uses a substitution matrix. Given N sequences, all $N(N-1)/2$ possible mutation pairs within an alignment column are scored using the substitution matrix, from which a vector of substitution scores is constructed. Each vector is then compared to every other vector using the Pearson correlation coefficient, where highly correlated columns correspond to coevolving positions. As generally implemented, the approach uses the McLachlan substitution matrix (65), which has led to it commonly being referred to as McBASC for *Mc*Lachlan *B*ased *S*ubstitution *C*orrelation. In a comparison of the four considered methods by Fodor and Aldrich (17), it was shown that McBASC is one of the two most accurate correlated mutation algorithms, as assessed by prediction of structural contacts.

Mutual information. Based in probability theory, *M*utual *I*nformation (MI) measures the mutual dependence of two random variables (37, 66). The measure of MI is supposed to measure what we know about column I given column J. Specifically, MI measures the amount of uncertainty in I that has been removed if the uncertainty in J is know. This is expressed by $\mathrm{MI}(I,J) = H(I) + H(J) - H(I,J)$, where $H(I)$ and $H(J)$ are common Shannon entropies. The Shannon entropy of column X is calculated by $H(X) = -\sum_x^{20} p_x \cdot \log p_x$. The joint entropy is calculated similarly using $H(X,Y) = -\sum_x^{20}\sum_y^{20} p_{x,y} \cdot \log p_{x,y}$, where the double sum accounts for all di-residue pairs. Approaches based on MI are attractive due to the intuitiveness of its interpretation; however, the approach is, without phylogenetic corrections, one of weakest correlated mutation algorithms in general use.

Phylogenetic perturbation. Phylogenetic perturbation methods quantify how much the residue distribution at column I is changed upon perturbation of the distribution at column J. Specifically, a subalignment is constructed that alters the distribution at column J. Thereafter, an updated residue distribution is determined within the subalignment. The statistical coupling energy is simply the difference of these two quantities. As originally implemented in Ranganathan et al.'s *S*tatistical *C*oupling *A*nalysis (SCA) software (2, 41, 67), the "phylogenetic perturbation" is based upon the most conserved residue within that position. The original SCA algorithm is similar to MI in its predictive power; however, a more powerful SCA algorithm has recently been developed (42).

Chi-square analysis. Finally, correlated mutations can be identified using an analysis partially derived from the chi-squared statistic, which is sometimes referred to as OMES for *O*bserved *M*inus *E*xpected *S*quared (68). The OMES score is calculated by $X_{I,J}^2 = \sum_i^L (N_i^{obs} - N_i^{ex})^2 / N_i^{valid}$, where L is a list of all possible nongapped residue pairs, N^{obs} is the number of times each pair is observed, N^{ex} is the number of times each pair is randomly expected to occur based on residue frequency, and N^{valid} is the number of sequences in the alignment that are do not have gaps at both positions. The predictive power of OMES is similar to that of McBASC.

2. Two background estimates were actually developed by Gloor et al. While closely related, the estimates vary depending upon whether the background MI is treated as additive or multiplicative, but those details are beyond the scope of this chapter.
3. In many instances, including the Lockless and Ranganathan paper, changes in substrate binding free energy are considered instead of changes in folding free energy. Nevertheless, the interpretation of the δ_{ij} as a reporter of thermodynamic coupling is unaffected.

Acknowledgments

The authors would like to thank Richard W. Aldrich and Gregory B. Gloor for helpful comments on the manuscript, and Donald J. Jacobs for numerous insightful discussions related to the correlated mutation algorithms, allostery, and the relationships therein.

References

1. Horovitz, A., Bochkareva, E. S., Yifrach, O., and Girshovich, A. S. (1994). Prediction of an inter-residue interaction in the chaperonin GroEL from multiple sequence alignment is confirmed by double-mutant cycle analysis. *J Mol Biol* **238**, 133–8.
2. Lockless, S. W., and Ranganathan, R. (1999). Evolutionarily conserved pathways of energetic connectivity in protein families. *Science* **286**, 295–9.
3. Fodor, A. A., and Aldrich, R. W. (2004). On evolutionary conservation of thermodynamic coupling in proteins. *J Biol Chem* **279**, 19046–50.
4. Chi, C. N., Elfstrom, L., Shi, Y., Snall, T., Engstrom, A., and Jemth, P. (2008). Reassessing a sparse energetic network within a single protein domain. *Proc Natl Acad Sci U S A* **105**, 4679–84.
5. Liu, Z., Chen, J., and Thirumalai, D. (2009). On the accuracy of inferring energetic coupling between distant sites in protein families from evolutionary imprints: illustrations using lattice model. *Proteins* **77**, 823–31.
6. Jensen, R. A., and Stenmark, S. L. (1970). Comparative allostery of 3-deoxy-D-arabino-heptulosonate-7-phosphate synthetase as a molecular basis for classification. *J Bacteriol* **101**, 763–9.
7. Jensen, A. A., and Spalding, T. A. (2004). Allosteric modulation of G-protein coupled receptors. *Eur J Pharm Sci* **21**, 407–20.
8. May, L. T., Avlani, V. A., Sexton, P. M., and Christopoulos, A. (2004). Allosteric modulation of G protein-coupled receptors. *Curr Pharm Des* **10**, 2003–13.
9. Hudson, J. W., Golding, G. B., and Crerar, M. M. (1993). Evolution of allosteric control in glycogen phosphorylase. *J Mol Biol* **234**, 700–21.
10. Royer, W. E., Jr., Knapp, J. E., Strand, K., and Heaslet, H. A. (2001). Cooperative hemoglobins: conserved fold, diverse quaternary assemblies and allosteric mechanisms. *Trends Biochem Sci* **26**, 297–304.

11. Royer, W. E., Jr., Zhu, H., Gorr, T. A., Flores, J. F., and Knapp, J. E. (2005). Allosteric hemoglobin assembly: diversity and similarity. *J Biol Chem* **280**, 27477–80.
12. Chakrabarti, S., and Panchenko, A. R. (2009). Coevolution in defining the functional specificity. *Proteins* **75**, 231–40.
13. del Sol, A., Tsai, C. J., Ma, B., and Nussinov, R. (2009). The origin of allosteric functional modulation: multiple pre-existing pathways. *Structure* **17**, 1042–50.
14. Cui, Q., and Karplus, M. (2008). Allostery and cooperativity revisited. *Protein Sci* **17**, 1295–307.
15. Formaneck, M. S., Ma, L., and Cui, Q. (2006). Reconciling the "old" and "new" views of protein allostery: a molecular simulation study of chemotaxis Y protein (CheY). *Proteins* **63**, 846–67.
16. Ashkenazy, H., and Kliger, Y. Reducing phylogenetic bias in correlated mutation analysis. *Protein Eng Des Sel* **23**, 321–6.
17. Fodor, A. A., and Aldrich, R. W. (2004). Influence of conservation on calculations of amino acid covariance in multiple sequence alignments. *Proteins* **56**, 211–21.
18. Wollenberg, K. R., and Atchley, W. R. (2000). Separation of phylogenetic and functional associations in biological sequences by using the parametric bootstrap. *Proc Natl Acad Sci U S A* **97**, 3288–91.
19. Noivirt, O., Eisenstein, M., and Horovitz, A. (2005). Detection and reduction of evolutionary noise in correlated mutation analysis. *Protein Eng Des Sel* **18**, 247–53.
20. Dimmic, M. W., Hubisz, M. J., Bustamante, C. D., and Nielsen, R. (2005). Detecting coevolving amino acid sites using Bayesian mutational mapping. *Bioinformatics* **21 Suppl 1**, i126-35.
21. Dutheil, J., Pupko, T., Jean-Marie, A., and Galtier, N. (2005). A model-based approach for detecting coevolving positions in a molecule. *Mol Biol Evol* **22**, 1919–28.
22. Ashkenazy, H., Unger, R., and Kliger, Y. (2009). Optimal data collection for correlated mutation analysis. *Proteins* **74**, 545–55.
23. Vicatos, S., Reddy, B. V., and Kaznessis, Y. (2005). Prediction of distant residue contacts with the use of evolutionary information. *Proteins* **58**, 935–49.
24. Kundrotas, P. J., and Alexov, E. G. (2006). Predicting residue contacts using pragmatic correlated mutations method: reducing the false positives. *BMC Bioinformatics* **7**, 503.
25. Dunn, S. D., Wahl, L. M., and Gloor, G. B. (2008). Mutual information without the influence of phylogeny or entropy dramatically improves residue contact prediction. *Bioinformatics* **24**, 333–40.
26. Dickson, R. J., Wahl, L. M., Fernandes, A. D., and Gloor, G. B. (2010). Identifying and Seeing beyond Multiple Sequence Alignment Errors Using Intra-Molecular Protein Covariation. *PLoS One* **5**, e11082.
27. Little, D. Y., and Chen, L. (2009). Identification of coevolving residues and coevolution potentials emphasizing structure, bond formation and catalytic coordination in protein evolution. *PLoS One* **4**, e4762.
28. Edgar, R. C., and Batzoglou, S. (2006). Multiple sequence alignment. *Curr Opin Struct Biol* **16**, 368–73.
29. Martin, L. C., Gloor, G. B., Dunn, S. D., and Wahl, L. M. (2005). Using information theory to search for co-evolving residues in proteins. *Bioinformatics* **21**, 4116–24.
30. Weil, P., Hoffgaard, F., and Hamacher, K. (2009). Estimating sufficient statistics in coevolutionary analysis by mutual information. *Comput Biol Chem* **33**, 440–4.
31. Buslje, C. M., Santos, J., Delfino, J. M., and Nielsen, M. (2009). Correction for phylogeny, small number of observations and data redundancy improves the identification of coevolving amino acid pairs using mutual information. *Bioinformatics* **25**, 1125–31.
32. Fernandes, A. D., and Gloor, G. B. Mutual information is critically dependent on prior assumptions: would the correct estimate of mutual information please identify itself? *Bioinformatics* **26**, 1135–9.
33. Brown, C. A., and Brown, K. S. Validation of coevolving residue algorithms via pipeline sensitivity analysis: ELSC and OMES and ZNMI, oh my! *PLoS One* **5**, e10779.
34. Burger, L., and van Nimwegen, E. Disentangling direct from indirect co-evolution of residues in protein alignments. *PLoS Comput Biol* **6**, e1000633.
35. Weigt, M., White, R. A., Szurmant, H., Hoch, J. A., and Hwa, T. (2009). Identification of direct residue contacts in protein-protein interaction by message passing. *Proc Natl Acad Sci U S A* **106**, 67–72.
36. Gobel, U., Sander, C., Schneider, R., and Valencia, A. (1994). Correlated mutations and residue contacts in proteins. *Proteins* **18**, 309–17.
37. Clarke, N. D. (1995). Covariation of residues in the homeodomain sequence family. *Protein Sci* **4**, 2269–78.
38. Mildvan, A. S., Weber, D. J., and Kuliopulos, A. (1992). Quantitative interpretations of

double mutations of enzymes. *Arch Biochem Biophys* **294**, 327–40.

39. Gloor, G. B., Martin, L. C., Wahl, L. M., and Dunn, S. D. (2005). Mutual information in protein multiple sequence alignments reveals two classes of coevolving positions. *Biochemistry* **44**, 7156–65.
40. Istomin, A. Y., Gromiha, M. M., Vorov, O. K., Jacobs, D. J., and Livesay, D. R. (2008). New insight into long-range nonadditivity within protein double-mutant cycles. *Proteins* **70**, 915–24.
41. Suel, G. M., Lockless, S. W., Wall, M. A., and Ranganathan, R. (2003). Evolutionarily conserved networks of residues mediate allosteric communication in proteins. *Nat Struct Biol* **10**, 59–69.
42. Halabi, N., Rivoire, O., Leibler, S., and Ranganathan, R. (2009). Protein sectors: evolutionary units of three-dimensional structure. *Cell* **138**, 774–86.
43. Kuriyan, J., and Eisenberg, D. (2007). The origin of protein interactions and allostery in colocalization. *Nature* **450**, 983–90.
44. Fenton, A. W. (2008). Allostery: an illustrated definition for the 'second secret of life'. *Trends Biochem Sci* **33**, 420–5.
45. Koshland, D. E. (1958). Application of a Theory of Enzyme Specificity to Protein Synthesis. *Proc Natl Acad Sci U S A* **44**, 98–104.
46. Yu, E. W., and Koshland, D. E., Jr. (2001). Propagating conformational changes over long (and short) distances in proteins. *Proc Natl Acad Sci U S A* **98**, 9517–20.
47. Ottemann, K. M., Xiao, W., Shin, Y. K., and Koshland, D. E., Jr. (1999). A piston model for transmembrane signaling of the aspartate receptor. *Science* **285**, 1751–4.
48. Swain, J. F., and Gierasch, L. M. (2006). The changing landscape of protein allostery. *Curr Opin Struct Biol* **16**, 102–8.
49. Kumar, S., Ma, B., Tsai, C. J., Sinha, N., and Nussinov, R. (2000). Folding and binding cascades: dynamic landscapes and population shifts. *Protein Sci* **9**, 10–9.
50. Monod, J., Wyman, J., and Changeux, J. P. (1965). On the Nature of Allosteric Transitions: a Plausible Model. *J Mol Biol* **12**, 88–118.
51. Gunasekaran, K., Ma, B., and Nussinov, R. (2004). Is allostery an intrinsic property of all dynamic proteins? *Proteins* **57**, 433–43.
52. Bruschweiler, S., Schanda, P., Kloiber, K., Brutscher, B., Kontaxis, G., Konrat, R., and Tollinger, M. (2009). Direct observation of the dynamic process underlying allosteric signal transmission. *J Am Chem Soc* **131**, 3063–8.
53. Schlegel, J., Armstrong, G. S., Redzic, J. S., Zhang, F., and Eisenmesser, E. Z. (2009). Characterizing and controlling the inherent dynamics of cyclophilin-A. *Protein Sci* **18**, 811–24.
54. Lee, A. L., Kinnear, S. A., and Wand, A. J. (2000). Redistribution and loss of side chain entropy upon formation of a calmodulin-peptide complex. *Nat Struct Biol* **7**, 72–7.
55. Mau, T., Baleja, J. D., and Wagner, G. (1992). Effects of DNA binding and metal substitution on the dynamics of the GAL4 DNA-binding domain as studied by amide proton exchange. *Protein Sci* **1**, 1403–12.
56. Forman, B. M., Umesono, K., Chen, J., and Evans, R. M. (1995). Unique response pathways are established by allosteric interactions among nuclear hormone receptors. *Cell* **81**, 541–50.
57. Conigrave, A. D., and Franks, A. H. (2003). Allosteric activation of plasma membrane receptors–physiological implications and structural origins. *Prog Biophys Mol Biol* **81**, 219–40.
58. Hardy, J. A., and Wells, J. A. (2004). Searching for new allosteric sites in enzymes. *Curr Opin Struct Biol* **14**, 706–15.
59. Pendergrass, D. C., Williams, R., Blair, J. B., and Fenton, A. W. (2006). Mining for allosteric information: natural mutations and positional sequence conservation in pyruvate kinase. *IUBMB Life* **58**, 31–8.
60. Fodor, A. A., Black, K. D., and Zagotta, W. N. (1997). Tetracaine reports a conformational change in the pore of cyclic nucleotide-gated channels. *J Gen Physiol* **110**, 591–600.
61. Mottonen, J.M., Jacobs, D. J., and Livesay, D. R. (2010). Allosteric response is both conserved and variable across three CheY orthologs. Biophys J **99**, 2245–2254.
62. Whitaker, R. J., Byng, G. S., Gherna, R. L., and Jensen, R. A. (1981). Comparative allostery of 3-deoxy-D-arabino-heptulosonate 7-phosphate synthetase as an indicator of taxonomic relatedness in pseudomonad genera. *J Bacteriol* **145**, 752–9.
63. Jensen, R. A., and Twarog, R. (1972). Allostery of 3-deoxy-D-arabino-heptulosonate 7-phosphate synthetase in Clostridium: another conserved generic characteristic. *J Bacteriol* **111**, 641–8.
64. Halperin, I., Wolfson, H., and Nussinov, R. (2006). Correlated mutations: advances and limitations. A study on fusion proteins and on the Cohesin-Dockerin families. *Proteins* **63**, 832–45.
65. McLachlan, A. D. (1971). Tests for comparing related amino-acid sequences. Cytochrome c and cytochrome c 551. *J Mol Biol* **61**, 409–24.

66. Atchley, W. R., Wollenberg, K. R., Fitch, W. M., Terhalle, W., and Dress, A. W. (2000). Correlations among amino acid sites in bHLH protein domains: an information theoretic analysis. *Mol Biol Evol* **17**, 164–78.

67. Hatley, M. E., Lockless, S. W., Gibson, S. K., Gilman, A. G., and Ranganathan, R. (2003). Allosteric determinants in guanine nucleotide-binding proteins. *Proc Natl Acad Sci U S A* **100**, 14445–50.

68. Kass, I., and Horovitz, A. (2002). Mapping pathways of allosteric communication in GroEL by analysis of correlated mutations. *Proteins* **48**, 611–7.

Chapter 22

The Advantage of Global Fitting of Data Involving Complex Linked Reactions

Petr Herman and J. Ching Lee

Abstract

In this chapter, we demonstrate the advantage of the simultaneous multicurve nonlinear least-squares analysis over that of the conventional single-curve analysis. Fitting results are subjected to thorough Monte Carlo analysis for rigorous assessment of confidence intervals and parameter correlations. The comparison is performed on a practical example of simulated steady-state reaction kinetics complemented with isothermal calorimetry (ITC) data resembling allosteric behavior of rabbit muscle pyruvate kinase (RMPK). Global analysis improves accuracy and confidence limits of model parameters. Cross-correlation between parameters is also reduced with accompanying enhancement of the model-testing power. This becomes especially important for validation of models with "difficult" highly cross-correlated parameters. We show how proper experimental design and critical evaluation of data can improve the chance of differentiating models.

Key words: Nonlinear least squares, Parameter correlation, Global fitting, Pyruvate kinase, Calorimetry, Two-state allosteric model, Monte Carlo analysis

1. Introduction

Since the discovery of allosteric regulation of proteins, the phenomena of cooperative ligand binding were recognized to play an important role in the regulation of biological functions in living cells. In considering the mechanism(s) of allostery, the investigator must be cognizant of interactions of at least two ligands with a macromolecule, which is capable of undergoing conformational change(s). Furthermore, these reactions are thermodynamically linked because each reaction is influenced by the presence or absence of the other. Decades of research have revealed highly diverse mechanisms of the allosteric regulation in different proteins. These mechanisms include the well-known concerted model proposed by Monod, Wyman, and Changeux (MWC) (1)

Aron W. Fenton (ed.), *Allostery: Methods and Protocols*, Methods in Molecular Biology, vol. 796,
DOI 10.1007/978-1-61779-334-9_22, © Springer Science+Business Media, LLC 2012

or a sequential model by Koshland, Nemethy, and Filmer (KNF) (2). Eigen developed yet another model, which is a combination of these two models (3).

Owing to a high variability of possible allosteric mechanisms, their detailed understanding often requires construction of alternative molecular or thermodynamic models. Quantitative parameterization as well as validation of such models depends on validation of mathematical expressions representing the model by experimental data. It is generally accepted that models cannot be proven but can only be disproven. Once conditions are found when a particular model cannot explain the experimental observations, the model has to be rejected or modified. During the validation process, a researcher is confronted by an important task of a thorough challenge of alternative models with data by varying different solution variables (temperature, pressure, pH, ionic strength, concentration, etc.), or technical details (excitation/emission wavelength in spectroscopy, angular velocity in sedimentation analysis, etc.). Objective comparison between the experimental data and the proposed model is often done by a nonlinear least-squares (NLSQ) analysis (4). Naturally, the mathematical description of the model should contain parameters that can account for changes of external conditions and that are adjusted during the NLSQ analysis. Then the parameterized models can also predict system behavior that further helps the researcher in model validation. Thus, Fig. 1 summarizes the general philosophy of studying complex linked reactions.

Global analysis (5–7) is a powerful method for discerning between models and for accurate recovery of model parameters. The method is based on an ordinary nonlinear least-squares minimization (4) and allows for simultaneous analysis of multiple datasets. Data for the analysis can be acquired under different experimental

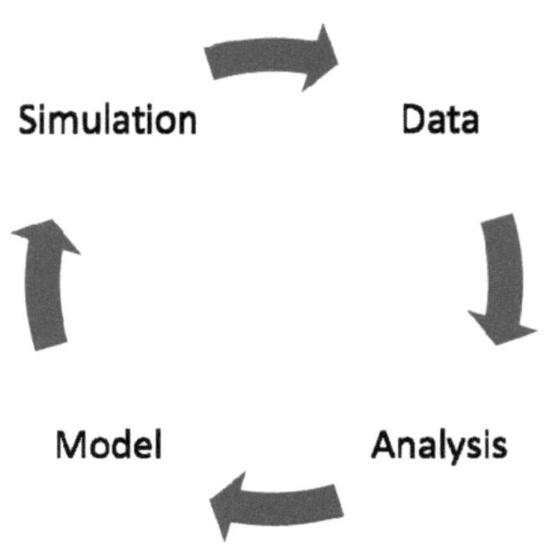

Fig. 1. Outline of general philosophy of studying complex linked reactions.

conditions and with different techniques. Consequently, some model parameters are common for multiple curves. During global fitting, the overall sum of weighed squared deviations of measured values and values calculated from a model that encompasses all experimental data is minimized. Global analysis, thus, allows determination of parameter values that are consistent with all datasets. Importantly, linkage between particular parameters and different datasets sharpens the minimum of the χ^2 surface with concomitant decrease of parameter correlation. Those parameters are recovered with higher accuracy (6). Overdetermination of parameters in the model, inherent to global analysis, helps to distinguish between alternative models by eliminating models inconsistent with data. The resolving power of global analysis is unequaled by a conventional NLSQ. Global analysis has been previously used for large variety of experimental data and techniques (5, 8–21). Recently, we have used global analysis of a dataset obtained by isothermal calorimetry (ITC), fluorescence binding experiments, and fluorescence temperature scans for evaluation of the functional energetic landscape in the allosteric regulation of rabbit muscle pyruvate kinase (RMPK) (22–24). In all cases, the global analysis approach provided performance far beyond the resolution of the conventional analysis.

In this chapter, we demonstrate the use of global analysis of multiple data curves acquired under different experimental conditions. We show that even a single properly chosen curve added to the dataset for analysis can significantly reduce correlation between parameters that often prohibits accurate recovery of parameter values and consequently lowering the chance to distinguish between alternative models.

2. Methods

2.1. Model

RMPK is an important allosteric enzyme of the glycolytic pathway catalyzing a transfer of the phosphate from phosphoenolpyruvate (PEP) to ADP:

$$\mathrm{PEP} + \mathrm{ADP} \overset{\mathrm{PK}}{\Leftrightarrow} \mathrm{Pyruvate} + \mathrm{ATP}$$

RMPK consists of four identical subunits (25), and its activity is subjected to a complex pattern of allosteric regulation. Although subject to recent challenges, numerous early experiments (13, 15, 16, 22–24, 26) suggested that all data can be described by a two-state concerted allosteric model (1) where protein assumes an active (R) and an inactive (T) form. In this model, differential affinities of ligands to the R and T-state perturb equilibrium between the states by shifting it to one or the other form. The general MWC model was customized for the tetrameric

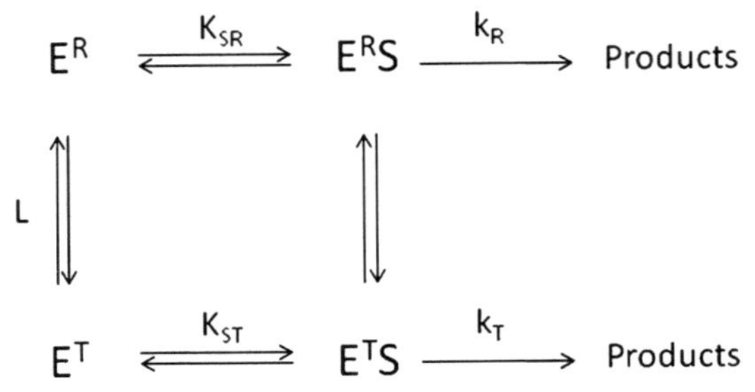

Fig. 2. Simplified model of the two-state allosteric enzyme. The enzyme equilibrates between an active R- and inactive T-state. Substrate binds to both the R- and T-states. Catalytic activities of the two states differ.

RMPK by Oberfelder et al. (13) and elaborated further by Herman and Lee (22–24).

For demonstration of the global analysis, we have chosen a model based on the reaction scheme from Fig. 2. It represents a simplified case of the general scheme published by Oberfelder (13). The substrate S binds to both the R and T-state of RMPK with dissociation constant K_{SR} and K_{ST}, respectively, which results in a catalytic activity. For the tetrameric RMPK, the steady-state reaction rate v can be described as (13):

$$v = \frac{k_R \cdot [S]/K_{SR} \cdot (1 + [S]/K_{SR})^3 + k_T \cdot L \cdot [S]/K_{ST} \cdot (1 + [S]/K_{ST})^3}{(1 + [S]/K_{SR})^4 + L \cdot (1 + [S]/K_{ST})^4} \tag{1}$$

where $[S]$ is concentration of substrate, $L = [T]/[R]$ is an equilibrium constant between the unliganded R- and T-state of the enzyme, and k_R and k_T are reaction rate constants for the R- and T-state, respectively. Assuming the T-state is fully inactive, $k_T = 0$ and binding of PEP to the T-state is negligible ($[S]/K_{ST} \rightarrow 0$ for any $[S]$) (23), Eq. 1 can be simplified to:

$$v = \frac{k_R \cdot [S]/K_{SR} \cdot (1 + [S]/K_{SR})^3}{(1 + [S]/K_{SR})^4 + L} \tag{2}$$

Temperature dependence of the equilibrium constant L can be expressed as:

$$L = \exp\left(-\frac{\Delta H_{R-T} - T \cdot \Delta S_{R-T}}{RT}\right) \tag{3}$$

where ΔH_{R-T} and ΔS_{R-T} are the enthalpy and entropy change of the $R \rightarrow T$ transition, respectively, R, the universal gas constant, and T, temperature in Kelvin (see Note 1). We use Eqs. 2 and 3 for simulation of reaction kinetics and their subsequent analysis.

2.2. Simulation of Reaction Kinetics

The task is to recover k_R, K_{SR}, and L at 40°C, i.e., close to the rabbit's physiological temperature, by fitting Eq. 2 to synthetic data supplemented with a realistic level of random noise. Then, confidence limits of the fitted values and correlation between parameters are evaluated for both the "standard" single-curve fit and the global fit of several curves of reaction kinetic data.

Employing Eq. 2 we have simulated reaction kinetics at 20, 30, 40, and 50°C using parameters values close to the ones published for RMPK (22–24) (see Note 2). Some adjustment of parameters was done to accommodate for the model simplification and to work with "pleasant round numbers". Qualitative behavior of the model, however, still closely resembles real RMPK. In particular, we used a normalized value of the rate coefficient $k_R = 1$ mol/s and the dissociation constant $K_{SR} = 100$ μM. Temperature dependence of the equilibrium constant between the R and T-state was calculated from Eq. 3 using $\Delta H_{R-T} = 44$ kcal/mol and $\Delta S_{R-T} = 150$ cal/mol/deg. Then a random Gaussian noise with standard deviation (SD) of 0.01 was added to each curve. The noise mimicked noise levels obtainable in real experiments (1% relative error at saturation). For illustration, the calculated temperature dependence of L is shown together with the inactive fraction of the enzyme $f_T = L/(1 + L)$ in Fig. 3. At 40°C, the value of L is 119.

2.3. Single-Curve Fit

The simulated reaction kinetics at 40°C together with the best fit is shown in Fig. 4. The fitting was done by a standard NLSQ method that can be found in most scientific graphing packages.

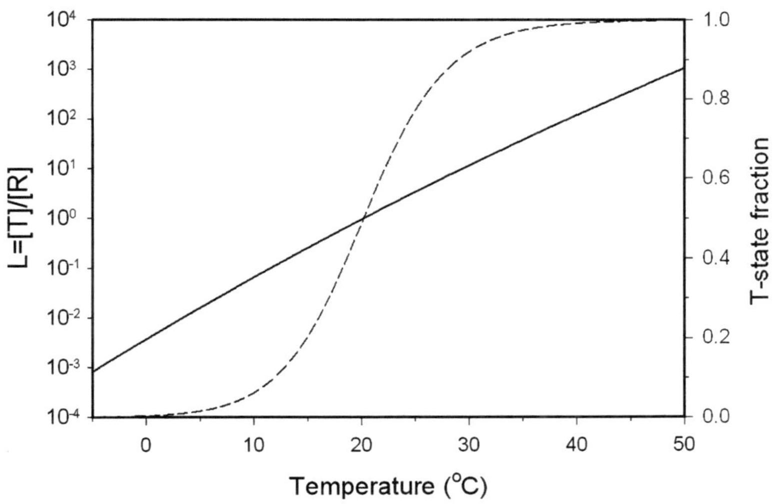

Fig. 3. Temperature dependence of the equilibrium constant between R and T state for $\Delta S_{R-T} = 150$ cal/mol/deg and $\Delta H_{R-T} = 44$ kcal/mol. The *dashed line* represents a fraction of the inactive T-state as a function of temperature.

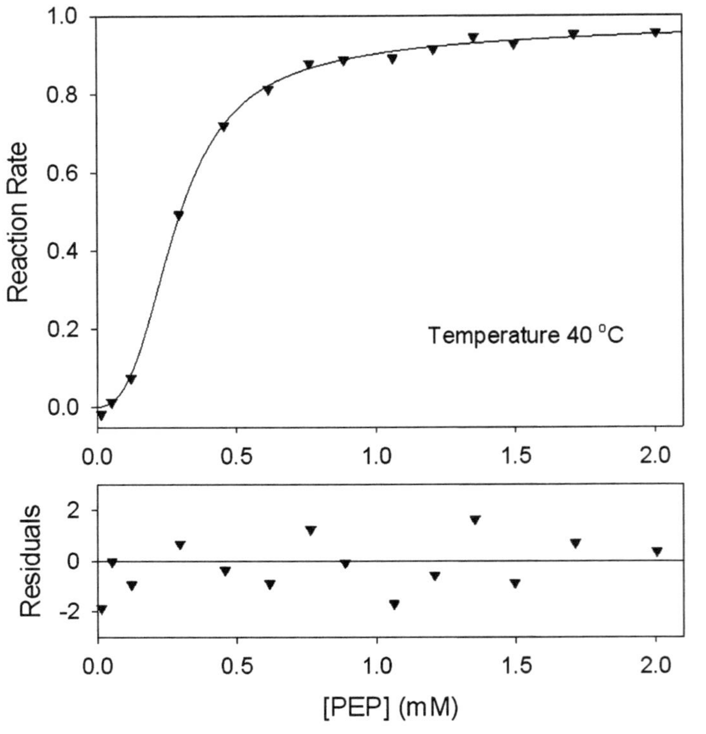

Fig. 4. Simulated steady-state reaction kinetics at 40°C. Standard deviation of the added Gaussian noise is 0.01. The *solid line* represents the best fit.

As described in numerous literatures, a function χ_R^2 called reduced chi-squared is minimized during the NLSQ fit (4):

$$\chi_R^2 = \frac{1}{(M - m - 1)} \cdot \sum_{i=1}^{n} \left[\frac{\Upsilon_{\text{Data}}(i) - \Upsilon_{\text{Model}}(i)}{\sigma(i)} \right]^2 \quad (4)$$

where M is the number of data points, m the number of fitted parameters, $\Upsilon_{\text{Data}}(i)$ the i-th experimental value, $\Upsilon_{\text{Model}}(i)$ the calculated value, and $\sigma(i)$ standard deviation of the corresponding data point. χ_R^2 is a quantity normalized for the number of data points and model complexity. When correctly used, minimization of χ_R^2 gives parameter values that have the highest probability of being correct (27) (see Notes 3 and 4). The NLSQ fit of the curve from Fig. 4 yielded parameter values $k_R = 0.998$ mol/s, $K_{SR} = 96.5$ μM, and $L = 146$. It is clear that we did not recover the exact values for parameters used in simulation. Especially, the recovered value for L is quite different from the correct value of 119. The sources of error are as follows: (a) the experimental noise that biases the true χ_R^2 minimum and (b) the high correlation between parameters that causes the minimum of the χ_R^2 surface to be shallow and insensitive to variations in parameters. As a consequence, a change of one parameter can be compensated by changes of other parameters without significant change of χ^2_R. When the magnitude of the correlation increases, the estimation of parameters by NLSQ becomes difficult or even impossible. One, therefore, always needs to know the degree of correlation and the confidence intervals (CI) of the fitted parameters. Without knowing the CI of the fitted results one cannot critically evaluate their significance and make comparison with other evidence.

2.4. Parameter Correlation and Confidence Intervals

There are several method for evaluation of the CIs for NLSQ analysis, see e.g., (27) for an overview. The simplest method is an asymptotic standard error (ASE) implemented in most commercial software packages. Unfortunately, this computationally simple method, which yields symmetrical CIs, uses implicit assumptions that can hardly be fulfilled for most NLSQ problems (27–30). In particular, ASE requires a linear fitting function, a large number of data points, and the absence of any correlation between parameters. As a consequence, the ASE confidence limits are usually far from reality. The preferred approaches are bootstrap (31) or Monte Carlo (29) methods. Although computationally extensive, they use fewer assumptions and provide more realistic estimates of the CIs. Parameter cross-correlation is also inherently and rigorously evaluated during calculation of CIs.

For estimation of CIs, we chose the Monte Carlo (MC) method. Its principle is rather simple:

1. Data were analyzed to obtain a set of the most probable model parameters, see Subheading 3.
2. A noiseless data curve was generated using parameters recovered in step 1 and independent variables x_i values from the original data.
3. Random Gaussian noise with the same level as contained in the original data was added to this noiseless curve ($SD = 0.01$).
4. The resulting data were analyzed and parameters tabulated.
5. Steps 3–4 were repeated many times to obtain sufficient statistics sampling the "best parameter values".
6. Histograms representing a confidence probability distribution for each particular parameter were constructed. As seen from the description of the MC procedure, the distributions originate from the noise contained in the data only. Accuracy of the recovered distribution depends mainly on a number of MC runs and how closely noise characteristics of the simulated curves match the original data noise. For a reasonable accuracy it is necessary to perform more than several hundred runs. In our example, we used 1,500 MC cycles.
7. Cross-correlation between the p_i and p_j parameter was visualized by plotting p_i against corresponding p_j value for each MC run.

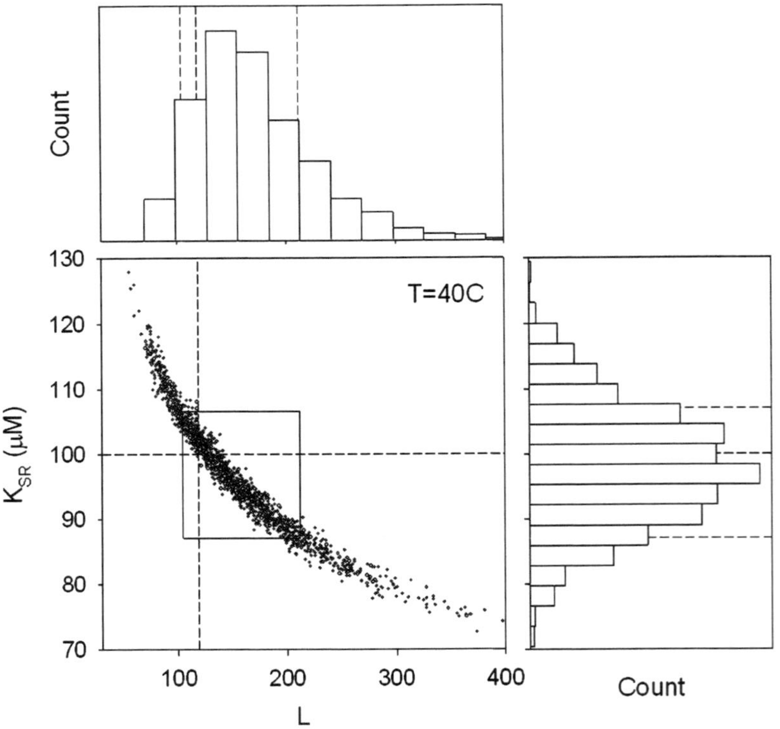

Fig. 5. Correlation between L and K_{SR} for a single curve analysis presented in Fig. 4. The correlation was obtained by 1,500 MC runs. Each point is a parameter pair from a single fit to data with unique Gaussian noise. *Black dashed lines* indicate original parameter values used for simulation. Histograms show resulting distribution of the fitted parameter values. *Gray dashed lines* border the region of ± 1 SD. This region is depicted by the *black rectangle* in the central graph.

Results of the MC analysis for the single-curve fit are visualized in Figs. 5 and 6 and quantitatively summarized in Table 1. Correlation between L and K_{SR} with corresponding confidence distributions are shown in Fig. 5. Each point in the central graph is a parameter pair from a single MC run. We can clearly see a strong negative correlation between L and K_{SR}. The parameter pairs are not randomly distributed in the parameter plane. Instead, they form a well-defined "sickle" shape indicating a dependence of K_{SR} on L. Black dashed lines mark the original parameter values used for the simulation of the "experimental" data from Fig. 4. The histograms present resulting distribution of the fitted parameter values. From the area of the histograms we evaluated confidence intervals for the 68% confidence level (1 SD) that are bordered by gray dashed lines. It is clear that the CI of the equilibrium constant L is not symmetrical and the distribution is significantly skewed to higher values. Similar pattern with slightly weaker correlation between L and k_R can be recognized in Fig. 6. Confidence limits

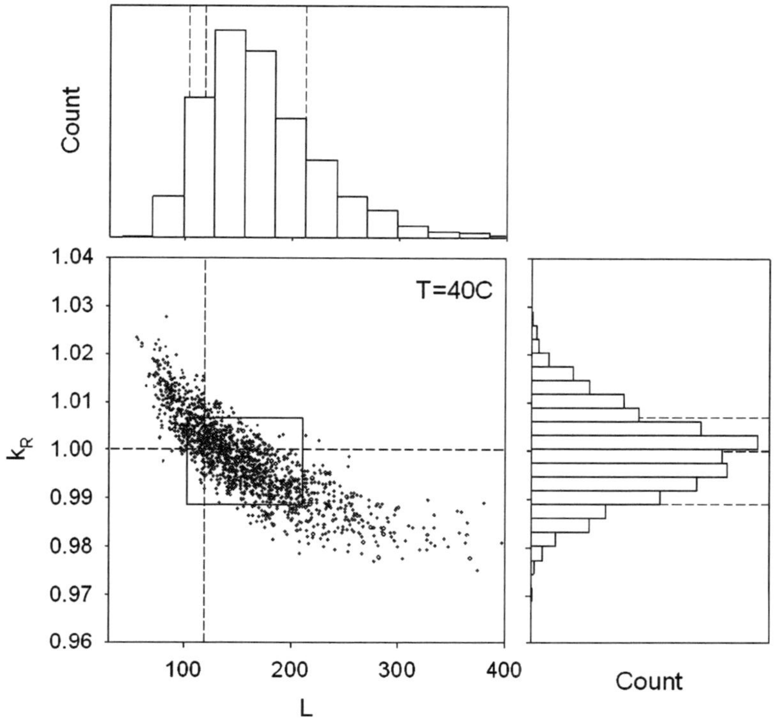

Fig. 6. Correlation between L and k_R for a single curve analysis presented in Fig. 4. The correlation was obtained by 1,500 MC runs. *Black dashed lines* indicate original parameter values used for simulation. Histograms show resulting distribution of the fitted parameter values. *Gray dashed lines* border the region of ±1 SD. This region is depicted by the *black rectangle* in the central graph.

calculated from the histograms are summarized in Table 1. The area defined by the CIs is depicted by the black rectangle in the central correlation graph. It is seen that the area is not randomly filled with points – a sign of cross-correlation.

2.5. Global Analysis: Background

From the MC analysis of the single-reaction kinetics, we learned that parameters of our two-state model describing allosteric regulation of RMPK are strongly cross-correlated. Consequently, the correlation also influences the widths of the CIs because deviation of any parameter can be compensated by an adjustment of other parameters without significant change of χ^2_R (see Note 5). Reduction of "covariance valleys" on the χ^2_R hypersurface would certainly improve both testing capability of the model and accuracy of recovered parameters. This can be done by simultaneous analysis of multiple curves covered by a single mathematical model where groups of curves share the same parameters.

Table 1
Comparison of fitting results at 40°C

Parameter	Lower limit[a]	Recovered value	Upper limit[a]	χ^2_R
L				
Single curve analysis	104	146	212	1.12
Global analysis	109	125[b]	137	0.86
Single curve, 4N analysis[c]	107	137	147	0.99
Global with calorimetry	117	121[d]	124	0.98
True value	–	*119*[e]	–	–
K_{SR}				
Single curve analysis	87	96.5	107	1.12
Global analysis	97	99.9	103	0.86
Single curve, 4N analysis	95	96.8	104	0.99
Global with calorimetry	99	100	101	0.98
True value	–	*100*	–	–
k_R				
Single curve analysis	0.989	0.998	1.007	1.12
Global analysis	0.999	1.0026	1.006	0.86
Single curve, 4N analysis	0.994	0.998	1.005	0.99
Global with calorimetry	1.000	1.0025	1.005	0.98
True value	–	*1.000*	–	–

[a]Confidence limits at 68% confidence level
[b]Derived value, calculated from Eq. 3 using $\Delta H_{R-T} = 42.8$ kcal/mol and $\Delta S_{R-T} = 146$ cal/mol/deg (the best fitted values)
[c]Single curve analysis with quadruple number of experimental points
[d]Derived value, calculated from $\Delta H_{R-T} = 43.998$ kcal/mol and $\Delta S_{R-T} = 150.03$ cal/mol/deg (the best fitted values)
[e]Calculated from Eq. 3 using the primary parameters $\Delta H_{R-T} = 44$ kcal/mol and $\Delta S_{R\text{-}T} = 150$ cal/mol/deg

The global analysis is conceptually simple. Instead of minimization χ_k^2, $k = 1, 2,\ldots, K$, separately for each curve form the set of K curves, a "global" reduced $\chi^2_{\mathrm{R,\,GLOB}}$ is minimized:

$$\chi^2_{\mathrm{R,\,Glob}} = \frac{1}{(N-m-1)} \cdot \sum_{k=1}^{K} \chi_k^2 = \frac{1}{(N-m-1)} \cdot \sum_{k=1}^{K} \sum_{i=1}^{M_k} \left[\frac{\Upsilon_{\mathrm{Data},k}(i) - \Upsilon_{\mathrm{Model},k}(i)}{\sigma_k(i)} \right]^2 \quad (4)$$

where N is the overall number of experimental points in all K analyzed curves, each of them containing $M_k$points, and m is the number of adjusted model parameters (see Note 6):

$$N = \sum_{k=1}^{K} M_k \quad (5)$$

A successful global fit requires all curves to be simultaneously well-fitted with one unique set of parameter values. Strict parameter linkages between different curves prevent inappropriate models to fit simultaneously all curves. On the contrary, with a correct model the parameter values will be determined with better accuracy (see Note 7).

2.6. Global Analysis of Multiple Enzyme Kinetics

To demonstrate the power of global analysis, we add to our reaction kinetics at 40°C, Fig. 4, additional three RMPK kinetics "measured" at 20, 30, and 50°C. The resulting set of curves is shown in Fig. 7. It is seen that as temperature increases, shape of the kinetic curves changes from near-hyperbolic to sigmoidal. One potential interpretation is that the effect is caused by a temperature-induced shift of the equilibrium between the R and T-state toward the inactive T-state, as shown in Fig. 3 (see Note 8). Recognizing this fact, we can set the first linkage between the curves. It is the temperature dependence of the equilibrium constant L. According to Eq. 3, we can calculate L at 20, 30, 40, and 50°C:

$$L(20^{\circ}\mathrm{C}) = \exp\left(-\frac{\Delta H_{R-T} - (273.15 + 20)\cdot \Delta S_{R-T}}{(273.15 + 20)R}\right) \tag{6}$$

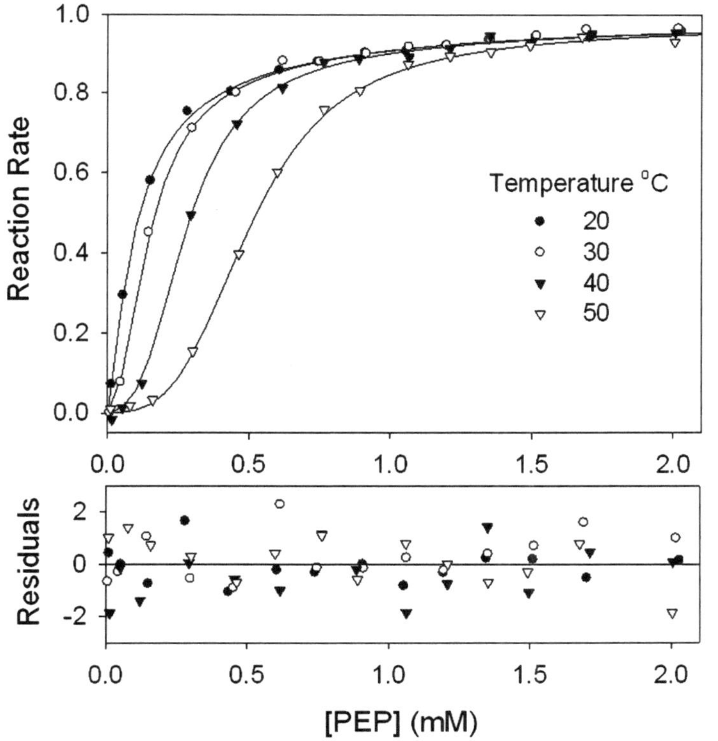

Fig. 7. Global analysis of steady-state reaction kinetics at different temperatures. Standard deviation of the added Gaussian noise is 0.01. The *solid lines* represent the best fit.

$$L(30^\circ\mathrm{C}) = \exp\left(-\frac{\Delta H_{R-T} - (273.15+30)\cdot\Delta S_{R-T}}{(273.15+30)R}\right) \quad (7)$$

$$L(40^\circ\mathrm{C}) = \exp\left(-\frac{\Delta H_{R-T} - (273.15+40)\cdot\Delta S_{R-T}}{(273.15+40)R}\right) \quad (8)$$

$$L(50^\circ\mathrm{C}) = \exp\left(-\frac{\Delta H_{R-T} - (273.15+50)\cdot\Delta S_{R-T}}{(273.15+50)R}\right) \quad (9)$$

At any temperature L can be expressed by two new parameters, ΔH_{R-T} and ΔS_{R-T}. Instead of four independent constants L, it is enough to fit ΔH_{R-T} and ΔS_{R-T}. As a consequence, the number of required model parameters decreases. Similar link can be made for k_R and K_SR. In real experiments, we would need to use equations similar to Eqs. 6–9 for temperature dependence of K_SR and the Arrhenius equation for temperature dependence of the rate constants $k_\mathrm{R.}$ In this simplified case, however, we neglected the temperature dependence as follows:

$$k_\mathrm{R}(20^\circ\mathrm{C}) = k_\mathrm{R}(30^\circ\mathrm{C}) = k_\mathrm{R}(40^\circ\mathrm{C}) = k_\mathrm{R}(50^\circ\mathrm{C}) = k_\mathrm{R} \quad (10)$$

and

$$K_\mathrm{SR}(20^\circ\mathrm{C}) = K_\mathrm{SR}(30^\circ\mathrm{C}) = K_\mathrm{SR}(40^\circ\mathrm{C}) = K_\mathrm{SR}(50^\circ\mathrm{C}) = K_\mathrm{SR} \quad (11)$$

In such a case, dimensionality of the NLSQ minimization problem is significantly reduced. Instead of 12 parameters required for independent fitting of the 4 individual curves, we need only 4 parameters, $k_\mathrm{R,}$ K_SR, ΔH_{R-T}, and ΔS_{R-T}, to describe all curves globally under a single model, Eq. 2. The more curves included in the global dataset, the higher the reduction of the overall number of parameters. Values of ΔH_{R-T} and ΔS_{R-T} extracted from the fit allow later reconstruction of L at any desired temperature according to Eq. 3.

The best global fit of the four enzyme kinetic curves is depicted in Fig. 7 by solid lines. Values of the corresponding parameters are summarized in Table 1. The primary adjusted parameters were ΔH_{R-T} and ΔS_{R-T} and the value of L was derived using Eq. 3. The fit yielded $\Delta H_{R-T} = 42.8$ kcal/mol and, $\Delta S_{R-T} = 146$ cal/mol/deg. Next we performed the same MC correlation and confidence analysis as described in Subheading 4. Results of 1,500 MC runs are shown in Figs. 8 and 9. In the correlation graphs, L was calculated from pairs of ΔH_{R-T} and ΔS_{R-T} as described (see Note 9). Besides the best parameter, values were closer to the expected ones, see Table 1, both the correlation graphs and the histograms indicated that addition of a few kinetic curves, i.e.,

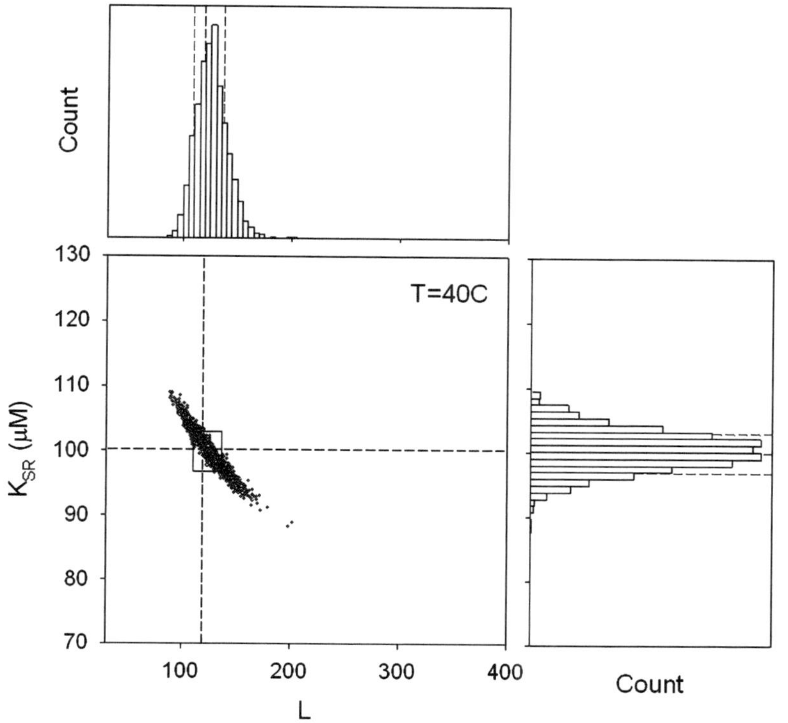

Fig. 8. Correlation between *L* and K_{SR} for global analysis of enzyme kinetics from Fig. 7. The correlation graph was obtained by 1,500 MC runs. *Black dashed lines* indicate original parameter values used for simulation. Histograms show resulting distribution of the fitted parameter values. *Gray dashed lines* border the region of ±1 SD. This region is depicted by the *black rectangle* in the central graph.

curves of the same character, to the dataset had a significant effect on accuracy and confidence intervals of all parameters. From a comparison of Figs. 5 and 8 and Figs. 6 and 9, it becomes obvious that although parameters are still correlated to some extent, the correlation is significantly reduced. In Table 1 we have also included results of a single curve fitting with quadruple number of data points that is comparable with number of data points used in the global analysis. As expected the confidence limits reduced, however, the reduction is inferior to the result of the global analysis. Addition of more curves of the same characters is but one of the first strategies to improve the accuracy of results and the limitation is due to the fact that these curves are defined by the same parameters. More significant improvement can be achieved by including data derived by alternative approaches that rely on different physical principles to monitor the changes, as shown in Subheading 7.

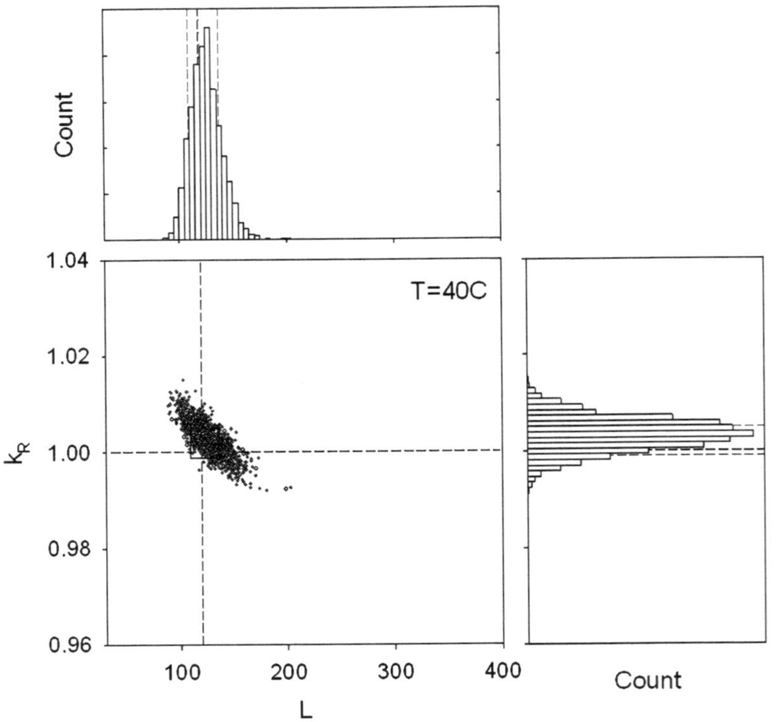

Fig. 9. Correlation between L and k_R for global analysis of enzyme kinetics from Fig. 7. The correlation was obtained by 1,500 MC runs. *Black dashed lines* indicate original parameter values used for simulation. Histograms show resulting distribution of the fitted parameter values. *Gray dashed lines* border the region of ±1 SD. This region is depicted by the *black rectangle* in the central graph.

2.7. Global Analysis of Data with Different Character

Many different strategies can be employed to improve the accuracy of parameters. One can lower the noise of the data by collecting more precise measurements or using more sensitive approaches. More similar curves can be included in the dataset for global analysis. One could also extend the temperature range to further modulate the free energy change $\Delta G = \Delta H_{R-T} - T \cdot \Delta S_{R-T}$ and partial contributions of ΔH_{R-T} and ΔS_{R-T} to L. By rearrangement of Eq. 3 we obtain:

$$L = \exp\left(-\frac{1}{R} \cdot \left(\frac{\Delta H_{R-T}}{T} - \Delta S_{R-T}\right)\right) \qquad (12)$$

Upon examining the relative contributions of ΔH_{R-T} and ΔS_{R-T} to L in Eq. 12, it is clear that at higher temperature the contribution of ΔS_{R-T} increases. If experiments were conducted at a single temperature only, one would not be able to extract ΔH_{R-T} and ΔS_{R-T} because the ΔG, not particular values of ΔH_{R-T} and ΔS_{R-T}, is important for quantification of L. Experimental temperature range, however, cannot be arbitrarily extended. At low temperature, we are limited by the freezing point of water and at high temperature by irreversible thermal processes degrading the enzyme structure. Using multiple curves of the same type (e.g., enzyme kinetics) has another inherent limitation. Mathematical description of all curves generated by the same approach has the same functional form and contains exactly the same parameters, Eqs. 2 and 3. For each curve, the correlation between parameters is similar. Significant improvement could be expected when a dataset includes data derived from a different approach in which only a limited subset of the globally linked parameters was needed to represent the data. When it is not necessary to include specific parameters in the mathematical description of the individual experimental curve, correlation among parameters would be reduced when multiple curves are analyzed. The extent of the effect depends on the "weight" that the curve has within the whole global dataset (see Note 10). The kind of experimental data to be included to the global fitting depends on the judicious choice of approach by the researcher.

In our next example, we include to the global fitting simulated data from isothermal titration calorimetry (ITC). Expression of these data naturally lacks the reaction rate constant $k_{R,}$ since the titration of RMPK by PEP is done in the absence of the second substrate ADP. Having in mind the rationale mentioned above, i.e., reduction of the parameter number, we do not analyze shapes of individual ITC curves because they obviously depend on the binding constant K_{SR}. Instead, we titrate RMPK with PEP to saturation and measure overall reaction heat as a function of temperature.

According to the two-state model, three major reactions contribute to the overall reaction heat Q_{tot} accompanying saturation of the enzyme by PEP. In particular, it is Q_R and Q_T that represent binding heats of PEP to the R and T state, respectively, and a heat Q_{R-T} accompanying the R → T (see Note 11):

$$Q_{tot} = f_{R,sat} \cdot Q_R + f_{T,sat} \cdot Q_T + (f_{T,sat} - f_T) \cdot Q_{R-T} \qquad (13)$$

Terms $f_{R,sat}$ and $f_{T,sat}$ are fractions of RMPK in the R and T state, respectively, at the end of the titration when the enzyme is completely saturated by PEP. The term f_T is a fraction of the unliganded T-state at the beginning of the titration. Neglecting linked proton reactions, the heats can be expressed as:

$$Q_{state} = 4 M_0 V_0 \cdot \Delta H_{state} \qquad (14)$$

and

$$Q_{R\to T} = M_0 V_0 \cdot \Delta H_{R\to T} \qquad (15)$$

where M_0 stands for molar concentration of RMPK and V_0 is a sample volume. The enthalpies ΔH_{state} and ΔH_{R-T} are calculated per mol of binding sites and per mol of tetrameric RMPK, respectively. The fraction f_{T} can be expressed as follows:

$$f_{\mathrm{T}} = L/(1 + L) \tag{16}$$

and

$$f_{\mathrm{T}} + f_{\mathrm{R}} = 1 \tag{17}$$

In our example we assumed that PEP does not bind to the T-state and $\Delta H_{\mathrm{R}} = 0$. As a consequence, $f_{T,\mathrm{sat}} = 0$ for RMPK saturated by PEP. Then,

$$Q_{tot} = -f_T \cdot Q_{R-T} = -M_0 V_0 \cdot \frac{L}{1 + L} \cdot \Delta H_{R-T} \tag{18}$$

$$\frac{Q_{tot}}{M_0 V_0} = -\frac{L}{1 + L} \cdot \Delta H_{R-T}[\mathrm{kcal/mol}] \tag{19}$$

Equations 19 and 3 describe the desired temperature dependence of overall reaction heats accompanying saturation of RMPK by PEP. Data simulated according to these equations with added random noise (SD = 0.44 kcal/mol) are shown in Fig. 10.

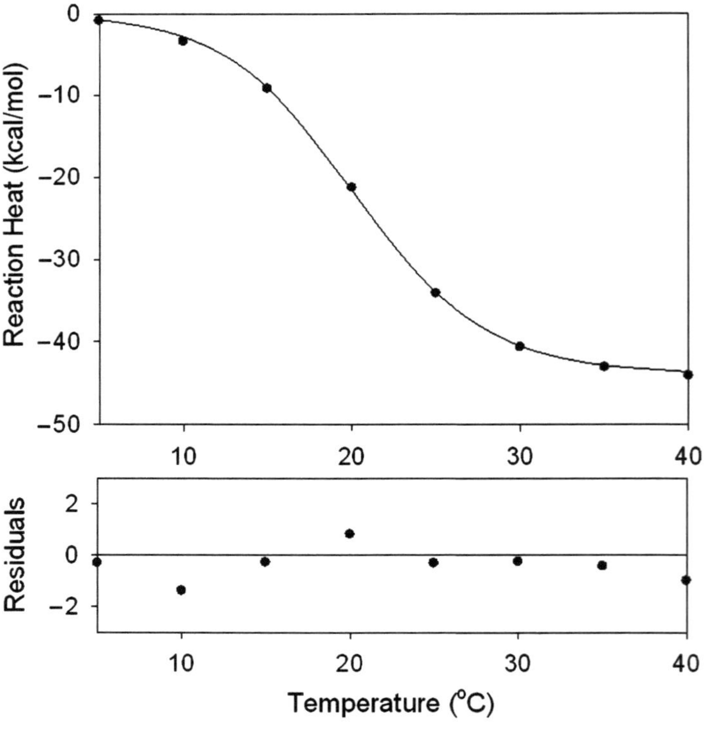

Fig. 10. Temperature dependence of overall reaction heat for PK titrated by PEP to the saturation. The *solid line* is a result of the best simultaneous fit of the calorimetric curve together with the set of 4 enzyme kinetics from Fig. 7. SD of noise is 0.44 kcal/mol.

This synthetic ITC curve depending only on ΔH_{R-T} and ΔS_{R-T} was used as the 5th dataset for the global analysis.

The result of the extended global fit is presented in Table 1, the best fit is shown in Fig. 10 by solid line. The fitted parameter values closely approached the true values used for the simulation, the ΔH_{R-T} and ΔS_{R-T} being 43.998 kcal/mol and 150.03 cal/mol/deg, respectively. Consequently, the value of equilibrium constant $L = 121$ (at 40°C) was determined with high accuracy. An outcome of the MC analysis is presented in Figs. 11 and 12, confidence limits in Table 1. Similar to the previous examples, 1,500 MC runs were performed for evaluation of correlation between parameters and confidence intervals. Figure 11 demonstrates a dramatic improvement of CI for both L and K_{SR} after addition of the ITC curve to the dataset, the highest effect being recorded for the equilibrium constant L. Although a correlation between L and K_{SR} can still be noticed from the direction of the main axis of the "ellipsoidal" scattergram, which is aligned with neither the x-axis nor the y-axis, the magnitude of the correlation

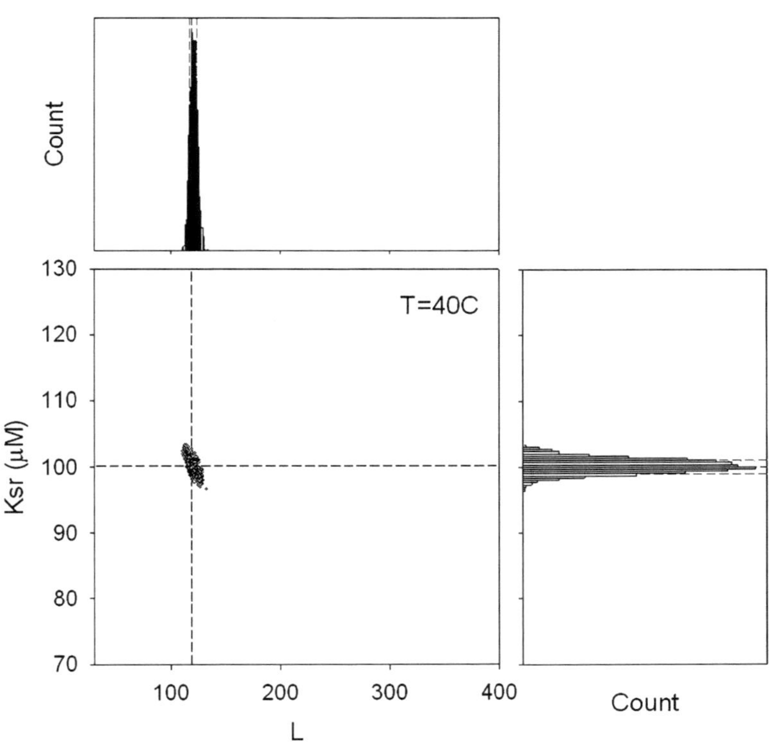

Fig. 11. Correlation between L and K_{SR} after extension of the dataset depicted in Fig. 7 by the calorimetric curve from Fig. 9. The correlation was obtained by 1,500 MC runs. *Black dashed lines* indicate original parameter values used for simulation. Histograms show resulting distribution of the fitted parameter values. *Gray dashed lines* border the region of ±1 SD. This region is depicted by the *black rectangle* in the central graph.

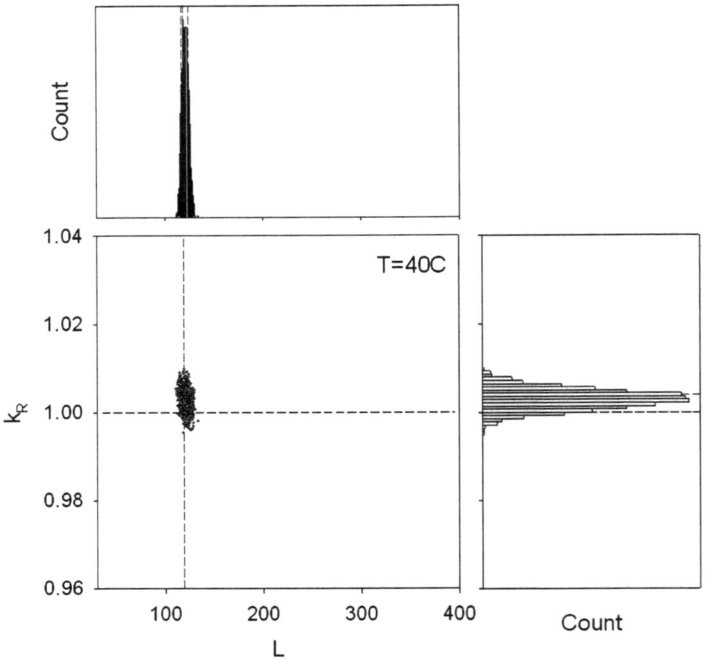

Fig. 12. Correlation between *L* and k_R after extension of the dataset depicted in Fig. 7 by the ITC curve from Fig. 10. The correlation was obtained by 1,500 MC runs. *Black dashed lines* indicate original parameter values used for simulation. Histograms show resulting distribution of the fitted parameter values. *Gray dashed lines* border the region of ±1 SD. This region is depicted by the *black rectangle* in the central graph.

is much smaller. This can be visually judged from the shape and rotation of the scattergram in Fig. 11, as compared to Figs. 8 and 5. Another indication of significantly weaker correlation is the confidence area, depicted by the black rectangle in the central graph, being fully filled by the points generated during the MC analysis (compare to Figs. 8 and 5). Correlation between L and k_R has almost disappeared, as shown in Fig. 12. Last but not the least; the true values of parameter used for data generation reside within the narrow confidence intervals obtained from the MC analysis.

There are unlimited numbers of ways to achieve the same goal as expressed in Subheading 7. One may employ spectroscopic techniques, such as fluorescence, to monitor ligand binding. Spectroscopic techniques provide another set of information on the conformation changes in the protein as a consequence of ligand binding (23).

In conclusion, when properly selected, even addition of a single curve to the dataset can significantly improve accuracy and confidence limits of recovered parameters with accompanying enhancement of the model-testing power. The benefit of the analytical power of global fitting is realized particularly when data obtained under different experimental conditions or using different experimental approaches are included for the analysis.

3. Notes

1. Recent detail studies have shown that there is a considerable isothermal heat capacity change ΔCp associated with the R → T transition. The K_{SR} is slightly temperature dependent as well (22–24). For the demonstration purposes, however, we neglect the ΔCp as well as temperature dependence of K_{SR} in this chapter. Another simplification in our example rests in the assumption of the "silent" second substrate ADP that has to be present in the reaction mix to achieve enzyme activity. Specifically, we assume that ADP binds with the same affinity to the R and T-state and does not shift the R-T equilibrium. For this reason, we do not explicitly see terms associated with ADP in the equation for the normalized reaction rate, Eq. 2. In reality, the role of ADP is much more complex, see Herman and Lee (24).
2. In reality the temperature of 50°C might be too high for completing measurement due to the onset of irreversible thermal denaturation of RMPK in these experimental conditions.
3. The following conditions must be satisfied for valid application of the least-squares parameter fitting (28): (a) Errors of the measured data should be assignable to the dependent variable (the *Y*-axis) only. Errors caused by an inaccurate assignment of the independent variable (*X*-axis) should be negligible and uncorrelated with the measured values. This could be, for example, inaccurate setting or measurement of experimental temperature, pH, concentration, etc. (b) Experimental noise in measured data should be random and described by the Gaussian distribution. (c) Data should not contain any systematic error. (d) The fitted model must be correct. If specifically this condition was not satisfied, the recovered parameter values would be without physical meaning, even though the data might be fitted well.
4. The standard deviation (SD) σ_i required for the correct weighing of the i-th data point can be evaluated during the experiment. One possible way is to measure each data point many times, calculate the mean and standard error of the mean (SEM) (4). Then, the SEM can be used as σ_i. This approach is particularly useful during automatic or semiautomatic data acquisition when data points can be sampled many times. For "counting experiments" when signal is measured by counters (e.g., photon counting or radioactivity), the noise follows the Poissonian distribution and the SD is known from the theory: $\sigma_i = \sqrt{\mu}$, where μ is a mean value of the signal. In practice, however, the mean can be often substituted for the

number of counts N measured and $\sigma_i = \sqrt{N}$. This is a rigorous and preferable method for setting the weighing factors. In other cases the researcher has to rely on specifications of the instrument or perform own calibration of the data noise.

5. The parameter cross-correlation rises largely from the functional form of the fitted model. It can be modulated to some extent only by changing the span and distribution of the sampling points along the x-axis.
6. There are a number of multipurpose and specialized commercial software packages containing NLSQ minimization routines with a global analysis option. However, global analysis can be easily performed in any "open model" NLSQ program that allows user to input own fitting formulas. In this case, multiple experimental curves can be integrated to form one longer data file with appropriate input format. Then a variable is introduced to the model with the only purpose to tell the program what formula to use for fitting of each particular segment of the input data corresponding to the particular curve. The "fooled program" then automatically performs the desired global analysis. This approach is used in Sigma Plot (Systat Software, Inc.) to derive the examples used in this chapter.
7. Errors other than random noise in the data can invalidate the results derived from global fit. Therefore, the analyzed dataset must to be very *carefully inspected* for the presence of any abnormalities. Illegitimate errors, i.e., obvious mistakes and outliers, should not create problem, since they can be easily discovered, data discarded and experiments repeated. Fundamental problem can create systematic errors that are more difficult to identify and correct. Systematic errors limit accuracy and precision of experimental data by a defined and often reproducible way. This could be, for example, a wrong calibration of the instrument or experimental conditions different from the assumed ones (mismatch in temperature, pH, pressure, concentrations, sample aging, etc). Systematic errors result in inaccurate data, even though precision could be excellent. This is actually the worst situation because the correct model cannot fit the wrong data and low data noise makes the mismatch clearly visible. Researcher then can erroneously discard the correct model due to its "inconsistency with the experimental data". Therefore, careful attention should be paid to experimental design, analysis of experimental conditions and techniques used. Minimum requirement for a successful global analysis is to have an internally consistent dataset where all data curves are reproducible and measured preferably within short period of time on the same sample using the same experimental and sample handling

protocols. For identification of potentially biased data files one can analyze data subsets where individual curves or group of curves are excluded from the analysis. A consistent dataset should always yield similar parameter values and about the same quality of fit.

8. Based on the model in Fig. 2, at low temperature, almost all RMPK is in the active R-state (Fig. 3), and the steady-state kinetics follows a simple hyperbolic Michaelis–Menten equation. With increasing temperature the equilibrium between the R and T-state shifts toward the inactive T-state that is accompanied by increasing sigmoidicity of the reaction kinetics. During the PEP titration, the increasing amount of substrate, which binds preferentially to the active R-state, progressively shifts the inactive RMPK fraction to the active one with corresponding cooperative increase of the reaction rate. The effect is a consequence of homotropic allosteric interaction between four equivalent RMPK subunits causing all subunits to change conformation in a concerted manner during the T–R transition.
9. Advantage of the Monte Carlo analysis is a straightforward generation of confidence distributions and correlation graphs for quantities mathematically derived from the primary fitted parameters. Uncertainties of the constituent parameters naturally propagate to the derived quantities without need for their complex mapping.
10. In some substantiated cases, a "weighing" of the curve within the global dataset, i.e., contribution of the curve to the overall chi-squared, can be artificially increased by scaling the weighing factors $1/\sigma_k(i)$ in Eq. 4. Not to bias noise distribution within the curve, all $\sigma_k(i)$ should be scaled by the same factor. After this weighing, however, one cannot expect the reduced chi-squared of the best fit to approach unity.
11. In a real experiment, we would also need to account for buffer ionization heats due to proton absorption or release accompanying both PEP binding and R-T transition. For details, see for example ref. (22).

Acknowledgments

Supported by NIH GM 775551 and the Robert A. Welch Foundation (JCL) and grant MSM 0021620835 of the Ministry of Education Youth and Sports of the Czech Republic (PH).

References

1. Monod, J., Wyman, J., and Changeux, J. P. (1965) On the Nature of Allosteric Transitions: a Plausible Model, *J Mol Biol 12*, 88–118.
2. Koshland, D. E., Jr., Nemethy, G., and Filmer, D. (1966) Comparison of experimental binding data and theoretical models in proteins containing subunits, *Biochemistry 5*, 365–385.
3. Eigen, M. (1967) Kinetics of reaction control and information transfer in enzymes and nucleic acids, in *Nobel Symposium* (Claesson, S., Ed.), pp 333–369, Almqvist & Wiksell, Stockholm.
4. Bevington, P. R., and Robinson, D. K. (2002) *Data reduction and error analysis for the physical sciences*, 3 rd ed., Mc Graw-Hill, New York.
5. Eisenfeld, J., and Ford, C. C. (1979) A systems-theory approach to the analysis of multi-exponential fluorescence decay, *Biophys J 26*, 73–83.
6. Knutson, J. R., Beechem, J. M., and Brand, L. (1983) Simultaneous Analysis of Multiple Fluorescence Decay Curves - a Global Approach, *Chemical Physics Letters 102*, 501–507.
7. Beechem, J. M., Ameloot, M., and Brand, L. (1985) Global and target analysis of complex decay phenomena, *Analytical Instrumentation 14*, 379–402.
8. Ackers, G. K., Johnson, M. L., Mills, F. C., Halvorson, H. R., and Shapiro, S. (1975) The linkage between oxygenation and subunit dissociation in human hemoglobin. Consequences for the analysis of oxygenation curves, *Biochemistry 14*, 5128–5134.
9. Johnson, M. L., Correia, J. J., Yphantis, D. A., and Halvorson, H. R. (1981) Analysis of data from the analytical ultracentrifuge by nonlinear least-squares techniques, *Biophys J 36*, 575–588.
10. Barisas, B. G., and Gill, S. J. (1979) Thermodynamic analysis of carbon monoxide binding by hemoglobin trout I, *Biophys Chem 9*, 235–244.
11. Gilbert, C. W. (1980) A vector method for non-linear least squares reconvolution and fitting analysis of polarized fluorescence decay data, in *Time-resolved fluorescence spectroscopy in biochemistry and biology (Nato ASI Series)* (Dale, R. E., and Cundall, R. B., Eds.), Plenum Press, New York.
12. Beechem, J. M., Knutson, J. R., Ross, J. B. A., Turner, B. W., and Brand, L. (1983) Global Resolution of Heterogeneous Decay by Phase Modulation Fluorometry - Mixtures and Proteins, *Biochemistry 22*, 6054–6058.
13. Oberfelder, R. W., Barisas, B. G., and Lee, J. C. (1984) Thermodynamic linkages in rabbit muscle pyruvate kinase: analysis of experimental data by a two-state model, *Biochemistry 23*, 3822–3826.
14. Consler, T. G., Woodard, S. H., and Lee, J. C. (1989) Effects of primary sequence differences on the global structure and function of an enzyme: a study of pyruvate kinase isozymes, *Biochemistry 28*, 8756–8764.
15. Consler, T. G., Jennewein, M. J., Cai, G. Z., and Lee, J. C. (1992) Energetics of Allosteric Regulation in Muscle Pyruvate Kinase, *Biochemistry 31*, 7870–7878.
16. Consler, T. G., Jennewein, M. J., Cai, G. Z., and Lee, J. C. (1990) Synergistic effects of proton and phenylalanine on the regulation of muscle pyruvate kinase, *Biochemistry 29*, 10765–10771.
17. Boo, B. H., and Kang, D. (2005) Global and target analysis of time-resolved fluorescence spectra of Di-9 H-fluoren-9-yldimethylsilane: dynamics and energetics for intramolecular excimer formation, *J Phys Chem A 109*, 4280–4284.
18. Ionescu, R. M., and Eftink, M. R. (1997) Global analysis of the acid-induced and urea-induced unfolding of staphylococcal nuclease and two of its variants, *Biochemistry 36*, 1129–1140.
19. Ucci, J. W., and Cole, J. L. (2004) Global analysis of non-specific protein-nucleic interactions by sedimentation equilibrium, *Biophys Chem 108*, 127–140.
20. Verveer, P. J., Squire, A., and Bastiaens, P. I. (2000) Global analysis of fluorescence lifetime imaging microscopy data, *Biophys J 78*, 2127–2137.
21. Bednarkiewicz, A., and Whelan, M. P. (2008) Global analysis of microscopic fluorescence lifetime images using spectral segmentation and a digital micromirror spatial illuminator, *J Biomed Opt 13*, 041316.
22. Herman, P., and Lee, J. C. (2009) Functional energetic landscape in the allosteric regulation of muscle pyruvate kinase. 1. Calorimetric study, *Biochemistry 48*, 9448–9455.
23. Herman, P., and Lee, J. C. (2009) Functional energetic landscape in the allosteric regulation of muscle pyruvate kinase. 2. Fluorescence study, *Biochemistry 48*, 9456–9465.
24. Herman, P., and Lee, J. C. (2009) Functional energetic landscape in the allosteric regulation of muscle pyruvate kinase. 3. Mechanism, *Biochemistry 48*, 9466–9470.

25. Larsen, T. M., Laughlin, L. T., Holden, H. M., Rayment, I., and Reed, G. H. (1994) Structure of rabbit muscle pyruvate kinase complexed with Mn2+, K+, and pyruvate, *Biochemistry 33*, 6301–6309.
26. Oberfelder, R. W., Lee, L. L., and Lee, J. C. (1984) Thermodynamic linkages in rabbit muscle pyruvate kinase: kinetic, equilibrium, and structural studies, *Biochemistry 23*, 3813–3821.
27. Johnson, M. L., and Faunt, L. M. (1992) Parameter estimation by least-squares methods, *Methods Enzymol 210*, 1–37.
28. Johnson, M. L., and Frasier, S. G. (1985) Nonlinear Least-Squares Analysis, in *Methods in Enzymology 117* (Hirs, C. H. W., and Timasheff, S. N., Eds.), pp 301–342.
29. Straume, M., and Johnson, M. L. (1992) Monte Carlo method for determining complete confidence probability distributions of estimated model parameters, *Methods Enzymol 210*, 117–129.
30. Johnson, M. L. (1994) Use of Least-Squares Techniques in Biochemistry, in *Methods in Enzymology, vol. 240* (Abelson, J. N., Simon, M. I., Johnson, M. L., and Brand, L., Eds.), New York.
31. Davison, A. C., and Hinkley, D. V. (1997) *Bootstrap Methods and Their Application* Cambridge University Press, Cambridge.

Chapter 23

Predicting Binding Sites by Analyzing Allosteric Effects

Dengming Ming and Michael E. Wall

Abstract

This chapter describes a method for analyzing the allosteric influence of molecular interactions on protein conformational distributions. The method, called Dynamics Perturbation Analysis (DPA), generally yields insights into allosteric effects in proteins and is especially useful for predicting ligand-binding sites. The use of DPA for binding site prediction is motivated by the following allosteric regulation hypothesis: interactions in native binding sites cause a large change in protein conformational distributions. Here, we review the reasoning behind this hypothesis, describe the math behind the method, and present a recipe for predicting binding sites using DPA.

Key words: Protein dynamics, Ligand binding, Allostery, Allosteric free energy, Allosteric regulation hypothesis, Functional site, Protein interaction, Dynamics perturbation analysis, Relative entropy, Kullback–Leibler divergence

1. Introduction

Prediction of small-molecule binding sites is a key aspect of protein function prediction (1) and can be an important step in identifying small-molecule interactions for drug discovery (2). It can also be used as a preprocessing step to reduce the search space in computational docking algorithms. There are many methods to predict binding sites, but the problem is not solved. Several methods use only information from analysis of protein structure and/or dynamics. We have developed a method called Dynamics Perturbation Analysis (DPA) which uses analysis of protein dynamics (3–7).

Before describing DPA, we mention some closely related methods for predicting binding sites. In the interest of brevity and direct relevance, we mention methods that are solely based on considering protein structure and/or dynamics; the reader is advised that there are many other methods that use sequence information either alone or in combination with other information. The methods mentioned

Aron W. Fenton (ed.), *Allostery: Methods and Protocols*, Methods in Molecular Biology, vol. 796,
DOI 10.1007/978-1-61779-334-9_23, © Springer Science+Business Media, LLC 2012

below are representative and convey the rich variety in this active and evolving field.

Existing protein structure analysis methods are based on diverse principles, including the following: associating binding sites with surface clefts (8, 9) and especially those that have extreme values of volume (10–13) or other shape descriptors (14–19); associating binding sites with buried volumes (20); using level surfaces of the van der Waals potential to predict ligand-binding envelopes (21); identifying spatial clusters of methyl probes (22) or using other probes (23) to locate sites that exhibit energetically favorable probe interactions with the protein; association of enzyme active sites with regions of high electrostatic potential (24) or charged surface residues either in unfavorable electrostatic environments (25) or with anomalous predicted pH titration curves (26); association of binding sites with anomalous surface distributions of hydrophobicity (27); defining structural features (e.g., motifs) associated with functional sites (28–33); and identifying residues that are on average close to other residues in the protein (closeness centrality) (34–36).

Principles of other methods that consider protein dynamics include association of functional sites with the following: hinge regions (37, 38); regions where the local harmonic vibrations are largely determined by high-frequency modes (39); intrinsically disordered regions that are highly mobile in the absence of a molecular interaction partner (40); and residues where mutations cause a large change in the couplings of local perturbations to remote, local changes in the distribution of folded vs. unfolded states of the protein (41).

The DPA method is motivated by a central hypothesis that protein functional sites tend to evolve at control points where interactions cause a large change in the protein conformational distribution (4). The reasoning behind the hypothesis is as follows. Cellular functions are regulated by molecular interactions that alter protein activity. To enable such control, protein activity, and thus protein conformational distributions, must be susceptible to alteration by molecular interactions at functional sites. In other words, protein activity should be controllable by allosteric effects (allostery). The importance of considering changes in the full conformational distribution to understand allostery, as opposed to considering mechanistic changes among discrete, well-defined structural states in earlier models of Monod, Wyman, and Changeux (42) and Koshland, Nemethy, and Filmer (43), was recognized by Weber in 1972 (44) and has been recently emphasized by Gunasekaran et al. (45). Given the above considerations, we hypothesized that protein functional sites might tend to evolve at control points where interactions cause a large change in the protein conformational distribution (4).

We used DPA to quantify changes in protein conformational distributions due to molecular interactions (4, 5), examined 305 protein structures from the GOLD (46) docking test set, and indeed found that interactions at small-molecule binding sites caused a relatively large change in a model of protein vibrations (6). Motivated by these results, we then used DPA to successfully predict small-molecule binding sites at locations where interactions cause a large change in protein vibrations (6) and developed a Fast DPA algorithm to accelerate analyses (3). Beyond the prediction of functional sites, which is the focus of this chapter, we also have extended the DPA methods to study allosteric linkages between remote sites in a protein (5).

Below, we describe both the original and Fast DPA methods for prediction of ligand-binding sites. The Materials section describes the key elements of DPA. The Methods section shows how these elements are assembled into a pipeline for functional site prediction and lists a method for validating predictions. Additional implementation details are provided in the Notes.

2. Materials

2.1. Relative Entropy D_x as a Measure of Allosteric Effect

In DPA (4, 6), a protein is decorated with M surface points that interact with neighboring protein atoms, as illustrated for Protein Data Bank entry 1JEF (47) in Fig. 1. The protein conformational

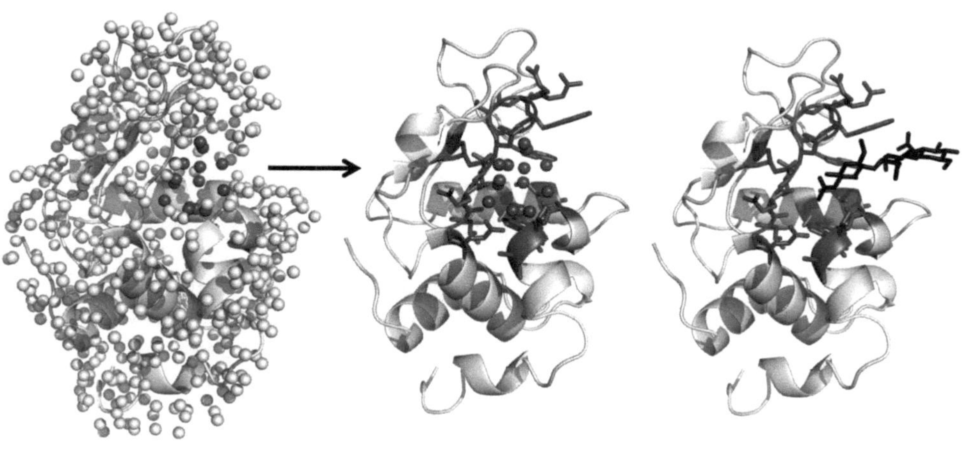

Fig. 1. Application of dynamics perturbation analysis (DPA) to predict ligand-binding sites. *Left*. In this example, the surface of lysozyme (PDB entry 1JEF (47), *white cartoon*) is decorated with 551 test points (*white spheres*), and D_x values are calculated as a measure of the degree to which the test points individually perturb the protein conformational distribution. The distribution of D_x values is fit to an extreme value distribution, and points that lie above the 96th percentile of the distribution are selected (*gray spheres*). *Center*. Points where the perturbation is largest are selected and clustered (*gray spheres*). Nearby residues are selected and predicted to lie within a functional site (*gray sticks*) *Right*. The predicted site (*gray sticks*) includes the two catalytic residues Glu35 and Asp52 and makes contact with a tri-NAG ligand in the crystal structure (*black sticks*).

distribution $P^{(0)}(\mathbf{x})$ is calculated in the absence of any surface points, and M protein conformational distributions $P^{(m)}(\mathbf{x})$ are calculated for the protein interacting with each point m. Any model may be used to generate conformational distributions; for ease of computation, we have used models of molecular vibrations. The distributions $P^{(m)}(\mathbf{x})$ are calculated from models of the protein in complex with each surface point. The relative entropy, or Kullback–Leibler divergence (48), $D_{\mathbf{x}}^{(m)}$ between $P^{(0)}(\mathbf{x})$ and $P^{(m)}(\mathbf{x})$ is calculated for each point m and is used as a measure of the change in the protein conformational distribution upon interacting with point m:

$$D_{\mathbf{x}}^{(m)} = \int \mathrm{d}^{3N}\mathbf{x}\, P^{(m)}(\mathbf{x}) \ln \frac{P^{(m)}(\mathbf{x})}{P(\mathbf{x})} \tag{1}$$

In the present case (unlike in other useful biological applications (49–53)), the relative entropy is not just an ad hoc measure; rather, it has real biophysical significance (7, 54, 55): $k_{\mathrm{B}} T D_{\mathbf{x}}^{(m)}$, where T is the temperature and k_{B} is Boltzmann's constant, is the free energy required to change the protein conformational distribution from an equilibrium distribution $P(\mathbf{x})$ to a nonequilibrium distribution $P^{(m)}(\mathbf{x})$.

2.2. Model of Protein Vibrations

The distributions $P(\mathbf{x})$ and $P^{(m)}(\mathbf{x})$ may be readily calculated using a simple model of protein vibrations – the elastic network model (ENM) (56–59). In the ENM, C_α atoms are extracted from an atomic model of a protein, and an interaction network is generated by connecting springs between all atom pairs (i, j) separated by a distance less than or equal to a cutoff distance r_c. Each spring has the same force constant γ, is aligned with the separation between the connected atoms, and has an equilibrium length equal to the distance d_{ij} between the atoms in the initial model. Thus, the potential energy is given by $U(\mathbf{x}) = \gamma/2 \sum_{i>j} \varepsilon_{ij} \left(\left|\mathbf{x}_i - \mathbf{x}_j\right| - d_{ij}\right)^2$, where $\varepsilon_{ij} = 1$ if atoms i and j are connected, and $\varepsilon_{ij} = 0$ otherwise. We previously found that coarse-grained vibrations were more accurately modeled using a backbone-enhanced ENM (BENM) in which neighboring C_α atoms on the polypeptide chain are connected by a stiffer spring (5). The interaction between the protein and a surface point m is modeled by connecting springs of force constant γ_s between the surface point and all protein atoms within a cutoff distance r_s of the surface point. The protein coordinates are not modified in modeling the interaction. The dynamics are defined using normal modes. In this model, the reference distribution $P(\mathbf{x})$ is given by

$$P(\mathbf{x}) = \prod_{i=1}^{3N, \lambda_i \neq 0} \left(\frac{\lambda_i}{2\pi k_{\mathrm{B}} T}\right)^{\frac{1}{2}} \mathrm{e}^{-\frac{1}{2k_{\mathrm{B}}T} \lambda_i \left|(\mathbf{x}-\mathbf{x}_0)\cdot \mathbf{v}_i\right|^2} \tag{2}$$

In Eq. 2, N is the number of atoms in the protein; $\mathbf{x}_0$ is the equilibrium structure; and λ_i and $\mathbf{v}_i$ are the ith eigenvalue and eigenvector of the Hessian $\mathbf{H} : h_{ij} = \partial U / \partial x_i \partial x_j |_{\mathbf{x}_0}$, where $U(\mathbf{x})$ is the potential energy function of the BENM. The perturbed distribution $P^{(m)}(\mathbf{x})$ is similar to Eq. 2, but substituting the eigenvalues and eigenvectors $\bar{\lambda}_i^{(m)}$ and $\mathbf{v}_i^{(\bar{m})}$ of the pseudo-Hessian $\bar{\mathbf{H}}^{(m)}$ for λ_i and $\mathbf{v}_i$. $\bar{\mathbf{H}}^{(m)}$ is derived from the full Hessian $\mathbf{H}^{(m)}$ for the protein model in the presence of the surface point m:

$$\mathbf{H}^{(m)} = \begin{pmatrix} \mathbf{H}_P & \mathbf{G} \\ \mathbf{G}^{\mathrm{T}} & \mathbf{H}_S^{(m)} \end{pmatrix}. \tag{3}$$

The submatrix $\mathbf{H}_P$ couples the protein coordinates, the submatrix $\mathbf{H}_S^{(m)}$ couples the test-point coordinates, and the submatrix $\mathbf{G}$ couples the protein to the test point. In terms of these matrices, $\bar{\mathbf{H}}^{(m)}$ is given by (5)

$$\bar{\mathbf{H}}^{(m)} = \mathbf{H}_P - \mathbf{G}\mathbf{H}_S^{(m)^{-1}}\mathbf{G}^{\mathrm{T}} \tag{4}$$

which is the Schur complement of block $\mathbf{H}_S^{(m)}$ of the matrix $\mathbf{H}^{(m)}$.

2.3. D_x Value Due to Perturbation of Protein Vibrations

Using expressions for $P(\mathbf{x})$ and $P^{(m)}(\mathbf{x})$, Eq. 1 becomes (4, 5)

$$D_{\mathbf{x}}^{(m)} = \frac{1}{2} \sum_{i=7\ldots3N}^{\bar{\lambda}_i^{(m)} \neq 0; \lambda_i \neq 0} \left(\ln \frac{\bar{\lambda}_i^{(m)}}{\lambda_i} + \sum_{j=7\ldots3N}^{\lambda_j \neq 0} \frac{\lambda_j}{\bar{\lambda}_i^{(m)}} \left| \bar{\mathbf{v}}_i^{(m)} \cdot \mathbf{v}_j \right|^2 - 1 \right). \tag{5}$$

Equation 5 is the central equation that enables DPA.

Empirically, the distribution of $y = D_{\mathbf{x}}^{(m)}$ values on the surface of a protein calculated using Eq. 5 is observed to obey an extreme value distribution (Fig. 2) (6),

$$\rho(y) = \frac{1}{\beta} \mathrm{e}^{\frac{y-\mu}{\beta} - \mathrm{e}^{\frac{y-\mu}{\beta}}}. \tag{6}$$

2.4. D_x^λ, A Surrogate Measure for Fast Calculations

The original DPA algorithm is a highly innovative approach that performs well in predicting functional sites (Fig. 3). However, it takes about an hour to analyze a 150-residue protein domain using DPA on a 2.75 GHz Linux workstation, and the method does not scale well to larger systems. These computational requirements limited the utility of the original method. We therefore developed a faster method for use in high-throughput scenarios and on web servers (3). The method, called Fast DPA, is based on a simple empirical observation: for dynamics defined by normal modes, the total value of $D_{\mathbf{x}}^{(m)}$ in Eq. 5 is highly correlated with just the first term,

$$D_{\mathbf{x}}^{\lambda,(m)} = \frac{1}{2} \sum_{i=7\ldots3N}^{\bar{\lambda}_i^{(m)} \neq 0; \lambda_i \neq 0} \log \frac{\bar{\lambda}_i^{(m)}}{\lambda_i}. \tag{7}$$

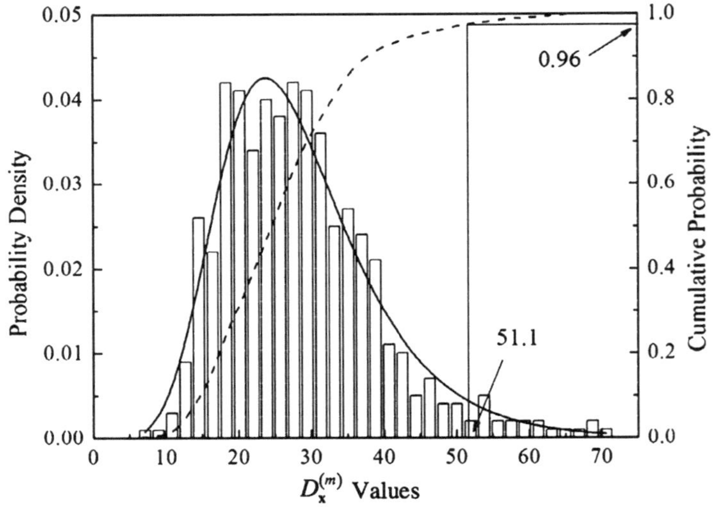

Fig. 2. Distribution of $D_{\mathbf{x}}^{(m)}$ values for 4,859 points on the surface of lysozyme 1JEF (the number of points was increased in this case to evaluate the fit). The distribution is well fit by an extreme value distribution Eq. 6 with parameters $\mu = 23.07$ and $\beta = 8.45$ (*solid line*). By examining the cumulative distribution (*dashed line*), the fit is used to find surface points that lie within the upper 96% of the distribution; these points are used to predict ligand-binding sites. Reprinted from Journal of Molecular Biology, Vol. 358, D. Ming and M. E. Wall, Interactions in native binding sites cause a large change in protein dynamics, pp. 213–223, Copyright (2006), with permission from Elsevier.

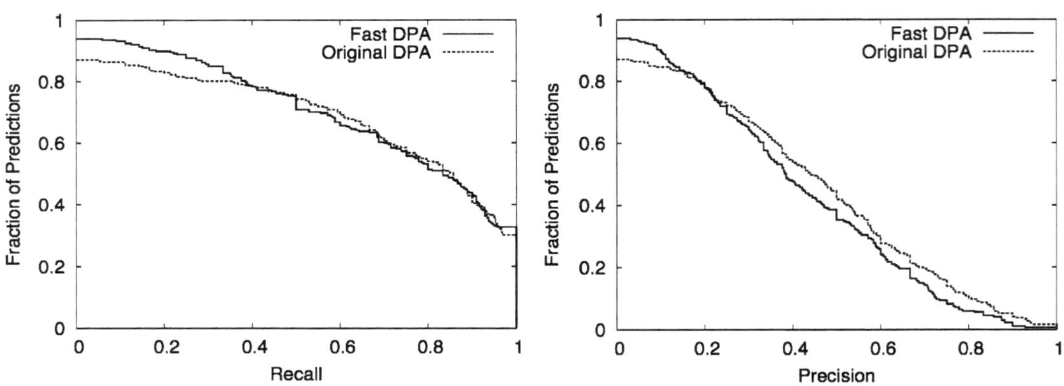

Fig. 3. Comparison of performance in predicting ligand-binding sites using DPA vs. Fast DPA for 287 (# of predictions using DPA) or 267 (# of predictions using Fast DPA) proteins in the 305-protein GOLD test set. *Left*. Recall of binding-site residues is slightly better using Fast DPA. *Right*. Precision of predicted residues is slightly worse using Fast DPA. The *y*-axis indicates the fraction of proteins with a recall or precision at least as high as the value on the *x*-axis (*y*-values should be read from the top of each step). Overall, the performance of Fast DPA is comparable to DPA.

Hereafter, we refer to $D_{\mathbf{x}}^{\lambda,(m)}$ simply as $D_{\mathbf{x}}^{\lambda}$. The left panel of Fig. 4 illustrates an example of the high correlation for lysozyme – other proteins exhibit a similar correlation. Observation of this correlation motivates the use of $D_{\mathbf{x}}^{\lambda}$ as a surrogate for $D_{\mathbf{x}}$ in DPA, and, because $D_{\mathbf{x}}^{\lambda}$ only involves eigenvalues, creates an avenue for using first-order perturbation theory. Indeed, we found that the

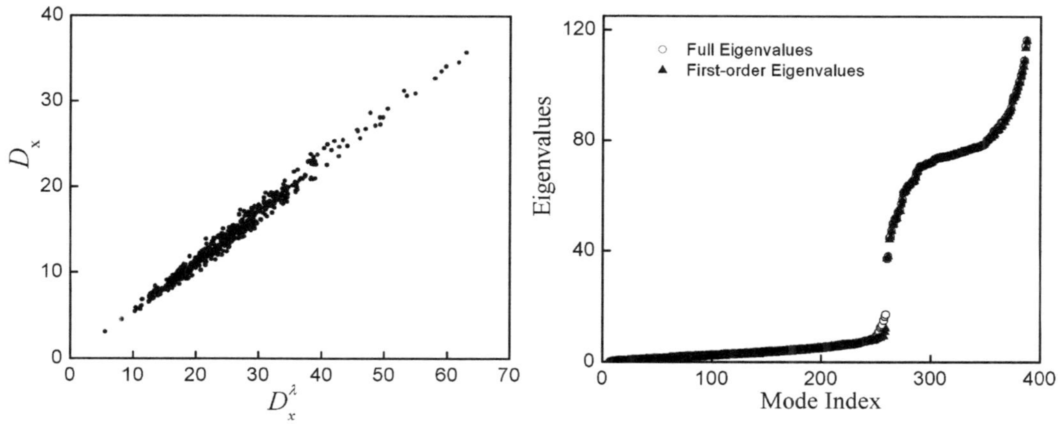

Fig. 4. *Left.* Values of D_x (*y*-axis) and $D_{\mathbf{x}}^{\lambda}$ (*x*-axis) are plotted for lysozyme DPA. The Pearson correlation coefficient between the two is 0.992. *Right.* Eigenvalues (used for calculation of $D_{\mathbf{x}}^{\lambda}$) that are estimated using perturbation theory (*filled triangles*) are a good approximation to the actual eigenvalues (*open circles*).

eigenvalues are well-approximated using first-order perturbation theory (Fig. 4, right panel). These results justified development of the Fast DPA algorithm which calculates $D_{\mathbf{x}}^{\lambda}$ using perturbation theory instead of $D_{\mathbf{x}}$ exactly (3). In this approximation, the pseudo-Hessian $\bar{\mathbf{H}}^{(m)}$ of the protein in the presence of point m is written as the Hessian $\mathbf{H}$ of the protein in the absence of the ligand plus a perturbation term $\delta\bar{\mathbf{H}}^{(m)}$:

$$\bar{\mathbf{H}}^{(m)} = \mathbf{H} + \delta\bar{\mathbf{H}}^{(m)}, \tag{8}$$

where $\bar{\mathbf{H}}^{(m)}$ is given by Eq. 4. To estimate the eigenvalues of $\bar{\mathbf{H}}^{(m)}$, we use the canonical first-order perturbation theory expression,

$$\lambda_i^{(m)} \approx \lambda_i + \mathbf{v}_i^T \delta\bar{\mathbf{H}}^{(m)} \mathbf{v}_i, \tag{9}$$

where λ_i is the ith eigenvalue of $\mathbf{H}$.

Use of Fast DPA for prediction of functional sites results in a N-fold speedup ($N =$ # of atoms in the model) compared to ordinary DPA (Fig. 5) while maintaining good performance in functional site prediction (Fig. 3) (3).

3. Methods

3.1. Dynamics Perturbation Analysis (DPA)

1. Generate surface points from a PDB structure. This is currently done using MSMS software (60) which locates points where a sphere rolling on the protein surface comes to rest in contact with three or more atoms. These points define what is called a

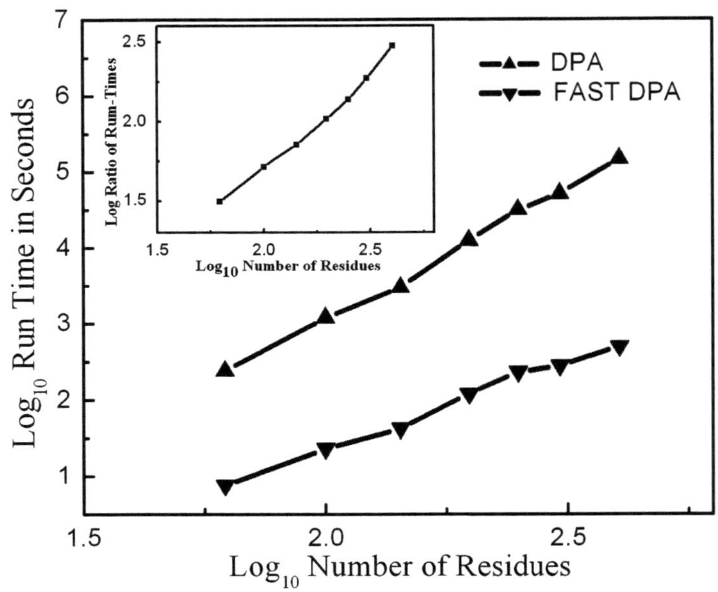

Fig. 5. Comparison of run times for DPA vs. Fast DPA.

reduced surface, and the spherical sections that contact the protein define faces that are used to construct more detailed analytic and triangulated surfaces. We run MSMS using a default probe radius of 1.5 Å and triangulation density of 1 vertex/Å^2 (see Note 1). We then select one surface point per spherical face from the vertex output file which has a .vert extension. These surface points are used as perturbation points (Fig. 1, see Note 2).

2. Generate the protein vibration model. As described above, we are currently using a custom implementation of a backbone-enhanced ENM (5). The default cutoff $\mathbf{r}_c$ is 8.5 Å in our previously published work (see Note 3) and the spring constant γ (whose value has no influence on the analysis) may be set to 1. Neighboring backbone C_α atoms are connected with a spring constant γ_b (= 40γ in published work). The interaction with a surface point m is added using a cutoff $\mathbf{r}_s$ (= $\mathbf{r}_c$ + 5.5 Å in published work) and a spring constant γ_s (= 12γ in published work). These parameter values have yielded satisfactory results for predicting functional sites in test sets, though they have not yet been optimized rigorously.
3. Calculate Hessian matrices (Subheading 2) for the protein model in the presence and absence of the surface points. We use our own code for this step.
4. Calculate normal modes from the Hessian matrices. We are currently using the DIAGQ module from CHARMM (61) for this step (see Note 4). In the Fast DPA method, special

attention should be paid to the construction of the perturbation matrix $\delta\bar{\mathbf{H}}^{(m)}$ (see Note 5).

5. Use Eq. 5 to calculate $D_{\mathbf{x}}$ (or Eq. 7 for $D_{\mathbf{x}}^{\lambda}$) from the normal modes. Here, a list of $D_{\mathbf{x}}$ values for all surface points is prepared for the following analysis. We have implemented this in our own code.
6. Fit the distribution of $D_{\mathbf{x}}$ values to an extreme value distribution, and select high-$D_{\mathbf{x}}$ points (e.g., above the $D_{\mathbf{x}}$ value corresponding to 0.96 in the cumulative distribution function). A minimum number of high-$D_{\mathbf{x}}$ points (e.g., 5) are required in the selection for input into the clustering step (see Note 6). We use a modified Levenberg–Marquardt algorithm for the fitting.
7. Spatially cluster the high $D_{\mathbf{x}}$ points, and rank them. We are currently using our own implementation of the OPTICS algorithm (62) for clustering and are ranking according to the mean value of $D_{\mathbf{x}}$ in a cluster. The default values of OPTICS parameters: generating distance $\varepsilon = 5$ Å and minimum number of objects MinPts $= 3$.
8. Select residues predicted to be in a ligand-binding site. Protein C_{α} atoms within a distance d_P ($= 6$ Å in our previously published work) of the points in each cluster are selected, the associated residues are used to define a set of predicted binding site residues P. Although we have used only the rank-1 cluster which has the highest mean D_{x} value in published validation studies, we have noticed anecdotally that in many cases the predicted sites found with lower ranked clusters are biologically significant.

3.2. Validation

1. Obtain known binding sites. We evaluate predictions using crystal structures that contain a protein in complex with one or more ligands. A set of residues L are selected whose C_{α} atoms are within a distance d_L ($= 6$ Å in previously published work) of any nonhydrogen atom of a ligand found in complex with the protein. Each ligand is associated with its own known binding site. The values of validation measures depend not only on the prediction algorithm but also on the precise definition of the known site.
2. Calculate binding site recall. The recall measures the degree to which a known site is recovered by the predictions. If there is predicted site P that overlaps a known site L by at least one residue (i.e., $\mathrm{P}\cap\mathrm{L}=\emptyset$), the recall of L with respect to P is defined as $|\mathrm{P}\cap\mathrm{L}|/|\mathrm{L}|$ where the notation $|\mathrm{A}|$ means the cardinality (number of elements) of set A. If there is no such predicted site then the recall of L is zero.

3. Calculate prediction precision. The precision measures the degree to which a predicted site overlaps a known site. Similar to the recall, if there is a known site L that overlaps a predicted site P by at least one residue, the precision of P with respect to L is defined as $|P \cap L|/|P|$. If there is no such known site then the precision of P is zero.
4. Calculate statistical significance of predictions. A simple null model may be used to assess the significance of predictions. In the null model, binding site predictions are made by randomly selecting surface residues. We define the set of surface residues S by selecting residues whose C_α atoms are within a distance d_S (= 6 Å in previously published work) of at least one MSMS surface point. The *P*-value P_{null} is defined as the probability of achieving an overlap of at least $|P \cap L|$ residues with a known site containing $|L|$ residues by randomly choosing $|P|$ surface residues from a total number $|S|$. This value is given by

$$P_{null} = \sum_{n=|P \cap L|}^{\min(|L|,|P|)} \binom{|L|}{n} \Big/ \binom{|S|}{|P|} = \sum_{n=|P \cap L|}^{\min(|L|,|P|)} \frac{|L|!|P|!(|S| - |P|)!}{n!(|L| - n)!|S|!} \quad (10)$$

P_{null} values were calculated for 250 proteins in GOLD test set where an overlap of at least one residue was found in both the predicted and ligand-binding sites (6). It was found that $P_{null} \leq 10^{-3}$ for 87% of proteins, indicating that the original DPA prediction had a statistically significant overlap with the real ligand-binding sites.

4. Notes

1. The algorithm can be run with alternative values for these parameters. In particular, MSMS occasionally fails to generate a surface for a protein using the default values but can sometimes be encouraged to do so by adjusting the probe radius.
2. Our experience showed that selecting one vertex point per reduced face yielded satisfactory results. In a few cases, the spatial distribution of the surface points thus derived might be too sparse to generate any predicted clusters. In these cases, it is possible to obtain predictions by increasing the number of vertices selected per face.

3. In some cases, the cutoff is too short to couple sufficiently the selected atoms, leading to additional zero-frequency modes, which foils the calculation. In these cases, r_c may be increased (e.g., in 1 Å increments) until the additional zero-frequency modes are eliminated, enabling the calculation to proceed. In some rare cases, e.g., when a thin structural feature protrudes from the surface, an alternative coarse-grained model might be preferred (63–65).
4. If more than 6 zero modes are found, the cutoff may be increased as indicated in step 2. In practice, it is good to establish an upper limit for r_c, e.g., 25 Å.
5. The perturbation Hessian $\delta\bar{\mathbf{H}}^{(m)}$ is calculated as

$$\delta\bar{\mathbf{H}}^{(m)} = -\mathrm{diag}\left\{\mathbf{G}_{1m}, \mathbf{G}_{2m}, \cdots, \mathbf{G}_{Nm}\right\} - \mathbf{G}\mathbf{H}_S^{(m)^{-1}}\mathbf{G}^{\mathrm{T}} \quad (11)$$

where the first part is due to the direct interaction between C_α and the perturbing surface points and the second term is due to the indirect interaction between C_α's induced by mediated by direct interactions with the surface point. In rare cases $\mathbf{H}_S^{(m)}$ can be singular and its inverse be undefined. This situation can be alleviated by increasing r_s. Note that $\mathbf{H}_S^{(m)}$ has the dimension of 3×3, since only one surface point interacts with the protein.
6. When the number of selected points is too small (e.g., <5), the threshold may be reduced (e.g., by 0.01 increments) until enough points are selected. In this case, a lower limit of the threshold may be established, e.g., 0.5. If this limit is reached, results might be obtained by increasing the density of surface points (see Note 2).

References

1. Ofran, Y., Punta, M., Schneider, R., and Rost, B. (2005) Beyond annotation transfer by homology: novel protein-function prediction methods to assist drug discovery, *Drug Discov Today 10*, 1475–1482.
2. Campbell, S. J., Gold, N. D., Jackson, R. M., and Westhead, D. R. (2003) Ligand binding: functional site location, similarity and docking, *Curr Opin Struct Biol 13*, 389–395.
3. Ming, D., Cohn, J. D., and Wall, M. E. (2008) Fast dynamics perturbation analysis for prediction of protein functional sites, *BMC Struct Biol 8*, 5.
4. Ming, D., and Wall, M. E. (2005) Quantifying allosteric effects in proteins, *Proteins 59*, 697–707.
5. Ming, D., and Wall, M. E. (2005) Allostery in a coarse-grained model of protein dynamics, *Phys Rev Lett 95*, 198103.
6. Ming, D., and Wall, M. E. (2006) Interactions in native binding sites cause a large change in protein dynamics, *J. Mol. Biol. 358*, 213–223.
7. Wall, M. E. (2006) Ligand binding, protein fluctuations, and allosteric free energy, *AIP Conf. Proc. 851*, 16–33.
8. Levitt, D. G., and Banaszak, L. J. (1992) POCKET: a computer graphics method for identifying and displaying protein cavities and their surrounding amino acids, *J Mol Graph 10*, 229–234.
9. Weisel, M., Proschak, E., and Schneider, G. (2007) PocketPicker: analysis of ligand

binding-sites with shape descriptors, *Chem Cent J 1*, 7.
10. Laskowski, R. A. (1995) SURFNET: a program for visualizing molecular surfaces, cavities, and intermolecular interactions, *J Mol Graph 13*, 323–330, 307–328.
11. Liang, J., Edelsbrunner, H., and Woodward, C. (1998) Anatomy of protein pockets and cavities: measurement of binding site geometry and implications for ligand design, *Protein Sci 7*, 1884–1897.
12. Dundas, J., Ouyang, Z., Tseng, J., Binkowski, A., Turpaz, Y., and Liang, J. (2006) CASTp: computed atlas of surface topography of proteins with structural and topographical mapping of functionally annotated residues, *Nucleic Acids Res 34*, W116–118.
13. Glaser, F., Morris, R. J., Najmanovich, R. J., Laskowski, R. A., and Thornton, J. M. (2006) A method for localizing ligand binding pockets in protein structures, *Proteins 62*, 479–488.
14. Peters, K. P., Fauck, J., and Frommel, C. (1996) The automatic search for ligand binding sites in proteins of known three-dimensional structure using only geometric criteria, *J Mol Biol 256*, 201–213.
15. Hendrix, D. K., and Kuntz, I. D. (1998) Surface solid angle-based site points for molecular docking, *Pac Symp Biocomput*, 317–326.
16. Coleman, R. G., Burr, M. A., Souvaine, D. L., and Cheng, A. C. (2005) An intuitive approach to measuring protein surface curvature, *Proteins 61*, 1068–1074.
17. Coleman, R. G., and Sharp, K. A. (2006) Travel depth, a new shape descriptor for macromolecules: application to ligand binding, *J Mol Biol 362*, 441–458.
18. Nayal, M., and Honig, B. (2006) On the nature of cavities on protein surfaces: application to the identification of drug-binding sites, *Proteins 63*, 892–906.
19. Xie, L., and Bourne, P. E. (2007) A robust and efficient algorithm for the shape description of protein structures and its application in predicting ligand binding sites, *BMC Bioinformatics 8 Suppl 4*, S9.
20. Brady, G. P., Jr., and Stouten, P. F. (2000) Fast prediction and visualization of protein binding pockets with PASS, *J Comput Aided Mol Des 14*, 383–401.
21. An, J., Totrov, M., and Abagyan, R. (2005) Pocketome via comprehensive identification and classification of ligand binding envelopes, *Mol Cell Proteomics 4*, 752–761.
22. Laurie, A. T., and Jackson, R. M. (2005) Q-SiteFinder: an energy-based method for the prediction of protein-ligand binding sites, *Bioinformatics 21*, 1908–1916.
23. Hernandez, M., Ghersi, D., and Sanchez, R. (2009) SITEHOUND-web: a server for ligand binding site identification in protein structures, *Nucleic Acids Res 37*, W413-416.
24. Bate, P., and Warwicker, J. (2004) Enzyme/non-enzyme discrimination and prediction of enzyme active site location using charge-based methods, *J Mol Biol 340*, 263–276.
25. Elcock, A. H. (2001) Prediction of functionally important residues based solely on the computed energetics of protein structure, *J Mol Biol 312*, 885–896.
26. Ondrechen, M. J., Clifton, J. G., and Ringe, D. (2001) THEMATICS: a simple computational predictor of enzyme function from structure, *Proc Natl Acad Sci U S A 98*, 12473–12478.
27. Brylinski, M., Prymula, K., Jurkowski, W., Kochanczyk, M., Stawowczyk, E., Konieczny, L., and Roterman, I. (2007) Prediction of functional sites based on the fuzzy oil drop model, *PLoS Comput Biol 3*, e94.
28. Wallace, A. C., Borkakoti, N., and Thornton, J. M. (1997) TESS: a geometric hashing algorithm for deriving 3D coordinate templates for searching structural databases. Application to enzyme active sites, *Protein Sci 6*, 2308–2323.
29. Shulman-Peleg, A., Nussinov, R., and Wolfson, H. J. (2004) Recognition of functional sites in protein structures, *J Mol Biol 339*, 607–633.
30. Stark, A., and Russell, R. B. (2003) Annotation in three dimensions. PINTS: Patterns in Non-homologous Tertiary Structures, *Nucleic Acids Res 31*, 3341–3344.
31. Stark, A., Shkumatov, A., and Russell, R. B. (2004) Finding functional sites in structural genomics proteins, *Structure 12*, 1405–1412.
32. Liang, M. P., Brutlag, D. L., and Altman, R. B. (2003) Automated construction of structural motifs for predicting functional sites on protein structures, *Pac Symp Biocomput*, 204–215.
33. Barker, J. A., and Thornton, J. M. (2003) An algorithm for constraint-based structural template matching: application to 3D templates with statistical analysis, *Bioinformatics 19*, 1644–1649.
34. Amitai, G., Shemesh, A., Sitbon, E., Shklar, M., Netanely, D., Venger, I., and Pietrokovski, S. (2004) Network analysis of protein structures identifies functional residues, *J Mol Biol 344*, 1135–1146.
35. Thibert, B., Bredesen, D. E., and del Rio, G. (2005) Improved prediction of critical residues

for protein function based on network and phylogenetic analyses, *BMC Bioinformatics 6*, 213.

36. Chea, E., and Livesay, D. R. (2007) How accurate and statistically robust are catalytic site predictions based on closeness centrality?, *BMC Bioinformatics 8*, 153.
37. Ma, B., Wolfson, H. J., and Nussinov, R. (2001) Protein functional epitopes: hot spots, dynamics and combinatorial libraries, *Curr Opin Struct Biol 11*, 364–369.
38. Yang, L. W., and Bahar, I. (2005) Coupling between catalytic site and collective dynamics: a requirement for mechanochemical activity of enzymes, *Structure 13*, 893–904.
39. Haliloglu, T., Keskin, O., Ma, B., and Nussinov, R. (2005) How similar are protein folding and protein binding nuclei? Examination of vibrational motions of energy hot spots and conserved residues, *Biophys J 88*, 1552–1559.
40. Radivojac, P., Iakoucheva, L. M., Oldfield, C. J., Obradovic, Z., Uversky, V. N., and Dunker, A. K. (2007) Intrinsic disorder and functional proteomics, *Biophys J 92*, 1439–1456.
41. Liu, T., Whitten, S. T., and Hilser, V. J. (2007) Functional residues serve a dominant role in mediating the cooperativity of the protein ensemble, *Proc Natl Acad Sci U S A 104*, 4347–4352.
42. Monod, J., Wyman, J., and Changeux, J. P. (1965) On the nature of allosteric transitions: a plausible model, *J Mol Biol 12*, 88–118.
43. Koshland, D. E., Jr., Nemethy, G., and Filmer, D. (1966) Comparison of experimental binding data and theoretical models in proteins containing subunits, *Biochemistry 5*, 365–385.
44. Weber, G. (1972) Ligand binding and internal equilibria in proteins, *Biochemistry 11*, 864–878.
45. Gunasekaran, K., Ma, B., and Nussinov, R. (2004) Is allostery an intrinsic property of *all* dynamic proteins?, *Proteins 57*, 433–443.
46. Jones, G., Willett, P., Glen, R. C., Leach, A. R., and Taylor, R. (1997) Development and validation of a genetic algorithm for flexible docking, *J Mol Biol 267*, 727–748.
47. Harata, K., and Muraki, M. (1997) X-ray structure of turkey-egg lysozyme complex with tri-N-acetylchitotriose. Lack of binding ability at subsite A, *Acta Crystallogr D Biol Crystallogr 53*, 650–657.
48. Kullback, S., and Leibler, R. A. (1951) On information and sufficiency, *Annals of Math Stats 22*, 79–86.
49. Sterner, B., Singh, R., and Berger, B. (2007) Predicting and Annotating Catalytic Residues: An Information Theoretic Approach, *J Comput Biol.*
50. Bhasi, K., Zhang, L., Brazeau, D., Zhang, A., and Ramanathan, M. (2006) Information-theoretic identification of predictive SNPs and supervised visualization of genome-wide association studies, *Nucleic Acids Res 34*, e101.
51. Igarashi, Y., Aoki, K. F., Mamitsuka, H., Kuma, K., and Kanehisa, M. (2004) The evolutionary repertoires of the eukaryotic-type ABC transporters in terms of the phylogeny of ATP-binding domains in eukaryotes and prokaryotes, *Mol Biol Evol 21*, 2149–2160.
52. Liu, X., Zhang, L. M., Guan, S., and Zheng, W. M. (2003) Distances and classification of amino acids for different protein secondary structures, *Phys Rev E Stat Nonlin Soft Matter Phys 67*, 051927.
53. del Sol Mesa, A., Pazos, F., and Valencia, A. (2003) Automatic methods for predicting functionally important residues, *J Mol Biol 326*, 1289–1302.
54. Donald, M. J. (1987) Free energy and the relative entropy, *J Stat Phys 49*, 81–87.
55. Qian, H. (2001) Relative entropy: free energy associated with equilibrium fluctuations and nonequilibrium deviations, *Phys. Rev. E 63*, 042103.
56. Atilgan, A. R., Durell, S. R., Jernigan, R. L., Demirel, M. C., Keskin, O., and Bahar, I. (2001) Anisotropy of fluctuation dynamics of proteins with an elastic network model, *Biophys J 80*, 505–515.
57. Bahar, I., Atilgan, A. R., and Erman, B. (1997) Direct evaluation of thermal fluctuations in proteins using a single-parameter harmonic potential, *Fold Des 2*, 173–181.
58. Hinsen, K. (1998) Analysis of domain motions by approximate normal mode calculations, *Proteins 33*, 417–429.
59. Tirion, M. M. (1996) Large amplitude elastic motions in proteins from a single-parameter, atomic analysis, *Physical Review Letters 77*, 1905–1908.
60. Sanner, M. F., Olson, A. J., and Spehner, J. C. (1996) Reduced surface: an efficient way to compute molecular surfaces, *Biopolymers 38*, 305–320.
61. Brooks, B. R., Bruccoleri, R., Olafson, B. D., States, D. J., Swaminathan, S., and Karplus, M. (1983) CHARMM: A program for macromolecular energy, minimization and dynamics calculations, *J Comput Chem 4*, 187–217.
62. Ankerst, M., Breunig, M. M., Kriegel, H. P., and Sander, J. (1999) OPTICS: ordering points to identify the clustering structure, *Proceedings of the ACM SIGMON International Conference on Management of Data 28*, 49–60.

63. Korkut, A., and Hendrickson, W. A. (2009) A force field for virtual atom molecular mechanics of proteins, *Proc Natl Acad Sci U S A 106*, 15667–15672.
64. Lu, M., Poon, B., and Ma, J. (2006) A new method for coarse-grained elastic normal-mode analysis, *J Chem Theory Compu 2*, 464–471.
65. Stember, J. N., and Wriggers, W. (2009) Bend-twist-stretch model for coarse elastic network simulation of biomolecular motion, *J Chem Phys 131*, 074112.

Index

Aron W. Fenton (ed.), *Allostery: Methods and Protocols*, Methods in Molecular Biology, vol. 796,
DOI 10.1007/978-1-61779-334-9, © Springer Science+Business Media, LLC 2012

T

V

MIX
Papier aus verantwortungsvollen Quellen
Paper from responsible sources
FSC® C105338

If you have any concerns about our products,
you can contact us on
ProductSafety@springernature.com

In case Publisher is established outside the EU,
the EU authorized representative is:
Springer Nature Customer Service Center GmbH
Europaplatz 3, 69115 Heidelberg, Germany

Printed by Libri Plureos GmbH
in Hamburg, Germany